Fundamentals of Telecommunication Networks

WILEY SERIES IN TELECOMMUNICATIONS AND SIGNAL PROCESSING

John G. Proakis, Editor
Northeastern University

Worldwide Telecommunications Guide for the Business Manager
Walter L. Vignault

Expert System Applications to Telecommunications
Jay Liebowitz

Business Earth Stations for Telecommunications
Walter L. Morgan and Denis Rouffet

Introduction to Communications Engineering, 2nd Edition
Robert M. Gagliardi

Satellite Communications: The First Quarter Century of Service
David W. E. Rees

Synchronization in Digital Communications, Volume 1
Heinrich Meyr and Gerd Ascheid

Telecommunication System Engineering, 2nd Edition
Roger L. Freeman

Digital Telephony, 2nd Edition
John Bellamy

Elements of Information Theory
Thomas M. Cover and Joy A. Thomas

Telecommunication Transmission Handbook, 3rd Edition
Roger L. Freeman

Digital Signal Estimation
Robert J. Mammone, Editor

Telecommunication Circuit Design
Patrick D. van der Puije

Meteor Burst Communications: Theory and Practice
Donald L. Schilling, Editor

Mobile Communications Design Fundamentals, 2nd Edition
William C. Y. Lee

Fundamentals of Telecommunication Networks
Tarek N. Saadawi and Mostafa H. Ammar with Ahmed El Hakeem

Optical Communications, 2nd Edition
Robert M. Gagliardi and Sherman Karp

Fundamentals of Telecommunication Networks

Tarek N. Saadawi
Department of Electrical Engineering,
City University of New York, City College

Mostafa H. Ammar
College of Computing,
Georgia Institute of Technology

with Ahmed El Hakeem
Department of Electrical and Computer Engineering,
Concordia University

A Wiley-Interscience Publication
JOHN WILEY & SONS, INC.
New York · Chichester · Brisbane · Toronto · Singapore

This text is printed on acid-free paper.

Library of Congress Cataloging in Publication Data:

Saadawi, Tarek N.

Fundamentals of telecommunication networks / Tarek N. Saadawi, Mostafa H. Ammar, Ahmed El Hakeem.

p. cm.—(Wiley series in telecommunications and signal processing)

"A Wiley-Interscience publication."

Includes bibliographical references.

ISBN 0-471-51582-5 (alk. paper)

1. Telecommunication systems. 2. Computer networks. 3. Data transmission systems. I. Ammar, Mostafa H. (Mostafa Hamed) II. El Hakeem, Ahmed. III. Title. IV. Series.

TK5101.S13 1994

621.382—dc20 94-6544

Printed in the United States of America

10 9 8 7 6 5

To Our Families

CONTENTS

PREFACE

In recent years, there have been dramatic changes and important successes in the field of telecommunication networks. Among the most significant of these are the penetration of data networks, the widespread use of local area networks, the deployment of the integrated services digital network (ISDN), the emergence of Broadband ISDN, and the development of fast packet switching.

As telecommunication networks evolved, we have witnessed the emergence of a body of knowledge that is clearly at the heart of our understanding and our capabilities for further innovations in the field. This book focuses, albeit not exclusively, on this fundamental knowledge, which consists of the concepts and mechanisms used in the design and operation of telecommunication networks.

This book is an outgrowth of our teaching, research, and consulting activities in telecommunication networks over the past 15 years. The material presented in this book is covered in a two-semester course at the City College of the City University of New York, at Concordia University, and a three-quarter course at the Georgia Institute of Technology.

The book is designed to meet the needs of a variety of audiences including students in electrical engineering, computer science, and computer engineering at their senior undergraduate or first-year graduate level as well as developers, practicing engineers, and managers who are dealing with telecommunication networks and systems.

The book is divided into six major parts. First we discuss the evolution of telecommunication networks and introduce existing networks and the principles of data communications: These are covered in Chapters 1 and 2. Chapter 2 deals with the physical layer of the Open System Interconnection (OSI) reference model. Performance evaluation of networks is covered in Chapter 3, which presents queueing theory fundamentals with detailed examples of its applications to networks. Layers 2 and 3 of the OSI reference model are covered in Chapters 4, 5, and 6. Chapter 4 addresses layer 2 and 3 protocol standards, Chapter 5 presents routing algorithms, and Chapter 6 discusses flow and congestion control techniques. Local area network (LAN) theory, standards, and technology are covered in Chapters 7 and 8. Chapter 7 presents multiple access communications techniques. Network interconnection and the transport layer are covered in Chapters 9 and 10. The final part covers integrated services digital networks and fast packet switching. Chapter 11 covers ISDN and Broadband ISDN and Chapter 12 discusses the theory and architectures of fast packet switching.

This book can be used as a text when offering a two-semester course on telecommunication networks. Each semester course would contain 2.5 hours of instruction per week for 15 weeks. We suggest teaching Chapters 1 to 6 and 11 in the first semester and Chapters 7 to 10 and 12 in the second semester. The first course is suitable for senior and first-year graduate students. The second is mainly for graduate students, with the first as a prerequisite for the second course. If the book is adopted for only a one semester offering, trading some material listed above for the first semester with added portions from the second semester is recommended. For example, Chapter 3 deals with mathematical techniques for the evaluation of networks and may be omitted or covered in a different order than it appears. Much of the coverage of the book is independent of the material covered in Chapter 3.

We greatly appreciate feedback from instructors, practitioners, and other readers of this book.

T. N. Saadawi
M. H. Ammar
A. El Hakeem

ACKNOWLEDGMENTS

The authors are grateful to a number of individuals whose encouragement and assistance have made this book possible. In particular, we wish to thank Professors D. L. Schilling, P. Enslow, J. F. Hayes, H. Mouftah, R. L. Pickholtz, S. Mahmoud, W. H. Tranter, B. Kraimeche, and G. Lundy.

We are also indebted to our former Ph.D. Advisors: Professor A. Ephremides of the University of Maryland, Professor J. W. Wong of the University of Waterloo, and Professor S. C. Gupta of Southern Methodist University. We are thankful to our students at City College of the City University of New York, the Georgia Institute of Technology, and Concordia University for their helpful comments and feedback.

Finally, this work would have never been completed without the encouragement and understanding of our respective families.

T. N. Saadawi
M. H. Ammar
A. El Hakeem

1

INTRODUCTION TO NETWORKS

1.1 THE GLOBAL VILLAGE

The term *global village* describes how modern technology has made it possible for any point on earth to communicate instantly with any other point. The information infrastructure we are now building is based on computers and the communication networks that interconnect them. This information infrastructure will drastically change our society, our economy, our working habits, and the way we live.

Computers have grown so powerful and cost effective that they are found nearly everywhere doing nearly everything. Thus, as the diversity and sophistication of computers and users' workstations have increased, so have the demands on the telecommunication networks transporting the information and knowledge generated. These networks must be able to provide communication paths to transport scientific and technological data and also serve diverse applications from entertainment to complex computational modeling. These networks must be able to handle and relay data at rates that range from a few characters per second to billions of bits per second (Gbps). In a broader sense, these networks must provide *flexible* information transport. Information should be transported with various degrees of speed, security, and reliability. This is vastly different from the capabilities of the telephone network built to carry voice signals at a fixed speed of 64,000 bps with uniform degrees of security and reliability.

Besides the flexible information transport mechanism, another important ingredient for telecommunication networks is a common language. It is essential that the communicating machines share conventions for representing the information in digital form and procedures for coordinating communication paths. These sets of conventions and the rules that determine how digital information will be exchanged are normally referred to as *communication protocols*. This marriage between two technological advances, namely, telecommunication technology and computing technology is the new challenge facing scientists, engineers, and designers. Below we briefly survey the evolution of both technologies.

1.2 EVOLUTION OF TELECOMMUNICATIONS NETWORKS

The first phase in the evolution of communication networks is dominated by the wide spread of telephony and covers over 90 years (see Fig. 1.1). However, the telegraph preceded the telephone by more than 30 years and the teletypewriter by half a century. In 1844, Samuel Morse, the telegraph's inventor, sent the first telegraph message from Washington D.C. to his colleague Alfred Vail in Baltimore, Maryland. Shortly after the invention of the telephone by Alexander Graham Bell in 1876, it was realized that telephone wires had to converge on central points where telephone-to-telephone connections could be made. These points were manual switchboards, staffed by operators. This was followed by the introduction of electromechanical switching and in 1889, Almond B. Strowger invented the first two-motion step-by-step switch. In a step-by-step switch, a call was established and routed in a set of progressive electromechanical steps, each under the direct control of the user's dialing pulses. It remained as the workhorse of central office switching well into the 1970s. The first crossbar switch was used in 1932 in Sweden. The Bell System introduced the No. 1 Crossbar System in 1938. Crossbar switching was carried out by a special circuit called a marker, which provided common control of number entry and line selection for all calls.

The golden age of automatic telephony occurred after World War II when the entire long-distance network was automated. The invention of the transistor spurred the application of electronics to switching systems and led to the deployment of the first electronic switching in the late 1950s. Electronic switching made it possible to design and build switches with greater capacity.

In the mid 1930s, analog point-to-point radio-relay transmission was possible by extending high-frequency (HF) communication technology into the very-high-frequency (VHF) range. During World War II, the operating frequency bands of the analog radio-relay systems were extended into the ultra-high-frequency (UHF) range. The first link with over 100 repeater stations was placed into commercial service in 1951 between New York and San Francisco, operating in the 4-GHz band with a 20-MHz band. This was followed by a growth of medium- and high-capacity analog radio-relay systems worldwide. By the early 1970s, digital microwave radio had gained importance and, by the 1980s, quadrature amplitude modulation (QAM) methods were widely used as the modulation scheme for digital radio-relay systems.

The second phase of networking, which occurred in the 1960s, included three major milestones: software switching, digital transmission, and satellite

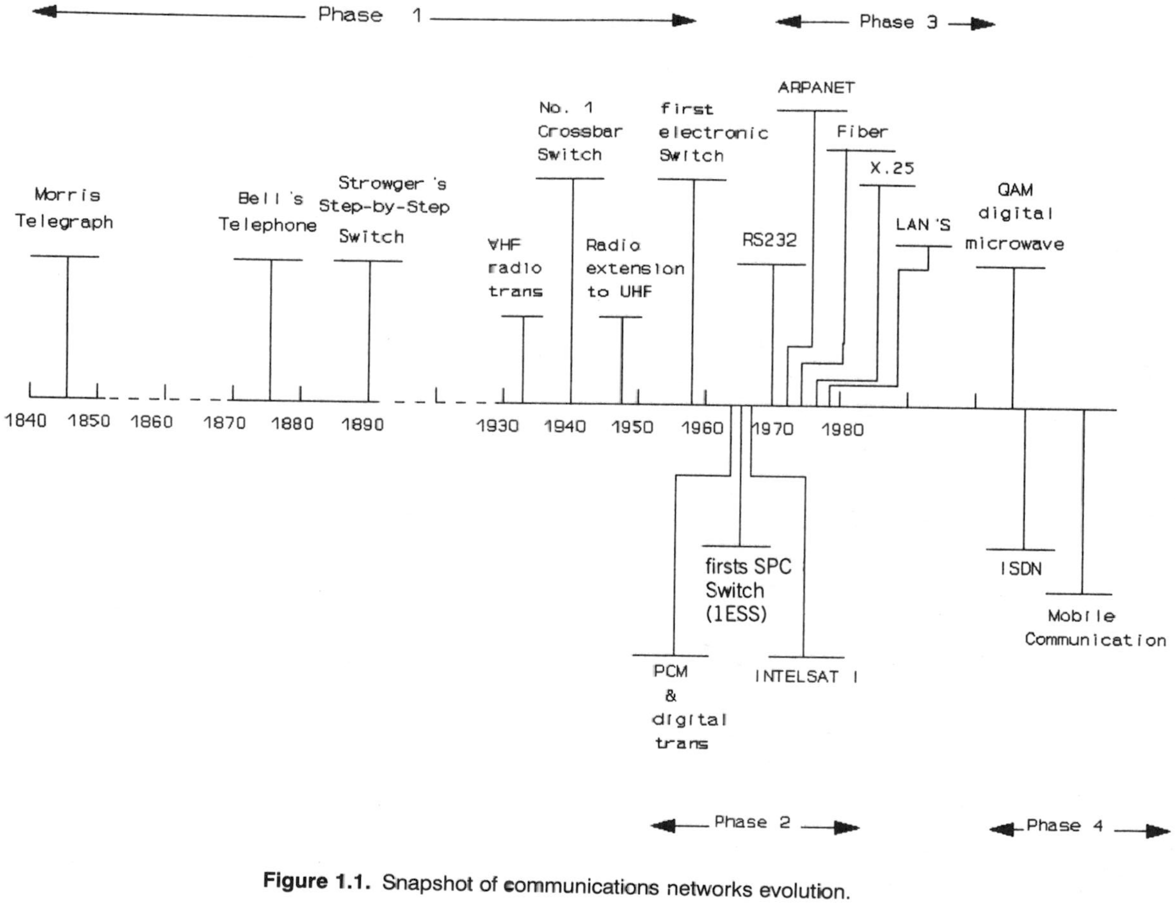

Figure 1.1. Snapshot of communications networks evolution.

deployment. In 1965, American Telephone and Telegraph (AT & T) introduced the first stored-program control (SPC) local switch, the 1 EES* switching system. With the use of software control, a family of custom calling services (speed calling, call waiting, call forwarding, and three-way calling, etc.) was made possible. The first software program for the 1 ESS switching system had approximately 100,000 lines of code; today switches may contain ten million lines of code.

The principle of converting analog signals to digital signals became popular with the introduction of pulse code modulation (PCM); consequently, the known rate for telephone-quality speech (4 KHz bandwidth) is 64 kbps. During the 1960s and 1970s, a hierarchy of digital transmission channels based on the 64 kbps channel was established and still forms the backbone of today's digital network. One of the popular digital transmission systems is called the *T1 carrier*, which supports 24 voice channels, each 64 kbps, resulting in a total of 1.544 Mbps.

Satellite communication was first proposed in 1945 by the British science fiction writer, Arthur C. Clarke. Satellite communications became a reality with the launching of the Russian satellite Sputnik (in 1957) and the American satellite Explorer (in 1958). The AT & T Telstar was the first experimental satellite capable of relaying television (TV) programs across the Atlantic Ocean. It was launched from Cape Canaveral (now Cape Kennedy) on July 1962. The first global civil communications satellite, INTELSAT I (Early Bird), was launched in April 1965.

The third phase of the evolution of communication networks occurred in the 1970s and is characterized by the introduction of data networks and packet switching technology. The concept of packet switching was first published in a 1964 report by Paul Baran of the Rand Corporation for the U.S. Air Force. In 1966, under the sponsorship of the Advanced Research Project Agency (ARPA) of the U.S. Department of Defense (DOD), an experimental packet-switching network, ARPANET, was set up and was put into service in 1971. ARPANET† led to the development and the widespread use of packet switching technology under the direction of Larry A. Roberts. Commercial offerings of packet-switched services rapidly followed the lead of the experimental networks, both in the United States and abroad. For example, in 1973, Bolt, Beraneck and Newman (BBN), Inc. founded TELENET, the first public packet-switched network, linking host computers and dial-up terminal users. The Canadian DATAPAC network was developed during the 1973–1977 period. The French Institute for information and automatic research, in 1973, set up the CYCLADES and CIGALE networks. As data networks advanced from terminal-oriented systems to packet-switched, computer-to-computer linkups, the protocols necessary to make networks function also grew more complex. Two fundamental standards were essential

*ESS is a trademark of AT & T.
†ARPANET was retired as of July 1990 after over 20 years of service.

for the advancement in data networks. The first is the American Standard Code for Information Interchange (ASCII), which was approved in 1964 and became the common method of coding data for communications. The second standard is the Electronics Industries Association (EIA) recommended standard (RS) RS-232D. Its first version, issued in 1969, specified how the encoded information would be transmitted by modem over a telephone network.

To achieve compatibility between computers and packet-switched networks, the International Telegraph and Telephone Consultative Committee (CCITT* from the French acronym) established a world-standard protocol, called X.25 in 1976. This led to several other international protocols for the interconnections between data networks. The widespread adoption of X.25 opened the door for the growth of packet-switched networks in the United States and other countries. Working closely with CCITT, the International Organization of Standardization (ISO) in 1978 approved its seven-layer framework of protocols for data communications, called the *open-system-interconnection* (OSI) reference model. The purpose of the OSI reference model is to allow any computer anywhere in the world to communicate with any other, as long as both obey the OSI standards.

This phase of networking is also characterized by the introduction and wide acceptance of local area networks (LANs). The oldest and best known of all LANs is the Ethernet, which started as a laboratory project in 1974 by R. M. Metcalfe and his colleagues at Xerox Corporation's Palo Alto Research Center. Ethernet was inspired largely by Alohanet, a packet-switched radio network developed by Norman Abramson at the University of Hawaii.

The invention of the laser (which stands for light amplification by simulated emission of radiation) in 1959 led to major technical developments in the field of optical communication. In 1970, Corning Glass Works reported on the first low-loss optical fiber (doped-silica clad fiber) that achieved 20 dB/km loss.

The fourth phase of communication networking, which began in 1980, is characterized by the availability of the integrated services digital networks (ISDN) and mobile communications. ISDN can be regarded as a general-purpose digital network capable of supporting a wide range of services such as voice, data, and image. A key ingredient in ISDN is the provision of a common integrated digital access between the customer's premises and the ISDN to support this multiplicity of services.

Mobile communications entered a new era with the introduction of the cellular concept. In 1981, the Federal Communications Commission (FCC) allocated 50 MHz of the spectrum (824–849 MHz and 869–894 MHz) to mobile radio cellular systems. By 1990, the U.S. cellular service has over five million subscribers.

*As of 1993, CCITT has become the International Telecommunications Union-Telecommunication Standardization Sector (ITU-TSS)

TABLE 1.1 The Four Phases of Communications Networks

	Telephony	Digital Network	Data Networks	Integrated Digital Networks
Year	1880s	1960s	1970s	1980s
Type of traffic	Voice	Voice	Data	Voice, data video, imaging etc.
Switching technology	Circuit switching (analog)	Circuit switching (digital)	Packet switching	Circuit, packet, and fast packet switching
Transmission media	Copper, then microwave	Copper, microwave, satellite	Copper, microwave, satellite	Copper, microwave, satellite, fiber optics

The four phases of the evolution of communications networks are shown in Table 1.1.

1.3 EVOLUTION OF COMPUTING TECHNOLOGY

The changes in computer technology can also be classified by four distinct phases. The first computer was invented in the late 1940s, and the programmable calculator which was designed as an engineering tool, became commercially available in the 1950s. The first phase of computing came in 1960s, when the computer was used as a data-processing machine. Computing power was expensive and data had to be processed in huge batches. Only large organizations could generate such batches. Jobs (i.e., programs with their accompanying data) were coded on punched cards or magnetic tapes and results were delivered as "Listings" printed on fanfold, perforated paper.

The second phase in the evolution of computing came in the 1970s, when computer services were shared among many subscribers. Thus, time-sharing made data processing more affordable by allowing many subscribers to split the cost of a computer. Computers became easier to use because they could be reached from a terminal and interrogated in real time.

The third phase, in the 1980s, transformed computers into a desktop productivity tool for individuals. Advances in microprocessor technology enabled manufacturers to fit a computer on a single chip, making it cheaper to buy a small computer than share a large one. Noticeable progress in super computer technology was also achieved during this phase.

The fourth phase, now underway, has two distinguishing features: (1) a reduction in the size of personal computers leading to an increase in their portability and mobility and (2) increased reliance on computer networking

TABLE 1.2 The Four Phases of Computing

	Batch	Time Sharing	Desktop	Network
Year	1960s	1970s	1980s	1990s
Location	Computer room	Terminal room	Desktop	Mobile
Technology	Medium-scale integration	Large-scale integration	Very large scale	Ultra large scale
Network architecture	None	Centralized	Centralized /distributed, LAN, WAN	Centralized /distributed, LAN, WAN, wireless

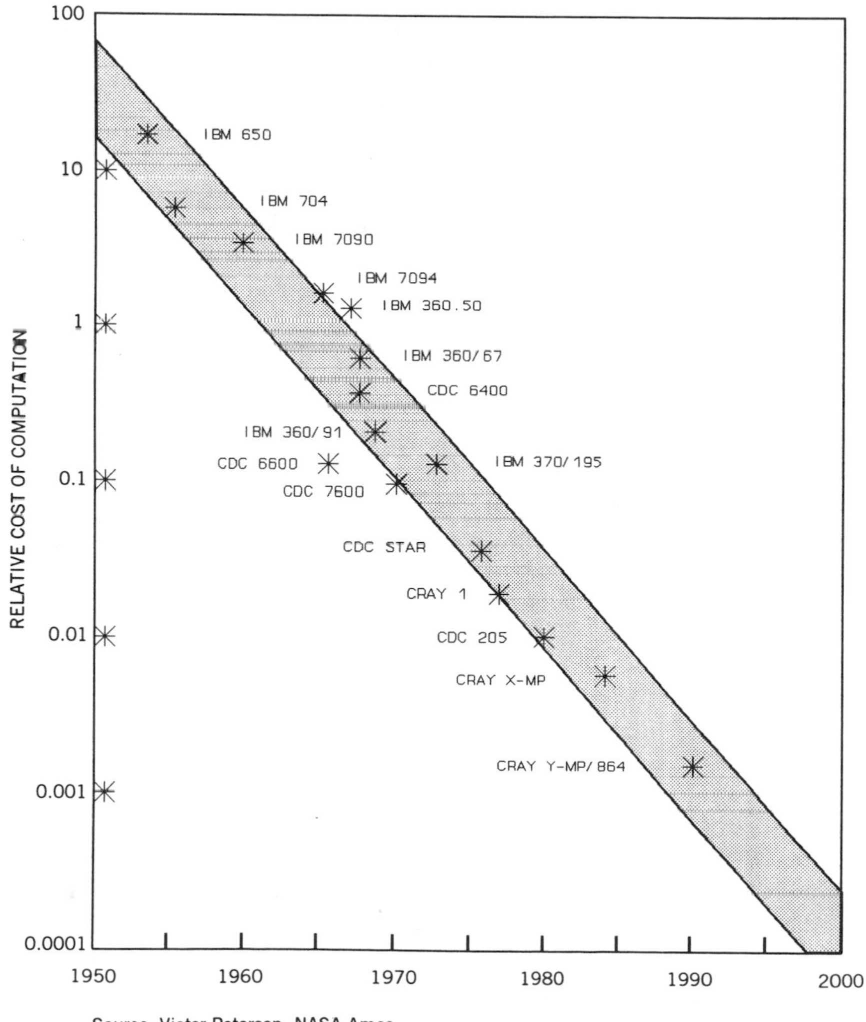

Source: Victor Peterson, NASA Ames

Figure 1.2. Relative cost of computing.

by computer operating systems and by user applications (e.g., client server technology). Innovations in multiprocessor technologies have also made large-scale supercomputers even faster and more powerful than before.

The four phases of the development and evolution of computers technology are summarized in Table 1.2 and illustrated in Figure 1.2, in which the cost for the most powerful machine of each era is plotted against a time line. The figure shows that the cost of computing is halved approximately every 3 years.

Networking became important as succeeding phases of computing increased the number of computers in service. When batch processors were few, there was no interest in coordinating their operation. Time-sharing, however, is unthinkable without networks; dedicated or dial-up lines must connect numerous terminals to a host computer, either a corporate mainframe or a departmental minicomputer. In the desktop computing era, computer networks have assumed new forms. Local-area networks connect personal computers to one another and to shared machines; both general purpose computers, called hosts or clients, and special purpose computers, called servers, provide communal files, high-quality printing, and institution-wide electronic mail. Wide-area networks (WANs) and metropolitan area networks (MANs) interconnect the various locations in an organization, linking mainframes and file servers to desktop machines. In the fourth phase of computing, where computers are becoming mobile, wireless connections and wireless networks will proliferate.

1.4 WHAT IS A COMMUNICATION NETWORK?

A *communication network* can best be defined as the set of devices, mechanisms, and procedures by which end-user equipment attached to the network can exchange meaningful information. The typical functions required of a communication network include:

1. A path by which electric signals (i.e., changes in voltage) can be transmitted.
2. A mechanism by which bits can be converted to and from electric signals.
3. The means to impart meaning to groups of bits. Bits are typically grouped in units called frames, packets or messages.
4. Methods to overcome deficiencies in the electric path that might cause electric signals (and therefore the bits they represent) to be misinterpreted.
5. Techniques for selecting and maintaining a path through the network to perform the above functions.

Networks can be classified as broadcast networks, switched networks, or hybrid networks. In a broadcast network, signals transmitted by one end-user's equipment are automatically heard by all other end-users' equipment. In switched networks, signals have to be "routed" through intermediate network points (called switches) to reach their desired destination. In the CCITT terminology, the switch or node is called the data terminal equipment (DTE), whereas the user equipment is called the data circuit-terminating equipment

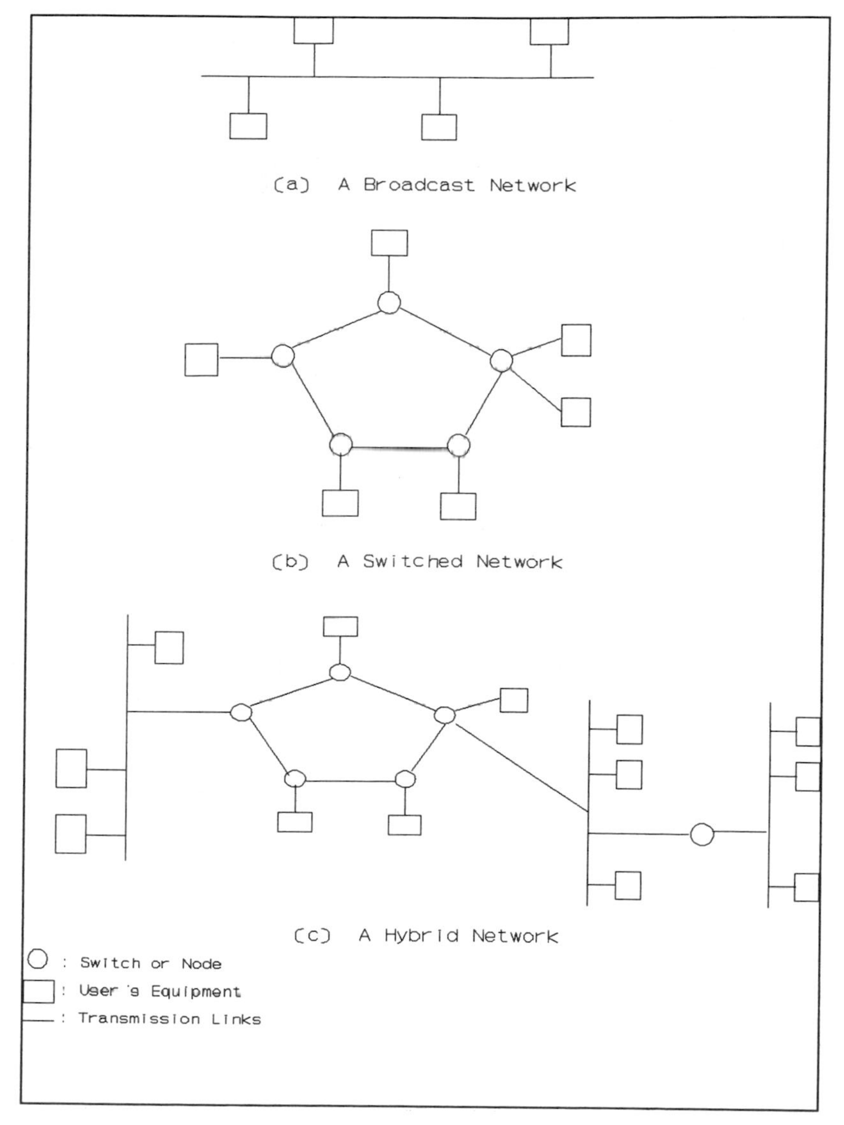

Figure 1.3. Communication Networks.

(DCE). Hybrid networks consist of a mixture of the other two types of networks, in that signals sometimes (but not always) need to pass through switches to reach their desired destination. Figure 1.3 shows the three types of networks.

1.5 SWITCHING TECHNIQUES

Switching techniques in telecommunication networks can be divided into three main types: circuit switching, message switching, and packet switching. *Circuit switching* is the oldest form of switching; it dates back to the telegraph era and is the common method for switching in the telephone network. In circuit-switched networks, there is a dedicated path, a connected sequence of links, between the calling and called stations for the duration of the call. A special signaling message (dialing tone) travels across the network and captures channels in the path as it proceeds to set up the circuit. After the path has been established, a return signal to the source indicates that information transmission may proceed.

Other forms of switching are typically used for data transmission where a continuous path is not established, rather memory in the switching node is used to store the message to be transmitted awaiting the availability of transmission facilities. This form of switching is known generally as *store-and-forward switching* and includes message switching and packet switching.

Both message switching and packet switching are more appropriate for data communication than circuit switching. In many data communication applications, data occur in bursts separated by idle periods, and the average data rate may be much lower than the peak rate (Fig. 1.4). This type of intermittant data can often be transmitted more economically by assembling the data into packets (or messages) and interspersing packets from several channels on one physical communications path.

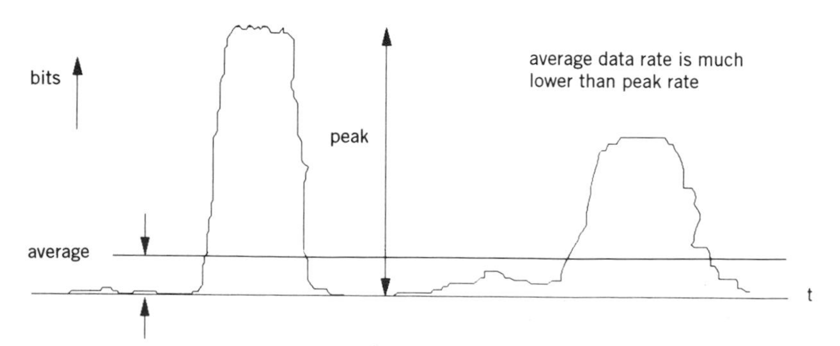

Figure 1.4. Data occur in bursts.

In *message switching*, the message is sent from one node to the next node in its path. The entire message is received at each node, stored, and then sent to the next node. Thus, the message "hops" from one node to the other along the path, with the possibility of being queued at busy channels.

Packet switching is the same as message switching, except a message is divided into a series of packets of limited lengths, each seperately addressed and sequentially numbered. The packets composing the message are then routed to their destination independently; many packets of the same message may be in transmission simultaneously. This pipelining effect results in reduced transmission delay. The first and still one of the best-known packet-switching network was the so-called ARPANET developed by the U. S. Department of Defense Advanced Research Projects Agency (ARPA) in 1969 to connect universities and Department of Defense operations involved in national defense research and development.

Figure 1.5 compares circuit-, message-, and packet-switching techniques. The transmission path shown in the Figure consists of four nodes (A through D) and three transmission links. For circuit switching (Fig. 1.5*b*), a dialing signal travels, from node A through D to set up a connection to the destination. There is a connection delay at each node. Also the transmission time and propagation time for a signal from one node to the other add to the total set-up delay. Note that the propagation time is the time it takes for a signal to propagate at the speed of light, typically 2×10^8 m/second in guided medium, from one node to the other. The transmission time is the time it takes for a transmitter to send out a block of data. For example, it takes 1 msecond to send 1000 bits of data over a 1 Mbps link. After the arrival of a return signal to the source, the path is essentially an open pipe from the source to the destination and the message is sent as a single block.

In message switching (Fig. 1.5*c*), there are no call set up procedures. However, there is a message header for identification and routing. Since the message is stored till the channel becomes available, considerable extra delay may result from storage at individual nodes. Thus, circuit set-up delays encountered in circuit switching are replaced by queueing (storage) delays in message switching.

For packet switching (Fig. 1.5*d* and *e*), the message is broken into three packets. The effect of pipelining is clear. For example, while packet 2 is being sent from A to B, packet 1 is being processed at node B and sent over link BC. This simultaneous transmission of packets results in a reduction of the total time needed to send data across the network. Also, since the individual packets are shorter compared to the complete message, they are less likely to have errors.

In summary, for a long continuous stream of data such as voice and video signals, circuit switched networks are probably an efficient approach. On the other hand, if the data flow comes in bursts, such as interactive data traffic, some technique of resource sharing, such as packet switching, should be used.

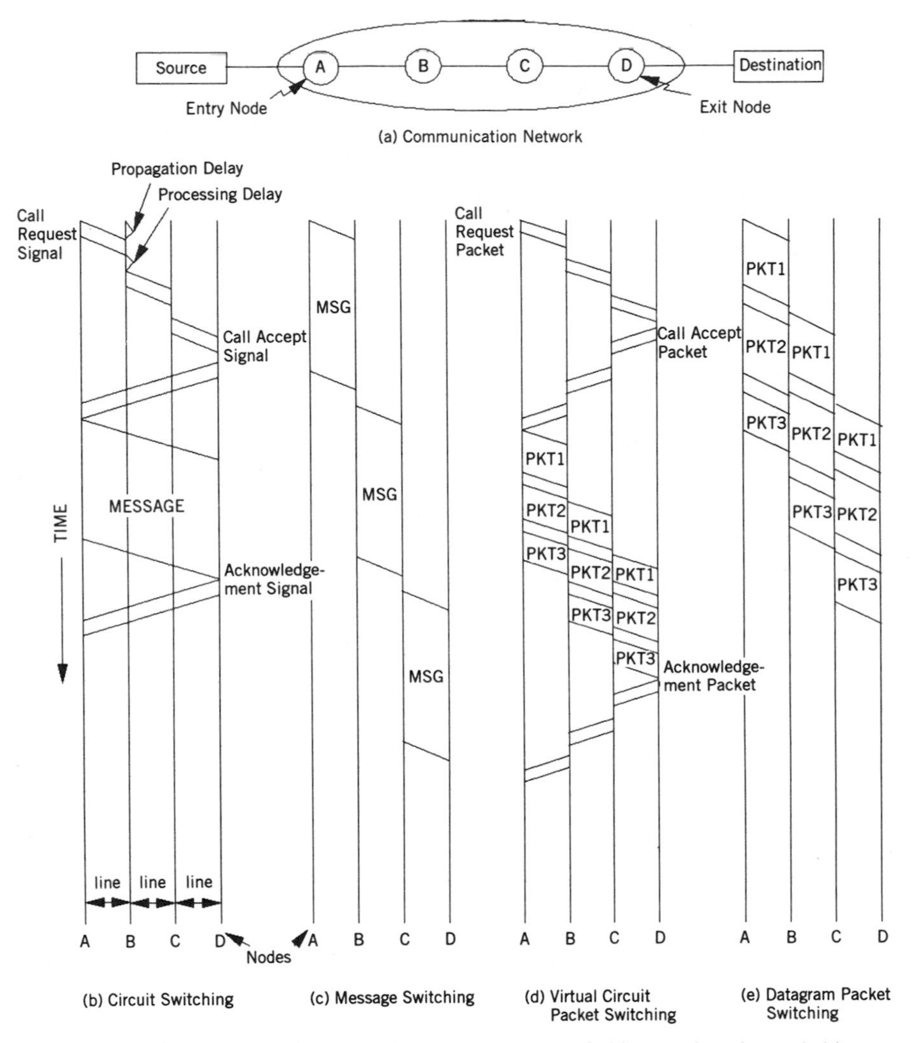

Figure 1.5. Comparison of circuit switching, message switching, and packet switching.

To better understand the difference between circuit switching and packet switching, we draw an analogy to a highway network. The capacity of the highway network is basically the number of lanes in a highway. One vehicle is analogous to one packet. Since vehicles, coming from different sources and heading to different destinations share a given lane, this is similar to packet switching. The toll-booth resembles the packet-switching nodes; vehicles wait in a queue and then advance to the toll-booth clerk (server), then follow the signs for the proper highway (routing function). An example of circuit switching in the highway network is when the President of the United States travels. Security and police forces first set up and reserve the path, and the highway (the channel) becomes physically dedicated to the President.

Packet switching can be classified into two basic techniques; virtual circuit and datagram. *Virtual circuit* (Fig. 1.5*d*) resembles the traditional circuit switching. To initiate a virtual-circuit call, the origin sends a call-request packet to node A, which relays it to B, then ultimately to the destination. A call-accept package is sent back to the origin. Thus, in this phase, a route has been set up prior to data transfer. Unlike circuit switching this route is only a logical connection and the path is not dedicated to this connection only and the packets are still stored at every node.

In *datagram* packet switching (Fig. 1.5*e*), there are no call-request and call-accept packets to establish a path between the source and the destination. The header of a typical datagram packet contains the packet's ultimate destination and its place in the message sequence. Unlike the header of a virtual-circuit data packet, it does not contain routing information; each node will make routing decisions as the packet moves through the network. Each node reads the destination information in the data packet's header, chooses the best route available at that moment, and sends the packet on to the next node. At the last node, the packet is stored until all other packets forming the message have arrived and sequenced.

1.6 NETWORK PROTOCOL ARCHITECTURE

The most basic reason for the existence of a communication protocol is to ensure that two communicating entities can send, receive, and interpret the information they wish to exchange. A protocol architecture is a framework into which protocol functionality is made to fit. Such architectures have been of great value in reducing the conceptual complexity inherent in the end-to-end communication task. Most protocol architectures are based on the concept of layering. In such architecture, an end-to-end (or application-to-application) communication task is accomplished by successively and incrementally "adding value" in each protocol layer.

To better understand the layered communication architecture, we make an analogy with human communications (Fig. 1.6). In the figure, the President of a Greek company is communicating with the President of a Spanish company, however neither speaks the other's language. Each employs an English translator and the translated messages are sent by the Fax operator over the physical communication channel. Thus, the two presidents are communicating with each other through three layers below them.

In this section, the layering concept is expanded by first discussing the open system interconnection (OSI) reference model and then discussing the other most prominent layered architectures: the Internet protocol architecture, IBM's system network architectures (SNA), and Digital's network architecture (DNA).

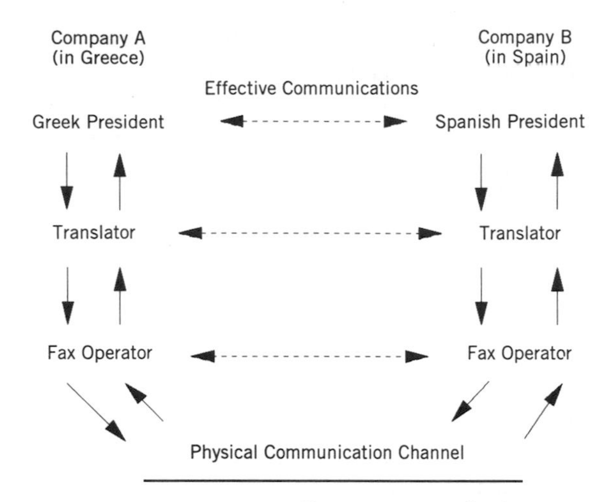

Figure 1.6. An example of human communications.

1.6.1 Open-System-Interconnection (OSI) Reference Model

As mentioned earlier, networking would be facilitated and the expense much reduced if computer manufacturers and customer equipment vendors could agree on a common communication interface protocol. Toward that goal, in 1978, the Geneva-based ISO set forth the OSI reference model.

The basic objective of the OSI reference model is to standardize the rules of interaction among interconnected systems. The choice of the term *open* clearly signifies that a system, conforming to those international standards, will be open to all other systems obeying the same standards throughout the world. Thus, only the external behavior of open systems must conform to OSI architecture. The internal organization, architecture, and functioning of each individual open system are outside the scope of OSI standards because they are not visible from any other system with which that individual system is interconnected. It is not the intent of the OSI to standardize the internal operation of a system. A system is considered to be one or more autonomous computer(s) and the associated software, prephirals, and users that are capable of information processing and/or transfer.

The OSI model divides the communications process into seven layers with each layer wrapping the lower layers and isolating them from the higher layers. Layering thus divides the total communications problem into smaller functions. It also ensures independence of each layer by defining services provided by a layer to the next layer, independent of how these services are performed.

Figure 1.7 shows the concept of layering. We refer to any layer as the N layer, and its next lower and next higher layers are referred to as the $N - 1$ layer and the $N + 1$ layer, respectively. There are one or more entities in

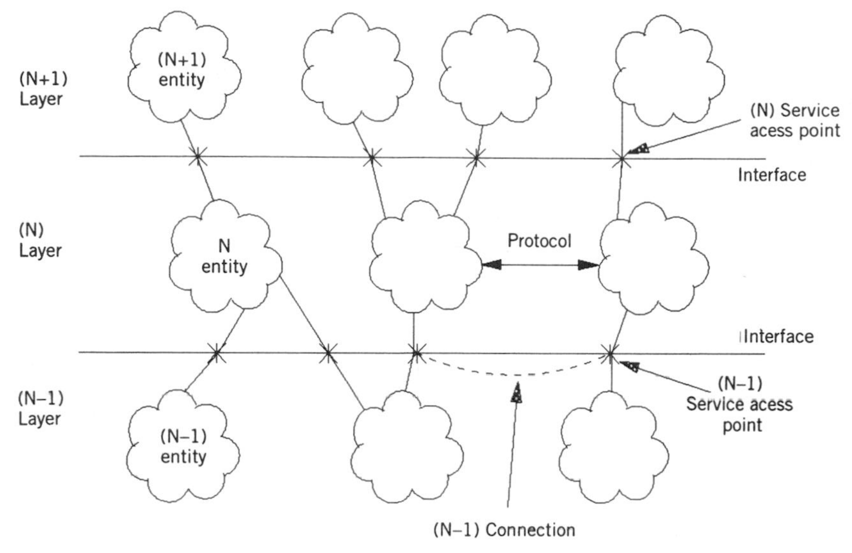

Figure 1.7. Concept of layering.

each layer. An N entity executes the functions of the N layer and also the protocol for communicating with N entities in other systems. A subroutine or a process is an example of entities. Communications with entities in the layers above and below a given layer is done across an interface. The logical interfaces between the N entities and the $N + 1$ entities are represented by the N service access points (SAP's). The $N - 1$ entity provides services to an N entity via the invocation of primitives. A *primitive* defines the function to be performed and passes data and control information.

Figure 1.8 shows the OSI seven-layer model. The bottom three layers (physical, data link, and network layers) provide telecommunications and networking functions. The highest three layers (session, presentation, and application layers) provide processing and dialog functions (session and presentation layers) and application control (application layer). The middle layer (transport layer) acts as a bridge between communications-oriented layers and processing-oriented layers.

The lowest layer (layer 1) in the model is the *physical layer*. It provides the physical medium for the information flow. In other words, it covers the physical interface between the devices and is concerned with transmitting raw bits over the communications channel. Thus, it is responsible for activating, maintaining, and deactivating the physical circuit between the sender and the receiver and providing the clocking signal. It also has a mechanism for informing layer 2 of the loss of a physical connection or electrical power. Examples of the physical interface are X.21, EIA-232-D, EIA-530, V.22 bis, and V.35.

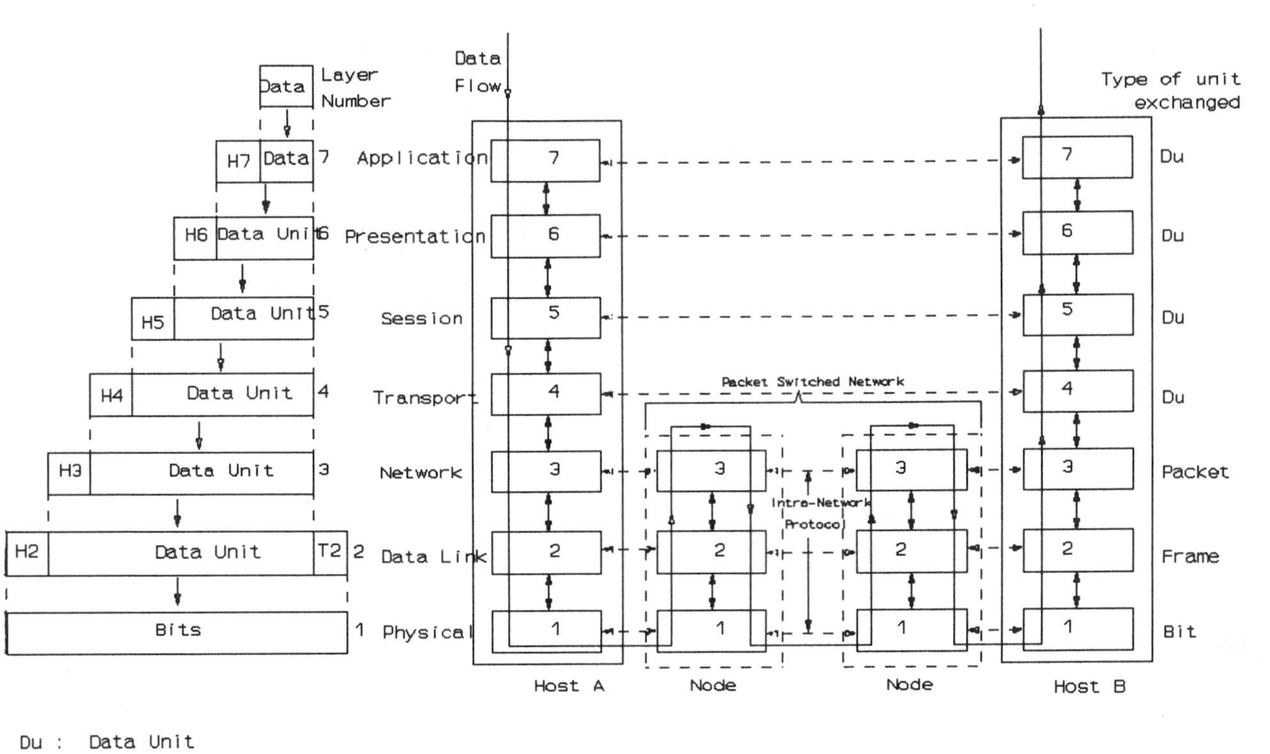

Figure 1.8. Open-system-interconnection reference model architecture.

The second layer is the *data link layer*, which is responsible for the transfer of data across the link. An important task for the data link layer is to take the transmission facility offered by the physical link and transform this into a link that appears free of transmission errors to the next layer above, the network layer. A layer 2 protocol achieves this transformation by breaking the data into frames, then transmitting these frames sequentially, and processing acknowledgment frames sent back by the receiver. Examples of data link layers for point-to-point connections found in switched networks are high-level data link control (HDLC), IBM's binary synchronous communications (BISYNC), and X.25 Link Access Procedures-Balanced (LAP-B); these will be discussed in Chapter 4. An example of a data link layer for a broadcast network is the IEEE 802 for local area networks (discussed in Chapter 8).

Layer 3 in the OSI model is called the *network layer*. The purpose of the network layer is to provide the functional and procedural means to set up and terminate a call, to route data, and to control data flow across the network. The transport layer is concerned only with the quality of service and its cost, not with whether optical fiber, packet switching, satellites, or local area networks are used. Thus, it is the function of the network layer to provide transport entities with independence from routing and switching considerations. Network layer protocols include level 3 of CCITT X.25 interface and the connectionless network layer protocol (CLNP).

The *transport layer* provides transparent transfer of data between end systems, thus relieving the upper layers from any concern with providing reliable and cost effective data transfer. Thus, the transport layer is an end-to-end layer; a process on the source end-system converses with a peer process on a destination end-system. Examples of the transport layer protocols are those contained in the ISO draft international standard 8703 and its CCITT equivalent X.224.

The *session layer* is concerned primarily with the characteristics of session connections. It provides the means for cooperating presentation entities (processes) to organize and synchronize their dialog and manage their data exchange. The session layer is the user's interface into the network.

The main function of the *presentation layer* is the provision of a mechanism for presenting data in a manner that can be understood by both the sending and receiving application processes or devices. The mechanism provided in the session layer allows for organizing and structuring the interactions between application processes.

The *application layer* is the highest layer in the OSI architecture. Application-layer protocols directly serve the end user by providing distributed information appropriate to an application, to its management, and to system management. It provides facilities for both connection-oriented and connectionless communication among application processes. Examples of connection-oriented applications are bulk file transfer, virtual terminal usage (i.e., long attachment of a terminal, workstation, or other device to a remote host), and stream-oriented access to distributed system components (such as print servers, and remote-job entry stations). Examples of connectionless applications are inward data collection (periodic active or passive sampling of a large number of data sources), outward data dissemination (the distribution of a single piece of information to a large number of destinations), broadcast and multicast (group-addressed) communication, and a variety of request–response applications, in which a single request is followed by a single response.

Figure 1.8 shows also the flow of information through the OSI reference model. User data messages pass down through the seven-layer model, each layer adding a control and communication capability and a protocol header to the message. This protocol header carries control information to the corresponding (peer) layer in the destination node for that layer. The message is then transmitted over the physical link to the first node (switch) in the network and back up through the first three layers; the protocol headers are removed along the way. This process is repeated in the network switches along the message's path. The message goes up the layer hierarchy to the destination user and the protocol headers are removed along the way. Adhering to the rule of protocol purity, each layer only examines its own protocol header.

Note that protocols at or above the transport layer are required for end-user functionality. Therefore, they are not required at intermediate points in the network.

1.6.2 The Internet Protocol Architecture

The Internet is perhaps the most successful and largest computer network in the world. As of 1993, it connects approximately 1.3 million (computers or hosts) and is experiencing an 80 percent annual growth rate. In the Internet, each host is typically connected to a particular network. These networks are in turn interconnected via gateways.

The Internet protocol architecture represents the evolution of the consensus of several software and hardware vendors, academic and industrial researchers, and various governmental agencies. The Internet standardization process is maintained by the Internet Architecture Board. The Internet protocol architecture defines the four layers listed below.

The Application Layer. It is the topmost layer of the architecture. A distinction is made between user application protocols and support application protocols. Among the most popular protocols are file-transfer protocols (FTP) and simple-mail-transfer protocols (SMTP). Support protocols include simple-network-management protocols (SNMP) and domain-name-system (DNS) protocols. The DNS set of protocols is used to map host names and internet addresses.

The Transport Layer. As in the OSI model, the Internet transport layer provides end-to-end communication services to the applications. The transmission control protocol (TCP) and user datagram protocol (UDP) are the primary Internet transport protocols.

The Internet Layer. One of the defining features of the Internet is the use of the Internet protocol (IP) to provide network layer services (as defined by the OSI model). Many management and control protocols such as the Internet control message protocol (ICMP) are used to support the services provided by IP.

The Link Layer. This is the lowest layer of the Internet architecture. It is the means by which hosts interface with the local network. Many protocols are used in this layer depending on the type of local network used.

1.6.3 Systems Network Architecture Protocol

Systems network architecture (SNA) provides the structure for creating networks of IBM computers and devices. The SNA protocol layers are (1) physical control (2) data-link control, (3) path control, (4) transmission control, (5) data-flow control, (6) presentation services, and (7) transaction services (Fig. 1.9).

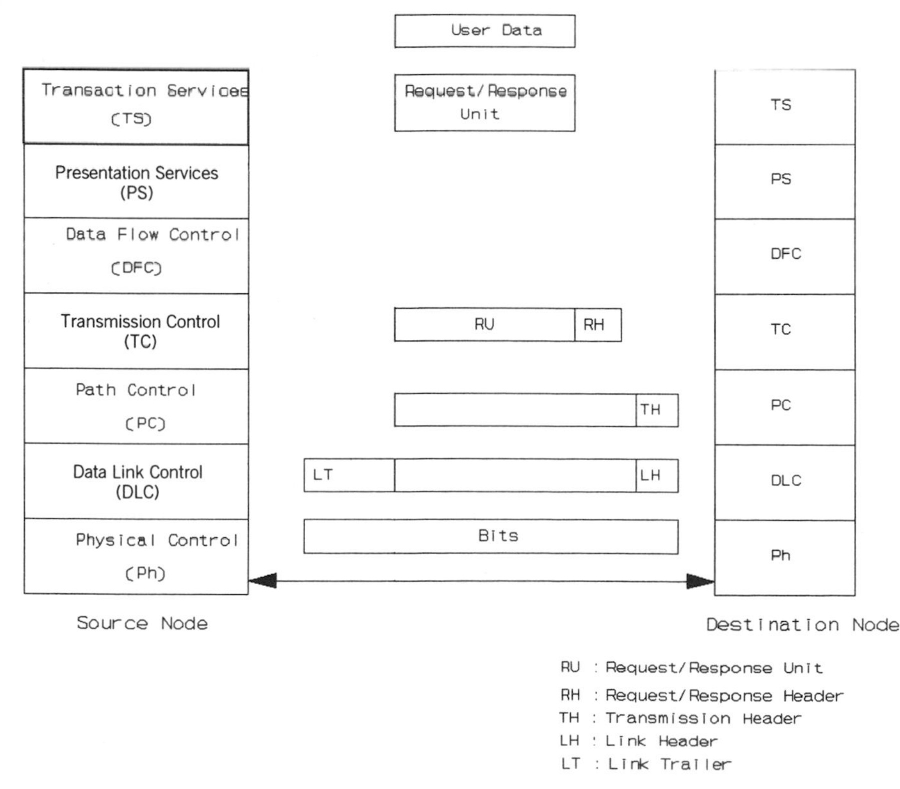

Figure 1.9. Systems network architecture protocol.

The physical control layer corresponds to OSI layer 1. The data-link control layer creates an error-free physical link and controls the flow of information on that link. The path control layer routes messages between sources and destinations. The transmission control layer coordinates a session and manages the flow of data on that session. The data flow control layer controls the user request–response flow and ensures flow integrity at the user level. Finally, the presentation services layer provides services such as data transformation (such as data and text compression), additions (such as column headings for display), and editing and translations (such as program commands into local terminal languages).

The transaction services layer provides network management services. These services include configuration services, which allow an operator to start up or reconfigure the network, operator services, such as network statistics gathering and display, maintenance services, and session services.

Figure 1.9 shows also SNA data encapsulation. Systems network architecture does not require a different header at each layer of the protocol hierarchy. User data network control information is converted into a request–response unit (RU) by the transaction services and the presentation services. The data flow control layer passes the information to the transmission control, which adds to the RU a request/response header (RH) on behalf of itself and the data-flow control. The path control adds a transmission header and the data-link control adds a link header (LH) and link trailer (LT). The resulting frame is passed to the physical control layer for transmission as a string of bits.

1.6.4 Digital Network Architecture

Digital network architecture is the architecture on which the Digital Equipment Corporation Network (DECNET) implementations are based. The DNA layers are (1) physical link, (2) data link, (3) routing, (4) end communications, (5) session control, (6) netwoks application, and (7) network management (Fig 1.10).

The physical layer has the same definition as OSI reference model. The data link layer creates error-free sequential channels using the Digital data communications message protocol (DDCMP), the Ethernet, and X.25. The routing layer (previously called the transport layer) implements a datagram service and provides the routing and congestion control functions. The end communications layer (previously called the network service layer) provides end-to-end error and flow control, reassembly, and segmentation. It is similar to the OSI transport layer. The session layer provides logical communications between users. The network application layer is similar to the OSI application and presentation layers. It provides functions such as remote file access and transfer, gateway access to other networks, and remote interactive terminal access. The network management layer allows for monitoring and

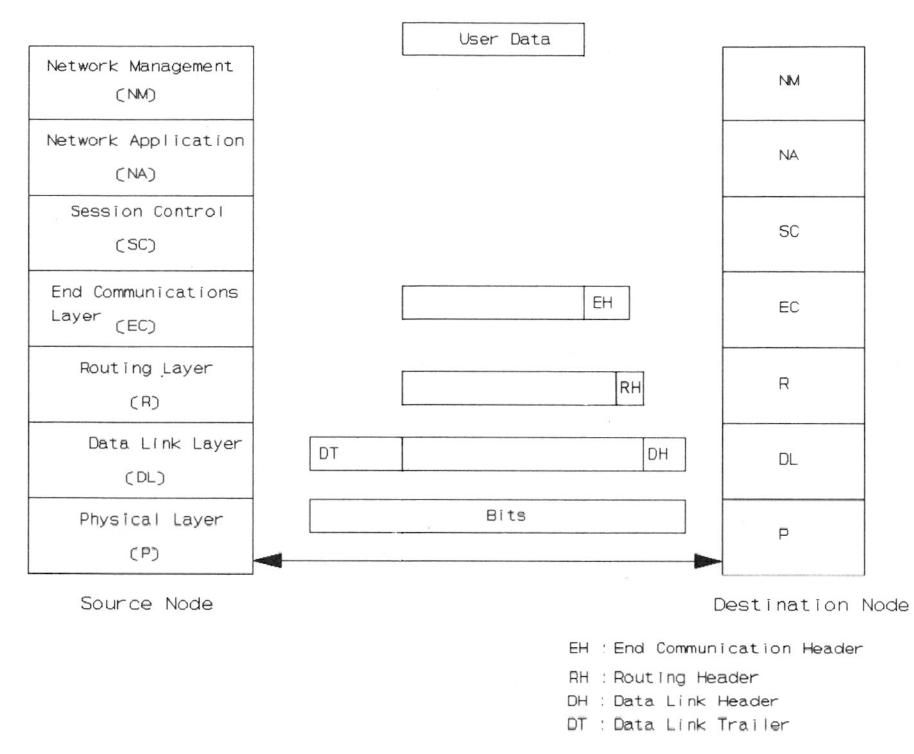

Figure 1.10. Digital-network-architecture protocol.

management of the network operations. Finally, the user layer provides a user interface to network management functions.

1.7 TYPES OF COMMUNICATION NETWORKS

Different types of communication networks are in existence today. The most common types of communication networks are the telephone network, wide area data networks, local area networks (LANs), metropolitan area networks (MANs), radio networks, satellite networks, mobile phone networks, and cable television networks. Below, we provide a brief description of these types of network.

1.7.1 The Telephone Network

The telephone network consists of three main components; the customer premises equipment (CPE), transmission facilities, and switching facilities. The *CPE* is the equipment located at the customer's location. Such equipment may include devices such as simple telephone instruments, modems, answering machines, and large private branch exchanges (PBXs).

Transmission facilities can be divided into the local loop and the trunk lines. The *local loop* (or subscriber loop) connects the CPE with the telephone company's switching office (central office or local exchange). Today most of the telephone local loops are wire-pair cables; however, a large percentage of the new installations are using fiber-optic cables. Local loops average more than 2 miles. It is estimated that there are several hundred million miles of telephone subscriber loops in the United States. *Trunk lines* (or circuits) connect two switching systems. Trunk lines carry traffic generated by many customers, whereas loops are dedicated to individual customers. The trunk lines range in length from less than a mile to several thousand miles. A variety of different transmission media is employed, including wire pairs, coaxial cables, microwave radio, satellites, and fiber optics.

The main function of the switching system is to interconnect circuits and route traffic through the network. Without switches, each subscriber would need a seperate direct line to every other subscriber. Switching systems in the telephone network can be classified into two groups: local and tandem. Local switching systems are called *central office* (CO) switches, and they connect customer loops directly to other customer loops or customer loops to trunks (Fig. 1.11). *Tandem* switches connect trunks to trunks or simply connect one CO switch to another CO switch. A *toll* switch is a tandem switch that serves the long distance network.

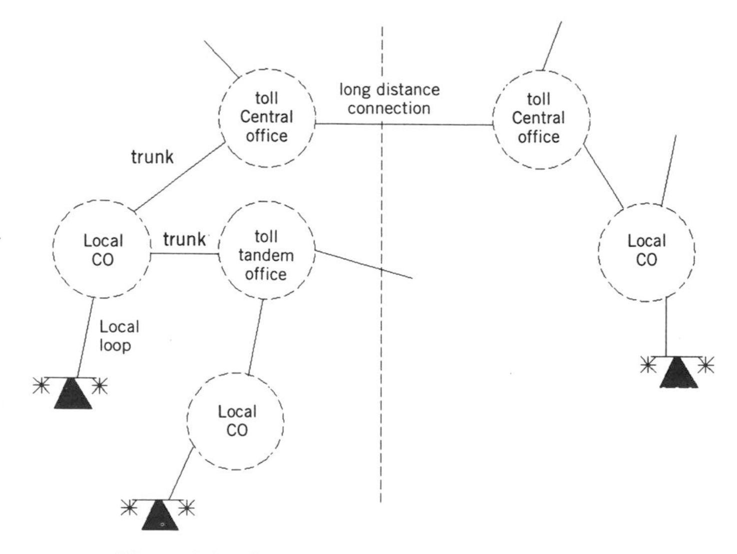

Figure 1.11. Basic elements of a telephone network.

1.7.2 Wide-Area Data Communication Networks

As mentioned earlier, the packet-switching technique is more appropriate for data transmission. *A packet-switched public data network*, similar to the telephone network, consists of customer-premises equipment (a computer or a terminal), network packet-switching nodes (a general-purpose computer for example), and transmission facilities. Telenet and Tymnet are examples of public-switched data network. Much of this text is devoted to issues dealing with this type of network.

1.7.3 Local Area Networks

A local area network (LAN) is a data communication network, typically a packet communication network, limited in geographic scope (ranging from a few meters up to a few kilometers). A local-area network generally provides high-bandwidth communication over inexpensive transmission media. A local-area network is composed of hardware elements and software elements. Hardware elements belong to three basic categories: a transmission medium (twisted pair, coaxial cable, fiber optics, or radio), a mechanism for control of transmission over the medium, and an interface to the network for devices (host computers, printers, servers, etc.) that are connected to the network.

The software elements are the set of protocols, implemented in the devices connected to the network that control the transmission of information from one device to another via the hardware elements of the network. These protocols function at various levels from data-link layer protocols to application layer protocols as previously mentioned in the discussion of the OSI model. Also, LANs are characterized by a large and often variable number of devices requiring interconnection. Thus, we face the situation in which a high bandwidth channel with short propagation delays is shared by independant users. Chapters 7 and 8 provide a detailed discussion of LAN principles and technology.

1.7.4 Metropolitan Area Networks

A metropolitan area network (MAN) is another example of data networks. They differ from LANs in two basic characteristics: geographic coverage and data rates. Metropolitan area networks span a geographic area ranging from a few kilometers to a few hundreds of kilometers. The data rate could vary from a few kilobits per second to a few gigabits per second.

They could be owned by an organization (private MANs) to enable the users to share efficiently widely distributed resources. A MAN could also serve as a backbone of a network that interconnects distributed LANs, covering distances of hundreds of kilometers. MANs could also be public and thus provide private organizations with a metropolitan access network as well as access into regional or wide-area public telecommunication networks.

1.7.5 Integrated Services Digital Networks

Integrated services digital networks (ISDN) provide an integrated digital access standard between the customer premises and the network for the transmission of a variety of services such as voice, data, image, and video. An ISDN should have most or all of the following functional elements: circuit switching, packet switching, common channel signaling, network operation and management databases, and information processing and storage facilities.

Common channel signaling is a key ingredient of ISDN. It allows for the control of multiple circuit-switched connections using a seperate common signaling path (common channel). Chapter 11 provides a detailed discussion of ISDN.

1.7.6 Radio-Based Networks

Radio-based systems, such as microwave radio systems, have the advantage that they provide communications over barriers such as mountainous terrain, bodies of water, and heavy forests. Furthermore, radio-based systems may be the only practical means of communicating when users are mobile.

Radio networks can be classified into two types: single-hop radio networks and multihop networks (Fig. 1.12). In single-hop radio network, all stations are within a line of sight of each other (i.e., there is full connectivity among all the users). Thus if one station is transmitting all stations receive the signal.

Multihop radio networks are more complex than single-hop networks. They are characterized by limited direct connectivity from geographic distance and obstacles to signal propagation. Thus, a transmitted signal is received only by a subset of stations, and the signal may have to be relayed if it is destined to a faraway station. As a result, global control of system operation and resource allocation are major issues in multihop radio networks.

The multipath problem affects the design of radio networks. It occurs when the transmitted signal propagates over differing path lengths between

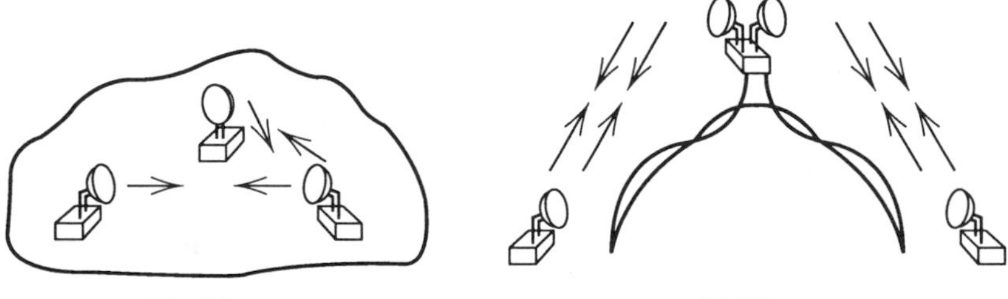

Figure 1.12. Radio-based networks.

the transmitter and the receiver. Multipath propagation may result in a reduction in the received signal level. The level of attenuation varies with frequency across the channel band. The complexity of radio networks further increases when users are mobile. The limited direct connectivity must be frequently updated to keep up with stations' mobility. Also, the radio propagation characteristics change accordingly. Finally, if the radio frequency spectrum is to be used fully, efficient algorithms must be devised to allow the dynamic allocation of the spectrum to a large population of intermittent mobile users.

1.7.7 Satellite Networks

The first communication satellite was placed in orbit in 1958, the first INTELSAT satellite was launched in 1965, and the INELSAT VI was launched in 1989. INTELSAT I had a 50-MHz bandwidth supporting only 240 voice channels, whereas INTELSAT VI has a 3.3 GHZ bandwidth supporting 120,000 voice channels and 3 TV channels.

Basically, a satellite is a microwave radio system with only one repeater; that is the satellite transponder in outer space. Earth stations communicate by sending signals to the satellite on an "uplink." The satellite then repeats those signals on a "downlink." However, the broadcast nature of the downlink makes it attractive for services such as the distribution of television programming.

Almost all of the communication satellites are in geosynchronous or geostationary orbits. For a geosynchronous satellite, the time to make one full revolution around the earth is the same as the time required for the Earth to make one full rotation, with the satellite moving in the same direction as the Earth's rotation. Thus, the satellite remains motionless with respect to earth stations. The satellite continuously views the same portion of the earth. In a geosynchronous satellite, the orbit may be inclined at any angle with respect to the earth's equatorial plane. In geostationary satellite, the orbit inclination is essentially zero.

Typically, satellites operate in the 6-GHz band for the uplink and the 4-GHz band for the downlink (6/4 GHz in the C-band). INTELSAT VI allows also for the 14/12 GHz (KU-band). A single geosynchronous satellite can cover approximately one third of the Earth's surface, thus worldwide coverage can be achieved with three satellites (Fig. 1.13). The distance from the surface of the earth to the geosynchronous satellite is 35,786 km, which results in a propagation delay of approximately 0.25 seconds.

Very small aperature terminal (VSAT) networks and direct broadcast satellite (DBS) are examples of the new applications to satellite communication. A VSAT is a small Earth station with antenna diameter typically less than 2.4 m and is suitable for easy installation on customer premises to provide a wide range of telecommunication services (e.g., voice, data, and video) with a large hub station or another VSAT. Direct broadcast satellites

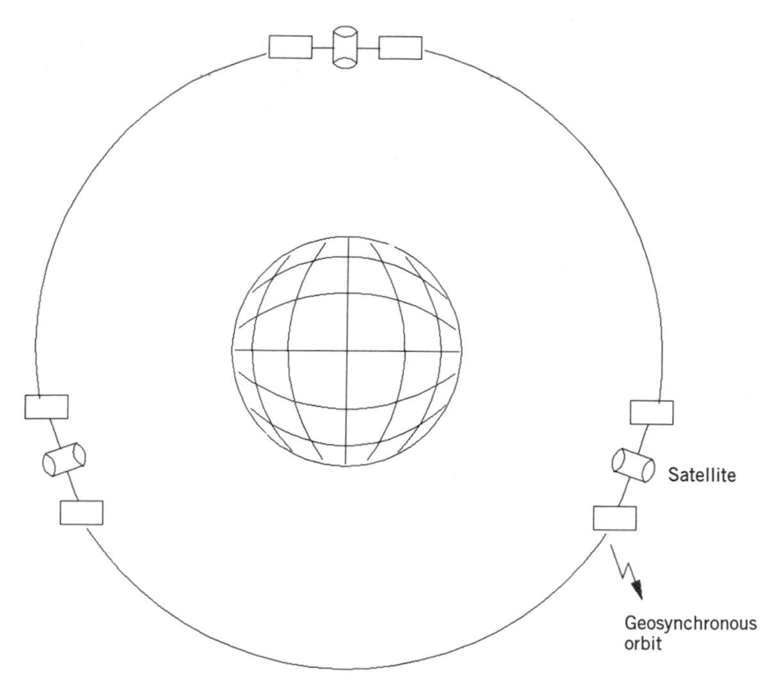

Figure 1.13. Satellite worldwide coverage.

provide multichannel television programming directly to users beyond the coverage areas of local television stations and cable television systems. These DBS systems are characterized by very high transmitting powers, thus allowing relatively simple and inexpensive receiving systems. In the western hemisphere, the frequency bands to be used are 17.3 and 18.8 GHz for the uplink and 12.1 to 12.7 GHz for the downlink.

Other applications for satellite communication include land mobile radio, position location, and communications for air traffic control, remote monitoring and tracking, and data communication using packet-switching techniques.

1.7.8 Mobile Communication Networks

Examples of mobile networks are mobile telephony, radio paging, the private land mobile radio, and personal communication networks (PCN).

1.7.8.1 Cellular Phone. Cellular phone, or the advanced mobile phone service AMPS, is a circuit-switched system. It is a means of providing mobile telephone service using radio frequency transmission. In a traditional mobile telephone system, a radio tower is placed at the center of a city, and serves mobile vehicles within a radius of 25 to 75 miles. In most areas the number of channels available is around 40, making it possible to serve only a limited

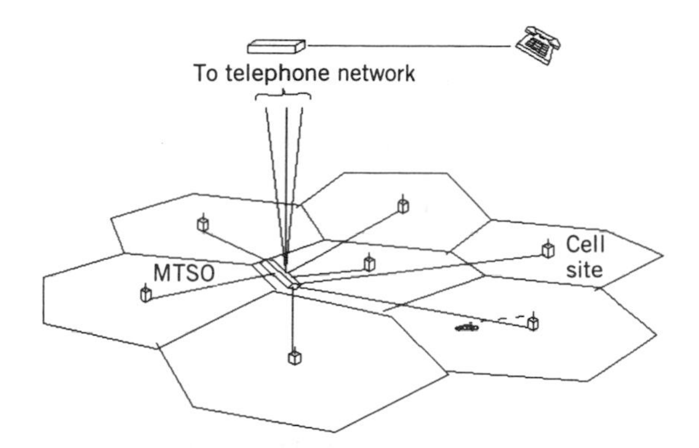

The AMPS system with cell sites located at the center of each cell

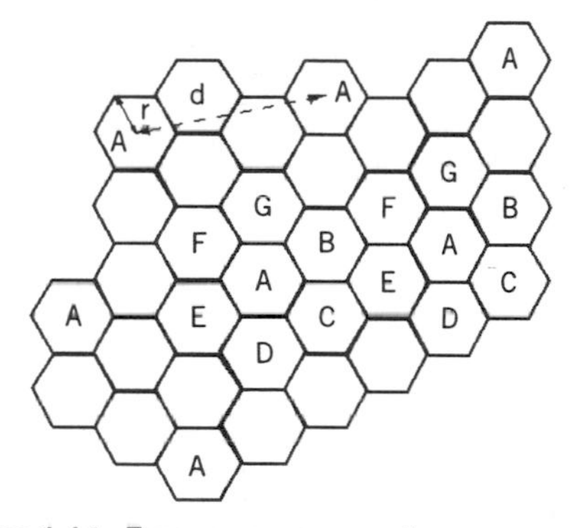

Figure 1.14. Frequency reuse pattern among cells.

number of customers. Moreover, the strong signal emitted from the radio tower prevents the use of the same frequencies by radio towers in nearby cities from interference. In the cellular phone system, the total area covered by the system is divided into cells. Each cell has the shape of a hexagon, or a circle (Fig. 1.14), and is served by a cell site. The cell site is an interface between the car phone and the terrestrial public telephone network. All cell sites are controlled by a large central controller, called the mobile telephone switching office (MTSO). All cell sites are wire connected to the MTSO.

Radio communications between the mobile vehicle and the cell site employs frequency modulation (FM). The capture effect of FM causes, when two seperate transmissions on the same frequency arrive at a receiver, the receiver to suppress the weaker (that is the interfering) signal and detect the stronger signal without significant quality degradation. The AMPS utilizes this FM capture effect to increase the efficiency in the use of the radio spectrum by repeating the use of channel frequencies in different cell sites. If two cell sites simultaneously transmit on the same radio channel, an FM receiver tuned to that channel will lock onto the transmitter with the stronger

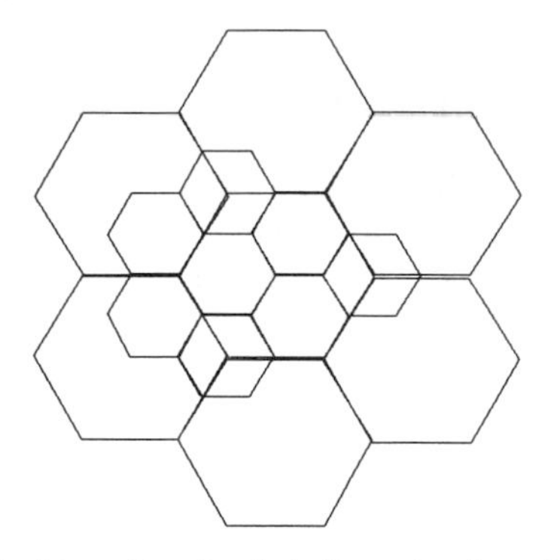

A mixture of small cells in the center city and large cells in the outskirts can coexist within a single system

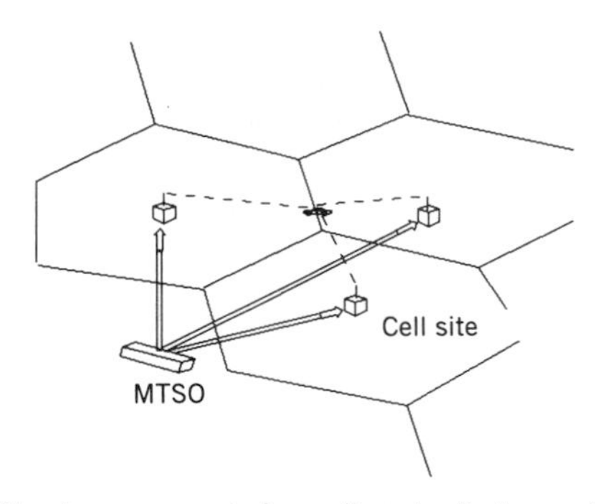

Signal measurements from adjacent cell sites provide the MTSO with the information necessary to hand off a mobile when another cell site can better serve the mobile

Figure 1.15. Cell splitting.

carrier. Thus, AMPS is an example of a space division multiple access (SDMA). Cellular systems using time division multiplexing has proven to be more efficient than frequency division multiplexing.

We define the parameter d/r as the ratio of the distance d between cell sites that reuse the same frequencies to the radius r of the cells. To guarantee that the degradation due to interference from cells using the same frequencies is kept to a minimum, d/r must be maintained to a minimum of 4.6. This value has been proven empirically. Figure 1.14 shows the frequency reuse pattern. A cluster of seven cells (A through G) repeats in a regular manner to provide continuous service across the coverage region. Each cell in a cluster has a distinct set of radio frequencies and thus will not interfere with frequencies assigned to any adjacent cell. Now, the same pattern is repeated in adjacent clusters with similar frequency allocations. The only criterion required is $d/r = 4.6$

The uniqueness of this system lies in the fact that the seperation between cells to avoid cochannel interference is not a specific distance in miles, but rather a distance large enough to make d/r equal to 4.6. Thus, the system can start up with relatively large cells. When more subscribers join the system and traffic density increases, we can add new cells and obtain a new pattern. Each small cell can support as many channels as a large cell. The whole idea is to maintain the ratio d/r equal to 4.6 (Fig. 1.15). This process of cell splitting increases the spectrum efficiency and allows for continuous growth, resulting in a lower cost to the customer for the service.

1.7.8.2 Personal Communications Networks. This is a digital microcellular system that operates at higher frequencies (1.7 to 2.3 GHz) allowing lower power and lower cost telephones than conventional cellular technology. It uses cells with a radius of 600 feet compared to the 2-mile radius of traditional cellular telephone.

1.7.9 Cable Television

Cable television started as CATV (community antenna television) in the late 1940s when it was used to provide television signals in areas that were underserved or unserved by standard broadcast stations.

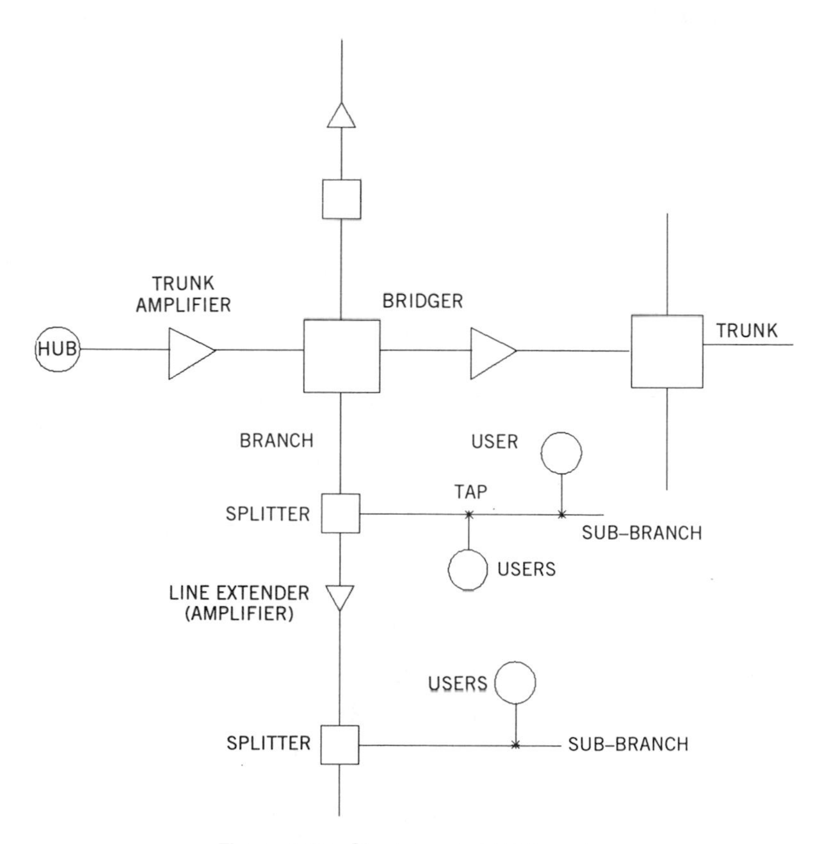

Figure 1.16. Single-trunk CATV layout.

A community antenna television (CATV) system provides a means for distributing television signals to individual homes. It can carry over 100 television channels at ranges up to a few tens of miles.

The central node of a CATV plant is the head-end. Signals from different sources, such as satellite and terrestrial broadcast, as well as local origination programming, are modulated onto radio frequency carriers and combined together for distribution over the cable system. Supertrunks (high-quality microwave, fiber optic, or cable links) connect the head-end to local distribution centers known as hubs. Several trunks may originate from a hub to provide coverage over a large contiguous area. Figure 1.16 shows a single trunk of a typical CATV plant. Trunk amplifiers are installed along the trunk to maintain the signal level and compensate for cable transmission characteristics. The bridge amplifier (or bridger) serves as a high-quality tap, providing the connection between the main trunk and multiple high-level branches. The line extender is a type of amplifier that maintains the signal level along the branch. The splitter connects a subbranch to a branch. Subscriber drops are connected to passive taps along the subbranch.

1.8 STANDARDS ACTIVITIES

Communication networks are designed to serve a wide variety of users with equipment from many different vendors. To design such networks effectively, standards are necessary to achieve interoperability, compatibility, and the required performance in a cost-effective manner.

Standards are even more crucial when a communication network is interconnected with other communication networks, because of the need to clearly define the network interfaces, maintain network functionality, and achieve expected customer service. Standardization of interfaces is necessary to achieve worldwide real-time connectivity. The lack of standards and/or the proliferation of proprietary interfaces often put a heavy burden on network providers, equipment manufacturers, and end users. Thus, the purpose of standards is to achieve the necessary or desired degree of uniformity in design or operation to permit networks to function properly for both providers and users. The scope of standards can vary; they may be internal within a company or they may apply to an entire country, a region, or the whole world.

Considerable activities are underway for the purpose of establishing national and international telecommunication standards. On the international scene, the ISO and the ITU with its two major committees; the CCITT and the International Radio Consultative Committee (CCIR) have full programs and regular meetings supported by representatives of carriers and equipment manufacturers. Standardization groups on the national scene are also very active in many countries. For example, within the United States, the American National Standards Institute (ANSI), the Electronic Industries Association (EIA), and the Institute of Electrical and Electronics Engineers (IEEE) are as active as their international counterparts. In Appendix 1A, we provide a detailed description of some of the important standards organiztions.

As discussed previously, the Internet Standardization process is overseen by the Internet Architecture Board.

SUGGESTED READINGS

[IEEE 84] chronicles some of the outstanding milestones in the field of telecommunications. [ROBE 78] and [GREE 84] cover the history of packet switching, and [JOE 84] discusses the evolution of telecommunications switching.

[SCIE 91] represents an overview look at the progress in telecommunications networks with articles written by M. Dertouzos, V. Cerf, L. Tesler, and N. Negroponte.

Text books such as [KLEI 75C], [HAMM 86], [SCHW 87], [TANN 88], [SPRA 91], [WALR 91], [HUI 91], [BERT 92], [HALS 92], [STAL 93], [ACAM 94], and [AIDA 94] provide a detailed look at telecommunication networks.

[ZIMM 80] and [FOLT 83] are excellent references for the OSI reference model. Internet standards are available in RFC (Request for Comments) documents. They are also available via anonymous FTP from the Internet Information Center with host address ds-internic.net.

[CERF 90] scans the history of ARPANET. The RFC in [POST 93], [BRAD 89a], and [BRAD 89b] contain the basics of the Internet architecture.

The work in [CLAR 93] presents a flexible model for inter-operability that relies on the use of multiple protocol standards.

[LEIN 87] provides an excellent treatment of packet radio networks and [MILL 87] provides additional reading on multipath effects. [HA 90], [WU 85], and [MORG 89] are good references on satellites. [LEE 93] is a good reference on mobile cellular telephony and [CHUA 93] and [STEE 92] discuss wireless personal communications. Examples of applications of CATV networks to data traffic can be found in [SAAD 85], [JAFF 86], and [TODD 93].

[RYAN 85] and [ABLE 94] provide different articles addressing telecommunication standards and the various organizations involved in standards making and their working methods and procedures. [SHER 92] presents various aspects of establishing telecommunication standards.

PROBLEMS

1.1 Examples of information signals are voice, video, image, data, and so on. The original information signal may be in a digital or analog form. Mention the original form (whether digital or analog) for voice, video, data, facsimile, and image signals.

1.2 Assume a fully connected network (every subscriber has a seperate line to every other subscriber). For N subscribers,

a. Determine the total number of lines in the network, L.

b. What is the value of L for $N = 2$, 10 100, 1000?

c. How many I/O (input/output) ports are needed at every station (subscriber)?

d. Based on the results of (b), do you recommend a fully connected network as a solution to the public switched telephone network? Why?

1.3 Before the wide spread of data networks in 1970s and 1980s, computer data had to be transmitted over the public telephone networks (in fact, a large percentage of data traffic is still being transmitted over the telephone network), what is the name of the device that allows computers to interface to the telephone network? Describe briefly its functions and characteristics.

1.4 Entertainment television programs are presently broadcasted using CATV, satellite, or terrestrial radio. Can entertainment television be sent to homes over the public telephone network? Explain why. What are the modifications needed in the telephone network to accomodate entertainment television? (Discuss from the two aspects of switching and transmission.)

1.5 The introduction of the cellular concept was a revolutionary step in mobile telephony. Explain the basic advantages of the cellular architecture.

1.6 CATV is a one-way network that broadcasts television signals to individual homes. What are the modifications needed in the CATV plant to allow home users to send data traffic to the head-end?

2

DATA COMMUNICATIONS PRINCIPLES

2.1 COMMUNICATION SYSTEM MODEL

A basic model for a communication system is shown in Figure 2.1. Regardless of the specific application and the system configuration, all information transmission systems invariably involve three major components: a transmitter, the communication channel, and a receiver.

The information $I(t)$ to be transmitted can be either analog (continuous) or digital (discrete). An analog signal varies continuously between a maximum and a minimum value. It has an infinite number of values between the two extremes. Examples of analog information include the human voice and television pictures. On the other hand, a digital signal has a limited set of discrete values, each of which represents a symbol like an alphabetical character or a number. Examples of digital information include telegraph signals, weather reports (rain, foggy, etc.), the results of throwing dice (1, 2, 3, etc.), data generated by computer devices, and pulses transmitted from a dial-pulse telephone.

The purpose of the transmitter is to couple the message to be exchanged to the channel, it must be capable of translating information from a form created by a human or a machine into a signal suitable for transmission over the transmission medium. For example, in some systems, it is necessary to modulate a carrier wave with the signal from the input device. *Modulation* is the systematic variation of some attribute of the carrier signal, such as amplitude, phase, or frequency, in accordance with a function of the information signal. In systems with a digital input device, it might be necessary to encode the input stream of bits into a code that is appropriate to the communications medium. In other systems, the transmitter must group the incoming bits into blocks of data and attach some additional control information. In addition to modulation (or encoding or grouping) of bits, the transmitter may provide filtering and amplification.

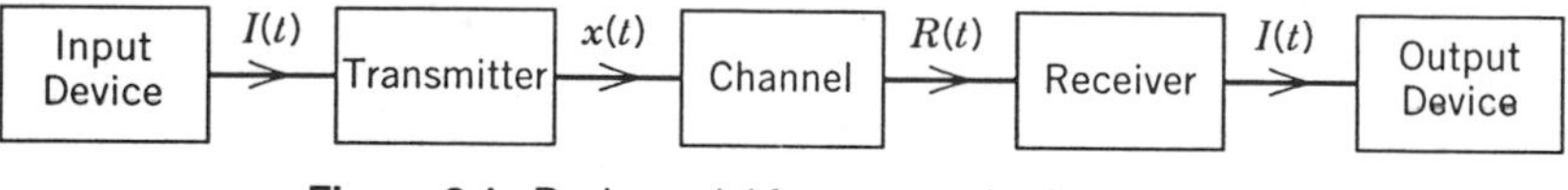

Figure 2.1. Basic model for communication system.

The information $I(t)$, whether analog or digital, can be transmitted in either analog or digital form as the signal $x(t)$ within the communication system. Hence the communication system itself can be either analog or digital (or a combination of both). It is important to note that signals may originate in one form, but may be converted to the other form for transmission over the network. It is obvious that analog signals can be transmitted over analog transmission systems and digital signals can be transmitted over digital transmission systems. However, the opposite is also true. Digital signals can be transmitted over analog communication systems by converting (modulating) the digital signals into analog signals (and then transmitting them over the analog communications facility). This can be done by using a device called a modem (see Section 2.4.2).

On the other hand, analog signals can be transmitted over a digital communications facility by converting the analog signal into digital forms (digitization). This, for example, can be done by using pulse code modulation (PCM), which is described in Section 2.2. The receiver must convert the received digital signal to its original analog form.

2.2 ANALOG INFORMATION OVER DIGITAL CHANNELS

It is increasingly common to transmit analog information, such as speech and video, as a digital rather than as a continuous waveform. There are several reasons for this including improved tolerance of noise, the possibility of regeneration, convenience of electronic implementation, and the capability of using the digital network as a bearer for combined analog and digital traffic.

2.2.1 Pulse Code Modulation (PCM)

The most common method of encoding analog signals into digital form is pulse code modulation (PCM), in which successive samples of the analog signal, say speech, are rounded to the nearest quantum level and expressed as an n bit binary number, with $N = 2^n$ where N is the number of quantum levels. Thus, the main functions of a PCM encoder (Fig. 2.2) are sampling, quantizing, and encoding. Sampling (Fig. 2.3) is the process of converting a continuous analog waveform $I(t)$ into a sequence of discrete analog samples, each sample being a very narrow pulse of width δt. Basically, sampling is the multiplication of the analog waveform $I(t)$ by a periodic train of pulses $s(t)$

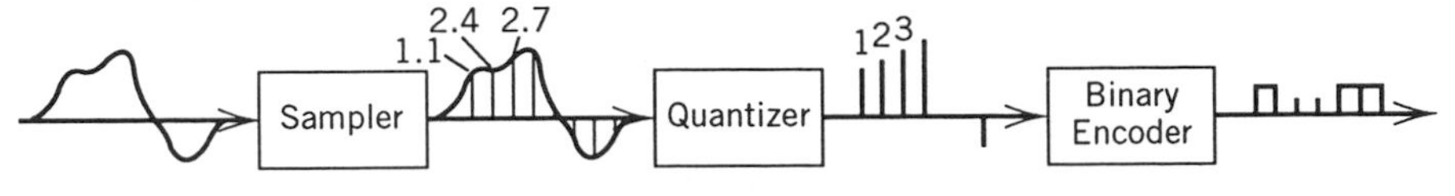

Figure 2.2. Pulse code modulation (PCM).

of unit amplitude. The time between the pulses is the sampling time T_s which equals the inverse of the sampling frequency (or rate), f_s.

The sampled output $I(t)S(t)$ is shown in Figure 2.3 to be the signal $I(t)$ sampled at the occurrence of each pulse. That is, when a pulse occurs, the output has the same value as does $I(t)$, and at all other times the output is zero. The sampling frequency is chosen according to the sampling theorem, which states that a waveform can in principle be reconstructed without distortion from accurate samples taken at a frequency at least twice the highest frequency present in the spectrum of the waveform. The reconstruction of the waveform can be obtained by passing the samples through a low-pass filter with a cut-off frequency at half the sampling frequency.

Since the sampled output $I(t)S(t)$ can take on an infinite number of nonzero amplitudes, the amplitude level has to be quantized. Each quantized sample is then represented (encoded) by a binary number with a finite number of digits (bits). In practical voice systems, each quantized sample is represented by an 8-bit word. Figure 2.4 shows an example of a quantized signal with a 4-bit encoding, for simplicity. A 4-bit encoder provides 16 different quantizing levels. The most significant bit is used to indicate the voltage polarity of the sampled signal; a "1" indicates positive and a "0" indicates negative. The samples are applied to a threshold detector (decision values) and binary encoded. The value of the encoded output is midway

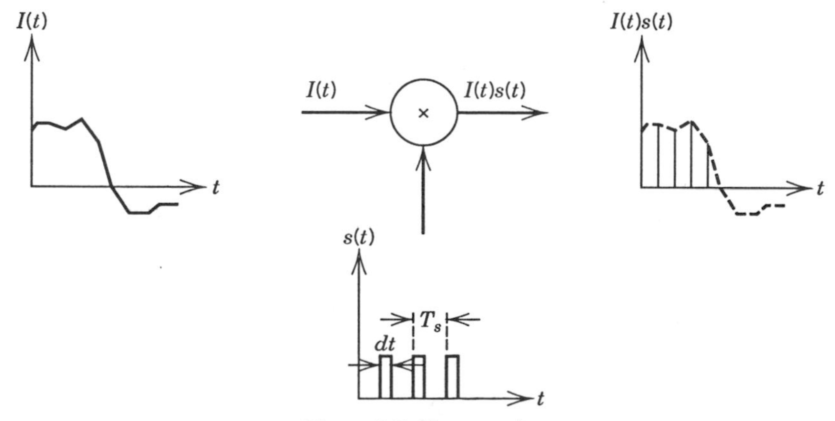

Figure 2.3. The sampler.

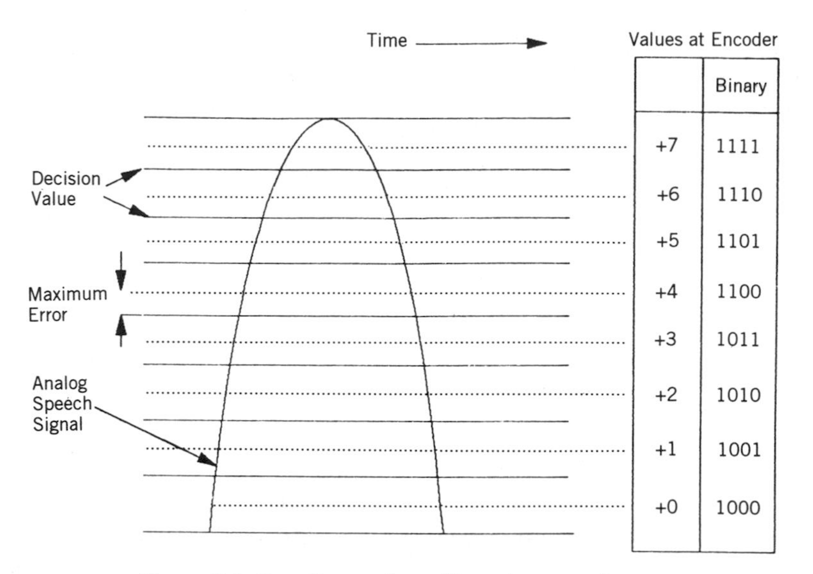

Figure 2.4. Quantizer; only positive values are shown.

between decision values. As a result, there may be quantizing errors* up to half the quantizing interval. The cumulative errors produce quantization distortion. The mean square quantization error, $\overline{e^2}$, for a uniform quantizer depends on quantization step (interval) d and has been shown equal to

$$\overline{e^2} = d^2/12 \tag{2.1}$$

Also, the output signal to quantization noise ratio (S/N)† depends on the input signal; that is, for larger input signals a larger S/N is obtained. Thus, a large S/N is not satisfied when the signal is small. Accordingly, before applying the input signal to the quantizer, we pass it through a device that has a nonlinear input–output characteristic known as compandor (*com*-pressing and ex*pand*ing). The combination of the compandor and the uniform-step quantizer results in a nonuniform-step quantizer. In the United States, the companding scheme is called the μ-Law transfer characteristic and is defined by

$$y = \pm \frac{\ln(1 + \mu|x|)}{\ln(1 + \mu)} \tag{2.2}$$

*The quantization noise e is the difference between the actual signal value and its quantized equivalent

†The amount of noise present is measured by the ratio of the signal power S to the noise power N and is referred to as the signal to noise ratio S/N.

where $x = v_i/V$ and $y = v_o/V$. v_i is the input voltage, v_o is the output voltage, and the range of allowable voltage is $-V$ to $+V$. The + sign applies when x is negative. The parameter μ determines the degree of compression. The Bell system (AT&T) uses $\mu = 255$.

In Europe, another companding scheme, called the A-Law characteristic, is used. It is defined by

$$y = \begin{cases} \pm \dfrac{A|x|}{1 + \ln A} & \text{for } |x| \preccurlyeq \dfrac{1}{A} \\ \pm \dfrac{1 + \ln A|x|}{1 + \ln A} & \text{for } \dfrac{1}{A} \preccurlyeq |x| \preccurlyeq 1 \end{cases} \tag{2.3}$$

Where, x and y are as defined in Eq. 2.2. The parameter A determines the degree of compression. A commonly used value is $A = 87.6$.

For example, the speech waveform in a telephone channel is bandwidth limited from 300 Hz to 3400 Hz. If we sample at twice the highest frequency component present, then $f_s = 2 \times 3400$ Hz $= 6800$ Hz. It is standard practice to use a sampling frequency of 8 kHz when encoding speech of telephone quality; this practice has been adopted internationally. Therefore, there is a margin between the highest speech frequency to be transmitted and half the sampling frequency, thus making it practical to design the low-pass filter at the receiver. It is convenient to think of the speech bandwidth fitting into a 4-kHz bandwidth with a sampling frequency of 8 kHz, that is, 8000 samples per second. Each sample is quantized into one of 256 quantization levels, requiring an 8-bit word per sample ($2^8 = 256$). Thus one voice channel will have a transmission rate of 64 kbps (8000 samples per second $\times$ 8 bits per sample).

The application of PCM can also be considered for other analog information. For high-fidelity music broadcasting with a sampling frequency of 32 kHz, some systems operate with uniform quantization at 16 bits per sample resulting in a 448-kbps transmission rate or with nonuniform quantization at 12 bits resulting in a 384-kbps transmission rate or at 10 bits resulting in a 320-kbps transmission rate. For digital music recording, typical sampling is 44.1 kHz with uniform quantizing at 16 bits resulting in a rate of 0.7 Mbps. When a large number of such PCM signals, say many voice calls, are to be transmitted over a common channel, multiplexing of these PCM signals is required. This technique, known as time division multiplexing (TDM), will be discussed in detail later.

Another example is color television signals, which have an analog baseband* bandwidth of approximately 5 MHz. Typical PCM used for video

*A baseband signal is the original basic signal generated by the source with no frequency translation carried out. A broadband signal is the baseband signal shifted to higher frequencies for efficient transmission.

encoding has a sampling rate f_s of 10^6 samples per second, other systems may use $f_s = 13.5$ MHz, with 9 bits per sample (or 8 bits), thus a transmission rate of 90 Mbps.

2.3 DATA TRANSMISSION

Transmitting a stream of bits (binary digits) is normally handled by grouping a number of bits into data units. These units could be a character, a packet, or a message. The number of bits in a character is fixed for a particular device or communication system. Typically, a character is 8 bits, yet it could be from 5 to 10 bits. Note that characters include not only letters, numbers, punctuation marks, but also control commands such as backspace, as well as special communications signals that mark, for example, the beginning and end of a message. Appendix 2A describes different types of character codes. A packet is another method of grouping bits together and typically includes more bits as overhead (contains information about addresses, type of packet, and error control). Packets could vary from a few bits to thousands of bits. Computer networks are usually designed with packets as the basic data units.

Basically, there are two approaches for transmission of data units between two devices; serial transmission and parallel transmission. In *serial transmission*, bits are transmitted one at a time (for example, the EIA-232 interface and modem transmission). In *parallel transmission*, some fixed number of bits, say 8 bits, are sent on different channels at the same time. Parallel transmission is faster than serial transmission, but requires a number of channels. Typical examples are the IEEE STD 488-1978 and microprocessor interfaces. The IEEE 488-1978 interface is designed for instruments and system components data transfer. The IEEE 488 is the standard method for connecting test and measurement instruments such as digital multimeters, signal generators, counters, printers, and plotters to computers. Within computer systems, transfer of bits is frequently done in parallel to minimize the time required. On the other hand, transmission across data networks is almost always serial because it requires only one channel.

Figure 2.5 shows the two methods of data transmission for an 8-bit character. A binary "1" is represented by positive voltage and a binary "0" is represented by negative voltage.

2.3.1 Transmission Modes

There are three modes of transmission; simplex, half duplex, and full duplex. In *simplex* transmission, data is transmitted in one direction only, for example a telemetry device transmits the collected data to a central computer. Another example in an analog environment is AM/FM radio, which receives signals from the radio station in a simplex communication mode. In *half-duplex* transmission, two devices exchange data alternately. In other

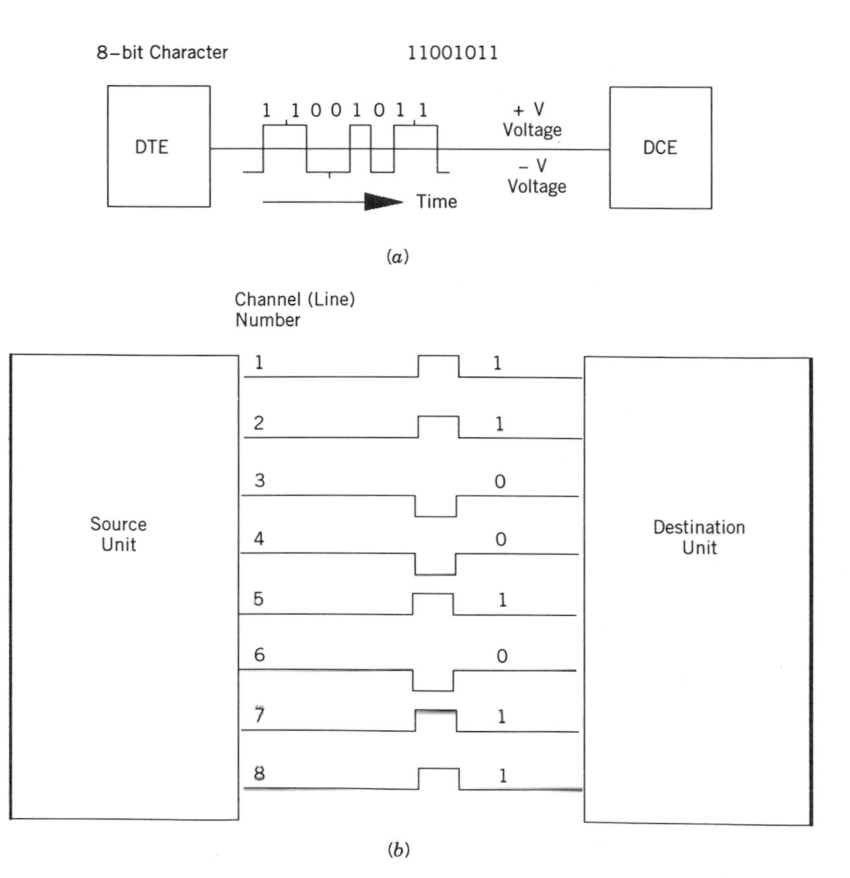

Figure 2.5. Data transmission. (*a*) Serial transmission. (*b*) Parallel transmission.

words, it permits transmission in either direction; however, transmission can occur in only one direction at a time. An example is human communication, in which person A talks while person B listens, then B talks and A listens. In *full-duplex* transmission, two devices exchange data simultaneously. For example, two computers exchange data files simultaneously. It is clear that the full-duplex mode can achieve higher rates of data flow but requires two channels to be available simultaneously between the communication end points.

2.3.2 Transmission Techniques

The transmission and reception of bits is based on the use of clocks to derive bit timing. A transmitter uses its clock to determine the beginning and end of each bit. At the receiver, a clock is used to determine when one bit ends and the next one begins. It is generally impossible to have two independent clocks running at precisely the same speed. Left by themselves, clocks tend to drift

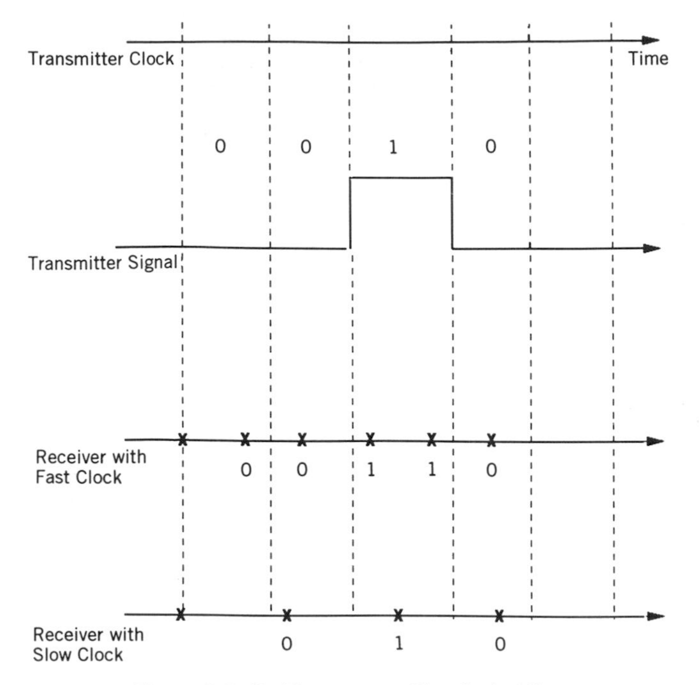

Figure 2.6. Problems caused by clock drift.

causing the interval between consecutive ticks to become longer or shorter than a prescribed nominal rate. For example, a clock used to generate a 100-bps data stream should tick once every 0.01 second. Left by itself the clock ticks will be anywhere between $0.01 - \epsilon$ and $0.01 + \epsilon$ apart, with ϵ depending on exactly how the clock is generated.

Clock drift can cause problems when a receiver uses its own independent clock to determine the beginning and end of bits in a received signal. Because of the differences between the receiver clock and the transmitter clock, the receiver might receive the same bit twice or might skip entirely over a bit. For example the bits 0010 might be interpreted as 00110 or 010 by the receiver as shown in Figure 2.6.

Asynchronous Transmission. There are two approaches to deal with this problem. The first, called *asynchronous transmission*, allows the transmitter and receiver to maintain independent clocks that are periodically synchronized with each other. In the second, called *synchronous transmission*, the receiver's clock is slaved to the transmitter's clock; that is, the receiver's clock is made to run at exactly the same speed as the transmitter's.

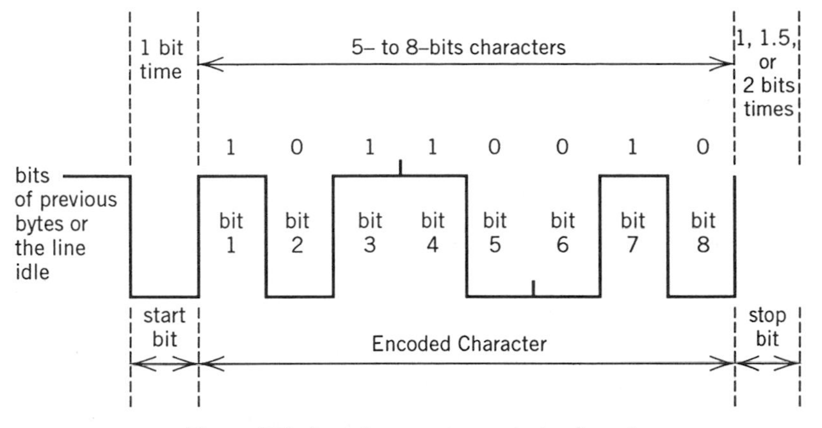

Figure 2.7. Asynchronous transmission format.

Asynchronous transmission relies on the fact that clock drift is limited to a certain range. This allows a receiver, that can somehow synchronize with one of the transmitter's clock ticks, to be able to correctly receive a number of bits correctly before it begins to loose or add any bits.

In asynchronous transmission, an idle transmission medium is placed in the high state, the state used to transmit a 1 when data is present. Before data is transmitted, the medium is placed in the low state for one bit time. This is known, as the start bit. The transition from the high to the low state is the indication used by the receiver to start its clock. The receiver then knows to expect a specified number of data bits (typically 5 to 10 bits) following the start bit. After the specified number of data bits have been transmitted, the transmitter then has to place the medium in the high state again for a minimum amount of time (typically ranging from 1 to 2 bit times) before transmitting more data. This is known as the stop bit (see Fig. 2.7).

The asynchronous technique just described achieves the desired objective of synchronizing the receiver's independent clock to the high-to-low transition event and then letting the receiver's clock run independently for a maximum number of bit times before requiring a new synchronization event. The primary drawback of asynchronous transmission is the overhead used in providing the start and stop bits, which can be quite high. In general, this makes its use only feasible in lower speed transmissions (110 bps to 19.2 kbps) where the start–stop overhead is minimal.

Synchronous Transmission. As discussed above, asynchronous transmission requires that bit synchronization be established at the start of each character, simply by using the start–stop bits. In synchronous transmission,

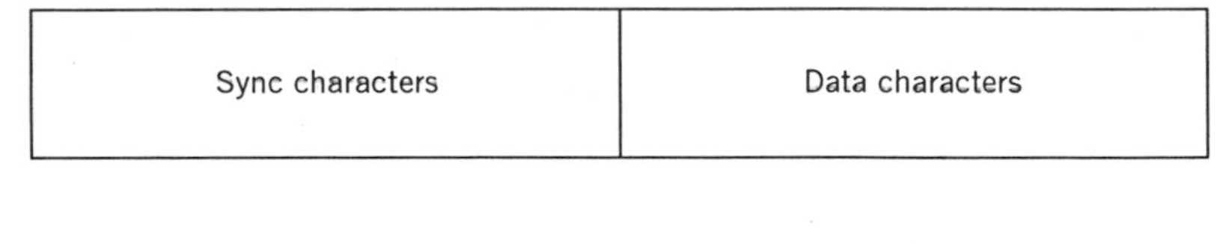

Figure 2.8. Synchronous transmission format.

bit timing is established at the very beginning of the transmission of a large number of bits and is maintained for the duration of transmission. This makes it possible to omit the start–stop bits needed in asynchronous transmission and thus increases transmission efficiency and speed.

To maintain synchronization of the receiving clock with the transmitting clock for the duration of a stream of bits, the transmission of the data is preceded by the transmission of one or more special characters (as shown in Fig. 2.8). The synchronization character is chosen to have a distinctive pattern from the data bits and is used to allow the receiver to acquire synchronization with the transmitter's clock. After synchronization is achieved, then actual data transmission can proceed. Synchronization is maintained through transitions in the transmitted data. This is achieved by using suitable signaling schemes such as Manchester or return-to-zero (RZ) encoding (see Section 2.4).

There are two different approaches to achieve synchronous data transmission: the character oriented and the bit oriented. The essential differences are the way the start and end of a frame, and the stream of contiguous information bits, are determined. Examples of synchronous transmission protocols are BISYNC, SDLC, HDLC, X.25 LAPB, and IEEE 802 standards. A

	ASYNCHRONOUS	SYNCHRONOUS
Serial	20 mA crrent loop EIA RS 232C EIA RS–422, 423, 499 EIA RS 485	BISYNC SDLC HDLC X.25 LAPB IEEE 202 Standards
Parallel	IEEE STP 488 – 1978	Microprocessor Interface

Figure 2.9. Typical data transfer interfaces.

detailed description of synchronous transmission is discussed under data-link protocols in Chapter 4.

Figure 2.9 provides examples of data transmission techniques discussed in this chapter.

2.4 DATA SIGNALING FORMATS

Before bits are sent over the transmission link, they must be encoded into a signaling format to satisfy the requirements of the medium or the receiver. These requirements may include avoiding direct current, clock extraction at the receiving end, overcoming noise interference, and transmission in the allocated bandwidth.

There are two basic signaling formats: digital signaling and analog signaling. In digital signaling (Fig. 2.10), the stream of bits to be transmitted $b(t)$ is mapped (encoded) into another stream of bits $d(t)$. The new stream of bits (signaling bits) is transmitted over the transmission link, and the receiver decodes the signaling bits to its original form $b(t)$. In analog signaling, referred to as digital modulation (Fig. 2.11), the stream of bits to be transmitted $b(t)$ is mapped (modulated) into an analog signal $a(t)$. The analog signal is then transmitted over the transmission link, and the receiver demodulates the analog signaling into its original bit stream form, $b(t)$.

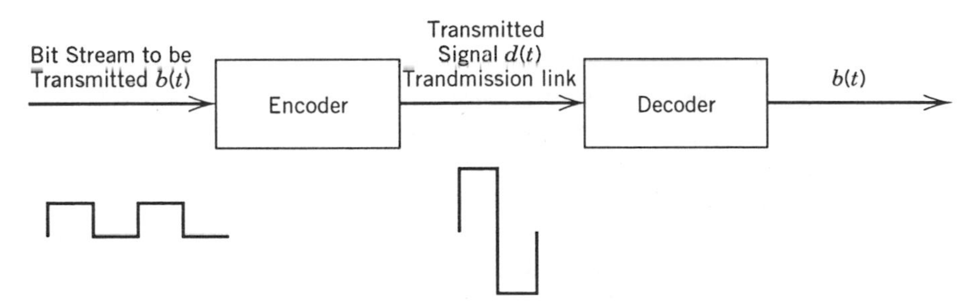

Figure 2.10. Bits are encoded into digital signaling.

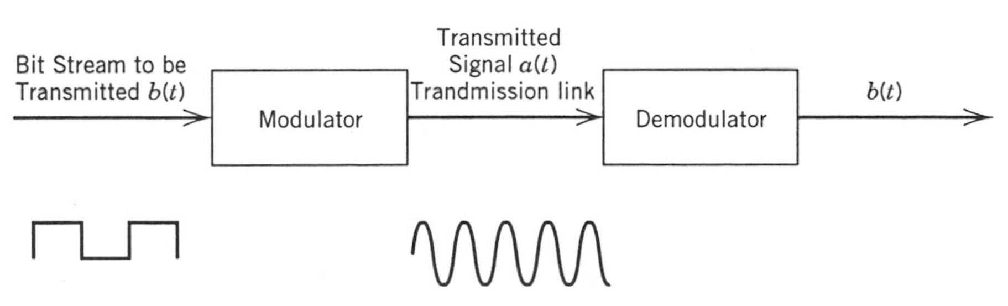

Figure 2.11. Bits are modulated into analog signaling.

2.4.1 Data Transmission over Digital Channels (Digital Signaling)

Some of the typical digital signaling formats used in communication networks are;

Return to zero (RZ)
Non return to zero (NRZ)
Non return to zero inverted (NRZI)
Manchester
Differential Manchester
Alternating mark inversion (AMI)
Duobinary

Reasons for using any of the above signaling format can be summarized as follows:

Signal frequency spectrum. It is important to minimize the high frequency components so that less bandwidth is required. Also, minimizing low frequency components, and especially d.c. component, is desirable, since along the communication path there may be transformers and coils.

Synchronization and clock extraction at the receiver. Some encoding techniques (such as Manchester encoding) can inherently provide the receiver with information regarding the beginning and end of each bit. Thus, alleviating the need for a separate clock channel (i.e., wire).

Other reasons. There are several other reasons, such as error detecting capability and noise immunity.

Return to Zero (RZ). In this scheme, the bit stream to be transmitted is encoded so that a binary 1 is represented by a positive pulse and a binary 0 as a negative pulse (Fig. 2.12). With this scheme the signal returns to zero after each encoded bit (whether a binary 1 or binary 0); therefore, it is referred to as a return to zero (RZ) encoding. The particular version shown in Figure 2.12 is called bipolar encoding. The RZ scheme is a simple one, yet it has some drawbacks. Because the signal changes faster than the bit rate, the bandwidth required is large. The RZ scheme is used in simple communication and recording equipment. Note that three distinct voltage levels are needed (positive, negative, and zero voltages).

Non Return to Zero (NRZ). In this scheme the voltage level is constant during the bit interval (there is no return to a zero voltage level). Hence, only two distinct voltage levels are needed (positive and negative voltages). The main disadvantage of NRZ is its lack of synchronization capability, because the signal level can be constant for long duration of 1s and 0s.

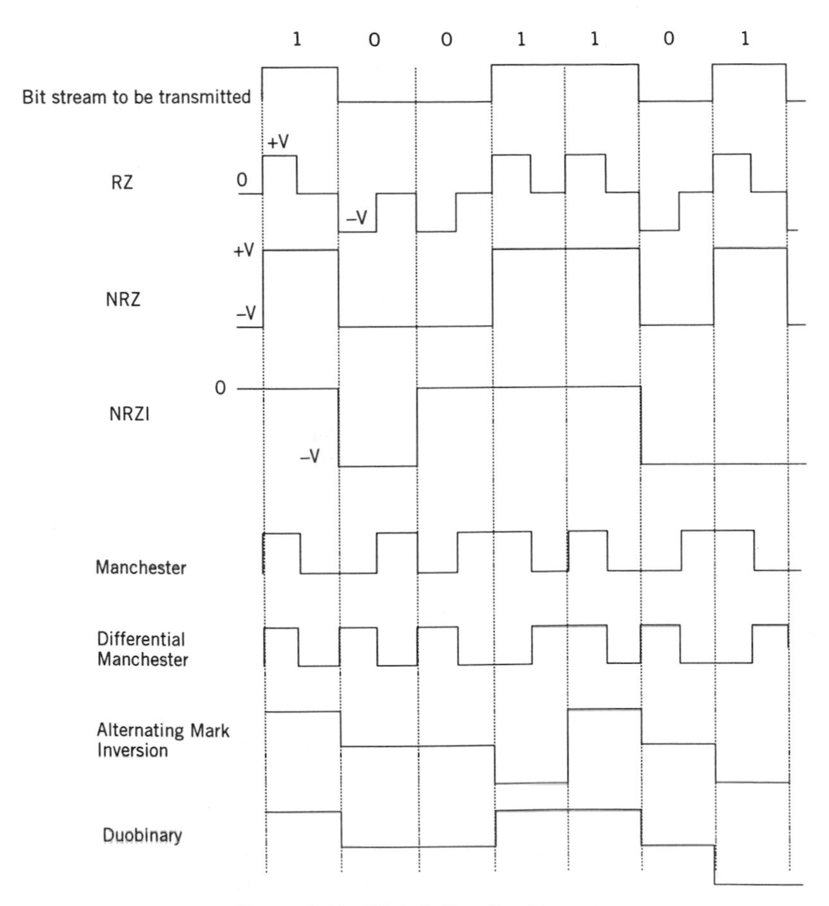

Figure 2.12. Digital Signaling Format.

Non Return to Zero Inverted (NRZI). NRZI is a form of differential encoding, in which the transmitted signal is determined by comparing the polarity of the adjacent bit, that is, the signal element (positive or negative) does not change if a binary 1 exists while a binary 0 causes change. This means that NRZI coding forces a transition whenever there is a binary zero in the bit stream to be transmitted. This scheme is specially attractive for bit-oriented protocols, since a binary 0 is always inserted after a sequence of five 1s, (bit-stuffing rule, see Section 4.6.1). Consequently the resulting waveform contains a guaranteed number of transitions, which enables the receiver to adjust its clock.

Manchester and Differential Manchester. In Manchester encoding, there is always a transition in the middle of the bit interval. The first half of the bit represents the bit value (i.e., the positive voltage for a binary 1 and negative voltage for a binary 0), and the second half of the bit is the complement of the first half.

In differential Manchester, there is also a transition in the middle of the bit interval. Like NRZI, it is a form of differential encoding. This means that there is a transition whenever there is a binary zero in the bit stream to be transmitted.

Alternating Mark Inversion (AMI). In alternating mark inversion, there are three voltage levels (positive, zero, and negative). A binary 0 is represented by zero voltage, and a binary 1 is alternately represented by positive and negative voltages.

Duobinary. Duobinary is similar to AMI, except that a binary 1 is either represented by positive or negative voltages according to the following rule:

1. A binary 1 is represented similarly to the previous binary 1 representation if the number of binary 0 between the two binary 1s is even.
2. A binary 1 is represented by the compliment of the previous binary 1 representation if the number of binary 0s between the two binary 1s is odd.

2.4.2 Data Transmission over Analog Channel (Analog Signaling)

To send digital signals, such as the output of a computer, over an analog channel, we must convert the digital signal into analog form to pass through the analog channel. For example, to log remotely onto a mainframe computer say from the home using the public telephone network, we attach a device called a modem (*mod*ulator/*dem*odulator) to our computer terminal. The modem is an electronic device used to convert the digital signal generated by a computer and its terminals into analog signals for transmission over the analog facilities. At the receiving end, a similar device accepts the transmitted analog signal, reconverts them to digital signals, and passes these signals to the destination device (see Fig. 2.13).

There are several modulation schemes to convert the digital signal into an analog signal. We scan briefly through some of these techniques.

In general, the modulation uses a sinusoidal signal referred to as the carrier of the form

$$A \cos(2\pi f t + \theta)$$

where f is the carrier frequency, A is the amplitude, and θ is the phase of the carrier signal. The modulation of the carrier is achieved by varying the amplitude, the phase, or the frequency of the carrier according to the data stream.

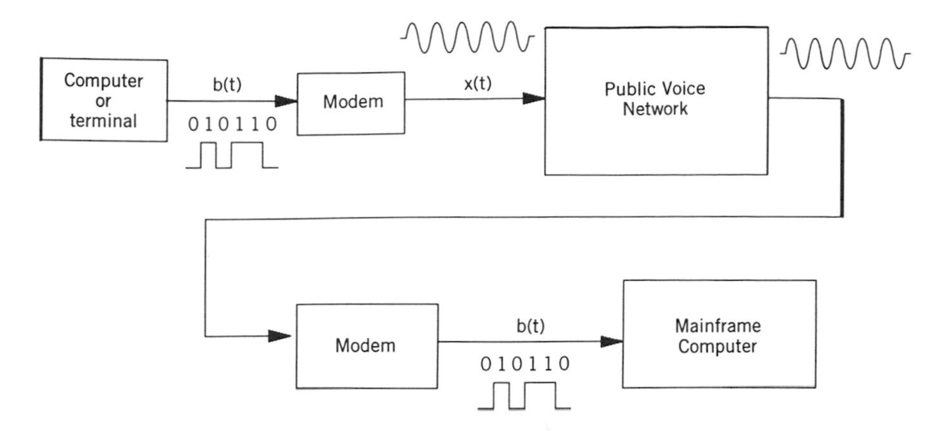

Figure 2.13. Data transmission over public voice network.

Amplitude-Shift Keying (ASK). In amplitude-shift keying the amplitude of the carrier signal is changed according to the binary stream of the data to be transmitted. This can be implemented by multiplying the carrier signal $A\cos(2\pi ft + \theta)$ by the data stream, $b(t)$, to be transmitted. Thus, the transmitted signal (i.e., the modem output, see Fig. 2.14), is

$$x(t) = b(t)A\cos(2\pi ft + \theta)$$

$$= \begin{cases} A_1 \cos(2\pi ft + \theta) & \text{if } b(t) = 1 \\ A_2 \cos(2\pi ft + \theta) & \text{if } b(t) = 0 \end{cases}$$

In frequency-shift keying (FSK), the frequency of a fixed amplitude carrier signal is changed according to the binary stream of data to be transmitted. Thus, the transmitted signal (i.e., the modem output) is

$$x(t) = \begin{cases} A\cos(2\pi f_1 t + \theta) & \text{if } b(t) = 1 \\ A\cos(2\pi f_2 t + \theta) & \text{if } b(t) = 0 \end{cases}$$

Frequency-Shift Keying (FSK). Frequency-shift keying is mostly used for low-bit-rate modems (300 bps to 1200 bps). For example, the Bell-103 modem (AT & T) is a full-duplex (FDX) modem that uses FSK and its data rate is up to 300 bps. In the originate mode, the modem transmits a 1070-Hz tone (f_1 = 1070 Hz) for binary 0 and a 1270-Hz tone (f_2 = 1270 Hz) for binary 1. It simultaneously

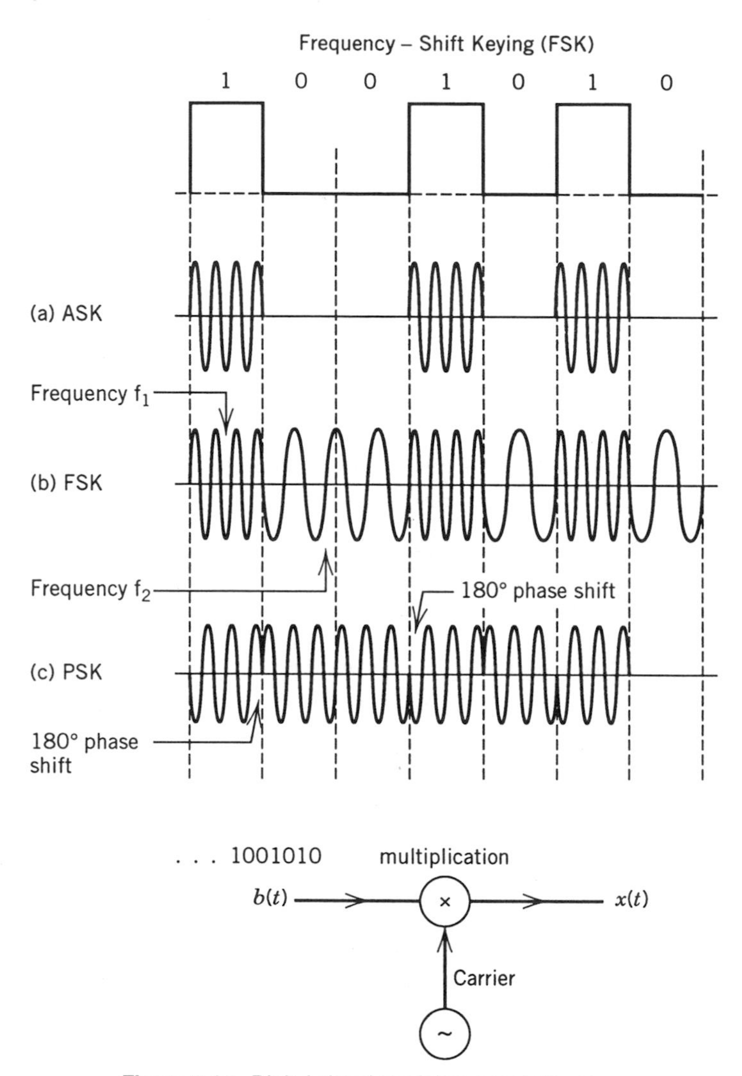

Figure 2.14. Digital signal modulation techniques.

receives a binary 1 in the high band at 2225 Hz and a binary 0 at 2025 on the same voice channel. The answer modem at the other end reverses the roles, so that it receives in the low band and transmits in the high band, thus allowing proper full-duplex operation. Figure 2.15 shows the frequency allocation of the Bell-103 modem.

Other types of modems that use FSK are the Bell-113 and the Bell-202 for voice channels. Frequency-shift keying is also used in high-frequency radio transmission and in broadband local area networks. Demodulation of the FSK signal is easy, since the information is contained in the frequency.

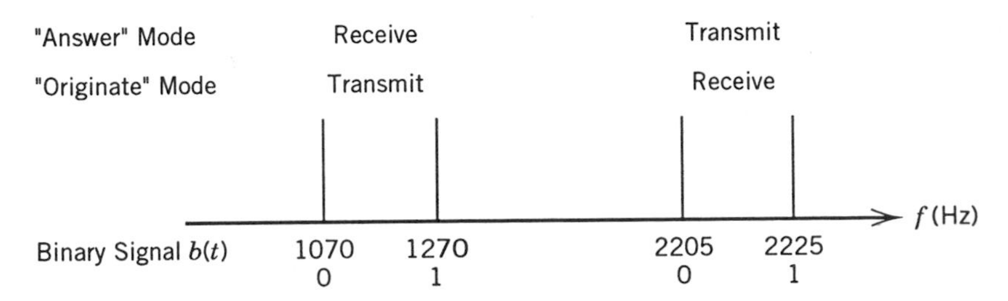

Figure 2.15. Modem using frequency-shift keying with full-duplex operation.

Phase-Shift Keying (PSK). In phase-shift keying, the frequency and amplitude of the carrier signal are kept constant, but the carrier is shifted in phase according to the input data stream. Thus, the transmitted signal (i.e., the modem output) is

$$x(t) = \begin{cases} A\cos(2\pi ft) & \text{if } b(t) = 0 \\ A\cos(2\pi ft + 180) & \text{if } b(t) = 1 \end{cases}$$

For synchronous data transmission, it is essential to enforce a phase change from the previously transmitted phase to help the receiver detect the beginning of every bit. This technique is called differential phase-shift keying (DPSK). In DPSK, at the end of every bit, the value of the next bit in the input data stream is as follows: if the bit is binary 1, insert a $+90$ degree phase shift, if the bit is binary 0, insert a -90 degree phase shift.

More efficient use of the bandwidth can be achieved if we squeeze more bits per second into a given bandwidth. For example, instead of a phase shift of 180-degree, as in PSK, another modulation technique allows phase shifts in multiples of 90 degree. This technique is known as quadrature phase-shift keying (QPSK). In QPSK, the incoming bit stream is broken into pairs of bits (these pairs are called dibits). Hence, there are only four possible values of these pairs (dibits) in any bit stream; 00, 01, 10, 11. These dibits change the carrier's phase in accordance with the digits values. The transmitted signal is

$$x(t) = \begin{cases} A\cos(2\pi ft) & \text{if } b(t) = 00 \\ A\cos(2\pi ft + 90) & \text{if } b(t) = 01 \\ A\cos(2\pi ft + 180) & \text{if } b(t) = 11 \\ A\cos(2\pi ft + 270) & \text{if } b(t) = 10 \end{cases}$$

An example of a QPSK modem is the Bell-type 212A, which operates at 300 bps or 1200 bps in a full-duplex mode and it has the ability to perform asynchronous or synchronous conversion. It is also compatible with the Bell type 103 and 113 modems.

TABLE 2.1 Examples of Information Signaling Format and the Devices Used

Information to be Transmitted	Type of Signal Transmitted	
	Analog	Digital
Analog	AM, FM, PM modulator	PCM
Digital	ASK, PSK, FSK, MSK (modems)	RZ, NRZ, NRZI (Codec, digital transmitter)

In general, we can break the incoming stream of bits into groups, each group is $\log_2 M$ bits long. These groups change the carrier's phase in accordance with the group value. This modulation scheme is referred to as M-ary PSK. The transmitted signal is

$$x(t) = A\cos\left(2\pi ft + \frac{2\pi K}{M}\right) \quad K = 0, 1, \ldots, M-1$$

Notice that when $M = 2$ it is PSK, and when $M = 4$ it is QPSK.

An example of M-ary PSK is the Bell-type 208 modem, which is a differential 8-PSK operating at 4800 bps half-duplex synchronous transmission. Table 2.1 summarizes the information signaling format and the devices used.

2.5 CHANNEL BANDWIDTH AND CAPACITY

Sending a message from one point to another involves the transmission of electromagnetic waves. Transmission media, such as twisted pair, coaxial cable, fiber optic, radio link, and satellite link, provide for the transport of the electromagnetic waves. These electromagnetic waves span a wide range of frequencies, from extremely low frequencies, say tens of cycles per second (hertz), to extremely high frequencies, as in microwave and higher as in optical (light) transmission at billions of cycles per second (giga hertz). The maximum information rate (called the channel capacity C in bits per second) that can be transmitted over a given bandwidth depends on the transmitted signal power and the noise characteristics of the channel (as well as the channel bandwidth B in hertz).

Claude Shannon, in the early 1940s, layed the foundation of a mathematical theory of communications, called information theory. Shannon's channel capacity theorem states that it is possible to find a technique (basically a coding technique) whereby it is possible to transmit information with arbitrarily low probability of error provided that the information rate R is less than or equal to a rate C (called the channel capacity). Conversely, it is not possible to find such a technique (code) when the information rate is greater than the channel capacity. One of Shannon's contributions is the following simple famous formula that relates the channel capacity and bandwidth.

$$C = B \log_2[1 + S/N]$$

where N is the total noise power within the channel bandwidth and S/N is the signal to noise power ratio.

The above formula, known as the Shannon-Hartley theorem, applies only to channels with additive Gaussian noise (i.e., noise power is evenly spread throughout the channel bandwidth). Fortunately, this is the most common situation in physical systems. Also, it turns out that the results obtained for a Gaussian channel are normally a lower bound on the performance of a non-Gaussian channel.

Clearly the Shannon-Hartley theorem provides for a trade-off between channel bandwidth B and signal to noise ratio S/N. For example, if $B =$ 4 kHz and $S/N = 200$ then the channel capacity $C = 30.6$ kbps. If B is reduced to 3 kHz, then to maintain the same channel capacity $C =$ 30.6 kbps, we must increase S/N to 1176.3.

2.6 TRANSMISSION MEDIA

Communication networks employ a variety of transmission media ranging from copper wires to satellite channels to transport user's information. The transmission media is the physical path for the communication signal. Transmission media can be classified into two major categories: guided media, which may constrain and guide the communication signal, and unguided media, which permit signals to be transmitted but not guide them. Examples of guided transmission media are open-wire lines, paired cable, coaxial cable, waveguides, and optical fibers. Examples of unguided transmission media are the atmosphere and outer space, which are used for transmitting microwave radio signals and satellite signals. An important characteristic of these different media is the bandwidth or simply the range of frequencies each can transmit. In general, the greater the bandwidth of a given media, the more it can carry.

2.6.1 Open-Wire Lines

Open-wire lines consist of bare (uninsulated) pairs of wires supported on poles spaced about 125 feet apart (Fig. 2.16*a*). The original telephone and telegraph transmission media for both local and long-haul transmission were open-wire lines. They have the advantage of low attenuation at voice frequencies; however, they are sensitive to external interference from storms and high-voltage power lines.

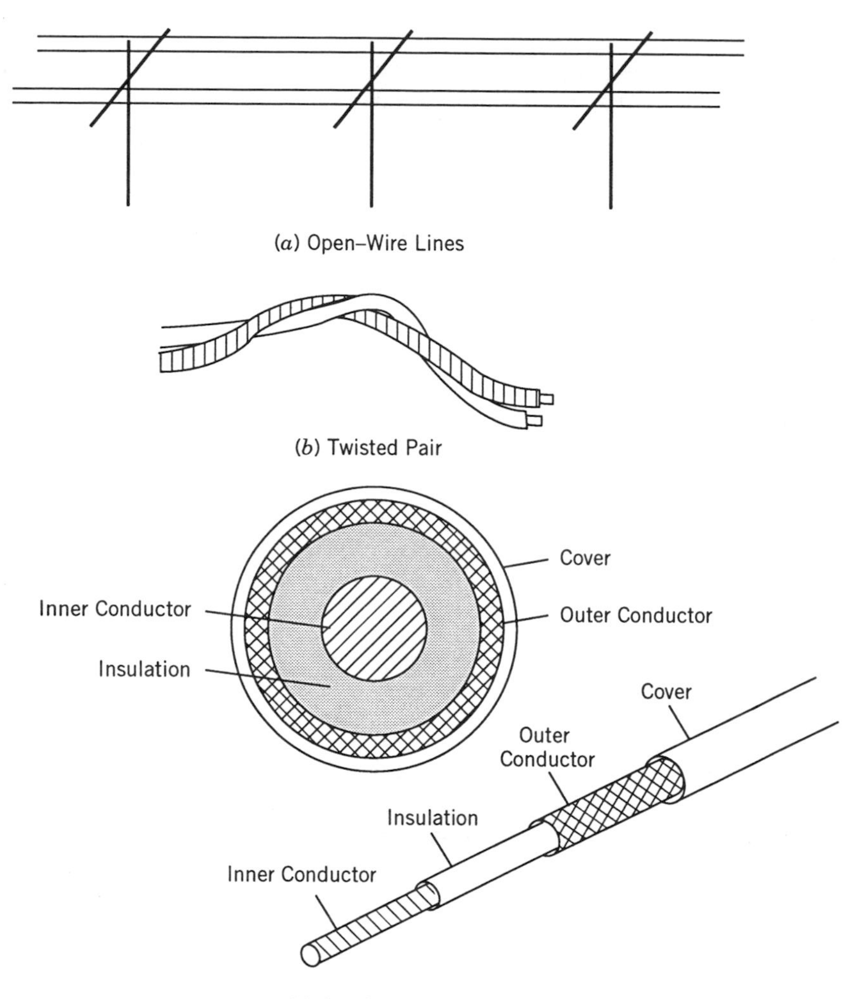

Figure 2.16. Different metallic transmission media (*a*) Open-wire lines. (*b*) Twisted pair. (*c*) Coaxial cable.

The wires range in diameter from 0.08 inches (12 gauge) to 0.165 inches (6 gauge) and are made of copper, copper-clad steel, bronze, or aluminum alloys. Note that the wire sizes are described by the American wire gauge system. The higher gauge number reflects thinner wire sizes.

2.6.2 Paired Cable

A pair of insulated wires is twisted together (twisted pair) in a continuous spiral at varying rates (Fig. 2.16*b*). The reason for twisting is to minimize the susceptibility of the cable to external interference and crosstalk.* A multipair cable consists of individually insulated wire pairs enclosed in a common protective sheath, and it may contain from 6 to 3600 wire pairs. Neighboring pairs are twisted with a different pitch (twist length) to limit crosstalk (electromagnetic interference) between them.

In most applications, wires range in diameters from 0.016 inches (26 gauge) to 0.036 inches (19 gauge). Paired cables are used for wiring buildings, between the customer's premises and the local central office (i.e., the local loop), and between the central offices themselves.

Long-distance lines typically use 19 gauge wires. The local telephone loops usually use 22 to 26 gauge, with the majority using the unshielded 24-gauge wire. Unshielded wire has no metallic cover around it. Shielded twisted pairs provide better resistance to external noise and crosstalk and are sometimes called data-grade media. Twisted-pair cable is relatively inexpensive, readily available, and can be installed fairly easily.

2.6.3 Coaxial Cable

A coaxial cable typically contains from 4 to 24 coaxial tubes. Each coaxial tube consists of a single wire centered in a tube of copper braid, which acts as the second conductor of the pair and electrically shields the center wire from external interference.

The two conductors are insulated from one another using various dielectric materials (e.g., plastic and gas). The outer conductor may consist of one or more layers of braided metal fabric (Fig. 2.16*c*). In addition, to coaxial tubes, coaxial cable may contain a small number of twisted wire pairs and single wires that are used for maintenance and alarm functions.

Coaxial cable can operate at very high frequencies, which permits it to carry a large number of voice channels, high-speed data, and television channels. Coaxial cables have been used extensively in the telephone network. Also, they have been largely used as the transmission media in the cable television industry. Another major application for coaxial cables is its use in local area networks (LANs).

2.6.4 Waveguides

A waveguide consists of a hollow metal tube. It may be rectangular, elliptical, or circular in cross section. Its main advantage is very low attenuation at microwave frequencies. Waveguides are used effectively to connect microwave transmitters to the microwave antenna.

*Crosstalk is a type of noise that results from having many wires running parallel to each other.

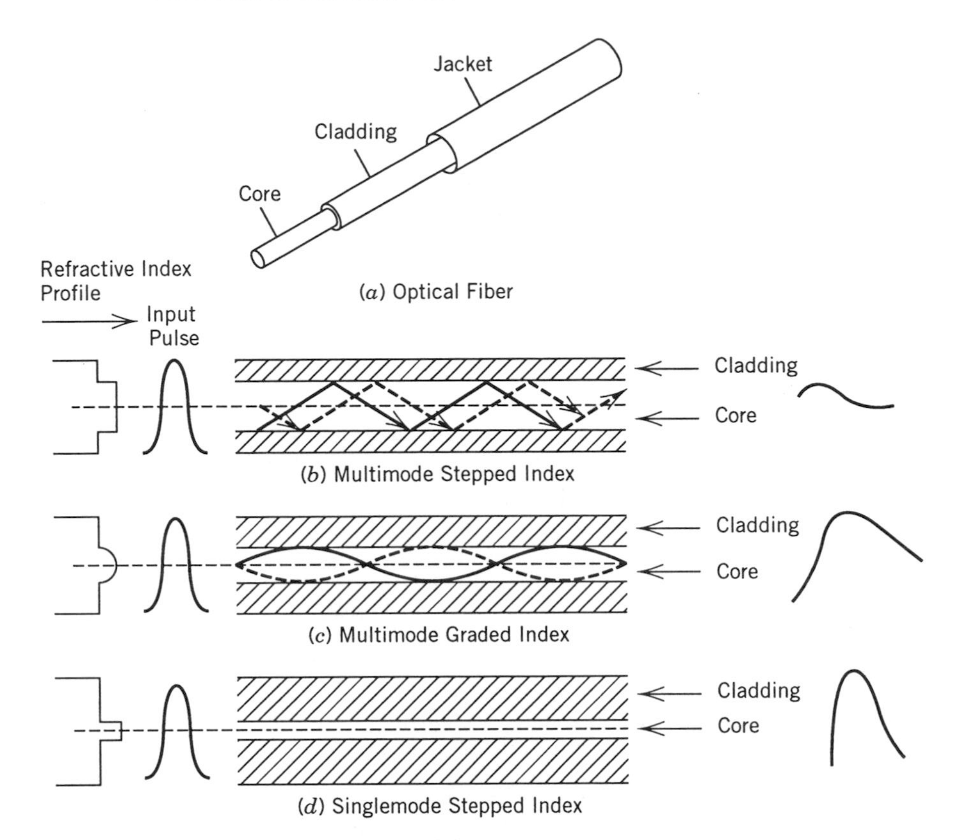

Figure 2.17. Basic types of optical fibers: (*a*) an optical fiber, (*b*) multimode stepped index, (*c*) multimode graded index, and (*d*) single-mode stepped index.

2.6.5 Optical Fibers

Optical fibers have unique characteristics when compared with other media. They have low transmission losses and high bandwidth. They provide immunity to electromagnetic interference and are small in size and lightweight.

Optical fibers have rapidly become the preferred method to transmit huge quantities of information from one place to another. All over the world, optical glass fiber, which as recently as 1970 were primarily objects for laboratory research, is superseding metallic wires in a wide variety of applications including LANs, metropolitan communication networks, long-haul systems, and undersea transmission systems.

Optical (lightwave) communication systems are typically digital in nature: information is encoded into a binary format of zeros and ones for transmission, with one represented by the transmission of a pulse of photons of light, and a zero represented by the absence of such a pulse. Optical fiber (Fig. 2.17*a*) consists of three concentric cylinders made of dielectric materials: the core, the cladding, and the jacket. The jacket is a plastic that absorbs light, prevents crosstalk, and protects the surface of the cladding. The core and the cladding are transparent glass that guide the light within the core and reflect it at the interface between them. The refraction index (defined as the ratio of the speed of light in a vacuum to its speed in a given media), of the core is greater than that of the cladding; hence light moves slower in the core than in the cladding. As a light wave travels in the core toward the cladding, it is reflected toward the core and is guided along the fiber.

There are two main types of optical fiber: multimode and single-mode (or monomode). *Multimode* fiber is subdivided into multimode step-index and multimode graded-index. Multimode fibers are used at modest information rates, tens of megabits per second to 100 Mbits/second, and for ranges up to 10 to 100 km or so. They are extensively used in optical-fiber–based LAN's. *Single-mode* fibers are mainly used in medium-to-long distance trunk lines and international systems, say to several gigabits per second. They are used in areas where multimode fiber is deployed due to falling prices and increasing bit-rate requirements.

The *step-index multimode fiber* (Fig. 2.17*b*) is the simplest type of fiber. They have a multitude of rays (modes) traveling within the fiber. Some of the rays that are reflected at very steep angles will travel a greater distance to reach the other end of the fiber than will the rays that hardly change direction at all. Thus, short pulses of light will be spread out or dispersed in time when they leave the fiber. *Single-mode fibers* provide a solution to this problem of modal dispersion since only the central mode (ray), which is actually the fundamental mode, can propagate in the core, so that the pulse spreading encountered in multimode fibers simply does not exist.

Graded-index multimode fibers are a compromise between the high bandwidth of single-mode fiber and the easy coupling of step-index multimode fibers. In graded-index fibers, the refractive index does not change abruptly between the core material and the cladding. Instead, the refractive index changes continuously and smoothly from a maximum value at the center of the fiber to a minimum value at the outer edge.

2.6.6 Microwave Radio

Microwave radio is one of the predominant long-haul transmission media in use today. The Earth's atmosphere is an effective way of transmitting electromagnetic waves and thus is used as the transmission media for terrestrial microwave radio systems (see Table 2.2 for the electromagnetic frequency spectrum). A microwave system usually consists of a microwave transmitter, a line-of-sight propagation path through the Earth's atmosphere, and receiving antenna. Repeaters are placed between the endpoints. They receive the signal, amplify it, and pass it on to the next repeater. The spacing between the repeater stations depends on the frequency used, the terrain,

TABLE 2.2 Electromagnetic Frequency Spectrum

Frequency	Wavelength (m)	Designation
3 Hz – 30 kHz	$10^8 - 10^4$	Very low frequency (VLF)
30 – 300 kHz	$10^4 - 10^3$	Low frequency (LF)
300 kHz – 3 MHz	$10^3 - 10^2$	Medium frequency (MF)
3 – 30 MHz	$10^2 - 10$	High frequency (HF)
30 MHz – 300 MHz	$10 - 1$	Very high frequency (VHF)
300 MHz – 3 GHz	$1 - 10^{-1}$	Ultrahigh frequency (UHF)
3 – 30 GHz	$10^{-1} - 10^{-2}$	Superhigh frequency (SHF)
30 – 300 GHz	$10^{-2} - 10^{-3}$	Extremely high frequency (EHF)
$10^3 - 10^7$ GHz	$3 \times 10^{-5} - 3 \times 10^{-9}$	Infrared, visible light, ultraviolet

the technology used, and the transmitter power. This spacing varies from 1 mile to 100 miles.

The main advantage of a microwave system is its ability to span natural barriers and obstructions, such as mountainous terrain, bodies of water, or heavily wooded areas. It is ideally suited for mobile communications.

Multipath fading represents a major technical issue for microwave system designers. It arises from the fact that the signal propagates along several paths, each of different electrical length. At the receiver, those relatively delayed signal components interfere with each other and this may lead to signal fading. Multipath fading can be minimized by using frequency-diversity or space-diversity techniques.

Another limitation to microwave systems is that it might be difficult to obtain frequency assignments in large cities because of the limited availability of channels in the radio spectrum.

2.6.7 Satellite

As mentioned in Section 1.7.6, communication systems use the electromagnetic spectrum. The frequencies used for satellite communications are allocated in superhigh frequency (SHF) and extremely high frequency (EHF) bands, which are broken down into subbands as shown in Table 2.3.

TABLE 2.3 Satellite Frequency Spectrum

Frequency Band	Range (GHz)
L	1 – 2
S	2 – 4
C	4 – 8
X	8 – 12
Ku	12 – 18
K	18 – 27
Ka	27 – 40
Millimeter	40 – 300

TABLE 2.4 The North America FDM Hierarchy

Multiplex Level	Number of Voice Circuits	Construction	Frequency Range (kHz)
Group	12	12 voice circuits	60 – 108
Supergroup	60	5 groups	312 – 552
Mastergroup	600	10 supergroups	564 – 3084
Mastergroup Mux	1200 – 3600	Various	312,564 – 17,548
Jumbogroup	3600	6 mastergroups	564 – 17,548
Jumbogroup Mux	10,800	3 jumbogroups	3000 – 60,000

2.7 MULTIPLEXING

Multiplexing combines and bundles together a number of communication channels and transmits them over one physical common broadband channel. At the receiving end, demultiplexing separates and recovers the original channels. The main purpose of multiplexing is the efficient use of the channel bandwidth. The two main multiplexing techniques are frequency division multiplexing (FDM) and time division multiplexing (TDM).

2.7.1 Frequency Division Multiplexing

In FDM systems, each transmitter is allocated a portion of the frequency spectrum (that is a frequency band). The transmitted signal spectral component must be confined to the allocated frequency band. Examples of FDM systems are the telephone network, cable television (that is, the community antenna television or CATV), voice frequency multiplexers, and satellite systems.

An illustration of this protocol is the North American FDM telephone hierarchy (Table 2.4), which is designed to transmit a large number of analog voice channels over a common transmission channel. The standard 0- to 4-kHz voice signal is modulated (frequency shifted) to occupy a specific frequency band; for example, with three successive levels of multiplexing, a total of 600 voice signals are multiplexed together over the common transmission media.

Figure 2.18 shows a schematic diagram for the North American FDM hierarchy up to the mastergroup. The lowest level building block in the hierarchy is a channel group of twelve 4-kHz voice channels resulting in a total bandwidth of 48 kHz. The channel group multiplexer uses twelve carriers spaced 4 kHz apart to generate 12 single-sideband signals that are summed to form the group. The second level is the 60-channel supergroup, which multiplexes five first-level channel groups. The supergroup occupies

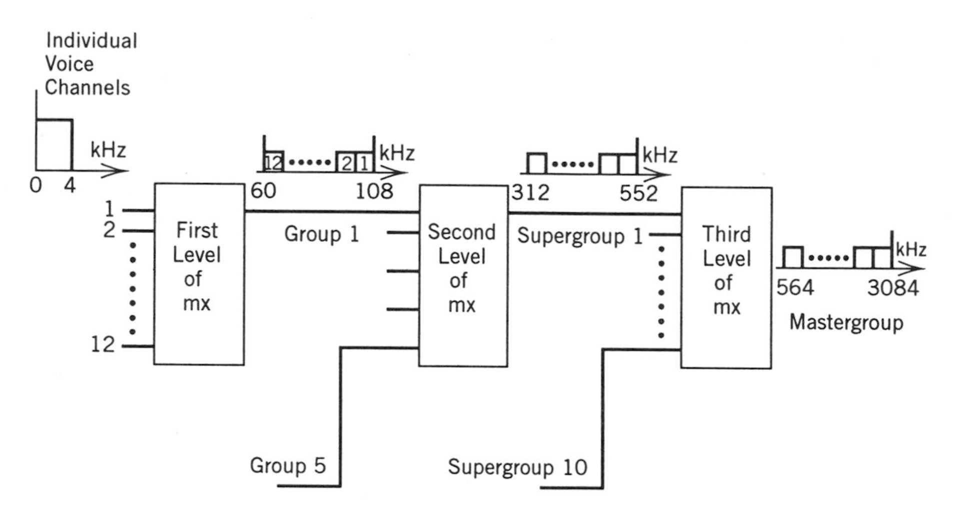

Figure 2.18. North American frequency division multiplexing hierarchy.

the frequency range 312 to 552 kHz. It uses five carrier frequencies 420, 468, 516, 564, and 612 kHz. The mastergroup consists of 10 supergroups.

2.7.2 Time Division Multiplexing (TDM)

In TDM systems, each station is allocated the entire frequency bandwidth of the transmission media, but only for a limited portion of the time, called a time slot. There are two basic techniques for TDM, synchronous time division multiplexing and asynchronous time division multiplexing.

In *synchronous TDM* (STDM), each source is repeatedly (and hence the word synchronous) assigned a portion of the transmission capacity. TDM is understood to imply STDM. Circuit switched telephone networks use (synchronous) TDM.

On the other hand in *asynchronous TDM* (ATDM), each source is assigned a portion of the transmission capacity only as it is needed (and hence the word asynchronous). Asynchronous is used in statistical multiplexers, known also as stat-Muxs or concentrators, packet switches, and asynchronous transfer mode switches proposed for broadband integrated services digital network. Asynchronous transfer model is discussed in Chapter 11; the remainder of this chapter will concentrate on (synchronous) TDM.

There are different TDM implementations: bit interleaving and word interleaving. In bit interleaving, each channel is assigned a time slot corresponding to a single bit. In word interleaving, each channel is assigned a longer time slot corresponding to some larger number of bits, referred to as a word. For example, lower levels of digital TDM tend to use word interleaving, whereas higher levels of digital TDM tend to use bit interleaving.

An example of TDM systems is the voice-TDM digital system, which allows multiple voice calls over the same wideband transmission media. There are two voice-TDM systems: the T1 system, used in the United States and the Conference of European Posts and Telecommunications Administrations (CEPT*) system. The American system was adopted by the CCITT in Recommendation G733; the European system appears in Recommendation G732.

The T1 Digital System. The T1 digital system time multiplexes 24 voice channels. As mentioned earlier in the discussion of PCM, each analog voice signal is band limited to approximately 3.4 kHz and is sampled at the 8-kHz rate (that is, the time between samples is 125 per microsecond). Each sample is then companded, quantized, and represented by 8 bits (that is, a time slot). Thus one voice call is represented by a 64-kbps data stream.

In the T1 system, 24 time slots are multiplexed together to form a frame (Fig. 2.19*a*). Each time slot is an 8-bit encoded word. In addition to the 24 time slots, one more bit is added for frame synchronization and alignement. The resulting PCM stream forms a rate of 1.544 Mbits per second, as follows:

$$\text{24 voice time slots} \times \text{8 bits per slot} = \text{192 bits}$$

$$\text{192 bits} + \text{1 framing bit} = \text{193 bits/frame}$$

$$\text{8000 frames/second} \times \text{193 bits/frame} = \text{1.544 Mbit/second}$$

This data rate of 1.544 Mbits per second and the frame structure is called T1 or DS1 (digital signaling 1). It is the primary level rate for long haul communication systems in the United States.

To multiplex and demultiplex the voice channels in an orderly fashion, the transmission system needs to identify the following:

1. The start and finish of each sampling sequence or frame, i.e., which is time slot 1, 2, and so on.
2. The initiation and termination of a channel and its routing information.

Framing bits essentially achieve requirement 1 (timing information), whereas signaling bits carry the information for requirement 2 (see below).

Frame Synchronization and Alignment. As mentioned earlier, one extra bit, the F bit, is added to each group of 192 bits to form a frame. Every twelve frames form a multiframe. The F bits in the odd numbered frames in a multiframe have the pattern 101010 and are used for frame synchronization. F bits in the even numbered frames have a pattern of 001110, and are used for multiframe synchronization.

*The conference of European Post and Telecommunication Administrations (CEPT) was a telecommunication standard organization before the establishment of ETSI (see Appendix A1).

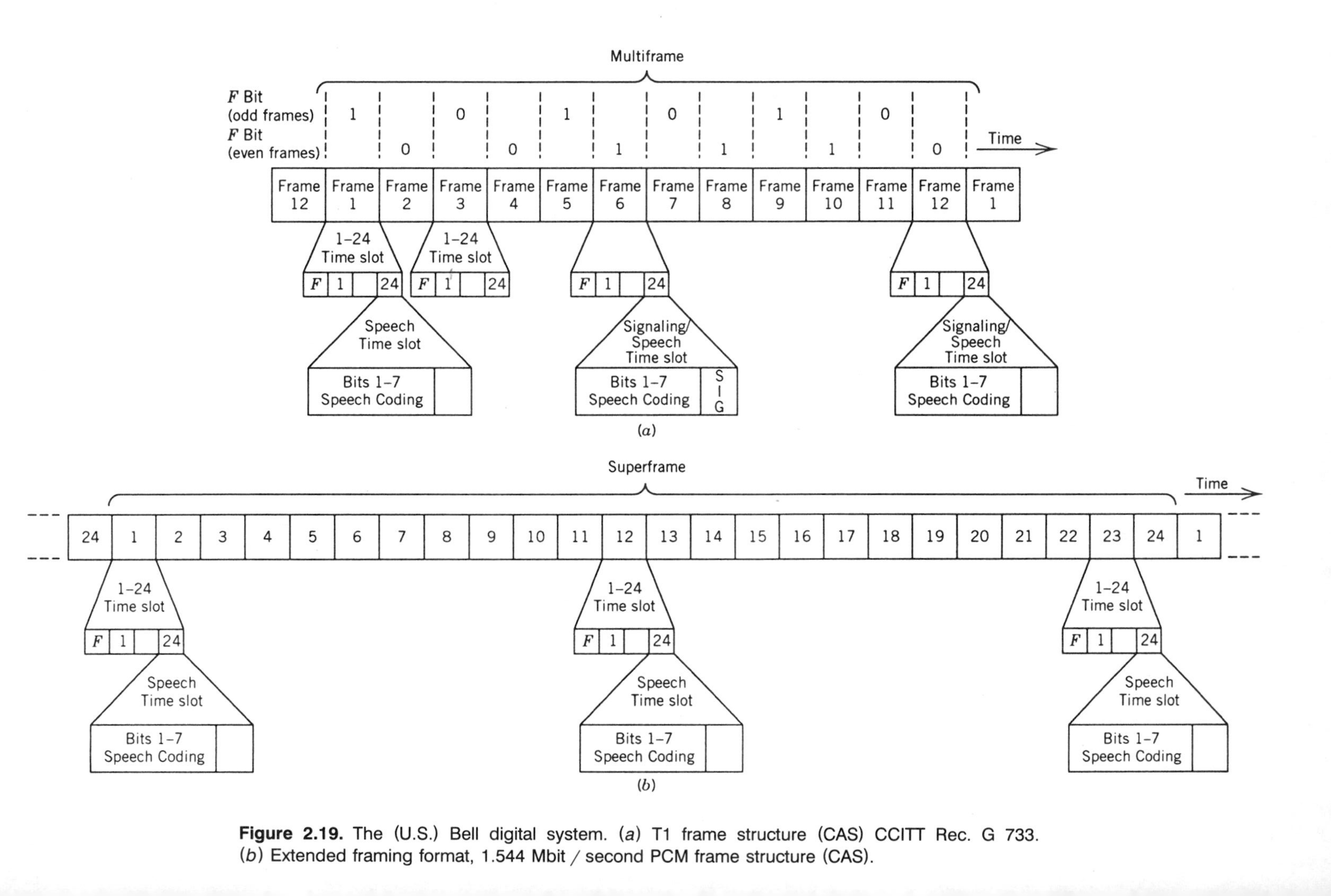

Figure 2.19. The (U.S.) Bell digital system. (*a*) T1 frame structure (CAS) CCITT Rec. G 733. (*b*) Extended framing format, 1.544 Mbit / second PCM frame structure (CAS).

Signaling. Signaling is achieved by one of two ways: channel-associated signaling and common-channel signaling.

Channel-Associated Signaling (CAS). In CAS, bits are stolen from the voice channel [the least significant bit (LSB) in all 24 time slots in frames 6 and 12]. The system basically overwrites the LSB of the speech signal with a signaling bit. This hardly affects speech quality. Bits stolen from frame 6 are called A signalling bits and those stolen from frame 12 are called the B signaling bit. Thus, the signaling rate is $\frac{1}{6}$th of the frame bit rate (8000 frames per second), that is $\frac{1}{6} \times 8000 = 1333$ bps.

Common-Channel Signaling (CCS). When the channel is carrying computer data, it is not adequate to steal bits. Therefore, with CCS, a separate line is used to carry the signaling information and bits are not stolen. Common-channel signaling may be used for voice and data traffic.

Extended Framing Format. To allow the transmission of actual data traffic, a variation to the standard framing of T1, called the extended framing format, is introduced (Fig. 2.19*b*). This new format redefines the framing bits, F, to allow the use of cyclic redundancy codes (CRC) and other features to accommodate in-service performance monitoring. Every consecutive 24 frames are grouped together to form a super frame. The framing bits, F, in every frame are defined as follows:

- F bits in frames 2, 6, 10, 14, 18, and 22 (total of 6 bits) are used for CRC bits to provide performance monitoring and error detection capabilities.
- F bits in frames 4, 8, 12, 16, 20, and 24 (total of 6 bits) are used to provide frame and superframe synchronization. The framing pattern sequence is 001011.
- F bits in all odd numbered frames (total of 12 bits) are used to provide a 4-kbit/second data channel link between DS-1 terminals for use in network maintenance, supervisory data, and so on.

For signaling, the least significant bits in frames 6, 12, 18, and 24 (called A,B,C, and D bits) are used (stolen) for channel-associated signaling.With the added features of error detection, monitoring, supervising, and so on, the extended framing format will become increasingly important in the ISDN.

High-Level Multiplexing. A hierarchy of further multiplexing is built up in stages from the basic T1 signal (Fig. 2.20*a*). Four T1 (or DS1) lines are multiplexed in a M12 multiplexer to generate the T2 (or DS2) transmission system. Seven T2 lines are multiplexed in an M23 multiplexer to form the T3 (DS3) transmission system. Six T3, through an M34 multiplexer, form the T4

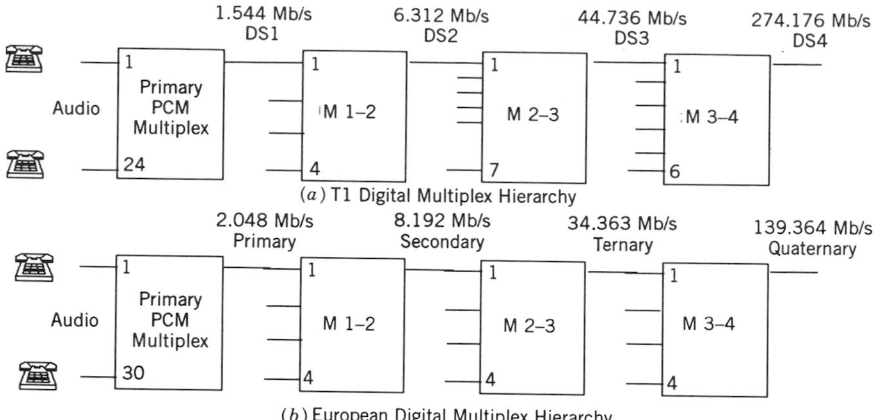

Figure 2.20. Digital hierarchy system. (*a*) T1 digital multiplex hierarchy (*b*) European digital multiplex hierarchy.

(or DS4) line. Additional frame synchronization bits must be added at every multiplexing stage.

The M12 Multiplexer adds 17 bits for frame synchronization and pulse stuffing. Hence, the number of bits per frame is $193 \times 4 + 17 = 789$ and the T2 bit rate is 789 bits/frame $\times$ 8000 frames/second = 6.312 Mbit/second.

The M23 multiplexer adds 69 bits for synchronization and pulse stuffing resulting in 5592 bits per frame ($789 \times 7 + 69$). Thus, the T3 bit rate is $5592 \times 8000 = 44.736$ Mbit/second.

The M34 multiplexer adds 720 bits for synchronization and pulse stuffing. Thus the T4 bit rate is 274.176 Mbit/second.

The European Digital System. The European system, referred to as CCITT G732, came into existence after the T1 system and benefited from the experience of the American system. The system, established by CEPT, resulted in a different digital system. Both systems coexist on the international level, although they are incompatible.

In the European system, 32 time slots (voice channels) are multiplexed together to form a frame (Fig. 2.21). Each time slot is an 8-bit encoded word. One entire time slot (slot 0) is devoted to frame synchronization and alignment. Another timeslot (slot 16) is devoted to signaling. Thus, the resulting bit rate is 2.048 Mbit/second as follows:

$$\text{30 voice timeslots} \times 8 \text{ bits} = 240 \text{ bits}$$
$$\text{1 frame synchronization timeslot} \times 8 \text{ bits} = 8 \text{ bits}$$
$$\text{1 signaling timeslot} \times 8 \text{ bits} = 8 \text{ bits}$$
$$\text{1 frame } (240 + 8 + 8 \text{ bits}) = 256 \text{ bits}$$
$$\text{8000 frame/second} \times 256 \text{ bits} = 2.048 \text{ Mbits/sec}$$

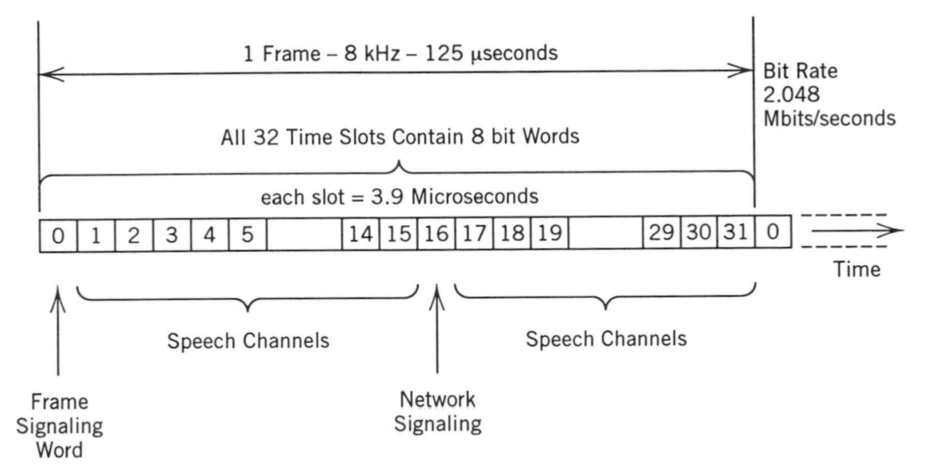

Figure 2.21. European digital system.

For CAS, a multiframe is composed of 16 frames, with timeslot 16 of all frames used for signaling. If the system utilizes common channel signaling (CCS) (that is signaling is not carried in timeslot 16), then this frees timeslot 16 to be used for carrying voice. In frame 0, time slot 16 is used for multiframe alignment. Figure 2.20*b* shows the higher level multiplexing of the CEPT system.

2.8 THE PHYSICAL LAYER OF THE OSI REFERENCE MODEL

As mentioned earlier in Chapter 1, the physical layer is the lowest layer in the OSI reference model. Its function is to interface with the transmission media and provide services to the data link layer such as the physical connection between data link entities. The OSI reference model defines the physical layer as the layer that provides mechanical, electrical, functional, and procedural functions to establish, maintain, and release physical connections between data terminal equipment (DTE), data circuit-terminating equipment (DCE), and/or data switches.

A number of physical level interfaces are used in the data communications industry. The most familiar is the Electronics Industry Association (EIA) RS-232. The latest version of this standard, introduced in 1987, is EIA-232-D which is now compatible with the similar standards CCITT V.24, V.28, and ISO 2110. Examples of other serial data DTE/DCE interfaces are EIA-422-A, EIA-423-A, and CCITT X.21.

2.8.1 EIA-232-D

The EIA-232-D standard interface is among the most common data communications interfaces. Originally developed for use with modems, it may also be used with digital networks. The formal title for the standard is "Interface between Data Terminal Equipment and Data Communications Equipment Employing Serial Binary Data Interchange".

The same interface is specified in the CCITT Recommendation V.24. The EIA-232-D specification actually describes the wires in a cable connecting data terminal equipment, such as a terminal or a computer, to a data circuit-terminating equipment, such as a modem. The EIA-232-D standard contains five important characteristics:

1. The interface mechanical characteristics
2. The electrical signal characteristics
3. A functional description of the interchange circuits (i.e., pins)
4. A list of standard subsets of specific interchange circuits for specific applications
5. Recommendations and explanatory notes.

The mechanical characteristics include the specifics of the connector, the assignment of interchange circuits to pins, the connector latching arrangement, mounting arrangements, and so on. Figure 2.22 shows the EIA-232-D connector which is a 25-pin connector. Table 2.5 shows the 25-pin assignments.

The electrical signal characteristics specify the signaling across the DCE/DTE interface. The signal on an interchange circuit represents a binary 1 (a marking condition or an OFF condition) when the voltage V_1 is more negative than -3 V with respect to the signal ground circuit AB (pin 7). The signal is considered to represent a binary 0 (a spacing condition or an

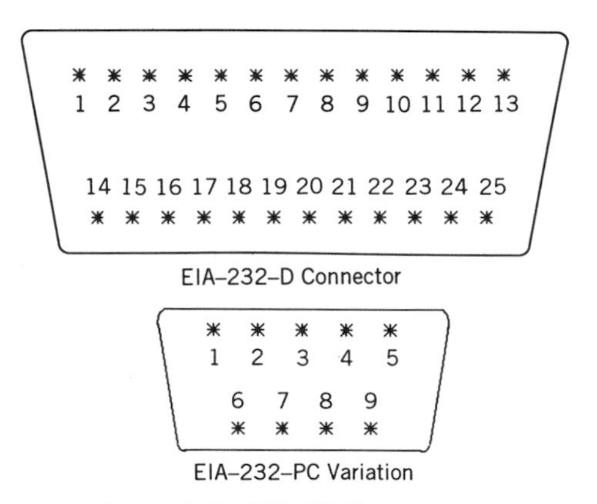

Figure 2.22. EIA-232-D connector.

TABLE 2.5 EIA-232-D Interchange Circuits and Pin Assignments

Inter-change Citcuit EIA Label	Pin Number	CCITT Equiv-alant	Signal Name	Function Group[a]
AA	1	101	Shield	G
AB	7	102	Signal ground / common return	G
BA	2	103	Transmitted data	D
BB	3	104	Received data	D
CA	4	105	Request to send	C
CB	5	106	Clear to send	C
CC	6	107	DCE ready	C
CD	20	108.2	DTE ready	C
CE	22	125	Ring indicator	C
CF	8	109	Received line signal detector	C
CG	21	110	Signal quality detector	C
CH	23	111	Data signal rate selector (DTE)	C
CI	23	112	Data signal rate selector (DCE)	C
DA	24	113	Transmitter signal element timing (DTE)	T
DB	15	114	Transmitter signal element timing (DCE)	T
DD	17	115	Receiver signal element timing (DCE)	T
SBA	14	118	Secondary transmitted data	D
SBB	16	119	Secondary received data	D
SCA	19	120	Secondary request to send	C
SCB	13	121	Secondary clear to send	C
SCF	12	122	Secondary received line signal detector	C
RL	21	140	Remote loopback	C
LL	18	141	Local loopback	C
TM	25	142	Test mode	C

[a]G, ground; D, data; C, control; T, timing.

ON condition) when the voltage V_1 is more positive than $+3$ V with respect to circuit AB. The open circuit voltage V_o should be less than or equal to 25 V in magnitude. The region between $+3$ and -3 V is a transition region. The functional description of the interchange circuits can be classified into four major groups: data, control, timing, and ground, as shown in Table 2.5. For example, in the data group function, the DTE transmits on pin 2 and receives on pin 3. The DCE does the opposite; it receives on pin 2 and transmits on pin 3. In cases where a DTE is communicating with another DTE, the connection must be altered (this arrangement is referred to as "null modem").

The distance coverage of EIA-232-D is 50 ft or less. In many applications, greater distances can be achieved. In general, EIA-232-D is limited to a 20 kbps and a few hundred feet of cable length. EIA-449 (formerly RS-449),

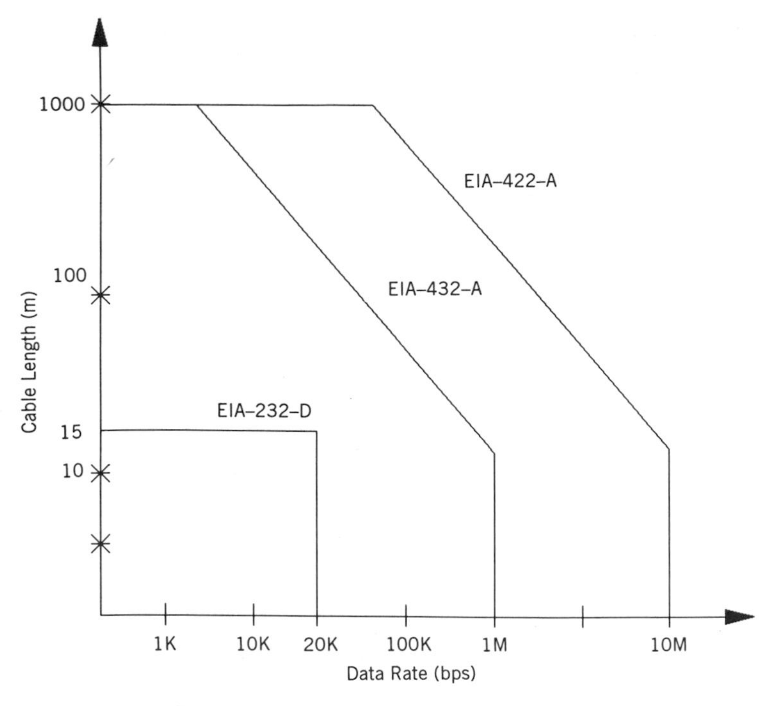

Figure 2.23. Performance of EIA interfaces.

introduced in 1977, attempts to overcome some limitations of EIA-232-D. It is a 37-bin connector and provides speeds up to 2 Mbps and is used with two other specifications, EIA-422-A and EIA-423-A. The EIA-422-A is a balanced electrical interface (that is, a differential signaling over a pair of wires for each circuit), and these can tolerate more noise to provide superior performance up to 10 Mbps at a distance of 10 m or to 100 kbps at 1 km. The EIA-423-A is an unbalanced electrical interface with a common return for each direction. It provides speeds of 1 kbps at 1 km or 100 kbps at 10 m. Figure 2.23 shows the cable distance–data rate curves for EIA-232-D, EIA-422-A, and EIA-423-A interfaces.

The EIA-530 standard has been introduced in 1987 with the intent of gradually replacing the 37-pin connector EIA-449. The EIA-530 is a 25-pin connector with a data rate from 20,000 bps to 2 Mbps.

2.8.2 X.21

Another important interface is the CCITT recommendation X.21 which is used in many European countries and Japan. The X.21 uses a 15-pin connector. A major objective in its design was to reduce the number of interchange circuits. Figure 2.24 shows the five basic interchange circuits of the X.21 interface: transmit, receive, control, indication, and signal timing. A transmit circuit and a receive circuit are used for user-data and network-

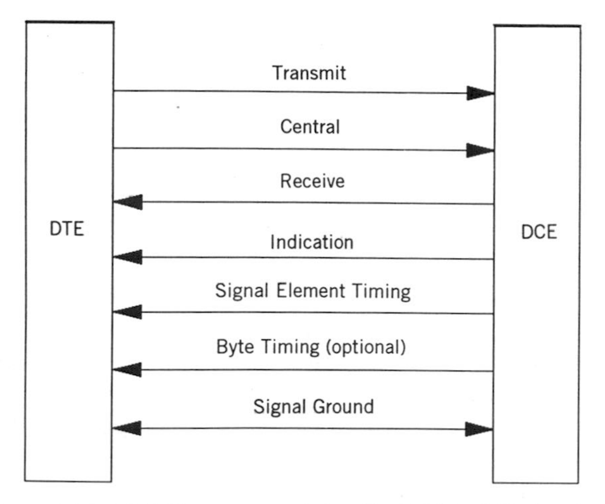

Figure 2.24. CCITT recommendation X.21 DTE / DCE interface.

control information depending on the state of the control circuit and indication circuit. Bit timing is provided by a signal-element timing circuit. A signal ground circuit and an optional interchange circuit for byte-timing information are provided.

2.9 NOISE AND ERROR DETECTION METHODS

At the receiver, the signal, whether it is transmitted over a wire channel or a wireless channel, is accompanied by noise. Also, as the received signal is processed by the various stages of the receiver (filtered, amplified, demodulated, decoded, etc.), each stage superimposes additional noise on the signal. The effect of the noise is to corrupt the desired signal and make it more difficult to determine correctly what the original data value was. The noise signal can have many shapes, causes, forms, and effects on system performance. One common type of noise is called *white noise* or Gaussian noise which is characterized by a relatively uniform amplitude in the frequency domain over the bandwidth of interest, while in the time domain it is very random in amplitude. Another type of noise is *impulse noise*, which is a burst of noise in the time domain, while in the frequency domain it is mostly a wideband of frequencies.

Crosstalk noise is another type of noise, which results from having many wires running parallel to each other. Intersymbol interference also contributes to data corruption. In general, it results from limited bandwidth. Noise sources, can be classified into two main groups; the first group is the noise generated by components within the communications system itself (sometimes referred to as internal noise), such as integrated circuit chips, leads, and resistors. The second group is the noise external to the communications system, such as atmospheric noise, man-made noise, and extraterrestrial noise.

In Section 2.5, we discussed Shannon's theorem which proved that a sufficiently sophisticated coding technique enables transmission at rates equal to or less than channel capacity with arbitrarily small error. Engineers are still faced with the task of finding a code with reasonable implementation complexity that satisfies the required error probability.

A number of codes have been invented for error detection and error correction. The basic idea in the design of these codes is to add redundant bits to the input data stream. Examples of error-detecting codes are parity-check codes. Error-detecting codes, as the name implies, can only detect the existence of one or more errors in part of the message. Yet they cannot tell exactly the error location. For some type of error patterns, error-correcting codes can detect with high probability the error location, and hence the error can be corrected. Examples of error-correcting codes are the Hamming, the Bose-Chaudhuri-Hocquenghem (BCH), the Reed-Solomon, and the convolutional code.

2.9.1 Parity-Check Codes

In parity-check codes, an additional bit, the parity bit, is appended to the binary-bit data stream of the encoded character to form a complex encoded character. The parity bit is chosen such that the entire character has an even number of binary 1s in it (even parity), or it is chosen such that the entire character has an odd number of binary 1s (odd parity).

For example, assume the ASCII character A which is represented by 1000001. With an even-parity check, the eighth bit will be 0 such that the total number of 1s is even. Thus the ASCII character A would be:

```
0   1 0 0 0 0 0 1
|
parity    data
 bit      bits
```

If one bit is in error, say the second bit is changed from 0 to 1, then the number of 1s will be three (i.e., an odd number of 1s), and the receiver will then detect the error in the bit stream. On the other hand, if there are two bits in error, say the second and third bits (i.e., the received pattern is 01000111). In this case, the total number of 1s is four (even) and the receiver will not detect the error. This example is shown below.

0 100 0001	ASCII character A with even parity bits
0 100 0011	with 1 bit in error
0 100 0111	with 2 bits in error

In conclusion, single-bit-parity check codes can detect all odd-number bit errors but it can not detect any even-number bit errors.

2.9.2 Block-Parity Check

Parity-check codes (even or odd) are mostly used in asynchronous transmission, whereas block check codes and cyclic codes are used for synchronous transmission. In block-parity check codes, a group of characters is transmitted as one block with no time gaps between them (synchronous transmission). A parity bit is formed on each individual character as well as on all the characters in the block. Thus, an additional character, referred to as block-check character (BCC) or the longitudinal-redundancy-check (LRC) character, is attached to the end of the block. Below we show an example to illustrate the use of block-parity checking for a block of 3 data characters (characters ABC). Note that the parity of the BCC is determined by the bits in the BCC itself and not on the parity bits of the text characters. Typical block size is approximately 256 characters.

Example. BCC Computation (even parity)

BCC C B A	⟶ direction of transmission

BCC is attached at the end of the message

	parity bit ↓	
A	0	100 0001
B	0	100 0010
C	1	100 0011
BCC	1	100 0000

BCC	C	B	A	
1100 0000	1100 0011	0100 0010	0100 0001	⟶ Bit stream transmitted after appending the check-bits

Now, if there are two bits in error in Character A, say in the 2nd and 3rd bits, the parity bit in character A will not detect the error (it is an even number of bit-errors), but the BCC will detect that error. This is shown below.

		errors ↓↓
A	0100	0111
B	0100	0010
C	1100	0011
BCC	1100	0000
		detect errors

Block-parity check is sometimes referred to as the vertical redundancy check/longitudinal redundancy check (VRC/LRC).

Still block-parity check fails to detect some errors. For example, assume character A has errors in bit 2 and bit 3 as before and, in addition, character B has errors also in bit 2 and bit 3. In this case, each character bit as well as the block parity character would fail to detect the errors. This is shown below:

	errors ↓↓
0100	0111
0100	$0\overline{10}0$
1100	$0\overline{01}1$
1100	0000

errors cannot be detected

2.9.3 Cyclic-Redundancy Check (Polynominal Codes)

The cyclic redundancy check (CRC), sometimes called polynominal codes, is a big improvement over the previous codes. It is by far the most widely used error-detection code. The basic idea behind CRC is first explained using decimal numbers. Assume that we want to transmit the message 57268. Both the transmitter and receiver agree on divisor, say 84. The transmitter divides the message sequence by the divisor 84:

$$\frac{57268}{84} = 681 + \frac{64}{84}$$

that is, the remainder of this division is 64. The transmitter attaches the remainder (64) to the message sequence and transmits the whole sequence 5726864.

Suppose that the transmission media introduces noise and as a result, the received sequence is 5754864. The receiver deals with (57548) and divides it by the divisor (84) and gets a remainder of (8) that is different from the transmitted remainder (64); thus the error is detected.

The CRC error detection follows the same principle operating on polynomials representing a binary data string. A binary string

$$b_{n-1}, b_{n-2} \cdots b_3, b_2, b_1, b_0$$

where $b_i = 0$ or 1, is represented as the polynomial

$$b_{n-1}x^{n-1} + b_{n-2} + \cdots + b_2x^2 + b_1x + b_0$$

For example, the string 101101 is represented as

$$x^5 + x^3 + x^2 + 1$$

We refer to the polynomial that represents the m data bits as the message polynomial $M(x)$. Similarly, the binary divisor is represented by the generator polynomial $G(x)$ and the remainder of dividing the message polynomial by the generator polynomial is represented by the remainder polynomial $R(x)$. In the following procedures, all divisions, multiplications, subtractions, and additions are done modulo-2 as follows:

a	b	EXCLUSIVE OR XOR $a+b$	Multiplications $a \cdot b$
0	0	0	0
0	1	1	0
1	0	1	0
1	1	0	1

Note that modulo-2 additions and subtractions are the same as using standard addition and substraction, with no carries for additions or borrows for substraction and both are identical to EXCLUSIVE OR. Also, multiplication rules are the same as standard multiplication.

Now we can summarize the procedure to generate the CRC for a given message as follows;

1. Both the transmitter and the receiver agree on the generator polynomial (i.e., the divisor), $G(x)$. Let r be of the order of $G(x)$, that is, the highest power of X. Both the high and low-order bits of $G(x)$ must be 1.
2. The transmitter appends r zero bits to the low-order end of the message to be transmitted, so the transmitted frame has $m + r$ bits and corresponds to the polynomial $x^r M(x)$.
3. The transmitter determines the remainder (which is always r or fewer bits) of the division of the bit string corresponding to $G(x)$, the divisor, into the bit string corresponding to the input m bits data stream plus the added r zero bits, represented by $x^r M(x)$.
4. The transmitter substracts the remainder from the bit stream corresponding to the $x^r M(x)$ using modulo-2 substraction.

Now, we will solve a binary example similar to the decimal example solved previously. The input data stream is 11101 with $m = 5$. The divisor is 1001 with $r = 3$:

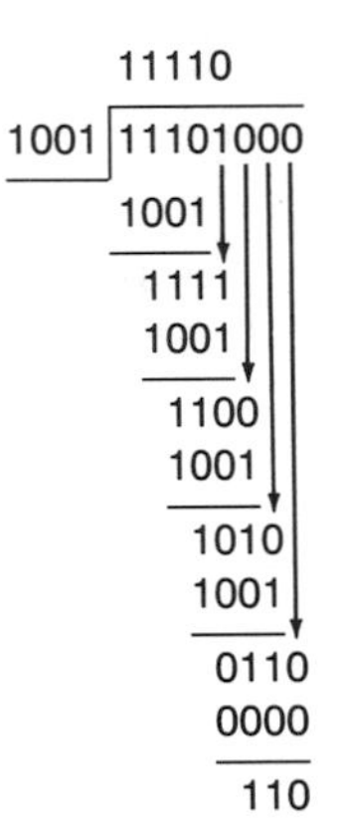

That is the remainder is 110. Now, the remainder is substracted from the dividend and the transmitted bit stream is

$$11101\underline{110}$$
remainder

The receiver uses the same divisor, and, if there are errors, the remainder will be different. Division of the polynomial uses the same rule as algebraic long-division. Now, the same procedures can easily be done using polynomials as follows:

$$M(x) = x^4 + x^3 + x^2 + 1 \qquad m = 5$$

$$G(x) = x^3 + 1 \qquad r = 3$$

Appending r zero bits to the input bit stream produces

$$x^r M(x) = x^7 + x^6 + x^5 + x^3$$

Dividing $x^r M(x)$ by $G(x)$ gives

$$\begin{array}{r|l}
 & x^4 + x^3 + x^2 + x \\
\hline
x^3 + 1 & x^7 + x^6 + x^5 + x^3 \\
 & x^7 + x^4 \\
\hline
 & x^6 + x^5 + x^4 + x^3 \\
 & x^6 + x^3 \\
\hline
 & x^5 + x^4 \\
 & x^5 + x^2 \\
\hline
 & x^4 + x^2 \\
 & x^4 + x \\
\hline
 & x^2 + x
\end{array}$$

Remainder $R(x) = x^2 + x$

$$T(x) = x^r M(x) + R(x)$$

$$= x^7 + x^6 + x^5 + x^3 + x^2 + x$$

the corresponding transmitted bit stream is

11101110
remainder

Note that $x^i + x^i = 0$ for polynomial additions

There is obviously one drawback to CRC, if the added error to the transmitted message is divisible by $G(x)$, the receiver will not detect the error. However, CRC has the following advantages:

1. It can detect all single errors
2. It can detect all double-bit errors when $G(x)$ has a factor with at least three terms.
3. It can detect any odd number of errors, when $G(x)$ has a factor $(x + 1)$
4. It can detect all burst errors for which the length of the burst is less than the length of CRC

CRC is very attractive when it comes to hardware implementation. The hardware is merely a number of shift register stages and EXCLUSIVE OR (XOR) gates (i.e., modulo two adders). Figure 2.25 shows an example of hardware implementation for CRC with $G(x) = x^3 + x^2 + 1$. The feedback lines with XOR gates are determined by coefficients of the generating polynomial g_i; for each nonzero g_i there is a feedback line with the XOR gate. Initially, the shift-register stages are all set to zero. The bits to be transmitted are entered one bit at a time, starting with the most significant bit. Assume the message is

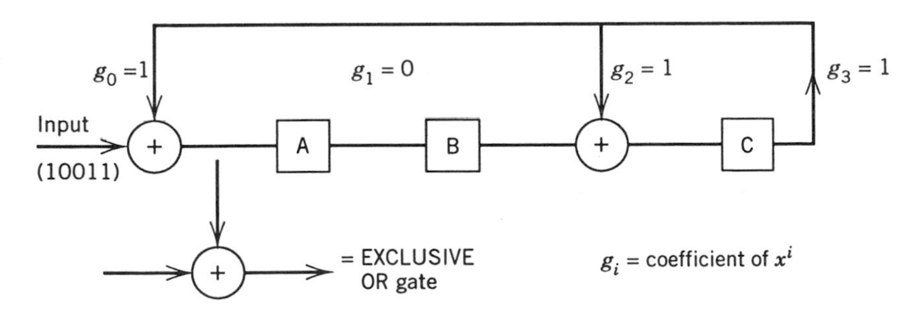

Figure 2.25. Shift Register Implementation for CRC with $G(x) = x^3 + x^2 + 1$.

TABLE 2.6 Shift Register Contents, $G(x) = x^3 + x^2 + 1$ and the Input Is 11001

Step Number	Input	Shift Register Contents A	B	C
0	0	0	0	0
1	1	1	0	0
2	1	1	1	0
3	0	0	1	1
4	0	1	0	0
5	1	1	1	0
6	0	0	1	1
7	0	1	0	0
8	0	0	1	0
			remainder	

11001, we append three zeros ($r = 3$) and input the message one bit at a time. Table 2.6 shows the content of the shift register at each step. The contents at the last step (step 8 in this example) shows a remainder of 010. At the receiver, the same circuit is used. The received message (i.e., 11001010 if there is no error) is fed into the shift register one bit at a time. The remainder is then the shift register contents. If there have been no errors, the register contents should be zero.

Some of the well known generator polynomials are listed in Table 2.7. The CRC-12 is used with 6-bit character codes and has now been replaced by 16-bit polynomials such as CRC-16. The CRC-16 (ANSI) is commonly used in the United States, whereas the CRC-CCITTT is commonly used in Europe. Both CRC-16 and CRC-CCITT V.41 are commonly used in character-oriented protocols, which are discussed in chapter 4. Both codes have 16 check bits, which are appended to the block as two 8-bit characters. As in the previous discussion on CRC, both codes will detect all errors in bursts up to 16 bits in length and over 99 percent of bursts longer than 16 bits. The CRC-32 is used in LAN standards

TABLE 2.7 Some of the Well Known $G(x)$

Standard	Generating Polynomial, $G(x)$
CRC-12	$x^{12} + x^{11} + x^3 + x^2 + x + 1$
CRC-16 (ANSI)	$x^{16} + x^{15} + x^5 + 1$
CRC-16	$x^{16} + x^{15} + x^2 + 1$
CRC-CCITT (V.41)	$x^{16} + x^{12} + x^5 + 1$
CRC-32	$x^{32} + x^{26} + x^{23} + x^{22} + x^{16} + x^{12} + x^{11} + x^{10} + x^8 + x^7 + x^5 + x^4 + x^2 + x + 1$

SUGGESTED READINGS

For detailed discussions on quantization and PCM, the reader is referred to such textbooks as [TAUB 86], [SKLA 88], [SCHW 90], and [PROA 94]. The reader is referred to [KRUT 88] for a detailed description of IEEE 488-1978. [LIND 73] and [STAL 91] address digital signaling formats. [PICK 87] presents an excellent treatment of modulation techniques and modems.

[BART 87], [REY 86], and [BLAC 89] provide good treatment for transmission media, and [TOWN 93] presents results of research on communications links. [BATE 86] is suggested for twisted-pair wires. [GLEN 85] and [NAGE 87] are excellent references for optical fibers, and [JONE 88] provides a mathematical treatment of optical communication systems. [IVAN 89] and [KEIS 89] are excellent books on radio communications.

[MCNA 88], [BLAC 88], [FOLT 80], [BERT 81], and [SPRA 91] provide good treatment of physical layer standards.

PROBLEMS

2.1 Derive Equation 2.1.

2.2 **(a)** The bit stream 001010 is transmitted using binary phase shift keying BPSK. Assume that the bit rate f_b is equal to the carrier frequency f, sketch the transmitted signal $X(t)$.

(b) Repeat (a), but with binary frequency shift keying BFSK used instead. Assume f_1 [carrier frequency corresponding to $b(t) = 1$] $= f_b$ and $f_2 = 2f_b$.

2.3 The bit stream 001010 is to be transmitted using QPSK. Assume that $f = f_b/2$, sketch the transmitted waveform.

2.4 Assume the public switched telephone network has a bandwidth of 3000 Hz per channel and assume a signal-to-noise ratio of 20 dB. What is the maximum theoretical information rate that can be obtained?

2.5 For a fixed signal power S and in the presence of white Gaussian noise with the channel-bandwidth total noise power N, the channel capacity C approaches an upper limit with increasing channel bandwidth B.

(a) Prove that this upper limit equals 1.44 S/n, where n is the noise power spectral density given by N/B.

(b) What is C for a noiseless Gaussian channel? Explain your results.

[Hint: use the formula $\lim_{x \to 0}(1 + x)^{1/x} = e$ (the naperian base).]

2.6 A block of N bits (using even parity check) is transmitted over a line with a bit error probability ϵ (assume bit errors are independent).

(a) Write an expression for the probability of K bits in error P_k in terms of N and ϵ

(b) Write an expression for the probability of undetected errors $P_{\text{undetected}}$

(c) Obtain P_k, $K = 1, 2, 3, 4$, and $P_{\text{undetected}}$ assuming only K errors occured for $\epsilon = 10^{-4}$, $N = 200$.

(d) Using (c), show that, for large N, the probability of undetected errors is approximately the probability of two errors.

2.7 In problem 2.6, and assuming single errors only, obtain the average time between errors, for $\epsilon = 10^{-4}$, $N = 200$, and $K = 1$. Assume a channel speed $C = 1$ Mbps and 100 blocks of bits are being transmitted. What is the average time between undetected errors?

2.8 A user inputs B bits of data that are transmitted using block parity check (VRC/LRC). The user's characters (i.e., horizontal lines) are L bits long. Prove that the rate of the block parity check code R (defined as the number of information bits to be transmitted divided by the total number of bits transmitted) is given by

$$R = \frac{LV}{(L+1)(V+1)}$$

where $V = B/L$. Compare with that of the single parity check of the ASCII code (i.e., $L = 7$)

2.9 The following table represents the 7-bit ASCII representation for letters D, E, and F. Using even-parity, obtain the parity-check bit and the block-check character BCC for the message DEF showing the transmitted bits.

	b_6	b_5	b_4	b_3	b_2	b_1	b_0
D	1	0	0	0	1	0	0
E	1	0	0	0	1	0	1
F	1	0	0	0	1	1	0

2.10 The information message 1011 is to be transmitted using the generator polynomial $g(x) = x^2 + 1$. What is the transmitted bit sequence? Draw a block diagram showing the hardware implementation. Show also the shift register contents at every step at both the sender and the receiver. Assume error-free transmission.

2.11 Show the CRC hardware implementation for the following generator polynomials $G(X)$:

(a) $X^{16} + X^{15} + X^5 + 1$ (CRC-16; ANSI)

(b) $X^{16} + X^{12} + X^5 + 1$ (CRC-CCITT; V.41)

2.12 Almost all subscriber loops in the public telephone network are implemented with a single pair of wires for both directions of transmission. If both ends of a voice connection talk simultaneously, their conversations are superimposed on the single pair and can be heard at the opposite end. Is this acceptable for data transmission over the analog loop? Show one technique to solve this problem.

2.13 Explain the reasons for the four-wire transmission (i.e., two pairs of wires: one pair for each direction) on a long-distance trunk in the telephone network. Draw a block diagram showing how conversion from the two-wire transmission of local loops to the four wire transmission is done.

2.14 Ten voice signals (each has a bandwidth from 0 to 4 kHz) are to be transmitted over a communication channel using time division multiplexing with binary PCM. Assume minimum sampling frequency and 256 quantization levels.

(a) Draw a block diagram for the PCM system and the time diagram showing the frame boundaries. What is the output bit rate?

(b) If a 240-kbits/second data source is to be time multiplexed with the PCM output and then the output is fed into a PSK modulator as shown in Figure 2.26, what is the input bit rate to the PSK? What is the PSK output bandwidth?

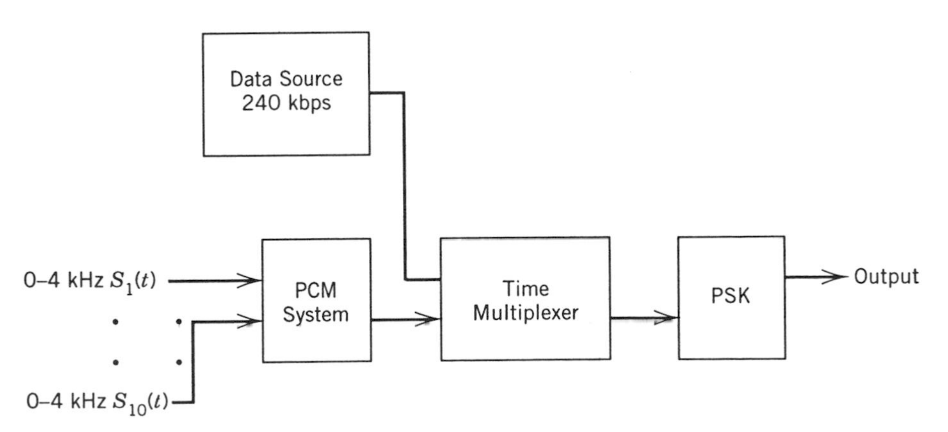

Figure 2.26. Problem 2.14.

2.15 Information bits are grouped into blocks of length N bits. Blocks are transmitted over a communication link with a bit error rate (BER) of ϵ. Errored blocks must be retransmitted.

(a) Obtain the expected number of transmission of a block E, in terms of N and ϵ.

(b) What is E for $\epsilon = 10^{-3}$ and $N = 100$? Interpret your results.

(c) Repeat (b) for $N = 300$ and 1000.

2.16 When information bits of a certain message are grouped into blocks, overhead bits are needed in every block to provide functions such as synchronization, error control, and addressing. Assume each block contains N bits including H overhead bits. The line bit error rate is ϵ and errored blocks must be retransmitted. A message with 1352 bits is to be transmitted.

Determine the optimum block length for this message N_{opt} such that the average number of total transmitted bits (including retransmissions) is minimum? Assume the overhead bits H are fixed and equal 168 bits regardless of the block size $\epsilon = 10^{-4}$? Repeat when $\epsilon = 10^{-3}$.

2.17 A multiplexer multiplexes the traffic from N computers and transmits it over one output line with a bit rate of C bps. Each of the N computers generates messages with an average of λ messages/second. Each message is L bits long. The multiplexer attaches a header of h bits long to each message for identification and control purposes.

(a) Obtain the maximum value of N that the multiplexer can support.

(b) Repeat (a) if the output line has a bit error rate of ϵ.

2.18 Explain why the balanced electrical interface (such as EIA-422-A) can tolerate more noise when compared with the unbalanced electrical interface.

3

QUEUEING SYSTEMS*

3.1 INTRODUCTION

Queueing theory studies are essential to analysis and design enhancement of telecommunication networks. Queueing disciplines arise not only in switched networks, local area networks, wide area data networks, multilink networks, and so on, but also in multiprocessor multimemory computing systems. In addition, queueing theory has been applied to oil pipelines, the shipping industry, different operations research problems, among others. In a wide-area data network, messages are queued at each node enroute to a destination, waiting, for example, for the appropriate link, because of congestion on the outgoing link, or due to node processing. The node processing time, to read the header, check for errors, or read routing information, for example, could take considerable time especially for internetworking of communication equipment from different vendors.

In this chapter, we neglect all the processing times and consider only the waiting time for the message to be served. This time depends on the arrival process in packets per second or messages per second (λ), and the capacity of the serving line or channel (C) (or alternately the service rate (μ),† which has the same units as λ). The arrival and service disciplines, number of servers, and buffer size in the queueing system with random arrivals are typically denoted by a string of letters and numbers. For example, $M/G/2/K$ denotes a queue whose arrivals obey a Poisson statistics (i.e., Markovian) denoted by M, G refers to general service time distribution, 2 is the number of servers, and K is the maximum buffer storage capacity (in packets, messages, calls, etc).

In the following sections, we introduce the basic queue types such as $M/M/1/\infty$, $M/G/1$, and their derivatives. We also discuss some of the

*The reader with little or no knowledge of stochastic processes and probability theory is encouraged to consult some of the references in the Suggested Readings section and Appendices 3A and 3B. The material presented in the following chapters is independent from the material presented here.

†Here we assume the capacity C already absorbed in the definition of μ.

applications of these queues to time division multiplexing (TDM) systems, frequency division multiplexing (FDM) systems, networks of queues, priority systems, and so on. For these systems, the main performance criteria are the average number in the buffer $E(n)$, expected waiting and total delay times $E(W)$ and $E(T)$ respectively, and the blocking probability (buffer overflow probability).

3.2 THE CONTINUOUS-TIME DISCRETE EVENT PROCESS

As mentioned earlier, for $M/M/1$, the first M denotes Poisson arrivals, the second M stands for exponential service time, and 1 stands for a single server. The absence of a fourth integer (∞) indicates an infinite buffer case. This celebrated queue (Fig. 3.1a) is an example of a one-dimensional birth–death* process, the behavior of which is described by a discrete valued process $n(t)$ turning into a discrete valued random variable n as $t \to \infty$ (under equilibrium). Messages arrive at a rate λ messages/second, served at a rate μ messages/second (Fig. 3.1). The service discipline is first-in-first-out (FIFO)† with no priority. The Poisson arrivals are independent with constant (but possibly state dependent) rate λ. Taking a small time interval (dt), the process $n(t)$ is a continuous time Markov chain, whose state probability at ($t = t + dt$) is related (Fig. 3.1b) to that at t by the formula:

$$\begin{aligned} P_n(t+dt) = {} & P_n(t) \cdot [\text{Probability of moving from state } (n) \\ & \qquad \text{at } (t) \text{ to state } (n) \text{ at } (t+dt)] \\ & + P_{n-1}(t) \cdot [\text{Probability of moving from state } (n-1) \\ & \qquad \text{at } (t) \text{ to state } (n) \text{ at } (t+dt)] \\ & + P_{n+1}(t) \cdot [\text{Probability of moving from state } (n+1) \\ & \qquad \text{at } (t) \text{ to state } (n) \text{ at } (t+dt)] \end{aligned} \tag{3.1}$$

In this equation, it is assumed that the probabilities of arrivals and services are independent of buffer state and that $n(t)$ also includes the message being serviced. Also, the probability of more than one arrival and/or more than one service is negligibly small ($O(dt)$). This leads to $P_n(t + dt)$ depending only on $P_n(t)$, $P_{n-1}(t)$, $P_{n+1}(t)$, and not $P_{n+2}(t)$ nor $P_{n-2}(t)$, and so on.

Inserting the different arrivals and service probabilities in Eq. 3.1, one obtains,

$$\begin{aligned} P_n(t+dt) = {} & P_n(t)[1 - \lambda\, dt][1 - \mu\, dt] + P_n(t)\mu dt\, \lambda\, dt \\ & + P_{n-1}(t)[\lambda\, dt][1 - \mu\, dt] \\ & + P_{n+1}[1 - \lambda\, dt][\mu\, dt] \end{aligned}$$

*A birth death process is a Markov process in which only two transitions are possible: the process decreases by one (a death) or increases by one (a birth).

†FCFS (First come first served) also called first in first out (FIFO)

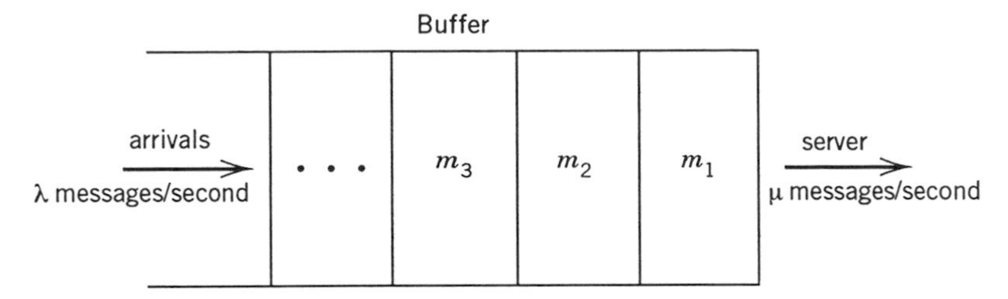

Figure 3.1a. An $M / M / 1$ queue.

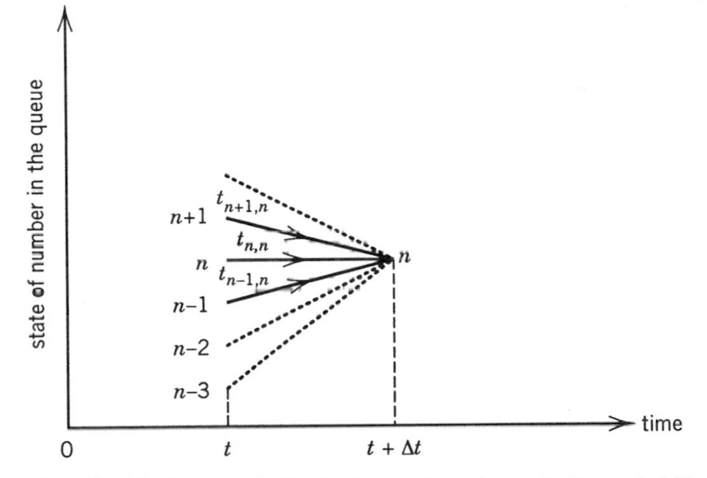

Figure 3.1b. The $M / M / 1$ queue in the buffer vs time. $t_{n+1,n}$ is the probability of buffer changing from state $(n+1)$ to state (n).

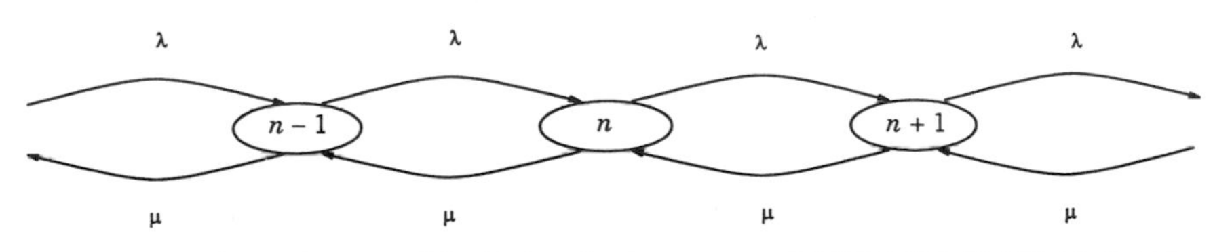

Figure 3.1c. State diagram of the birth – death process underlying the $M / M / 1$ queue.

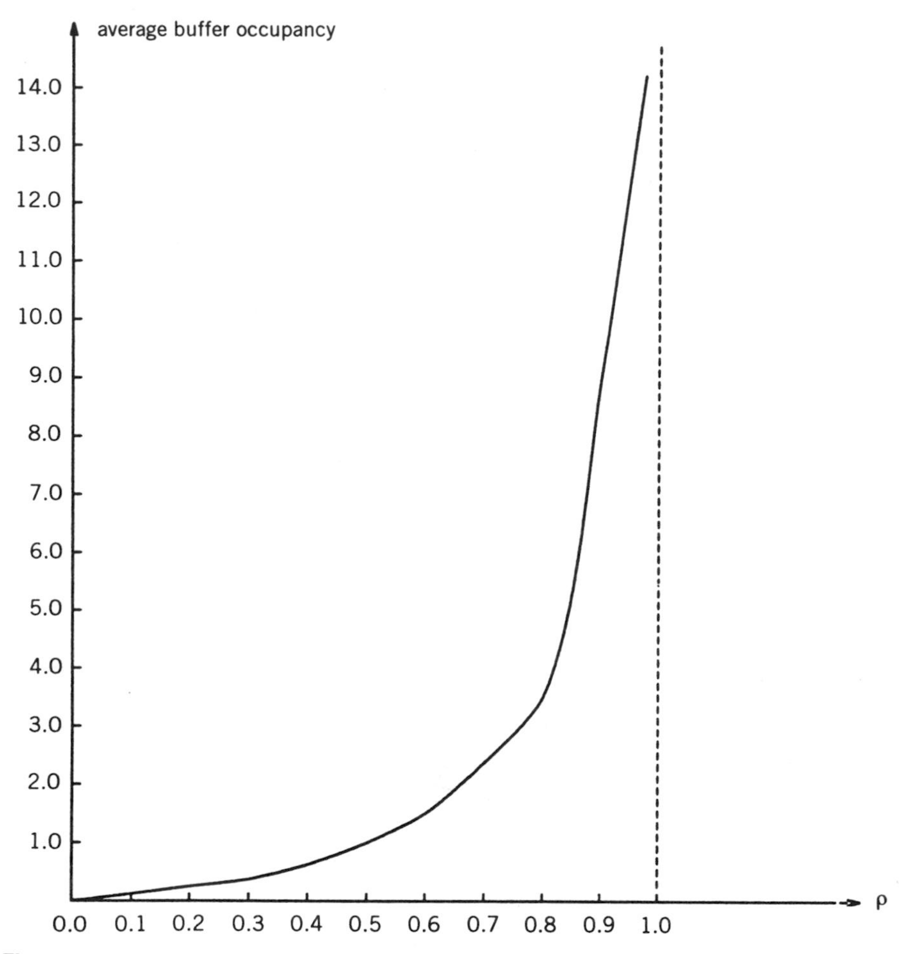

Figure 3.1d. Average number in the buffer vs traffic intensity in an $M/M/1$ queueing system.

Neglecting all $(dt)^2$ terms, taking the limit as $dt \to 0$ and observing that

$$P_n(t + dt) = P_n(t) + \frac{dP_n(t)}{dt}(dt),$$

then

$$\frac{dP_n(t)}{dt} = -(\lambda + \mu)P_n(t) + \lambda P_{n-1}(t) + \mu P_{n+1}(t) \tag{3.2a}$$

Similarly, at the boundary (when $n = 0$), we have

$$\frac{dP_0(t)}{dt} = -\lambda P_0(t) + \mu P_1(t), \qquad n = 0 \tag{3.2b}$$

The transient solution of the stochastic differential equation (3.2) is a function of the initial number in the system (m) and the solution is well-documented in the literature (The interested reader may refer to DAIG86), that is,

$$P_{n^{(t,m)}} = e^{-(\lambda+\mu)t}$$

$$\times\left\{(\lambda/\mu)^{n-m/2} I_{n-m}\left[2\left(\sqrt{\lambda\mu t}\right) + (\lambda/\mu)^{(n-m+1)/2} I_{n+m+1} 2\sqrt{\lambda\mu t}\right]\right.$$

$$\left. + (1-\lambda/\mu)(\lambda/\mu)^n \sum_{j=m+n+2}^{\infty} (\lambda/\mu)^{-j/2} I_j\left(2\sqrt{\lambda\mu t}\right)\right\} \quad (3.3)$$

Where $I(\cdot)$ is the Bessel function of first kind and second order. Aside from the computational aspects of Eq. 3.3, emphasis is usually put on the steady-state solution. This is the solution in Eq. 3.2 as it becomes independent of time (t) and the initial state m. To obtain such a solution, one should substitute $dP_n(t)/dt = 0$ and drop the dependence on t in Eq. 3.2 leading to,

$$\mu P_{n+1} = (\lambda+\mu)P_n - \lambda P_{n-1}, \qquad n = 1,2,\ldots,\infty \qquad (3.4a)$$

and

$$\mu P_1 = \lambda P_0 \qquad (3.4b)$$

One could also write upon observing Fig. 3.1c and applying the principle of flow conservation,

$$(\lambda+\mu)P_n = \mu P_{n+1} + \lambda P_{n-1}$$

This equation (same as Eq. 3.4a) says that the rate at which the queue leaves state n (the left-hand side of the equation) should be equal to the rate at which it enters that state (the right-hand side) if steady-state conditions are to prevail.

Now substituting $n = 1,2,\ldots$ in Eq. 3.4a, we get,

$$\begin{aligned} P_2 &= (\lambda/\mu)P_1 = (\lambda/\mu)^2 P_0 = \rho^2 P_0 \\ P_3 &= (\lambda/\mu)P_2 = (\lambda/\mu)^3 P_0 = \rho^3 P_0 \\ &\vdots \\ P_n &= \rho^n P_0 \end{aligned} \qquad (3.5)$$

where $\rho = \lambda/\mu$ is the traffic intensity in message arriving/message served (i.e., Erlangs), which is sometimes called the utilization factor, the intensity factor, or simply the load.* Now since $\Sigma_{i=0}^{\infty} P_i = 1$, P_0 can be obtained from Eq. 3.5 as follows:

$$P_0 = \frac{1}{\sum_{i=0}^{\infty} \rho^n} = 1 - \rho \tag{3.6a}$$

$$P_0 = (1 - \rho)$$

and substituting into Eq. 3.5 we get

$$P_n = (1 - \rho)\rho^n, \qquad n \geq 0 \tag{3.6b}$$

Evaluating the expected value of the random variable n, we obtain the average number of messages in the queueing system in the steady state, that is

$$\overline{N} = E(n) = \sum_{n=0}^{\infty} nP_n = \frac{\rho}{1 - \rho} = \frac{\lambda}{\mu - \lambda} \tag{3.7}$$

We note that $\lim_{\lambda \to \mu} \overline{N} = \infty$, that is, an infinite number of messages exist in the buffer, which is said to be unstable in this case, which sets the limit ($\rho < 1$) for the system to be stable (Fig. 3.1d).

To find the expected delay time in seconds, we use Little's formula, which gives the relation between the average number of messages in a queueing system, the average arrival rate, and the mean delay time $E(T)$ as follows

$$E(T) = \frac{\overline{N}}{\lambda} = \frac{1}{(\mu - \lambda)}$$

This equation indicates that under steady-state equilibrium conditions, the average number in the system is nothing but the accumulation of arriving packets during the total delay time of a typical message. Note also that the total expected delay time is also given by the average waiting time plus the average service time (i.e., message transmission time).

$$E(T) = E(w) + 1/\mu$$

where $1/\mu$ is the average message transmission time of one message and $E(w)$ is the expected waiting time.

*The load ρ is a dimensionless quantity, however, like decibels that also measure dimensionless quantities, it is usually measured in units called "Erlang."

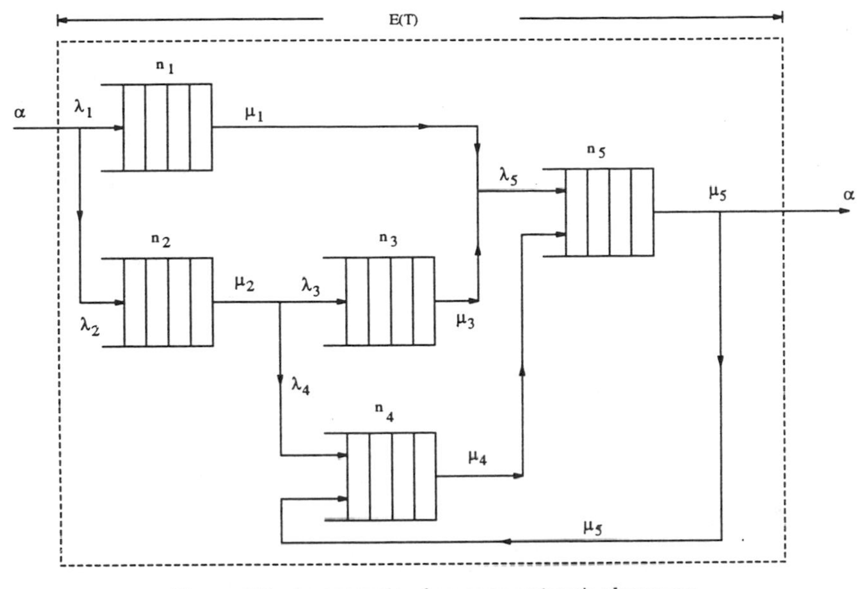

Figure 3.2. An example of an open network of queues.

Actually Little's formula is so general that it can be applied to a network of queues to find the network-wide delay over the M links. For the jth link, Little's result gives (Fig. 3.2)

$$E(T_j) = E(n_j)/\lambda_j$$

where $E(n_j)$ is the average number of messages in service or queued at link j and λ_j is the average traffic flow over link j. If the combination of m buffers in the network is assumed to be one, we get the total average delay of the whole network T, that is,

$$T = \frac{E(n)}{\alpha} = \frac{1}{\alpha}\sum_{j=1}^{M} E(n_j) = \frac{1}{\alpha}\sum_{j=1}^{M} \frac{\lambda_j}{\mu_j - \lambda_j} \tag{3.8}$$

where α is the aggregate arrival rate into the network, which is the same as the external traffic to the network (Fig. 3.2).

Another quantity of interest is the probability that n exceeds a certain threshold ξ, that is,

$$P(n > \xi) = \sum_{n=\xi+1}^{\infty} P_n = \rho^{\xi+1} \tag{3.9}$$

This becomes of particular importance in the case of a finite $M/M/1$ queue as per the following example.

Example 3.1. Consider an $M/M/1/K$ queue, that is, an $M/M/1$ queue operating with a finite buffer of size K, find the probability of buffer overflow (blocking), that is, $P_B = P(n = K)$, and the buffer size given that the required blocking probability is equal to χ.

Solution. Proceeding as in the case of the infinite buffer $M/M/1$, queue up to Eq. 3.6 and summing over the range $(0 \leq n \leq K)$, we get,

$$\sum_{n=0}^{K} P_0 \rho^n = \frac{P_0(1 - \rho^{K+1})}{1 - \rho} = 1$$

yielding

$$P_0 = \frac{(1 - \rho)}{(1 - \rho^{K+1})} \quad \text{and} \quad P_n = \frac{(1 - \rho)}{(1 - \rho^{K+1})} \rho^n$$

The blocking probability P_B becomes

$$P_B = P(n = K) = P_K = \frac{(1 - \rho)\rho^K}{(1 - \rho^{K+1})}$$

To obtain the buffer size K, given the blocking probability χ, we solve for K,

$$K = \frac{\log[\chi/(1 - \rho + \chi\rho)]}{\log \rho}$$

where $[\chi]$ is the smallest integer greater than or equal to x. For $\chi = 10^{-2}$ and $\rho = 0.5$, we get $K = 6$, whereas, for $\chi = 10^{-4}$ and $\rho = 0.9$, we get $K = 66$, and, for $\chi = 10^{-6}$, $\rho = 0.9$, we get $K = 110$.

The extension of $M/M/1$ queueing to the case where the arrival and service rates (λ_n and μ_n, respectively) are dependent on the state of the system under equilibrium is now given. In this case, the equilibrium difference equation becomes

$$\begin{aligned} (\lambda_n + \mu_n)P_n &= \mu_{n+1}P_{n+1} + \lambda_{n-1}P_{n-1}, \qquad n \geq 1 \\ \lambda_0 P_0 &= \mu_1 P_1 \end{aligned} \tag{3.10}$$

(see Fig. 3.3). Starting with P_0 and solving iteratively, we obtain,

$$P_n = \frac{\lambda_0 \lambda_1 \cdots \lambda_{n-1}}{\mu_1 \mu_2 \cdots \mu_n} P_0 \tag{3.11}$$

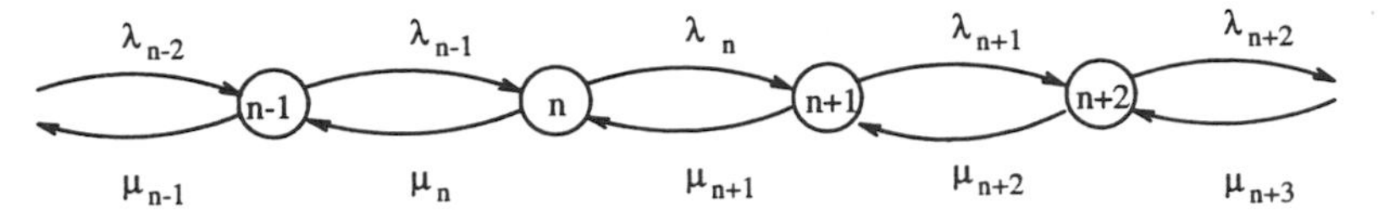

Figure 3.3. General birth – death process (continuous time, discrete state). Transitions only to neighboring states. λ_n, μ_n are rates and no self-loops exit.

For the case of equal arrivals, that is, $\lambda_n = \lambda$, $\mu_n = n\mu$,* we sum over all probabilities, $\Sigma_{n=0}^{\infty} P_n = 1$, to obtain P_0, that is,

$$P_0 = e^{-\rho} \quad \text{and} \quad P_n = \rho^n e^{-\rho}/n! \tag{3.12}$$

The service grows with the state so that the probability of higher states becomes less than that of the $M/M/1$ (for the same ρ), as can be easily seen by comparing Eq. 3.12 and Eq. 3.6.

The moment-generating function (MGF) (see also Appendix 3A) provides a systematic way for solving the recurrence equations (3.4) and the like. This can be easily achieved by multiplying both sides of Eq. 3.4 by z^n and summing over all n $(1 \leq n \leq \infty)$,

$$\sum_{n=1}^{\infty} \mu z^n P_{n+1} = \sum_{n=1}^{\infty} (\lambda + \mu) z^n P_n - \sum_{n=1}^{\infty} \lambda z^n P_{n-1}$$

From the definition of $G(z)$, in Appendix 3A, we get,

$$\frac{\mu[G_n(z) - P_0 - zP_1]}{z} = (\lambda + \mu)[G_n(z) - P_0] - \lambda z G_n(z)$$

Noting that $\lambda P_0 = \mu P_1$ and $P_0 = 1 - \lambda/\mu$, we obtain

$$G_n(z) = \frac{\mu(1-z)(1-\lambda/\mu)}{\lambda z^2 - (\lambda + \mu)z + \mu} = \frac{\mu(1-\lambda/\mu)}{\mu - \lambda z} = \frac{1-\rho}{1-\rho z} \tag{3.13}$$

As in Appendix 3A, for $G_n(z)$ to be stable, that is, for the coefficients of z^n in $G_n(z)$ to be less than 1, the roots $G_n(z)$ should be inside the unity circle, thus implying $\rho < 1$. The average number of customers (or messages) in the buffer is given by

$$E(n) = G_n'(1) = \frac{\rho}{(1-\rho)}$$

For finite buffers and state-dependent arrivals and services (Eq. 3.10), the evaluation of the blocking probability is quite different from the simple $M/M/1$ case, but it is easily derived as in the next example.

*This might be the case at some voice multiplexing devices, where the service increases with the load or the number of calls.

Example 3.2. Find the blocking probability, $P_b(K)$, of the finite buffer of size K of the generalized birth–death process of Eq. 3.10, prove that this obeys the $P_b(K)$ recurrence equation:

$$P_b(K) = \frac{\rho_K P_b(K-1)}{1 + \rho_K P_b(K-1)}$$

where $P_b(0) = 1$. Specialize to the case of the $M/M/1/K$ queue and compare $P_b(K)$ of this case to the probability that $P(n > K)$ in the $M/M/1/\infty$ queue.

Solution. In the steady state, all arrivals while the system is at state K (with steady-state probability P_K) are blocked, hence the average number of blocked messages in time T seconds is again given by $\lambda_K P_K T$. On the other hand, the total average arrivals during the same time is given by $\sum_{l=0}^{K} P_l \lambda_l T$, giving rise to a blocking probability equal to

$$P_b(K) = \frac{\lambda_K P_K}{\sum_{l=0}^{K} \lambda_l P_l}$$

Noting that $P_1 = \lambda_0 P_0 / \mu_1$, starting with $P_b(0) = 1$ (i.e., no buffer, all are blocked), we get

$$\begin{aligned} P_b(1) &= \frac{P_b(0)\lambda_1 P_1}{\lambda_0 P_0 + \lambda_1 P_1 P_b(0)} \\ &= \frac{P_b(0)\lambda_1 P_1}{\mu_1 P_1 + \lambda_1 P_1 P_b(0)} = \frac{\lambda_1/\mu_1 P_b(0)}{1 + \lambda_1/\mu_1 P_b(0)} \end{aligned}$$

Also

$$P_b(2) = \frac{\lambda_2 P_2}{\lambda_0 P_0 + \lambda_1 P_1 + \lambda_2 P_2}$$

Substituting $P_2 = (\lambda_1/\mu_2)P_1$ and $P_1 = (\lambda_0/\mu_1)P_0$, by induction,

$$P_b(K) = \frac{\lambda_K/\mu_K P_b(K-1)}{1 + \lambda_K/\mu_K P_b(K-1)} = \frac{\rho_K P_b(K-1)}{1 + \rho_K P_b(K-1)}$$

For a finite buffer, $M/M/1/K$ queue, $\rho_K = \rho$ for all K, hence

$$P_b(K) = \frac{\rho P_b(K-1)}{1 + \rho P_b(K-1)}$$

In an ordinary $M/M/1/\infty$ queue, the probability $P(n \geq K)$ is equal to ρ^K (see Eq. 3.9), which is typically higher than the finite $P_b(K)$ case for the same ρ, K. The latter equals $(1-\rho)\rho^K/(1-\rho^{K+1})$, whose proof is clear from Example 3.1.

3.3 THE GENERAL BIRTH–DEATH AND POISSON PROCESSES

These are defined as a special case of Markov chains (see Appendix 3B), where the state transitions from state n take place only to neighboring states, that is, $n+1$, $n-1$. Transitions from n to n (self-loops) exist only in discrete Markov chains. λ_n and μ_n represent the birth (i.e., arrival) and death (i.e., departure) rates at state n, respectively, and they depend on the number of customers, n, in the system. At steady state, and equating the flow rate out of state n to the flow rate into this state, the dynamic equations for the general birth–death process in the continuous case are obtained. Also one might substitute state-dependent arrivals and departures in Eq. 3.2, thus obtaining,

$$\frac{dP_n(t)}{dt} = -(\lambda_n + \mu_n)P_n(t) + \lambda_{n-1}P_{n-1}(t) + \mu_{n+1}P_{n+1}(t), \qquad n > 0$$

$$\frac{dP_0(t)}{dt} = -\lambda_0 P_0(t) + \mu_1 P_1(t), \qquad n = 0$$

$$(3.14a)$$

As was previously outlined, solution of this stochastic differential equation is tedious except in simple cases such as the pure birth process (also called Poisson process) where $\lambda_n = \lambda$ and $\mu_n = 0$ for all n. In this case, we obtain

$$\frac{dP_n(t)}{dt} = -\lambda P_n(t) + \lambda P_{n-1}(t), \qquad n > 0$$

$$\frac{dP_0(t)}{dt} = -\lambda P_0(t), \qquad n = 0 \qquad (3.14b)$$

Setting $n = 1$ in Eq. 3.14b and assuming $P_n(0) = 1$ for $n = 0$ and $P_n(0) = 0$ for $n \neq 0$, it follows from Eq. (3.14b) that $P_0(t) = e^{-\lambda t}$ and

$$\frac{dP_1(t)}{dt} = -\lambda P_1(t) + \lambda e^{-\lambda t}$$

$$P_1(0) = 0$$

which yields the solution

$$P_1(t) = \lambda t e^{-\lambda t}$$

Substituting recursively, $n = 2, 3, \ldots,$ we obtain for any n

$$P_n(t) = \frac{(\lambda t)^n e^{-\lambda t}}{n!}, \qquad n \geq 0, \quad t \geq 0 \tag{3.14c}$$

This Poisson distribution has mean, variance, and interarrival times shown in the following example.

Example 3.3. Find the mean, variance of the pure birth process (i.e., Poisson process). What is the distribution function of the interarrival times of this process?

Solution. The MGF of the Poisson distribution is given by,

$$G_n(z) = E(z^n) = \sum_{j=0}^{\infty} z^j P_j(t) = \sum_{j=0}^{\infty} \frac{(\lambda tz)^j}{j!} e^{-\lambda t} = e^{-\lambda t + \lambda t z}$$

The mean and variance of the distribution are (see Appendix 3A),

$$E(n) = G_n'(1) = (\lambda t) e^{\lambda t(z-1)}|_{z=1} = \lambda t$$

$$\sigma_n^2 = G_n''(1) - [G_n'(1)]^2 + G_n'(1) = \lambda^2 t^2 e^{\lambda t(z-1)}|_{z=1} - (\lambda t)^2 + \lambda t = \lambda t$$

Now defining $X(t)$ as the probability that the time between adjacent arrivals t' is less than or equal to t

$$X(t) = P(t' \leq t) = 1 - P(t' > t)$$

$P(t' > t)$ is the probability of no arrival in the time interval $\{0, t\}$, so,

$$X(t) = 1 - P_0(t) = 1 - e^{-\lambda t}$$

$X(t)$ is the cumulative distribution function (CDF) of the interarrival times. This celebrated exponential distribution for the inter arrival times has the property that the past history plays no role in prediciting the future, (so called memoryless property or the Markovian property). The probability density function (pdf) is given by

$$f_x(t) = \lambda e^{-\lambda t}, \qquad t \geq 0$$

The expected value of this interarrival time is $(1/\lambda)$.

The application of Little's theorem to many generalized birth–death queues may not be that straightforward as in the following example.

Example 3.4. In a discouraged arrivals birth–death process defined by

$$\lambda_n = \frac{\beta}{n+1}, \qquad n = 0, 1, 2, \ldots, \qquad \mu_n = \mu, \qquad n = 1, 2, 3, \ldots$$

find the distribution of the number in the system, the average number in the system, and the expected waiting time, $\beta \leq 1$

Solution. In the steady state, Eq. 3.14a becomes Eq. 3.10 and by substituting $n = 1, 2, \ldots$ we obtain,

$$P_n = P_0 \prod_{l=0}^{n-1} \frac{\beta/(l+1)}{\mu} = P_0 \left(\frac{\beta}{\mu}\right)^n \frac{1}{n!}$$

where P_0 (obtained from, the condition $\sum_{n=0}^{\infty} P_n = 1$)

$$P_0 = \frac{1}{\left[1 + \sum_{m=1}^{\infty} (\beta/\mu)^m / m!\right]} = e^{-\beta/\mu}$$

The distribution of the number in the buffer now becomes

$$P_n = \frac{e^{-\beta/\mu}}{n!} (\beta/\mu)^n$$

Again we see another Poisson distribution as that of Eq. 3.12. The traffic intensity ρ is calculated as $\rho = 1 - P_0 = 1 - e^{-\beta/\mu}$, ($\rho = 1 - P_0$) is true for any general queue (general arrival, general service with one server, that is, a $G/G/1$ queue). The average number in the system is,

$$E(n) = \sum_{n=0}^{\infty} nP_n = \beta/\mu$$

The expected waiting time is

$$E(w) = E(T) - 1/\mu \tag{3.15}$$

whereas the average time spent in the system is given by

$$E(T) = E(n)/\bar{\lambda} = E(w) + 1/\mu$$

$$= \frac{E(n)}{\bar{\lambda}} = \frac{\beta}{\bar{\lambda}\mu} = \frac{\beta\mu}{\mu^2\bar{\lambda}} = \frac{\beta}{\mu^2\rho} = \frac{\beta}{\mu^2(1 - e^{-\beta/\mu})} \tag{3.16}$$

Where $\bar{\lambda} = \sum_{n=0}^{\infty} \lambda_n P_n$.

Substituting Eq. 3.16 into Eq. 3.15, the average waiting time becomes

$$E(w) = \frac{E(n)}{\bar{\lambda}} - 1/\mu = \frac{\beta}{\mu^2(1 - e^{-\beta/\mu})} - \frac{1}{\mu}$$

Other special cases of the general birth–death processes such as $M/M/m/m$ queues can be found in the problems.

3.4 THE *M/G/1* QUEUEING SYSTEM

This denotes a system having Poisson arrivals, general service time distribution, and single server. The memoryless property of the exponetial interarrival and service times of the $M/M/1$ queue, implies the future of an exponentially distributed process is independent of its past. This enables the analyst to model the state of such a system by only one variable (buffer content) to represent the system.

In the $M/G/1$ system, however, the service distribution is general, so the system's past history has to include the service time already received by the message in service as well as the above-mentioned buffer content. If the embedded points between adjacent arrivals is extremely small, there will be only one arrival or one departure in the case of the $M/M/1$ queue; however many messages may arrive during the general message length of the $M/G/1$. This implies that a two-dimensional Markov chain is an appropriate representation of such a queue. It is possible, however, to model this as a one-dimensional embedded Markov chain, where the state variable is the number of messages in the system immediately following the service of the ith message, and the transition takes place only at the embedded points. (Fig. 3.4)

The service discipline is first come–first served. n_i, n_{i-1}, and w_i denote, respectively, the number of messages in the buffer following the departure of the ith and $(i - 1)$th messages and number of messages that have arrived during the service time of the ith message (with statistics to be found shortly).

It is easy to see that the number in the buffer is reduced by 1 from the service of the ith message and is incremented by w_i (the number of messages arriving) if the buffer is nonempty, $n_{i-1} > 0$. If the buffer is empty (i.e.,

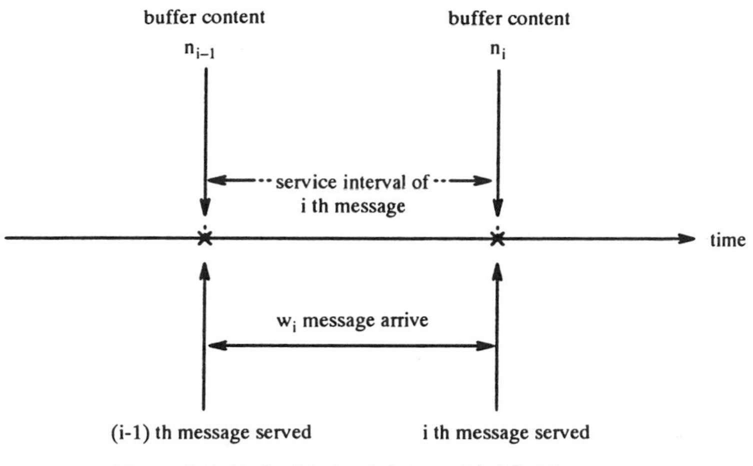

Figure 3.4. Embedded points in an $M/G/1$ queue.

$n_{i-1} = 0$), the number in the buffer at the end of service of the ith message (n_i) is the same as the number that arrived, that is, w_i. Hence, one can write

$$n_i = \begin{cases} n_{i-1} - 1 + w_i, & n_{i-1} > 0 \\ w_i, & n_{i-1} = 0 \end{cases}$$

These two equations can be combined to give

$$n_i = (n_{i-1} - 1)^* + w_i, \qquad n_{i-1} \geq 0$$

where $(\chi)^* = \chi$ if $\chi \geq 0$ and $(\chi)^* = 0$ otherwise. Employing the definition of the MGF, and realizing that,

$$E\{Z^{\chi^* + y}\} = E(Z^{\chi^*})E(Z^y)$$

Hence,

$$G_n(z) = G_{\chi^*}(z)G_w(z) \tag{3.17}$$

Note that in the steady state (equilibrium) the MGF $G_{\chi^*}(z)$ is readily obtained, by

$$G_{n_i}(z) = G_{n_{i-1}}(z) = G_n(z),$$

and is independent of time. Also $\chi^* = 0$ in the two cases $(n_i - 1) = 0$ or when $n_{i-1} = 0$, leading to,

$$G_{\chi^*}(z) = G_{(n_{i-1}-1)}(z) = P_0 + P_1 + P_2 z + P_3 z^2 + \cdots = P_0 + \frac{G_n(z) - P_0}{z} \tag{3.18}$$

The MGF $G_w(z)$ can be obtained by recalling that $w_i = w$ under equilibrium as the above is the number of messages received during the service time τ. This is a continuous random variable with PDF $f_\tau(\chi)$ and the arrival process is Poisson. It follows,

$$P(w = j) = \int_0^\infty P(w = j|\chi) f_\tau(\chi)\, d\chi$$

where the conditional Poisson arrivals are given by

$$P(w = j|\tau) = \frac{(\lambda\tau)^j e^{-\lambda\tau}}{j!}$$

The moment generating function $G_w(z)$ is then given by

$$G_w(z) = \sum_{j=0}^{\infty} P(w = j) z^j = \sum_{j=0}^{\infty} \left\{ \int_0^\infty \frac{(\lambda\chi)^j e^{-\lambda\chi}}{j!} f_\tau(\chi)\, d\chi \right\} z^j$$

Interchanging the order of summation and integration, we get

$$G_w(z) = \sum_{j=0}^{\infty} P(w = j) z^j = \int_0^\infty \left\{ \sum_{j=0}^{\infty} \frac{(\lambda\chi)^j e^{-\lambda\chi}}{j!} f_\tau(\chi) z^j \right\} d\chi$$

$$= \int_0^\infty e^{-\lambda(1-z)\chi} f_\tau(\chi)\, d\chi = F_\tau[\lambda(1 - z)] \tag{3.19}$$

where
$$F_\tau(s) = \int_0^\infty e^{-s\chi} f_\tau(\chi)\, d\chi \tag{3.20}$$

is the Laplace transform of $f_\tau(\chi)$. Substituting Eqs. 3.19 and 3.18 into 3.17 and reducing, we obtain

$$G_n(z) = \frac{P_0(z - 1) F_\tau[\lambda(1 - z)]}{z - F_\tau[\lambda(1 - z)]} \tag{3.21}$$

which is the celebrated Pollaczek-Khinchin (PK) equation.

To find P_0 we reexpress Eq. 3.21 in terms of $G_w(z)$, that is,

$$G_n(z) = \frac{P_0(z-1)G_w(z)}{z - G_w(z)} \tag{3.22}$$

Expanding $G_w(z)$ around the point $z = 1$ leads to

$$G_w(z) = G_w(1) + G_w'(1)(z-1) + \frac{G_w''(1)}{2!}(z-1)^2 + \cdots$$

Now since

$$G_w'(1) = E(w), \quad G_w''(1) = E(w^2) - E(w), \quad \text{and} \quad G_w(1) = 1$$

$$G_w(z) = G_w(1) + (z-1)E(w) + \left[E(w^2) - E(w)\right]\frac{(z-1)^2}{2!} + \cdots$$

and

$$z - G_w(z) = (z-1)\left\{[1 - E(w)] - \left[E(w^2) - E(w)\right]\frac{(z-1)}{2!} \cdots\right\} \tag{3.23}$$

Substituting Eq. 3.23 into Eq. 3.22 canceling the $z - 1$ term and setting $z = 1$, yields $P_0 = 1 - E(w) = 1 - \rho$. Now to prove that $E(w) = \rho$, we see from Eq. 3.19 that

$$E(w) = G_w'(1) = -\lambda \left.\frac{dF_\tau(s)}{ds}\right|_{s=0} \tag{3.24}$$

From Eq. 3.20, this is equal to

$$E(w) = -\lambda \int_0^\infty \chi f_\tau(\chi)\, dx = \lambda E(\tau) = \rho \tag{3.25}$$

Finally we obtain

$$G_n(z) = \frac{(1-\rho)(z-1)F_\tau[\lambda(1-z)]}{z - F_\tau[\lambda(1-z)]} \tag{3.26}$$

Now, specializing to the case of an $M/M/1$ queue, we see that the exponential service time (PDF) is given by

$$f_\tau(\chi) = \mu e^{-\mu\chi} \tag{3.27}$$

so

$$F_\tau(s) = \frac{\mu}{\mu + s} \tag{3.28}$$

and

$$F_\tau[\lambda(1 - z)] = \frac{\mu}{\lambda(1 - z) + \mu} = \frac{1}{\rho(1 - z) + 1}$$

Substituting Eq. 3.27 into Eq. 3.26 and reducing, yields

$$G_n(z) = \frac{1 - \rho}{1 - \rho z}$$

as found in Eq. 3.13. Expanding, one obtains

$$G_n(z) = (1 - \rho)\sum_{j=0}^{\infty}(\rho z)^j = \sum_{j=0}^{\infty}(1 - \rho)\rho^j z^j$$

The steady-state probability distribution of the number in the buffer readily follows:

$$P_n = (1 - \rho)\rho^n$$

In the $M/D/1$ case, the PDF of the constant service time (τ_0) is given by

$$f_\tau(\chi) = \delta(\chi - \tau_0)$$

The distribution of the number of messages arriving during the service of one message is given by

$$P_w(j) = P(w = j) = \frac{e^{-\lambda\tau_0}(\lambda\tau_0)^j}{j!}$$

The MGF of w is given by

$$G_w(z) = \sum_{j=0}^{\infty} P_w(j) z^j = \sum_{j=0}^{\infty} \frac{e^{-\lambda\tau_0}}{j!}(\lambda\tau_0 z)^j = e^{\lambda\tau_0(z-1)} \tag{3.29}$$

evaluating $E(w) = G_w'(1) = \lambda\tau_0 < 1$ with the later condition imposed to guarantee convergence of the series (in the $M/M/1$ case the equivalent condition is $\rho = \lambda/\mu < 1$).

Trying to invert Eq. 3.26 does not yield a closed-form solution for the $M/G/1$ buffer case. One can still find useful entities such as mean and variance of n from Eq. 3.22 for the $M/G/1$ case, which can be then easily extended to the $M/D/1$ and $M/M/1$ cases. Straight differentiation of Eq. 3.22 to evaluate $E(n)$ yields (0/0). Using Eq. 3.23, however, in Eq. 3.22 gives

$$G_n(z) = \frac{[1 - E(w)]G_w(z)}{\{1 - E(w) - (z-1)/2![E(w^2) - E(w) - \cdots]\}}$$

Now differentiating this and evaluating at $z = 1$ yields

$$E(n) = G_n'(1) = \frac{1}{1 - E(w)}\left[\frac{E(w)}{2} - E^2(w) + \frac{E(w^2)}{2}\right]$$

$$= \frac{1}{2(1-\rho)}\left[\rho(1-\rho) + \sigma_w^2\right]$$

Now substituting σ^2 as in Appendix 3.A

$$E(n) = \frac{1}{2(1-\rho)}\left[\rho(1-\rho) + G_w''(1) - G_w'^2(1) + G_w'(1)\right]$$

From Eq. 3.19, we obtain

$$G_w''(1) = \lambda^2 \frac{d^2F_\tau(s)}{ds^2}\bigg|_{s=0} = \lambda^2 E(\tau^2) = \lambda^2\left[\sigma_\tau^2 + (1/\mu)^2\right]$$

which leads to

$$E(n) = \frac{1}{(1-\rho)}\left[\rho - \frac{\rho^2}{2}(1 - \mu^2\sigma_\tau^2)\right] \tag{3.30}$$

In the $M/M/1$ case,

$$\sigma_\tau^2 = 1/\mu^2$$

yields

$$E(n) = \left[\frac{\rho}{1-\rho}\right]$$

whereas in the $M/D/1$ case, $\sigma_\tau^2 = 0$ (fixed message size τ_0),

$$E(n) = \frac{\rho(2-\rho)}{2(1-\rho)}, \qquad \rho = \frac{\lambda}{\mu} = \lambda\tau_0$$

The $M/G/1$ average waiting time is one service time less than the average queueing time, that is,

$$E(w) = E(T) - \frac{1}{\mu} = \frac{E(n)}{\lambda} - \frac{1}{\mu}$$

$$= \frac{1}{\mu}\left[\frac{2-\rho(1-\mu^2\sigma_\tau^2)}{2(1-\rho)} - 1\right] = \frac{\lambda E(\tau^2)}{2(1-\rho)} \tag{3.31}$$

In the $M/M/1$ case, $\sigma_\tau^2 = 1/\mu^2$; so

$$E(w) = \frac{\lambda}{\mu^2(1-\rho)} = \frac{\rho}{\mu(1-\rho)} \tag{3.32}$$

whereas, for $M/D/1$ queues, $\sigma_\tau^2 = 0$, $E(\tau^2) = 1/\mu^2 = \tau_0^2$

$$E(w) = \frac{\rho}{2\mu(1-\rho)}$$

We notice that $E_{M/D/1}(w) < E_{M/M/1}(w)$. Also the total queueing delay $E_{M/D/1}(T) < E_{M/M/1}(T)$ is due to the variability of the exponential service time of the $M/M/1$ queue.

3.5 APPLICATIONS OF THE *M/G/1* QUEUE

Service on Channels with Errors and Automatic Repeat Request

In the STOP AND WAIT* protocols typically used at the data link and transport layers of the OSI hierarchy to mask the effects of the channel and all physical layer errors, the head-of-line packet is transmitted on error-prone channels (Fig. 3.5) and waits for acknowledgment (ACK) to come in the reverse direction. If T_0 seconds (called time out) elapse without receiving the ACK, the message is transmitted again and so on, until ACK is received. This implies that the message service time is higher than that of an errorless system, such as the one we treated earlier ($M/G/1$ queue). To find the effective transmission time, we assume j transmissions until successful transmission, each costing τ seconds to transmit and propagate and an extra T_0 seconds (timeout period). Also, the acknowledgment channel is assumed to

*The operation of the stop and wait protocol is discussed in detail in Chapter 4.

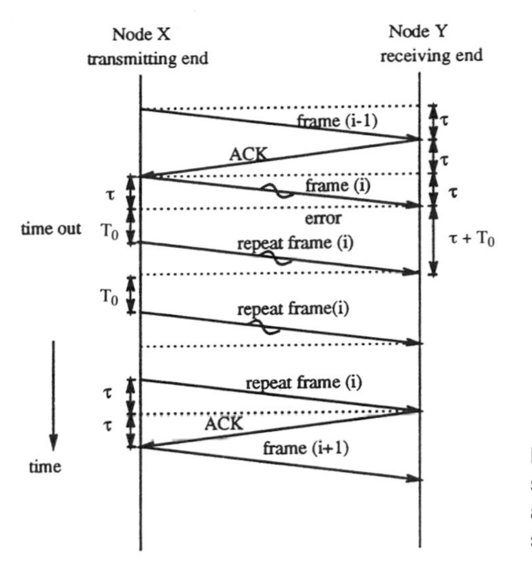

Figure 3.5. Transmission discipline in the stop and wait protocol case using positive acknowledgement and time out. T_0 is assumed $\gg 2\tau$; ~ denotes frame in error.

be perfect, such that acknowledgment packets are error free. The number of transmissions $(j-1)$ is a random variable with a geometric distribution (each user tries $j-1$ times but fails with probability $(1-\alpha)^{j-1}$ and finally succeeds in the final trial with probability α,* that is,

$$P_j = (1-\alpha)^{j-1}\alpha, \quad j = 1, 2, \ldots, \infty \tag{3.33}$$

where α is the probability of message transmission errors on the channel.

Thus, the total effective transmission time t equals $j(T_0 + \tau)$, obtaining the Laplace transform of its conditional probability density function and averaging over the random variable j, we obtain

$$T'(s) = \sum_{j=1}^{\infty} (1-\alpha)^{j-1}\alpha B(s)C(s) \tag{3.34}$$

where

$$B(s) = E\{e^{-sj\tau}\} = \int_0^{\infty} e^{-sy}\delta(y - j\tau)\, dy = e^{-j\tau s} \tag{3.35}$$

and

$$C(s) = e^{-jT_0 s} \tag{3.36}$$

*The last transmission trial takes only 2τ seconds but is assumed here to take the same time as the other trials, i.e., $T_0 + \tau$ for convenience of the analysis.

Substituting Eqs. 3.35 and 3.36 into Eq. 3.34

$$T'(s) = \sum_{j=1}^{\infty} (1-\alpha)^{j-1} \alpha e^{j\tau s} e^{-jT_0 s} = \frac{\alpha e^{-s(\tau+T_0)}}{\left[1-(1-\alpha)e^{-s(\tau+T_0)}\right]}$$

The average and mean-square value of these transmission times are,

$$\overline{T}' = \frac{-dT'(s)}{ds}\bigg|_{s=0} = \frac{\tau+T_0}{\alpha} \tag{3.37}$$

$$\overline{T'^2} = \frac{d^2T'(s)}{ds^2}\bigg|_{s=0} = \left[\frac{2-\alpha}{\alpha^2}\right](\tau+T_0)^2 \tag{3.38}$$

The variance of T' is given by

$$\sigma_{T'}^2 = \overline{T'^2} - \overline{T}'^2 = \frac{(1-\alpha)}{\alpha^2}(\tau+T_0)^2$$

which is nothing but the variance of the aforementioned geometric distribution multiplied by a factor $(\tau + T_0)^2$.

This implies that one could have just approximated the continuous server of the ordinary $M/G/1$ queue by another intermittently available one whose service probability is geometrically distributed with parameter α and average service time $[(\tau + T_0)/\alpha]$. With the knowledge of $\sigma_{T'}^2$, one can easily find the average buffer content from Eq. 3.30.

3.6 PRIORITY QUEUEING

The presence of messages with different priority levels is not uncommon in many applications. The ACK packets in the automatic repeat request (ARQ) technique just presented are usually given higher priority than normal data packets. Expedited data units in layer 4 transport protocols are treated similarly. So is emergency and management data and voice traffic in integrated voice-data systems.

Conceptually, we have n types of traffics feeding n buffers but only one server. Traffics arrive with rates $\lambda_1, \lambda_2, \ldots, \lambda_n$ arranged in descending priority. The service discipline of the various traffics depends on the priority structure, which can take one of the several types:

1. Nonpreemptive Priority. In this type, arriving messages of users with priorities $1, 2, \ldots, (k-1)$ higher than a certain level k will have to wait till user k finishes his current transmission. Upon completion of the kth user message, service moves to the 1st, 2nd, $\ldots, (k-1)$th buffers which will be emptied in succession.

2. Preemptive Resume Priority. Here, transmission of messages with lower priority could by interrupted by arriving messages of a higher priority. However, upon completion of service to all higher priority messages, the server returns to continue service of all lower priority messages (where it left). In this case, there is no loss of work, but implementation is harder than the nonresume priority case (to follow).
3. Preemptive Nonresume Priority. In this case, upon completion of all higher priority messages that interrupted lower priority services, service returns to the very start of the lower priority messages. Here we have a loss of work, but implementation is easier (no counters are necessary at the interrupted low priority buffers to record the messages portion which was finished upon interruption). In the following analysis we find the mean waiting time for the various priority levels.

We assume in all cases independent Poisson arrivals to all buffers with rates $\lambda_1, \lambda_2, \dots, \lambda_k$ (for k priority classes arranged in descending order), and mean service times of $1/\mu_1, 1/\mu_2, \dots, 1/\mu_k$ time units. In the following, we start with the case of an $M/G/1$ *queueing discipline at all buffers and a nonpreemptive priority structure.*

The interaction between the service of various priority users is clear in this case. For example, higher priority users have to wait for a lower priority buffer to completely transmit its message. Also, lower priority users will have to wait for the higher priority ones to empty their buffers. In the following, we find the total average waiting time of a user of priority level i that has the following three components:

1. Average time to finish the current service (residual time S_r).
2. Average time to serve all queued messages with higher priorities $(i, i-1, i-2, \dots, 1)$ when the subject message arrives, S_q.
3. Average time to serve all messages with higher priorities $(i-1, i-2, \dots, 1)$ that arrive during the period W_i seconds and are transmitted first before our subject message, S_a.

Summing, we obtain

$$W_i = S_r + S_q + S_a$$

$$= S_r + \sum_{l=1}^{i} E(S_l) + \sum_{l=1}^{i-1} E(S_i)$$

where $E(S_l)$ is the average time of the lth priority level waiting message. This is given by

$$E(S_l) = \frac{E(n_l)}{\mu_l}$$

where $E(n_l)$ is the difference between the average number of messages of priority level i (waiting and in service) and the average number in service ($\rho_l = \lambda_l/\mu_l$). From Little's formula, we obtain the expected total number of messages in the lth buffer, that is,

$$E(m_l) = \lambda_l[E(w_l) + 1/\mu_l] = E(n_l) + \rho_l$$

and

$$E(S_l) = \frac{\lambda_l}{\mu_l} \cdot E(W_l) = \rho_l E(W_l)$$

The S_a component is similarly defined as

$$S_a = \sum_{l=1}^{i-1} \rho_l E(W_i)$$

representing the services of all higher priority messages that have arrived during the waiting period of customer i (i.e., W_i seconds),*

$$W_i = S_r + \sum_{l=1}^{i} \rho_l E(W_l) + \sum_{l=1}^{i-1} \rho_l E(W_i)$$

solving iteratively for W_i, $i = 1, 2, \ldots, k$

$$E(W_i) = S_r[(1 - \xi_{i-1})(1 - \xi_i)] \tag{3.39}$$

where $\xi_i = \sum_{j=1}^{i} \rho_j$. To find S_r, we compare Eq. 3.39 when $i = 1$ to Eq. 3.32; this reveals that

$$S_r = \lambda \overline{E}(\tau^2)/2$$

For k priority classes, we average over such classes and note that $\lambda = \sum_{l=1}^{k} \lambda_l$ and

$$S_r = \tfrac{1}{2} \sum_{l=1}^{k} \lambda_l E(\tau_l^2)$$

Substituting into Eq. 3.39 and evaluating at $i = 1, 2, \ldots, k$ yields the various

*Queued messages of priority i are served before our subject message, so the first sum goes to (i), whereas arriving messages with priority i during the waiting time of the subject message are served after and so the second sum goes only to ($i - 1$).

total delays of the k priority classes, that is,

$$E(T_i) = E(W_i) + \frac{1}{\mu_i} = \frac{\sum_{l=1}^{k} \lambda_l E(\tau_l^2)}{2\left(1 - \sum_{j=1}^{i-1} \rho_j\right)\left(1 - \sum_{j=1}^{i} \rho_j\right)} + \frac{1}{\mu_i} \tag{3.40}$$

Tuning our attention now to preemptive nonresume priority, we see less interaction between the priority classes, because the user of a certain priority does not have to wait for the completion of service of the lower class user. In other words, the user does not see this lower class, and the delay analysis for a certain class i considers this as the lowest in the system and is independent of classes $i + 1, i + 2, \ldots, k$. Also the unfinished work (the sum of all service times of all users of classes $1, 2, \ldots, i$) is the average waiting time in an overall $M/G/1$ queue with no priorities whose traffic utilization ρ is

$$\rho = \rho_1 + \rho_2 + \cdots + \rho_i$$

This waiting component (similar to Eq. 3.31), is given by

$$\frac{\sum_{l=1}^{i} \lambda_l E(\tau_l)^2}{2\left(1 - \sum_{l=1}^{i} \rho_l\right)}$$

where i was used rather than k (as explained). Accounting for the service time of the class i message and waiting time for the higher priorities arrivals to be served (similar to the nonpreemptive service component S_a), we obtain

$$T_i = \left[\sum_{l=1}^{i} \lambda_l E(\tau_l^2) \Big/ 2\left(1 - \sum_{l=1}^{i} \rho_l\right)\right] + \left(\sum_{l=1}^{i-1} \rho_l\right) T_i + 1/\mu_i$$

For $i = 1$, we get

$$T_1 = \frac{\lambda_1 E(\tau_1^2)/2}{(1 - \rho_1)} + 1/\mu_1 \tag{3.41}$$

For $i > 1$, one can find by successive substitution that

$$T_i = \frac{\left[\sum_{l=1}^{i} \lambda_l E(\tau_l^2)\right] \Big/ 2 + \left(1 - \sum_{l=1}^{i} \rho_l\right) \Big/ \mu_i}{\left(1 - \sum_{l=1}^{i-1} \rho_l\right)\left(1 - \sum_{l=1}^{i} \rho_l\right)}$$

The difference in performance from using preemptive or nonpreemptive policies are reflected in the following example.

Example 3.5. Assume only two priority classes ($k = 2$), $M/M/1$ queue with service rates μ_1, μ_2, respectively, and arrival rates λ_1, λ_2. Compare the service delays of the two classes to the average nonpriority delay in the two cases of preemptive nonresume as well as nonpreemptive priority.

Solution. For an $M/M/1$ queue with nonpreemptive priority, we substitute $E(\tau_1^2) = 2/\mu_1^2$ and $E(\tau_2^2) = 2/\mu_2^2$* into Eq. 3.40 to obtain

$$E(T_1) = \frac{\rho_1/\mu_1 + \rho_2/\mu_2}{1 - \rho_1} + \frac{1}{\mu_1}$$

$$E(T_2) = \frac{\rho_1/\mu_1 + \rho_2/\mu_2}{(1 - \rho_1)(1 - \rho_1 - \rho_2)} + \frac{1}{\mu_2}$$

The waiting delay component of the average waiting time of an ordinary $M/M/1$ queue with no priority (averaging Eq. 3.32 over the two service times) is

$$E(W) = \frac{\rho_1/\mu_1 + \rho_2/\mu_2}{1 - \rho}$$

where $\rho = \rho_1 + \rho_2$. Comparing the waiting time components of the three cases above, it is clear that priority reduces the waiting delay of the priority user (1) at the expense of the nonpriority user (2). In the preemptive case, we obtain the total delay plus service times from Eq 3.41.

$$E(T_1) = \frac{\rho_1/\mu_1}{1 - \rho_1} + \frac{1}{\mu_1}$$

$$E(T_2) = \frac{\rho_1/\mu_1 + \rho_2/\mu_2}{(1 - \rho_1)(1 - \rho_1 - \rho_2)} + \frac{1}{\mu_2(1 - \rho_1)}$$

*See Appendix 3A for details.

whereas the average nonpriority case gives

$$E(T) = \frac{\rho_1/\mu_1 + \rho_2/\mu_2}{1-\rho} + \frac{\lambda_1}{(\lambda_1+\lambda_2)}\frac{1}{\mu_1} + \frac{\lambda_2}{\lambda_1+\lambda_2}\frac{1}{\mu_2}$$

Comparing $E(T_2)$ in the two cases above (nonpreemptive and preemptive), we notice that the waiting delay component is the same in both preemptive and nonpreemptive priority cases, however the service time has gone up in the preemptive cases for the lower class.

3.7 NETWORKS OF QUEUES

Store-and-forward networks, interconnected local area networks, large computer systems, and so on, can be modeled as a network of queues. This model is useful not only for evaluating the delay and the throughput of different queues but also for capacity allocation to different links in circuit switched networks, for studying the effects of flow and congestion control. As an example, Figures 3.6 and 3.7 show typical examples of open and closed networks of queues, respectively. In the open networks of queues cases, we notice the existence of a finite relationship between the input and output arrivals λ_s and λ_d and the various internal traffics. From flow conservation, we obtain $\lambda_s = \lambda_d$ (the source traffic goes to the destination). Also, since the incoming traffic to a certain node j emanates with probability α_{lj} from all other nodes as well as from the source node,

$$\lambda_j = \alpha_{sj}\lambda_s + \sum_{l=1}^{J} \alpha_{lj}\lambda_l = \lambda'_j + \sum_{l=1}^{J} \alpha_{lj}\lambda_l, \qquad 1 \le j \le J \tag{3.42}$$

where λ_j is the total input traffic to the jth buffer, λ'_j is the percentage of source (external) traffic directed to node j, and J is the number of buffers. α_{lj} is the probability that a packet has been served at node l is routed to node j (in Fig. 3.6 $q_{lj} = \alpha_{lj}\lambda_l$).*

$$\alpha_{jd} + \sum_{l=1}^{J} \alpha_{jl} = 1, \qquad 1 \le j \le J \tag{3.43}$$

which stems from the fact that the sum of all traffics from node j to its destination and to all other nodes adds to 1. As an example of Eq. 3.42, we see from Figure 3.6 that $\lambda_1 = \alpha_{s1}\lambda_s + \alpha_{41}\lambda_4$. The assumption of Poisson

*For example $q_{57} = \alpha_{57}\lambda_5$.

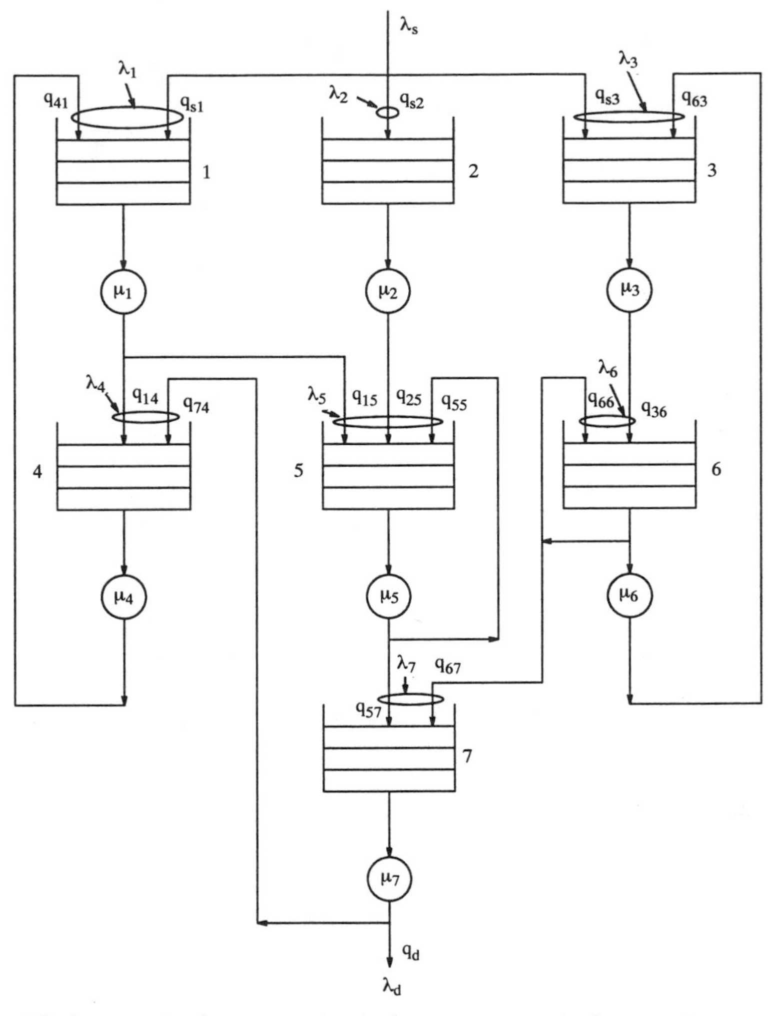

Figure 3.6. An example of an open network of queues. $q_{ij} = \alpha_{ij}\lambda_i$, for example $q_{15} = \alpha_{15}\lambda_1$.

arrivals,* exponentially distributed with independent service times at all buffers and random routing are typical; moreover, this kind of queue network is typically called a Jackson Network.

A product form of the solution has been proposed. After this product form is substituted into the balance equation, it is possible to prove its validity. In the product form, the joint distribution of the buffer contents of

*Arrivals to each nodes are the sum of the input as well as internal traffic, which is not Poisson distributed.

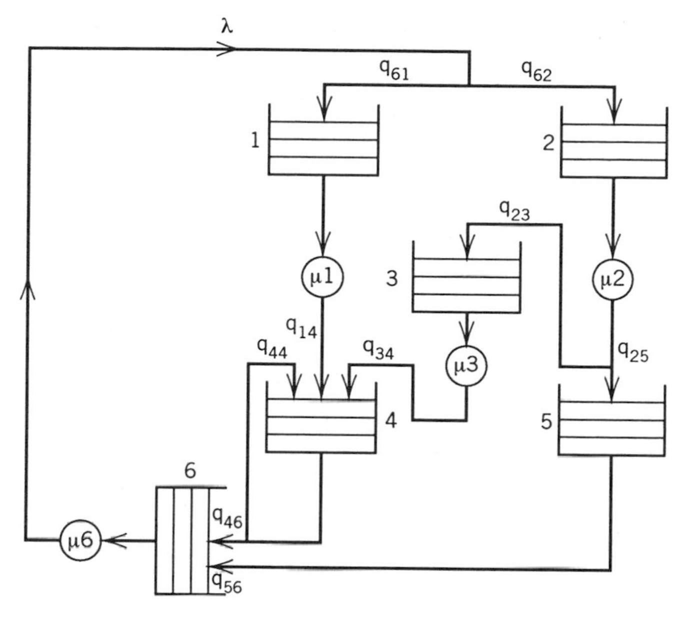

Figure 3.7. An example of a closed network of queue.

all J queues is given by

$$P(\bar{n}) = P(n_1, n_2, \ldots, n_J) = \prod_{l=1}^{J} P_l(n_l)$$

where $P_l(n_l)$ is the probability that the lth buffer has n_l messages. The total arrival rate at the overall state $\bar{n}$ is equal to the total departure rate from that state, that is,

$$\begin{aligned}
&\sum_{l=1}^{J} \alpha_{sl}\lambda_s P(n_1, n_2, \ldots, n_{l-1}, n_l - 1, n_{l+1}, \ldots, n_J) \\
&\quad + \sum_{j=1}^{J} \alpha_{jd}\mu_j P(n_1, n_2, \ldots, n_{j-1}, n_j + 1, \ldots, n_J) \\
&\quad + \sum_{j=1}^{J} \sum_{l=1}^{J} \alpha_{jl}\mu_j P(n_1, n_2, \ldots, n_l - 1, \ldots, n_j + 1, \ldots, n_J) \\
&\quad = \left[\sum_{l=1}^{J} \lambda'_l + \sum_{j=1}^{J} \mu_j \right] P(n_1, n_2, \ldots, n_J)
\end{aligned} \tag{3.44}$$

The right-hand side, represents rate of flow out of state $\bar{n}$. The first term on the left-hand side covers arrivals (with rates $\alpha_{sl}\lambda_s$) at state $\bar{n}$ from the source to each node. The second term represents departures (with rates $\alpha_{jd}\mu_j$) at node n_j to destination. The third term represents internal departures from node j and arrivals to node l (with rates $\alpha_{jl}\mu_j$).

Invoking at this point the local balance equations at each node (assuming they hold),

$$\lambda_j P(n_1, n_2, \ldots, n_{j-1}, n_j - 1, n_{j+1}, \ldots, n_J)$$
$$= \mu_j P(n_1, n_2, \ldots, n_{j-1}, \ldots, n_j, n_{j+1}, \ldots, n_J) \tag{3.45}$$

Similarly,

$$\lambda_j P(n_1, n_2, \ldots, n_{j-1}, n_j, n_{j+1}, \ldots, n_J)$$
$$= \mu_j P(n_1, n_2, \ldots, n_{j-1}, \ldots, n_j + 1, \ldots, n_J) \tag{3.46}$$

Substituting from Eq. 3.42 into Eqs. 3.45 and 3.46 and summing both sides of the resulting equations, one obtains

$$\left[\sum_{j=1}^{J} \mu_j + \sum_{j=1}^{J} \lambda'_j\right] P(n_1, n_2, \ldots, n_{j-1}, n_j, n_{j+1}, \ldots, n_J)$$
$$= \sum_{j=1}^{J} \lambda'_j P(n_1, n_2, \ldots, n_{j-1}, n_j - 1, n_{j+1}, \ldots, n_J)$$
$$+ \sum_{j=1}^{J} \mu_j P(n_1, n_2, \ldots, n_{j-1}, n_j + 1, n_{j+1}, \ldots, n_J)$$
$$+ \sum_{j=1}^{J} \sum_{l=1}^{J} \alpha_{lj}\lambda_l P(n_1, n_2, \ldots, n_{j-1}, n_j - 1, n_{j+1}, \ldots, n_J)$$
$$- \sum_{j=1}^{J} \sum_{l=1}^{J} \alpha_{lj}\lambda_l P(n_1, n_2, \ldots, n_{j-1}, n_j, n_{j+1}, \ldots, n_J)$$

Substituting from Eq. 3.43, we obtain

$$\left[\sum_{j=1}^{J}\mu_j + \sum_{j=1}^{J}\lambda'_j\right]P(n_1, n_2, \dots, n_{j-1}, n_j, n_{j+1}, \dots, n_j)$$

$$= \sum_{j=1}^{J}\lambda'_j P(n_1, n_2, \dots, n_{j-1}, n_j - 1, n_{j+1}, \dots, n_J)$$

$$+ \sum_{j=1}^{J}\mu_j\alpha_{jd}P(n_1, n_2, \dots, n_{j-1}, n_j + 1, n_{j+1}, \dots, n_J) \quad (3.47)$$

$$+ \sum_{l=1}^{J}\sum_{j=1}^{J}\alpha_{jl}\{\mu_j P(n_1, n_2, \dots, n_{j-1}, n_j + 1, n_{j+1}, \dots, n_J)$$

$$+\lambda_j P(n_1, n_2, \dots, n_l - 1, n_{l+1}, \dots, n_J)$$

$$-\lambda_j P(n_1, n_2, \dots, n_{j-1}, n_j, n_{j+1}, \dots, n_J)\}$$

Comparing Eqs. 3.47 and 3.44, we see that all corresponding terms are equal, except for the double summation. However, recalling that Eq. 3.45 implies that,

$$\lambda_j P(n_1, \dots, n_{l-1}, n_l - 1, n_{l+1}, \dots, n_J)$$

$$= \mu_j P(n_1, n_2, \dots, n_{l-1}, n_l - 1, \dots, n_j + 1, \dots, n_J)$$

Substituting this and Eq. 3.46 into Eq. 3.47 yields exactly Eq. 3.44. This proves that Eq. 3.45 is a sufficient condition for satisfying the flow equation (3.44). Effectively, it was proven that the local balance equations (3.45) are an equivalent substitute for the global balance equation (3.44). To find our multiplicative solution for the buffer state, we substitute iteratively n_j times into Eq. 3.45, thus obtaining

$$P(n_1, n_2, \dots, n_J) = \left\{\frac{\lambda_j}{\mu_j}\right\}^{n_j} P(n_1, n_2, \dots, n_{j-1}, 0, n_{j+1}, \dots, n_J) \quad (3.48)$$

Repeating the same procedure for each of the J queues,

$$P(n_1, n_2, \dots, n_J) = \prod_{j=1}^{J}(\rho_j)^{n_j}P(0) \quad (3.49)$$

where $P(0)$ is the probability of having all the queues empty. This is obtained by summing over all possible values for buffer contents, that is,

$$\sum_{n_1=0}^{\infty} \sum_{n_2=0}^{\infty} \cdots \sum_{n_j=0}^{\infty} \prod_{j=1}^{J} \rho_j^{n_j} P(0) = 1$$

This yields

$$P(0) = \left\{ \sum_{n_1=0}^{\infty} \sum_{n_2=0}^{\infty} \cdots \sum_{n_j=0}^{\infty} \prod_{j=1}^{J} (\rho_j)^{n_j} \right\}^{-1}$$

For a solution to exist, $P(0)$ should be finite, so that the sums and multiplications in the denominator can be exchanged, leading to

$$P(0) = \frac{1}{\prod_{j=1}^{J} \left\{ \sum_{n_1=0}^{\infty} \sum_{n_2=0}^{\infty} \cdots \sum_{n_j=0}^{\infty} (\rho_j)^{n_j} \right\}} = \frac{1}{\prod_{j=1}^{J} 1/(1-\rho_j)} = \prod_{j=1}^{J} (1-\rho_j)$$

The complete overall solution (the joint probability distribution of the buffer occupancies at equilibrium) is

$$P(n_1, n_2, \ldots, n_J) = \prod_{j=1}^{J} (1-\rho_j)\rho_j^{n_j} \tag{3.50}$$

There is nothing special here, the right hand side is only the multiplications of the solutions of J $(M/M/1)$ independent queues! One must recall, however that the λ_j are uniquely related and defined by the system of equations (3.42 and 3.43). The product in Eq. 3.50 facilitates the analysis and design of interconnected networks. For example, the treatment of routing, congestion, and flow control problems in these networks is made much simpler by the application of Eq. 3.50, as can be seen in the Suggested Readings section.

3.8 SYNCHRONOUS-TIME DIVISION MULTIPLEXING

In synchronous time division multiplexing (STDM) a population of U users each having an infinite $M/G/1$ buffer with Poisson arrival rate λ_u messages/second and general message length distribution with mean and mean square values equal to $\bar{l}$ packets and $\overline{l^2}$, respectively, share equally the capacity of a transmission line. Transmission time can be looked upon as divided into frames of equal size T_F, where each user is allocated one slot of width $T_s = (T_F/U)$. During each slot (Fig. 3.8), the server removes exactly

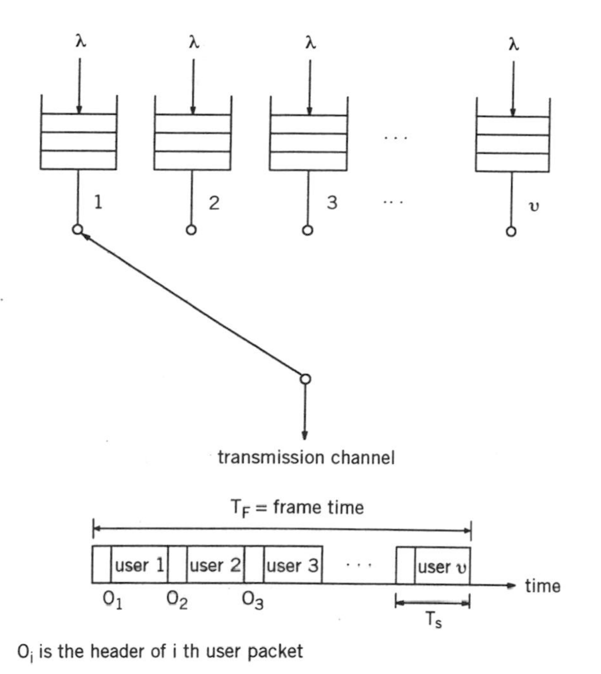

Figure 3.8. Synchronous service to v users in synchronous-time division multiplexing (STDM).

one packet from one of the buffers. To each packet, guard bits, synchronization bits, and so on are usually added; this overhead is less than that of asynchronous time division multiplexing (ATDM) (introduced in Section 3.10) in which the source or destination addresses must also be included in each packet.

The average number of packets per message $\bar{l}$ may exceed 1. Also messages arriving after the start of the frame have to be buffered too. This gives rise to a waiting time delay component equal to (Eq. 3.31)

$$D_1 = \frac{\lambda_u T_F^2 \overline{l^2}}{2(1 - \rho_u)}$$

where ρ_u is the user traffic utilization, which is defined as

$$\rho_u = \lambda_u T_F \bar{l}$$

The transmission time component of the user message on the frame (once it becomes the head of line on its queue) is given by

$$D_2 = (\bar{l} - 1/2)T_F + T_s \text{ seconds}$$

It is easy to see that a message of one packet ($\bar{l} = 1$) takes $D_2 = (T_F/2 + T_s)$ seconds to transmit. The first term, $T_F/2$, assumes a user whose assigned slot is located (on average) midway on the frame, and the second term, T_s, is the time necessary to transmit one packet on the multiplexer TDM line. If $\bar{l} = 2$, $D_2 = 3/2T_F + T_s$ with the first packet taking T_F seconds to transmit and the second takes $(T_F/2 + T_s)$ seconds as before, and so on. Adding D_1 and D_2, one obtains the total message delay D;

$$D = (D_1 + D_2) = \frac{\lambda_u T_F^2 \overline{l^2}}{2(1 - \rho_u)} + (\bar{l} - 1/2)T_F + T_F/U \qquad (3.51)$$

This is the simplest case; a more rigorous queueing analysis can be found in the suggested reading section.

3.9 FREQUENCY DIVISION MULTIPLEXING

In frequency division multiplexing (FDM), each user is allocated a frequency band equal to W_s/U, where W_s is the total bandwidth available to all users. All users transmit simultaneously but at a split bit rate approximately equal to (W_s/U). (The rate is approximate because of the effects, for example, of overhead.) However, the user packet transmission rate remains the same as in STDM, that is, $\mu = 1/T_F$, because they transmit for the whole frame in FDM, but at a reduced rate (W_s/U).* Frequency guard bands are usually incorporated but are ignored in the analysis. Only the $M/G/1$ queue plus transmission components exist in FDM, that is,

$$D = \frac{\lambda_u \overline{l^2} T_F^2}{2(1 - \rho_u)} + \bar{l}T_F \qquad (3.52)$$

where $T_F = UT_s$, the reference TDM frame. With all parameters remaining the same as in the TDM case, the first component is the $M/G/1$ queueing component and the second is the transmission time of $\bar{l}$ packets per message.

Comparing Eqs. 3.51 and 3.52 reveals that STDM systems have a delay advantage of $(T_F/2) - (T_F/U)$ over FDM systems. Synchronization, packet header size, and channel error effects were ignored in both derivations and a more detailed comparison usually takes the channel environment into consideration (mobile network, satellite networks, indoor communications, etc.).

*Recall STDM, each user transmits only in a fraction $1/U$ of the frame (slot) but at full channel rate (W_s); so, effectively in both systems, $\mu = 1/T_F$, and each user transmits only one packet per frame in both systems.

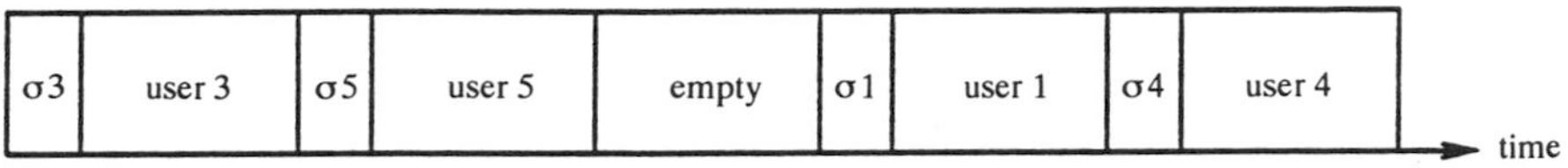

σ3	user 3	σ5	user 5	empty	σ1	user 1	σ4	user 4

time

Figure 3.9. ATDM frame service to U users. The packet header σ_i is larger than that of STDM (i.e., $\sigma_i > O_i$)

3.10 ASYNCHRONOUS-TIME DIVISION MULTIPLEXING

At low-traffic conditions, many of the users in STDM are idle and so their assigned slots are unutilized while the active users encounter unnecessarily larger delays, especially with larger message sizes. This is the motivation behind using ATDM (Fig. 3.9) where the U users still share the total line capacity, but there is no slot allocation. Head-of-line packets of the various users buffers are served first come first serve. The frame definition does not apply here, and all users are thought of as being virtually combined into a single $M/G/1/\infty$ buffer with aggregate total arrival from all users equal to λ_t messages/second.

These aggregate arrivals are served at the slot rate ($1/T_s$ rather than the $1/T_F$ in STDM) and the resulting average number in the overall $M/G/1$ buffer (from Eqs. 3.31 and 3.32) is given by

$$\bar{n} = \rho + \frac{\lambda_t^2 \overline{l^2} T_s^2}{2(1-\rho)}$$

where $\rho = \lambda_t \bar{l} T_s$ is the overall utilization factor. From Little's formula, we obtain the ATDM delay,

$$D = \frac{\bar{n}}{\lambda_t} = \bar{l} T_s + \frac{\lambda_t \overline{l^2} T_s^2}{2(1-\rho)} \tag{3.53}$$

To compare the STDM and ATDM systems on fair basis, one should substitute a value $\lambda_t = U\lambda_u$ for λ_t of ATDM, and keep the same $T_s, \rho, \bar{l}, \overline{l^2}$, [Note that $\rho = \rho_{\text{ATDM}} = \lambda_t T_s \bar{l} = U\lambda_u (T_F/U)\bar{l} = \rho_u$.] However even in this case, the comparison ignores the fact that ATDM consumes more overhead than TDM, leading to an artificially superior ATDM performance over a wide range of traffic. If this overhead is taken into consideration (say it is 10% in ATDM and 0% in TDM) the ATDM traffic will effectively rise by 10% and $\rho_{\text{ATDM}} = 1.1\rho_{\text{TDM}}$, as in Problem 3.19.

SUGGESTED READINGS

The reader is encouraged to look further at the classic works in [KLEI75C] which detail the theoretical basis and models for the various queues that arise in computer networks. [SCHW77] and [SCHW88] have two chapters dealing with queueing problems and their application to congestion control and network of queues in Telecommunication Networks. [HAYE84] deals with the general treatment of queueing problems and their application to local area networks. [HAMM86], and [STUC85] mainly deal with the application of queueing theory to telecommunication problems. [COOP81], [DAIG92], [BERT92], [COHE69], [JACK57], [RUBI79], [DAIG86], [KOBA78], and [JAIS88] provide other treatments and applications of queueing systems

PROBLEMS

3.1 Eight data terminals statistically share the capacity of an outgoing link, traffic is combined from the eight terminals and then served first-come first-served. An $(M/M/1)$ queue with infinite buffer is assumed. Find the expected number of packets in the buffer and the expected waiting time in the following cases:

a. A terminal generates a packet every 10 seconds, the link speed is 1024 bits/second, and packet length = 512 bits.

b. Repeat (a) using finite buffer $k = 10$ and compare your results to the infinite buffer case.

3.2 For the $M/M/1$ queue with buffering capacity K

a. Verify that the blocking probability $= 1/(K + 1)$ if $\rho = 1$.

b. The queue above multiplexes the traffic of ten users each generating 256 bits/second of traffic. The link service rate is 16 packets/second, each message consists of four packets on average, each packet is 512 bits length. Find the blocking probability and the expected waiting time if $K = 4$

3.3 Show that the time between adjacent arrivals in an $M/M/1$ queueing system is given by $(1/\lambda)$. Check using Laplace transform techniques. Prove that the interdeparture time is $(1/M)$ by conditioning on each of the two cases of empty and nonempty queue.

3.4 For a continuous time general birth–death process with finite buffer size k, the solution of the buffer occupancy (before reaching the steady state) is given by $\overline{P}_{(t)} = \overline{P}_{(0)}e^{\Delta t}$, where $\overline{P}_{(t)}$ is the transitional probabilities of buffer occupancy $= [P_0(t), P_1(t), \ldots, P_k(t)]^T$.

$\bar{P}_{(0)}$ is the initial state vector. Find all elements of the matrix Δt in terms of λ_i, μ_i, and k, that is, for arrival and service rates from state i and k. Find the average number of messages in the buffer and the stability criteria (equilibrium condition).

3.5 Repeat problem 3.4 in the special case where $k = 1$, $\lambda_0 = \lambda$, $\mu_1 = \mu$.

3.6 In a finite population with M customers, m servers, and K finite storage queue, the arrival and service rates are given by

$$\lambda_n = \begin{cases} \lambda(M-n), & 0 \le n \le K \\ 0, & \text{otherwise} \end{cases}$$

$$\mu_n = \begin{cases} n\mu, & 0 \le n \le m \\ m\mu, & n \ge m \end{cases}$$

Assume Poisson arrivals, exponential service times, and $M \ge K = m$. Sketch the state transition diagram. Find the equilibrium distribution of the number of customers in the queue, the average delay, and the blocking probability.

3.7 The $M/M/m/m$ queue arises in circuit switching applications where Poisson call arrivals are served by a maximum of m exponentially distributed servers. The $(m+1)$ call is blocked. The arrival and service rates are given by

$$\lambda_n = \begin{cases} \lambda, & n < m \\ 0, & n > m \end{cases}$$

$\mu_n = n\mu$, $n = 1, 2, \ldots, m$. Find the steady-state distribution of the number of active calls on the system P_n, the blocking probability B_l, and the expected call waiting time $E(W)$.

3.8 In a circuit switched public network, calculate the number of circuits such that the call-blocking probability is ≤ 0.2, if the Poisson traffic has an arrival rate of 120 calls/hour and exponentially distributed call duration of 2 minutes on average. Assume that blocked calls are cleared (i.e., lost).

3.9 The output of a voice source is digitized. Voice packets are formed and transmitted over a link with capacity 150 Mbps; each packet is fixed and is of length T_p (Fig. 3.10). Let α be the probability that in the next

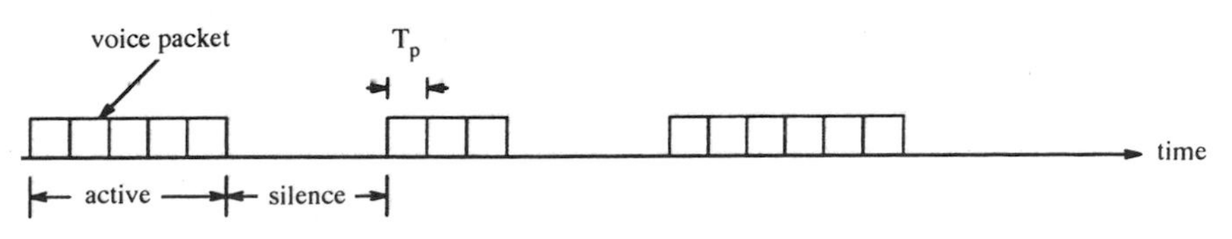

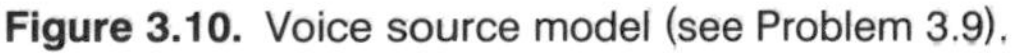
Figure 3.10. Voice source model (see Problem 3.9).

packet transmission time T_p the source will switch from the active period to the silence period. Let β be the probability that in the next packet transmission time T_p the source switches from the silence period to the active period.

a. Draw a discrete time two-state Markov chain representation of the above model. Find the steady-state probabilities in each state.

b. If the mean active period = 350 mseconds, and the mean silence period = 650 mseconds, find α and β. Assume packet length = 400 bits.

3.10 Figure 3.11a depicts a four-node packet switched network using full duplex lines. The capacity of the lines are in kilobits per second. The matrix of Figure 3.11b represents the traffic matrix $[\gamma_{ij}]$, where γ_{ij} is the number of packets sent from node i to node j (assume a Poisson arrival process), and the route to be used for $i - j$ traffic. Given these routing and traffic matrices, determine the average packet delay for every link T_i, the link flows λ_i, the average number of hops n, and the average end-to-end packet delay. Assume that the packet length is exponentially distributed with an average of 1000 bits per packet.

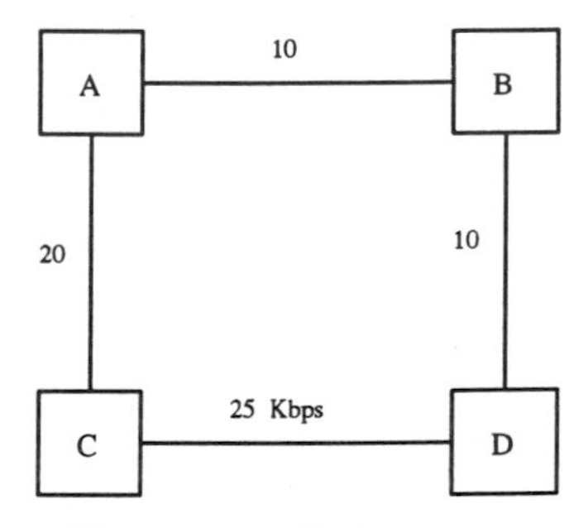

Figure 3.11a. Problem 3.10.

i \ j	A	B	C	D
A		5 AB	3 A C	8 ACD
B	5 B A		4 BDC	1 B D
C	3 C A	4 CDB		3 C D
D	8 DCA	1 D B	3 D C	

Figure 3.11b. Problem 3.10.

3.11 In Figure 3.12, a concentrator in each of the three colleges of the City University of New York is to be connected to the main computerin downtown Manhattan. Terminals (with buffering capability) are in turn connected to the concentrator. Each terminal transmits at a rate of 100 bps (line speed). Messages are, on the average, 1000 bits long. A typical terminal transmits a message on the average of once a minute. The number of terminals in each college is as follows:

CCNY	20
Brooklyn	15
Queen's	10

Determine the average message delay in this network (one-way delay to the main computer, only; and assume $M/M/1$ queueing).

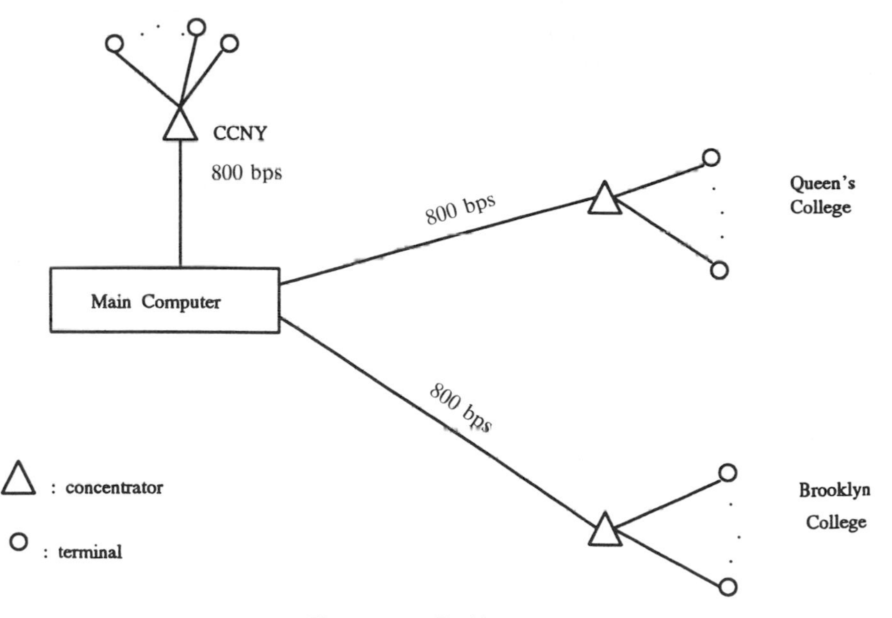

Figure 3.12. Problem 3.11.

3.12 A generalized *M/M/m* queue has infinite storage capacity and the following parameters,

$$\lambda_n = \lambda, \qquad n = 0, 1, 2, \ldots$$

$$\mu_n = \min\begin{cases} n\mu, & 0 \le n \le m \\ m\mu & n \ge m \end{cases}$$

Sketch the state diagram; find the distribution of the number in the buffer P_n and the probability of an arriving customer being forced to join the queue.

3.13 An $M/G/1$ queue has as its input traffic rate the aggregate sum of the following inputs; 5 users at 0.01 messages/second, 10 users at 0.08 packets/second, 15 users at 10.24 bps. Depending on link condition the server alternates between two states (with probabilities α and $1 - \alpha$). With probability α, the normal server state is defined by $\bar{\tau}_1 = 0.8$ second, $\overline{\tau_1^2} = 16$. With probability $(1 - \alpha)$, the congested server state defined by $\bar{\tau}_2 = 2$ seconds, $\overline{\tau_2^2} = 20$. Find the expected number in the buffer and the expected waiting time if $\alpha = 0.4$, the packet length is 256 bits, and the message length is five packets on average.

3.14 For the $M/G/1$ queueing system find the probability of an empty buffer, average time between busy periods and the average length of the busy period, all in terms of λ and $\bar{l}$, that is, the average arrival rate in messages per second and the average message length in packets, respectively.

3.15 Stop-and-wait ARQ protocols are typically used at the data link (and/or) transport layers to mask the errors of lower layers. Using positive acknowledgement and time out, the server tries transmitting the head-of-line packet and repeats the trial if he does not receive (ACK) within T_0 seconds. The effective transmission time now becomes $T' \cong j(T_0 + \tau)$, where τ is the packet length plus the one-way propagation delay in seconds and j is the total number of trials until success. Assuming a geometric distribution for j, let P_e be the probability of bit errors and R the data rate in bits per second. Find the mean and variance of T'. Substitute into the $M/G/1$ equations to find $E(n)$ and $E(w)$ if the traffic rate is in packets per second.

3.16 Expedited and normal data packets are served from an $M/G/1$ buffer, the average expedited higher priority and normal packets length ratio is $(1/10)$, the arrival rates are $\lambda_1 = 2$ packets/second, $\lambda_2 = 4$ packets/second, respectively, and the total traffic utilization $\rho = 0.6$. The service times are exponentially distributed. Find $E(W_2)$, that is, the waiting time of the lower priority class (i.e., the normal data packets) in the following two cases.

a. Nonpreemptive priority service

b. Nonpriority service

3.17 In the packet network of Fig. 3.13, all arrivals at all nodes are assumed Poisson distributed and all packets are exponentially distributed with

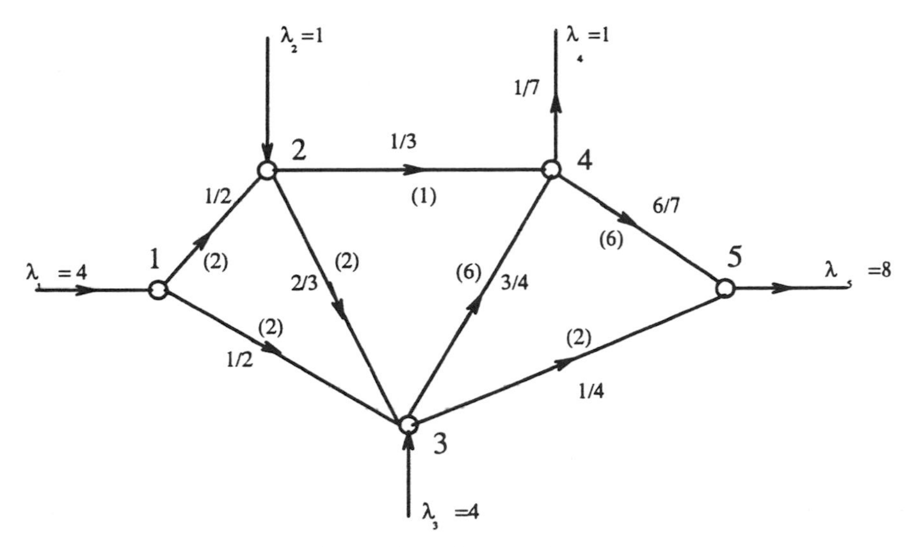

Figure 3.13. Problem 3.17.

an average service rate of $\mu = 8$ packets/second. Traffic on each link is enclosed by parenthesis. The figure also shows on each link the portion of input traffic forwarded on that link.

a. Verify and find the values of traffic on each link (shown in parenthesis) and out of each node.

b. Find the average number of messages (transmitted and queued) between nodes (1, 5) and (2, 4), assuming that the routes are (1-2-4-5) and (2, 3, 4) respectively.

3.18 In the packet network of Fig. 3.13, assuming $\lambda_2 = \lambda_3 = \lambda_4 = 0$, and retaining all remaining parameters. Use Little's formula to find the networkwide delay $E(T)$ and the probability that the sum of the average number of messages in the first and second nodes is larger than, or equal to, two messages, $(n_1 + n_2) \geq 2$.

3.19 Use Eqs. 3.51 and 3.53 to compare the total transfer delay of the TDM and ATDM systems if both systems have the same number of users U, arrival Poisson rate per user λ_u, and constant message length $= \bar{\tau}$ packets. Because of overhead, the ATDM-to-TDM slot ratio is $T_s'/T_s = \alpha > 1$. Find the value of α that makes $D_{ATDM} > D_{TDM}$ in terms of λ_u if $T_s = \bar{l} = 1$ and $U = 10$.

4

THE DATA LINK LAYER

4.1 INTRODUCTION

This chapter discusses the data link layer and the network access layer protocol X.25. The purpose of the data link control (DLC) protocol is to provide error-free transfer of information between senders and receivers over a given communication link. An important consideration is the integrity of the information being transferred. Information in error or information lost has no value to the users and can be disastrous. It is the nonideal nature of the data communications link and environment that necessitates the presence of the data link layer.

4.2 Structure of Data Link Control Protocols

The fundamental structure of any data link control protocol must include the following:

1. The incoming input bit stream must be segmented into blocks or frames. The beginning and end of each frame/block must be clearly identified, thus providing for frame/block synchronization.
2. The protocol must provide a means of identifying and addressing a particular sender or receiver among the many present on a multipoint facility or among the very large number of users connected through the network.
3. The protocol must provide a technique for checking for errors (see Section 2.9) and must have the mechanism to initiate recovery action to correct the error to maintain a high degree of message integrity. It must also provide a mechanism for flow control. The sending station must not send frames at a rate faster than the receiving station can absorb them.

4.3 FUNCTIONS OF DATA LINK CONTROL PROTOCOLS

Besides the fundamental structure of data link control protocols, there are a basic set of protocol functions that are common to most of these protocols. The basic protocol functions are described below.

1. *Frame control* delimits the beginning and end of the transmission blocks or frames by the use of delimiting characters or flags.
2. *Error control* provides for the detection of errors, the acknowledgment of correctly received blocks, and the requests for retransmission of incorrectly received blocks. As discussed in Chapter 2, the most commonly used error-detection techniques are vertical and longitudinal checks and cyclic redundancy checks.
3. *Initialization control* guarantees the establishment of an active data link. It requires the exchange of messages to establish identification and readiness to receive or transmit.
4. *Link management capabilities* are used to supervise the link by controlling transmission direction, establishing and terminating logical connections, and so on. Link management responsibility may reside in a master station or in all user stations.
5. *Transparency* allows the link control to be independent of the pattern or the code of the transmitted information. In particular, all bit patterns should be allowable in the content of transmitted messages.
6. *Flow control* governs the flow of information across the data link and is achieved through coordination between the transmitter and the receiver.
7. *Abnormal recovery controls* supervise the actions to be taken to recover from abnormal occurrences. For example, time-outs discussed in Section 4.5 are common methods to detect abnormal conditions.

4.4 ERROR CONTROL

Error control is an area of utmost importance in communication systems. Errors in certain data cannot be tolerated. It is easy to imagine, for example, the effect of an undetected data error on a weapons control system or on an important financial data transfer.

There are two basic categories of error-control schemes for data communications: automatic-repeat-request (ARQ) and forward-error correction (FEC) schemes. Automatic-repeat-request error-control systems employ error-detecting codes, such as those described in Section 2.9. In ARQ schemes, the receiver discards the block or frame received in error, and requests a retransmission of the same frame of data via a feeback channel (return channel). Retransmission continues until the frame is successfully received.

This process is widely used in data communication systems for error control because it is simple and provides high system reliability. However, the performance of ARQ schemes depends on the channel error rate and the round-trip delay.

In an FEC error-control system, an error-correcting code (such as block codes or convolutional codes) is used. When the receiver detects the presence of errors in a received frame of data, it attempts to locate and correct the errors. After error correction has been performed, the corrected frame is then delivered to the user. Since no retransmission is required in an FEC error-control system, no feedback channel is needed. Powerful error-correcting codes must be used which makes decoding difficult and expensive to implement. This is why ARQ schemes are often preferred over FEC schemes for error control in data communication systems, such as packet-switched data networks. On the other hand, in communication systems where feedback channels are not available or retransmission is not suitable for some reason (such as in satellite and space communications), the use of FEC is necessary.

Another approach to error control is through the use of hybrid ARQ schemes, which incorporate both FEC and retransmission techniques. In hybrid ARQ schemes, the FEC scheme corrects the error patterns that occur most frequently leading to a reduction in the frequency of retransmission. This increases the system throughput performance. However, when a less frequent error pattern occurs and is detected, the receiver requests a retransmission, thus increasing the system reliability.

4.5 AUTOMATIC-REPEAT-REQUEST SCHEMES

There are two basic types of ARQ schemes: stop-and-wait and continuous ARQ. We describe their operation next using the following assumptions. (1) A single transmitter is sending information to a single receiver. (2) The receiver can send acknowledgements back to the transmitter. (3) Information frames and acknowledgments contain error-detection codes. (4) Information frames and acknowledgments received in error are ignored and discarded.

Later we generalize the operation of our ARQ schemes to include two-way information transfer.

4.5.1 Stop-and-wait

4.5.1.1 Error-Free Operation. The basic error-free operation of the stop-and-wait ARQ is shown in Figure 4.1. The transmitter sends a single information frame and waits to receive an acknowledgment from the receiver. The next frame can only be transmitted after this acknowledgment is received.

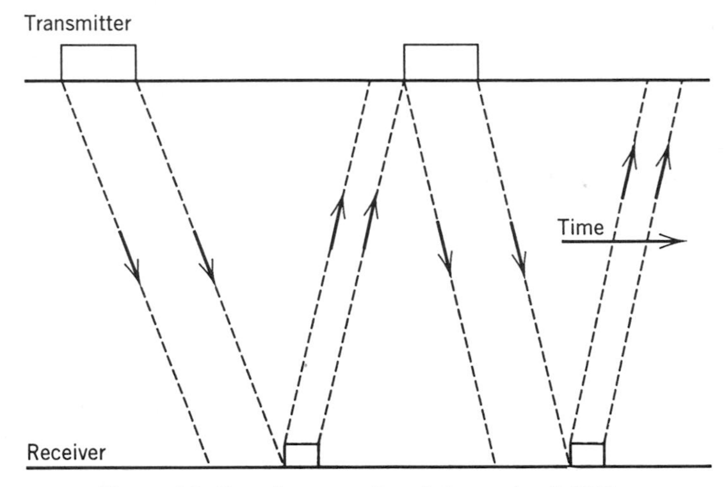

Figure 4.1. Error-free operation of stop-and-wait ARQ.

4.5.1.2 Recovery from Lost Information Frame. When an information frame is corrupted, the error is detected by the receiver and no acknowledgment is returned. To recover from this, the transmitter starts a timer immediately after the conclusion of a frame's transmission. Once this timer exceeds a threshold called the *timeout value*, the transmitter retransmits the frame. The process continues until the transmitter receives an acknowledgment. This is shown in Figure 4.2.

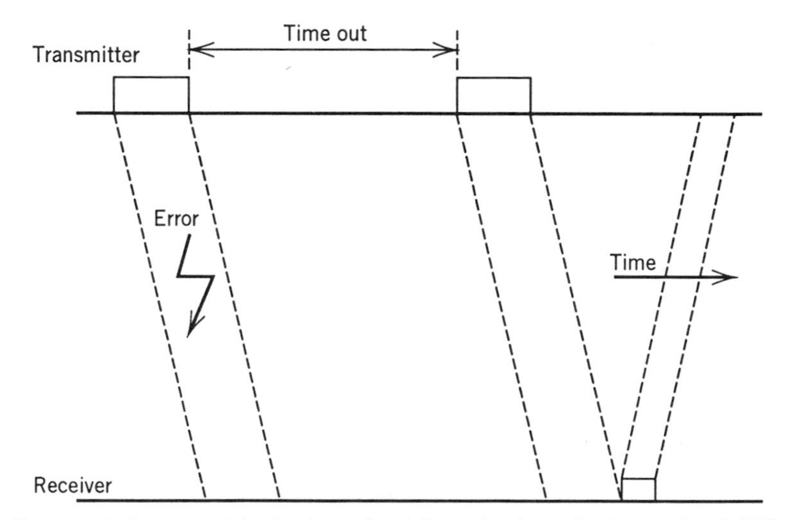

Figure 4.2. Recovery from the loss of an information frame in stop-and-wait ARQ.

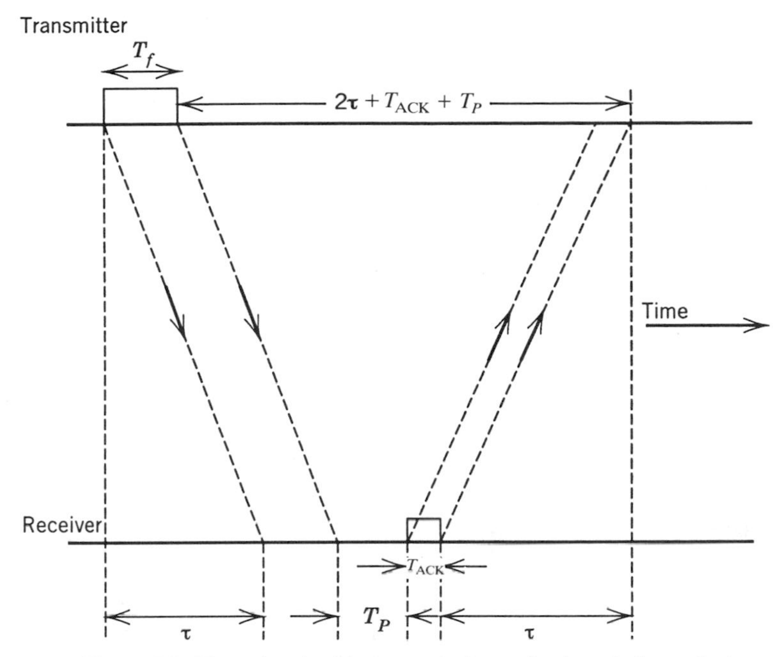

Figure 4.3. Times involved in transmission and acknowledgment.

The selection of an appropriate value for the timeout threshold is important. The transmitter should give the receiver a chance to return an acknowledgment before transmitting again. Consider Figure 4.3 which shows the timing involved in transmission and acknowledgment. There are 4 components: T_f is the information frame transmission time; T_{ACK} is the acknowledgment transmission time; T_p is the processing time of the frame at the receiver; and τ is the one-way signal propagation delay between transmitter and receiver.

Given these times an acknowledgment can be expected $2\tau + T_{ACK} + T_p$ after the transmitter concludes its transmission. It is, therefore, necessary to have a timeout that is greater than $2\tau + T_{ACK} + T_p$. Because there is some variability in the frame processing time T_p, the timeout value is often made several times larger than the minimum.

4.5.1.3 Recovery from Lost Acknowledgment. When an acknowledgment is corrupted, it is ignored by the transmitter, which now acts as if no acknowledgement has been received. The timeout mechanism will then cause a retransmission of the frame. This is shown in Figure 4.4

Observe that in this scenario, the receiver receives two copies (duplicates) of the same frame. It is important for the receiver to be able to determine that it is actually receiving the same frame twice. It is important to emphasize that it is not appropriate for the receiver to compare the data content of the

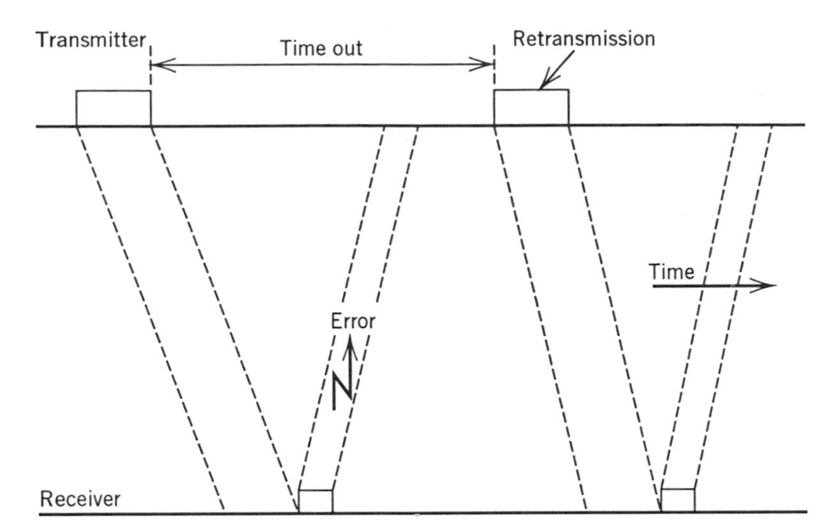

Figure 4.4. Recovery from lost acknowledgment in stop-and-wait ARQ protocols.

received frames to determine whether they are duplicates, because the data content of two consecutive and distinct frames can indeed be identical.

Duplicate detection is achieved through the use of *sequence numbers*. Each frame is given a number by the transmitter. The receiver compares this number with that of the previously received frame. A duplicate frame is detected if its sequence number is identical to that of the previous frame. A duplicate frame is normally discarded by the receiver, but it is still acknowledged in order to inform the transmitter that the frame has been received. In stop-and-wait ARQ, it is sufficient to use a single bit to indicate a frame's sequence number. Consecutive frames are thus numbered $0, 1, 0, 1, 0, 1 \ldots$.

4.5.1.4 Numbering of Acknowledgments. Consider the scenario shown in Figure 4.5. Frame 0 is transmitted, but it requires an unusually long time to process because the receiver is slow. However finally an acknowledgment is returned. Meanwhile the transmitter takes a timeout and retransmits frame 0. After receiving the acknowledgment from the first frame 0 transmission, the transmitter continues with frame 1, which is corrupted and received in error. The transmitter, however, thinks that frame 1 has been received correctly, because it receives an acknowledgment that is actually returned for the retransmission of frame 0.

Clearly an error in the operation of the protocol has occurred and frame 1 was not delivered correctly to the receiver. To avoid this and similar erroneous operations it is necessary to label acknowledgments with the sequence number of the frame being acknowledged. Figure 4.6 shows how this helps with the situation just described.

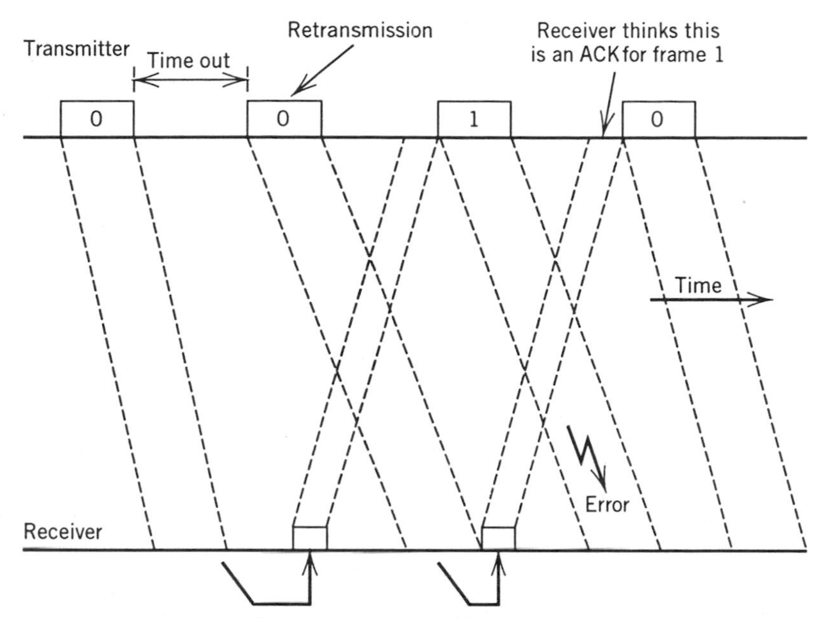

Figure 4.5. The need for numbering acknowledgments.

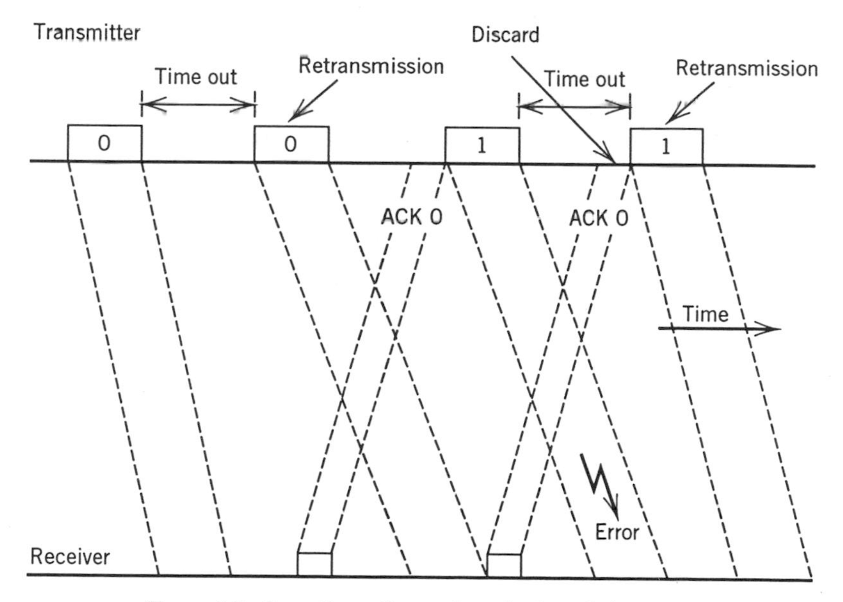

Figure 4.6. Operation with numbered acknowledgment.

4.5.1.5 The Trouble with Stop-and-Wait ARQ. As discussed earlier in Section 4.5.1.2, it takes at least $2\tau + T_p + T_{\text{ACK}}$ time units for the acknowledgment to return in stop-and-wait ARQ. During this time, the transmitter has to remain idle even if it has other packets to transmit.

How significant is this idle time? This depends on several factors. Let us define transmitter maximum utilization, η, as the proportion of time it is not idle, assuming it always has packets to transmit. This is given by (assuming error-free transmission and reception)

$$\eta = \frac{T_f}{T_f + 2\tau + T_p + T_{\text{ACK}}} \tag{4.1}$$

Notice that both T_f and T_{ACK} increase for lower transmission data rates. T_f is also proportional to the size of the information frame, typicaly $T_f \gg T_{\text{ACK}}$. However, the round-trip propagation delay 2τ and the frame processing time T_p are independent of data rates and are typically constant for a given hardware configuration.

Consider the case where information frames are 10,000 bits each and acknowledgment frames are 100 bits each. For a connection between transmitter and receiver that is 10,000 m long, signal propagation delay is approximately $\tau = 50$ μsecond (assuming that the signal speed is $0.6 \times$ speed of light . A typical processing rate is $T_p = 10$ μsecond. Now, we evaluate Eq. 4.1 for two typical scenarios:

1. *Data Rate = 10 Kbps:*

$$T_f = \frac{10000 \text{ bits}}{10 \text{ Kbps}} = 1 \text{ sec}, \qquad T_{\text{ACK}} = \frac{100}{10 \text{ Kbps}} = 10^{-2} \text{ second}$$

$$\eta = \frac{1}{1 + 2 \times 50 \times 10^{-6} + 10^{-2} + 10 \times 10^{-6}} \approx 1$$

2. *Data Rate = 100 Mbps:*

$$T_f = \frac{10000 \text{ bits}}{100 \text{ Mbps}} = 100\ \mu\text{sec} \qquad T_{\text{ACK}} = \frac{100}{100 \text{ Mbps}} = 1\ \mu\text{sec}$$

$$\eta = \frac{100}{100 + 100 + 10 + 1} \approx 0.5$$

It is clear from the above calculations that the transmitter idle time is negligible when data rates are low and stop-and-wait ARQ is perfectly suited to this scenario. However, when data rates are high, the stop-and-wait protocol can result in wasting a significant amount of the transmitter's time due to waiting for acknowledgments.

4.5.2 Continuous ARQ

4.5.2.1 Error-free Operation. Continuous ARQ is used in situations where stop-and-wait ARQ results in very low utilizations. The basic principle is simple. The transmitter continuously transmits information frames without waiting for acknowledgments. As before, frames and acknowledgments are numbered. The basic error-free operation of continuous ARQ is shown in Figure 4.7

It is clear that, under error-free conditions, a transmitter utilization of 1 can be achieved with continuous ARQ. As will be seen next, this is achieved at a price of more complex error-recovery techniques. It should be noted, however, that in typical low-error-rate environments, error recovery is infrequently invoked. However the improvement in the error-free operation of a protocol always outweighs any complication to the error recovery technique.

4.5.2.2 Error Recovery Techniques In Continuous ARQ. Two error recovery approaches are commonly used in continuous ARQ: *selective-reject* and *go-back-N*. *Selective-reject* operates under the principle that only lost or unacknowledged frames should be retransmitted. A typical information frame-loss scenario is shown in Figure 4.8.

Lost acknowledgments are handled in the same manner, that is, by timeout and retransmission. As in stop and wait, a lost acknowledgment can

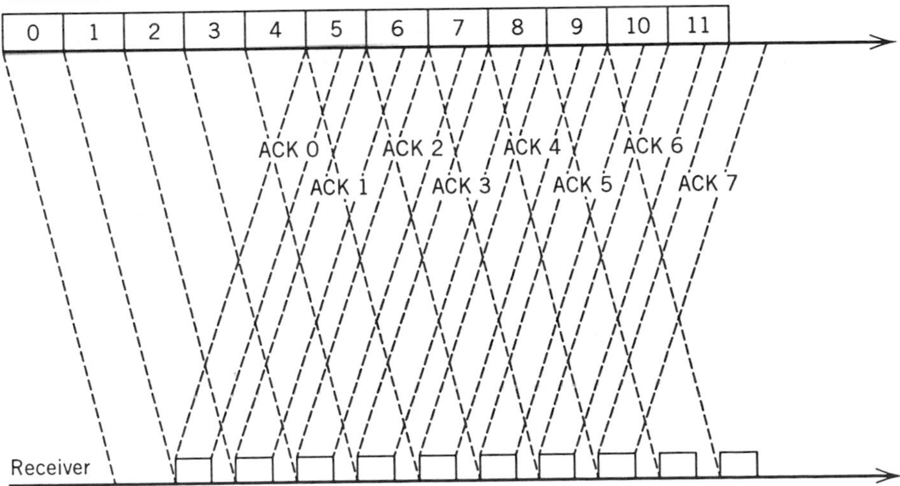

Figure 4.7. Continuous ARQ error-free operation.

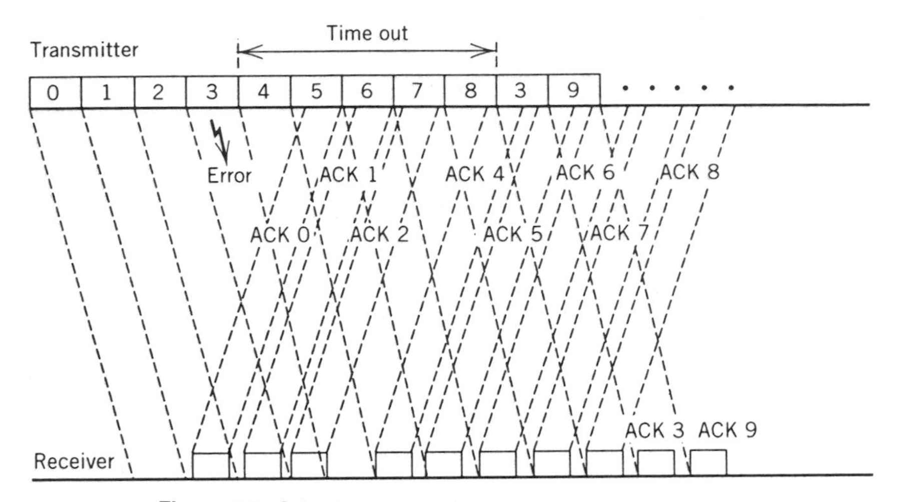

Figure 4.8. Selective-reject lost information frame recovery.

cause duplicate frames to be received. Sequence numbers are used to detect duplicates. It should be clear that a single bit sequence number is not sufficient in this case, because a frame may duplicate one that was transmitted several frames ago.

Although the selective-reject method sounds like a reasonable approach, it has a major drawback in certain instances; namely, the packets may be delivered to the receiver in a different sequence than the one in which they were transmitted. This is illustrated in Figure 4.8, where the receiver gets the frames in the order 0 1 2 4 5 6 7 8 3 9. . . . Why is the order or out of sequence phenomena a problem? It would not be a problem if each frame is a distinct entity unto itself. This, however, is typically not the case. Each frame is normally a piece of a longer message such as a record in a file or a field in a data base query. The receiver, typically needs to reconstruct the original message (file, query, etc.) before it can be processed. This is done at a higher layer, which normally expects to receive the data from the data link layer in sequence.

When selective-reject is used, the receiving data link layer passes on information frames to its higher layer as long as they are received in sequence. Once an out-of-sequence packet is received, it should be buffered until missing packets are received. For example, in the scenario of Figure 4.8, the receiving data link layer will buffer frames 4, 5, 6, and 7 until the retransmitted frame 3 is received. It will then pass frames 3 through 7 to its higher layer. This buffering can become quite extensive especially for channels having long propagation delays. It is for this reason that another continuous ARQ error recovery technique known as go-back-N was developed. In *go-back-N* the receiver insists on receiving information frames in sequence. A frame received out-of-sequence is discarded. However, an

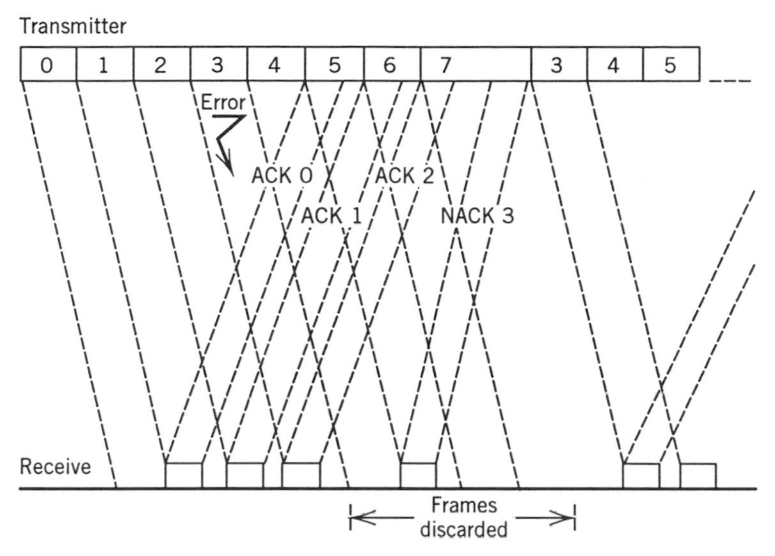

Figure 4.9. Go-back-N recovery through out-of-sequence frame reception.

indication of the out-of-sequence frame is sent back to the transmitter. This indication is in the form of negative acknowledgment (NAK) which contains the sequence number of the next frame that the receiver is expecting. An example is shown in Figure 4.9.

In the scenario illustrated in the figure, when frame 3 is received in error, it is ignored. When frame 4 is received, the receiver realizes that frame 3 is missing and sends a NAK3 to tell the transmitter that an out-of-sequence frame has been received.

In go-back-N protocols, once the transmitter realizes that a frame has been lost it "goes-back" and retransmits all frames starting with the lost frame. In the example of Figure 4.9, the transmitter receives NAK3 while transmitting frame 7, at which point it retransmits frames 3 through 7 again. Comparing this with the selective-reject approach, we find that more bandwidth is required for error recovery. However, since the receiver ignores all out-of-sequence frames, it receives all frames in the sequence they were transmitted and the buffering required in the selective-reject approach is eliminated.

In go-back-N (as illustrated in Fig. 4.9), only the first out-of-sequence frame received generated a NAK. The subsequent frames are simply ignored until the correct frame is received. Because of this, timeout recovery is still needed. If a NAK is received in error, the transmitter will continue until a timeout threshold is exceeded and then start retransmitting from the frame that caused the timeout. Timeout recovery is also needed if the lost information frame is the last one transmitted, in which case there is no subsequent frame to cause a NAK to be returned.

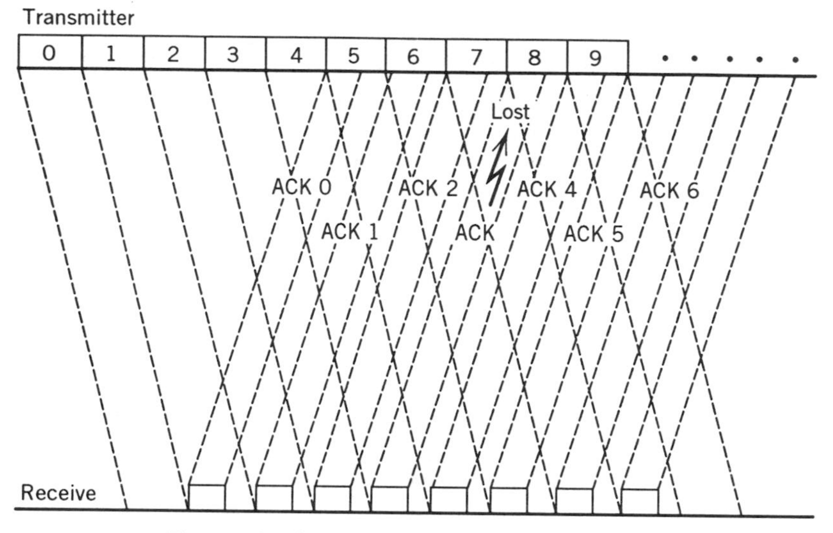

Figure 4.10. Lost acknowledgments in go-back-N.

Finally, we consider how lost acknowledgments are recovered in go-back-N. In the scenario illustrated in Figure 4.10, the transmitter does not receive ACK3 but receive ACK4. This latter acknowledgment indicates that the receiver has accepted frame 4, which implies that it has already received frame 3. So although ACK3 was not received, ACK4 served to acknowledge frame 3 as well. Thus in go-back-N, an acknowledgment serves to confirm the reception of all frames up to and including the one being acknowledged. Sometimes, this fact is used to group acknowledgments together and have the receiver send one ACK for every few messages it receives, this could occasionally save precious channel capacity.

4.5.2.3 Sequence Numbers. As we discussed earlier, a single bit is sufficient to indicate the sequence numbers in stop-and-wait ARQ. For continuous ARQ protocols, more bits are required. Exactly how many depends on several factors. Consider, for example a go-back-N protocol using 2-bit sequence numbers. The sequence numbers will wrap around every four frames. This can cause confusion as illustrated in Figure 4.11. When NAK3 is received, the transmitter will go back to the most recent frame 3 and not to the appropriate frame. In this scenario a 4-bit sequence number is required.

The basic rule is that the sequence number should not repeat within the time an ACK or NAK for a frame is expected. This dictates a large number of bits for sequence numbering over long-propagation-delay channels (e.g., satellite channels). In our discussion of ARQ flow control, we explain another constraint on the sequence numbering of frames.

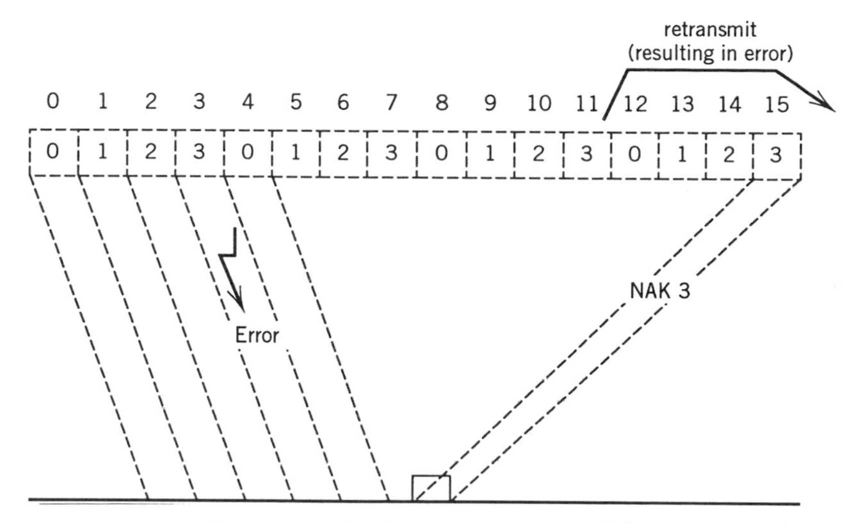

Figure 4.11. Two-bit sequence number ARQ.

4.5.2.4 Two Way Information Transfer. So far we have been discussing ARQ protocols for information transfer in one direction. ARQ protocols can be easily used for two-way information transfer by maintaining two independent exchanges one in each direction.

It is possible, however, to have the added efficiency of "piggybacking" acknowledgments. In such a situation an ACK or NAK message and associated sequence number are made part of the information frame. Piggybacking is used whenever possible, and explicit ACKs or NAKs are used whenever they need to be sent and no information frames are available.

4.5.3 Flow Control in ARQ Protocols

4.5.3.1 Window Mechanism. We have seen that information frames received by a data link layer are passed on to a higher layer. Normally, the rate at which this higher layer can receive new information is limited. If this rate is lower than the rate at which the receiver is receiving information frames, then the receiver needs to buffer frames. This buffering will solve the problem up to a point, because if the rate mismatch continues the receiver's buffers will fill up and it will start discarding information frames due to the lack of buffer space.

To avoid this, it is necessary to provide a mechanism for the receiver to control the flow of information frames from the transmitter. Automatic repeat request protocols normally use the *sliding window* mechanism to accomplish this. A window size W is defined as the number of information frames that can be transmitted before receiving an acknowledgment from the

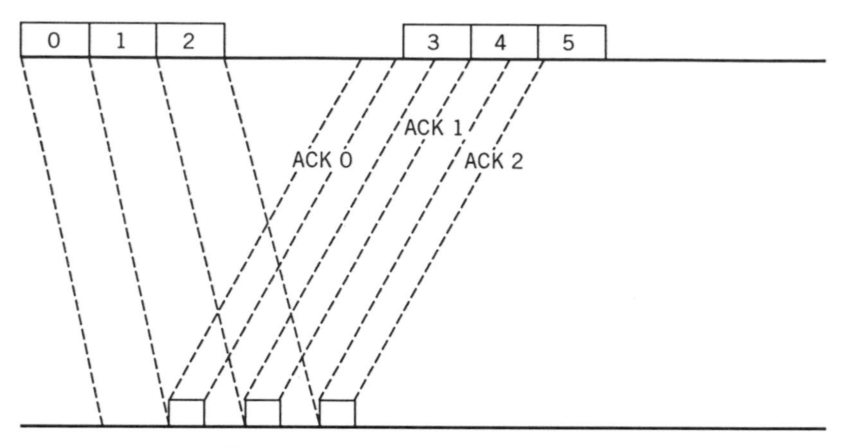

Figure 4.12. Continuous ARQ with $W = 3$.

receiver. The operation of continuous ARQ with window size $W = 3$ is illustrated in Figure 4.12. The transmitter has to stop after transmitting three consecutive frames. The transmitter can transmit an additional frame after receiving ACK0. ACK1 and ACK2 also allow the transmission of new frames. After frame 5 is transmitted, it has to stop because there are now 3 unacknowledged frames.

Note that the receiver can, by withholding acknowledgments, slow down the transmitter. The rate of new frame transmission can match the rate of frame accomodation at the receiver. We discuss the window mechanism further in Chapter 6.

4.5.3.2 Sequence Numbers and the Window Mechanism. The use of the window flow-control mechanism imposes certain limitations on the use of sequence numbers. To illustrate this, consider the use of go-back-N with a 2-bit sequence number and window size $W = 4$. In Figure 4.13, we illustrate a scenario where frames 0 to 3 are received correctly but their ACKs are all lost.

Since 2-bit sequence numbers are being used, the receiver expects the next new frame to have number 0. If the transmitter has not received the ACKs, it will enter a timeout phase and then retransmit frames 0 to 3. These frames will be treated as new frames by the receiver, even though they duplicate previous frames.

This problem can be remedied by making $W = 3$. As illustrated in Figure 4.14, new and duplicate frames now are clearly distinguished. In general, for go-back-N, when n bits are used for the sequence number, W must be less than 2^n for correct operation. Defining a similar value for selective-reject is left as an exercise to the reader.

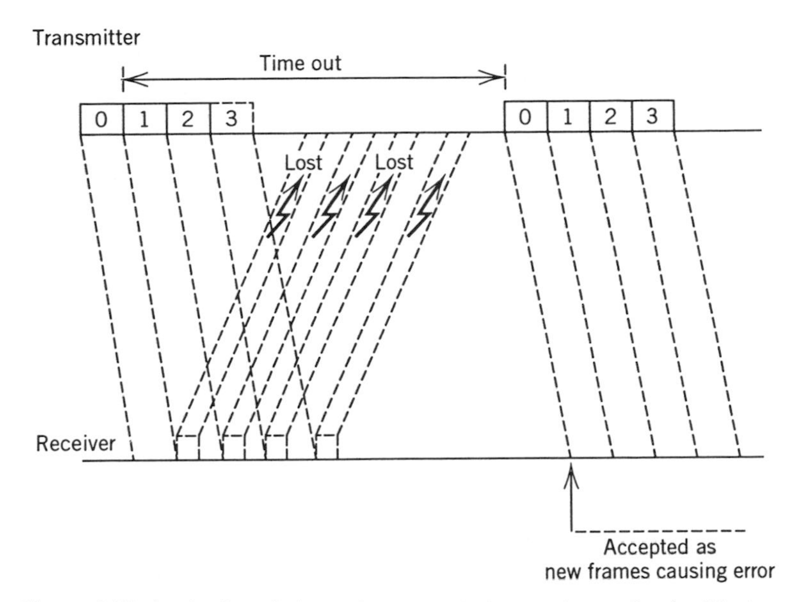

Figure 4.13. Lost acknowledgments can cause incorrect operation for $W = 4$.

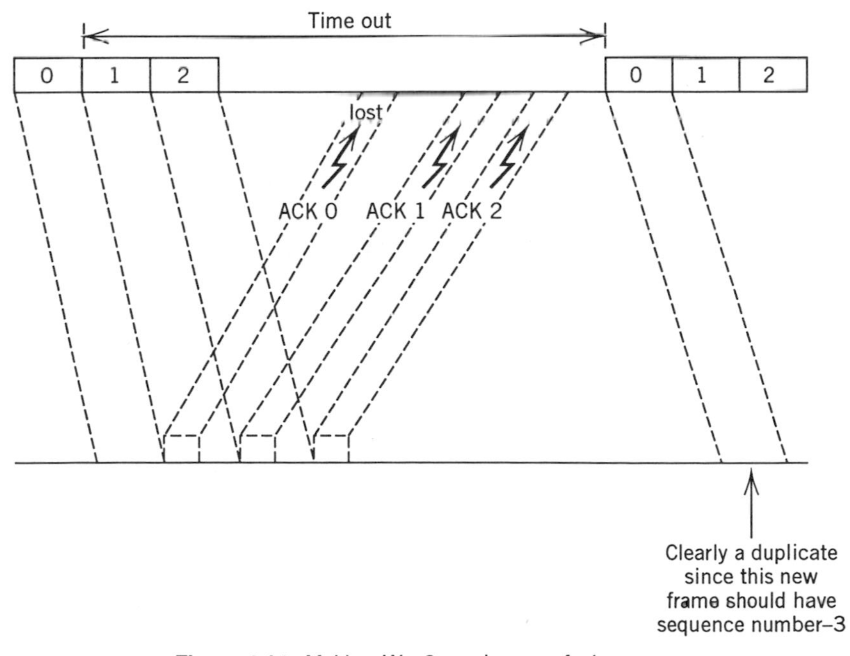

Figure 4.14. Making $W = 3$ resolves confusion.

4.6 HIGH-LEVEL DATA LINK CONTROL — A BIT-ORIENTED PROTOCOL

Data link protocols can be classified into two major categories: character-oriented and bit-oriented protocols. Below, we discuss the high-level data link control (HDLC) protocol as an example of the bit-oriented protocol.

Although our discussion of bit-oriented protocols centers around HDLC, there are many bit-oriented protocols:

The synchronous data link control (SDLC) developed by IBM as the data link protocol for Systems Network Architecture (SNA).

The advanced data communication control procedure (ADCCP) standardized by the American National Standards Institute (ANSI).

The high-level data link control (HDLC) adopted by the International Organization of Standardization (ISO).

The link access procedure–balanced (LAP-B) adopted by the International Telephone and Telegraph Consultative Committee (CCITT) for its X.25 packet-switched network standard.

Most of these bit-oriented protocols (except for minor differences) have the same format and protocol structure. Below, we present the basic structure of the bit-oriented protocol using HDLC to show the protocol format.

The HDLC protocol attempts to satisfy the requirements of a general data link control procedure that would have widespread applicability. As a result, these protocols can be implemented over point-to-point and multipoint configurations using half-duplex and full-duplex operation over switched and nonswitched transmission lines. Also the peculiarities of the underlying channel such as terrestrial or satellite connections should be taken into account. To satisfy the above requirements, three different data-transfer modes of operation and three different types of stations are defined in HDLC:

1. *The primary station* controls the operation of the link, issues commands, and receives expected responses. In many instances, the primary station is a mainframe host computer.
2. *The secondary station* receives commands and issues responses that depend on the nature of the command received and the mode of operation used. The secondary stations are subservient to the primary station at the data link level. Their role is generally passive, and they have little or no capability for recovery from system errors. Examples of secondary stations are terminals, simple data collection, or data display devices.
3. *The combined station* is capable of initializing the link, activating the other combined station, and logically disconnecting the link. It can both

issue commands and responses and receive commands and responses. A combined station may be a host computer or a packet-switching node.

Two link configurations are defined: the unbalanced and the balanced. In the unbalanced configuration, one primary station is connected to one or more secondary stations in a point-to-point or multipoint operation. In the balanced configuration, one combined station is connected to another combined station in a point-to-point operation.

The three data transfer modes are normal response mode, asynchronous response mode, and asynchronous balanced mode:

1. *Normal response mode* (NRM) is an unbalanced configuration and is ideally suited for polled multipoint operation where a single primary station is connected to a number of secondary stations. In this configuration, the secondary stations can only transmit after receiving permission (through poll messages) from the primary stations.
2. *The asynchronous response mode* (ARM) is also an unbalanced configuration. Here a secondary station is not required to receive explicit permission from the primary station to initiate transmission (responses) of its own. Asynchronous-response-mode operation, therefore, is less disciplined than NRM operation. Because of the asynchronous nature of secondary station transmissions, only one secondary station can be activated (on-line) at a time. Other secondary stations on the multipoint link must be kept in a disconnected mode (off-line) so as not to interfere with any transmission.
3. *Asynchronous balanced mode* (ABM) provides a balanced type of data transfer mode between two logically equal stations (i.e., two combined stations).

For a point-to-point configuration, the asynchronous modes (ARM and ABM) are usually more efficient than the normal response mode (NRM), because there is no polling overhead required.

4.6.1 Frame Format

The basic transmission unit is called a frame. The frame format is given in Figure 4.15 and consists of the following fields:

$$F, A, C, \text{Info}, \text{FCS}, F$$

where F is the flag sequence, A is the address field, C is the control field, Info is the information (data) field, and FCS is the frame check sequence.

The flag (F) is a unique 8-bit pattern (a 0-bit followed by six 1-bits and ending with a 0-bit, i. e., F = 01111110) used to synchronize the receiver with

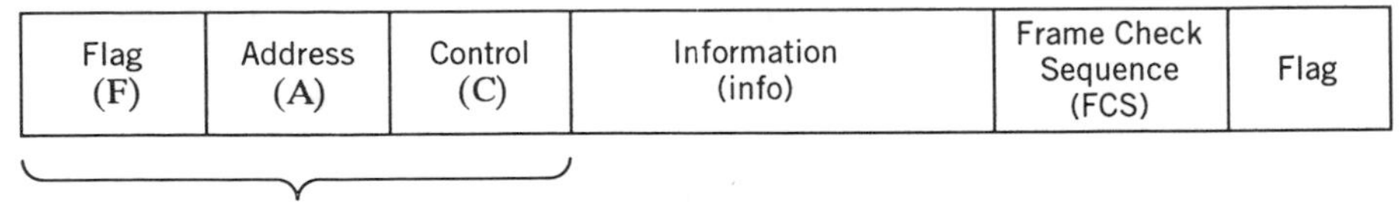

Figure 4.15. The basic bit-oriented data link format.

the incoming frame. It identifies the start and end of each transmitted frame and is also used by the sender to fill the time between frames.

To assure transparency and uniqueness of this flag in the transmitted frame, a sending station always monitors the bit stream that is being transmitted; whenever five consecutive 1s are transmitted, an additional 0 is inserted. This technique is known as bit stuffing (or zero insertion). The receiver similarly monitors the incoming bit stream and after five consecutive 1s, the sixth bit is dropped if it is 0 to insure data integrity. However if the sixth bit is a 1, either the flag (the next bit must then be 0) or the sequence for abortive termination of the frame (at least seven 1s) is received. The closing flag for one frame may also serve as the opening flag for the next frame.

The address field (A) contains the station address (secondary or primary). The address may be for the receiving station or the sending station depending on the particular class of procedure being used. Command frames are always sent with the receiving station's address. Response frames are always sent with the sending station's address. The length of the address field is 1 byte or multiple bytes (extended addressing). Multiple-byte addressing provides for more than 256 addresses. The low-order bit of each 8-bit field indicates whether this is the last (bit = 1) or there is more to follow (bit = 0). A global (broadcast) address is specified as the "all 1s" address.

The control field (C) identifies the function and purpose of the frame. There are three different types of frames (Fig. 4.16):

1. Information frames (I frames) are used for data transfer. The data may be of any length and may consist of any code or grouping of bits.
2. Supervisory frames (S frames) are used to control the flow of data.
3. Unnumbered frames (U frames) are used to provide additional control functions. These are not included in the send and/or receive sequence counter.

Figure 4.17 shows the control field format. The first bit of the control field (bit C_1) is set to 0 to denote an information frame (I) and to 1 otherwise. The first and second bits (C_1 and C_2) are set to 10 to identify supervisory frames (S) and set to 11 to identify unnumbered frames (U) (also shown in Fig. 4.16).

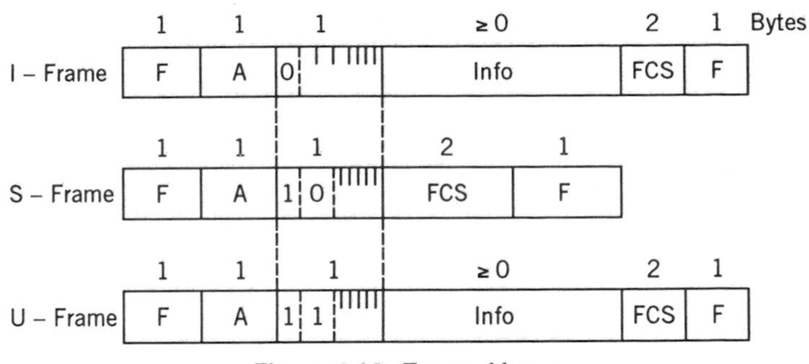

Figure 4.16. Types of frames.

Only an I frame has a send sequence number $N(S)$, bits 2, 3, and 4, (C_2, C_3, C_4) to uniquely identify the frame. Both the I and S frames have a receive sequence number $N(R)$ [bits 6, 7, and 8 (C_6, C_7, and C_8)] to acknowledge the I frames received. $N(R)$ is used for acknowledgment and represents the number of the next frame expected to be received.

The P/F bit (poll/final) [bit 5 (C_5)] in all three types of frames is considered to be the P bit if the frame is a command and the F bit if the frame is a response. In the normal response mode, the primary station sets the P bit to 1 to poll the addressed secondary station, whereas the secondary station, in response to a received P bit set to 1 , sets the F bit to 1 to identify its final sent frame.

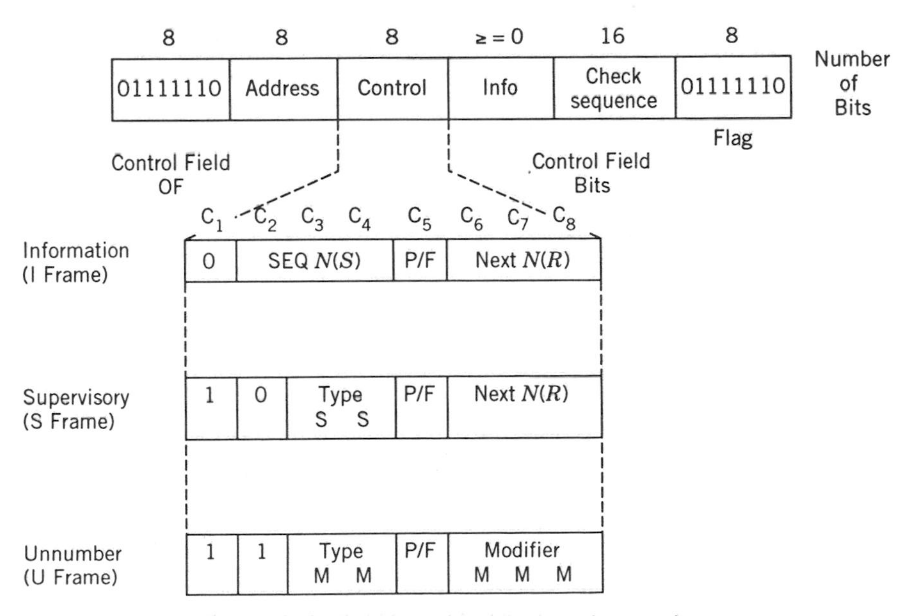

Figure 4.17. Field format for bit-oriented protocols.

In the ARM and ABM, the receipt of a frame with the P bit set to 1 will cause the secondary station (ARM) or combined station (ABM) to set the F bit equal to 1 in the next appropriate frame transmitted. Bits 3 and 4 in the supervisory frame control field specify the following supervisory functions;

1. *Receive ready* (RR): Bits 3 and 4 (C_3 and C_4) are set to 00. RR frames are used to acknowledge I frames received from the other station and to indicate readiness to receive.
2. *Reject* (REJ): Bits 3 and 4 (C_3 and C_4) are set to 01. REJ frames are used to request retransmission of all I frames starting from a designated point in the numbering cycle and to acknowledge I frames already received from the other station.
3. *Receive not ready* (RNR): Bits 3 and 4 (C_3 and C_4) are set to 10. RNR frames are used to indicate a temporary busy condition and to acknowledge I frames already received from the other stations.
4. Selective reject (SREJ): Bits 3 and 4 are set to 11. SREJ frames are used to request retransmission of a single designated I frame previously transmitted.

Bits 3, 4, 6, 7, and 8 (C_3, C_4, C_6, C_7, C_8) in the U frame provide for the specification of up to 32 commands and 32 responses to cover the remaining control functions required. Normally, not all 32 unnumbered frames are used.

Some of the unnumbered commands are:

Unextended numbering set mode commands. These frames are used to establish the particular Modulo 8 sequence numbering mode of operation to be used. Three unextended numbering set mode commands are defined:

SNRM: Set normal response mode

SARM: Set asynchronous response Mode

SABM: Set asynchronous balanced mode

Extended numbering set mode commands. These frames are used to establish the particular Modulo 128 sequence numbering mode of operation to be used. When extended numbering is used, the control field is extended to two bytes to accommodate the 7-bit sequence number. Three extended numbering set mode commands are defined as follows:

SNRME: Set normal response mode extended

SARME: Set asynchronous response mode extended

SABME: Set asynchronous balanced mode extended

Disconnect Command. The disconnect (DISC) command is used to logically terminate a previously established operational mode.

Unnumbered Information Command/Response. Unnumbered information (UI) frames are used to send data to one or more stations, independent of the normal flow of information frames.

Reset Command. The reset (RSET) command is used to reset the send state variable at the transmitting station and the receive state variable at the receiving station to zero.

Set Initialization Mode Commands. The set initialization mode (SIM) command is used to establish the initialization mode of operation.

Unnumbered Acknowledgment Response. The unnumbered acknowledgment response is used to acknowledge receipt and execution of a mode setting, initializing, resetting, or disconnecting command.

Frame Reject Response. The frame reject (FRMR) response is used to indicate that invalid conditions are detected, such as (1) receipt of a control field that is invalid or not implemented, (2) receipt of an information-bearing frame with an information field that exceeds the maximum established length, or (3) receipt of an N(R) which either points to an I frame which has been transmitted and acknowledged or to an I frame that has not been transmitted and is not the next sequential I frame awaiting transmission.

All three type of frames (I, S, and U frames) include a 16-bit frame check sequence (FCS) field prior to the closing flag sequence to assist in the detection of transmission errors.

4.6.2 Examples of Typical Operations

In this section, we describe some typical on-line operations using the shorthand described here: A transmitted frame is described by the fields: A, YN(S)N(R), and P/F, where the fields are defined as follows:

A represents the address associated with the frame. Primary stations use the address of the secondary station for which the frame is intended. Secondary station transmissions include the address of the secondary station that is transmitting. The combined station uses the remote station address when a command frame is sent and the local station address when a response frame is sent.

Y represents the abbreviation for the command or response. N(S) and N(R) are the send and receive sequence number values, respectively, that are an integral part of I and S frames. If only a single number is present, it represents N(R). For example I34 represents an information frame with N(S) = 3 and N(R) = 4, and RR2 represents a S frame (receive ready) with N(R) = 2.

P/F when present, indicates that P and F bit is set to "1" in that frame. When not shown it means that the value of the P or F bit is set to equal 0.

Example 4.1 Link Set-up and Sequence Numbers. Figure 4.18 shows the physical configuration of a primary station connected to a number of secondary stations (multidrop line). Half-duplex operation is assumed in this example. The primary station A activates the secondary stations by addressing SNRM (set normal response mode) to each of them individually with the P bit set to 1, because it is a command frame and thus grants the addressed secondary station the right to transmit. Note that SNRM is an unextended mode, hence the sequence numbers are 3 bits long. B responds with an unnumbered acknowledgment (UA) frame with the F bit set to 1 identifying that this UA is a final frame in this response transmission. Similarly, C responds by sending C, UA, F. After setting up the link between A and B, and C, the primary station A polls the secondary station B for information (traffic). B responds by sending four information frames, with the F bit set to 1. Each I frame has a different send sequence number (first digit), plus a receive sequence number (second digit) that identifies the next I frame expected from the primary station A.

In fact, B can send up to seven frames (it has a 3-bit sequence number field) before it stops and waits for an acknowledgment from station A. Note that the primary station always inserts the address of the intended secondary station whenever it sends a frame. All secondary stations listen to the frames on the line and the station picks up the frame addressed to it. On the other hand, the secondary station always inserts its own address in the frames sent out to A, thus enabling A to recognize the frame sender.

Upon the receipt of frame B, I30, F from B, the primary station acknowledges the receipt of the four frames by issuing frame B, RR4, where 4

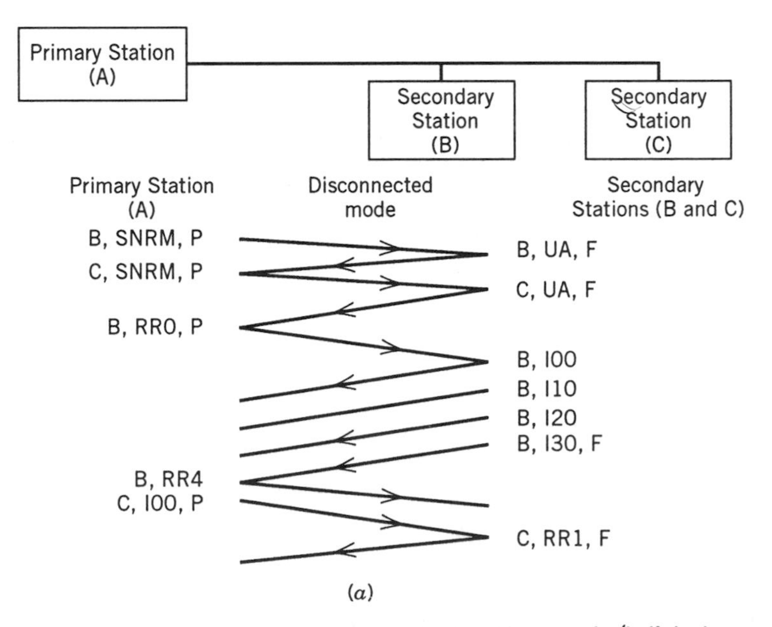

Figure 4.18. Examples of typical operations of bit-oriented protocols (half-duplex, error-free line). (*a*) Link set / up and sequence numbers. (*b*) RNR, RR frames and global addressing. (*c*) Error recovery (SREJ, go-back-N, timeout), ~ is the error in the transmitted frame.

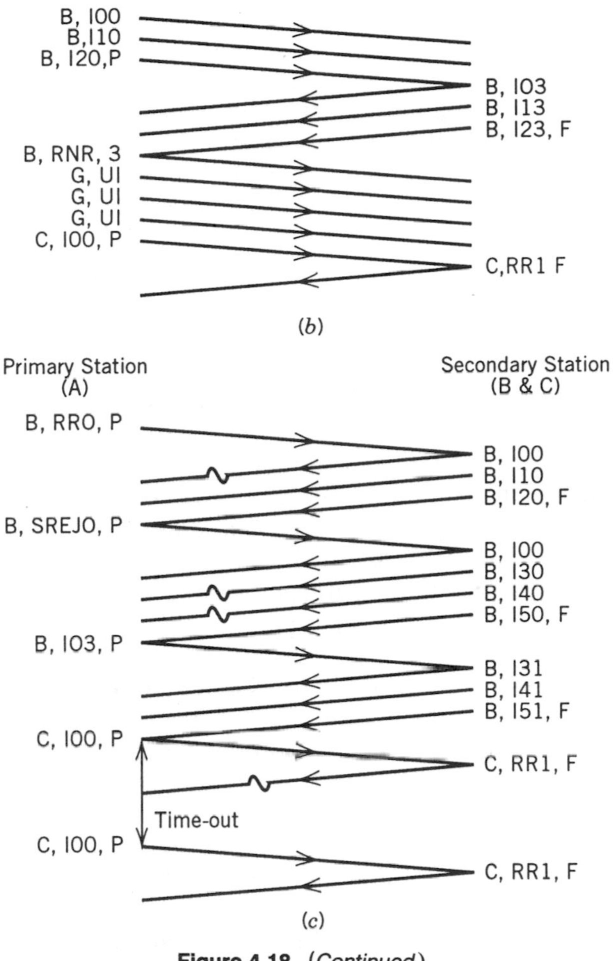

Figure 4.18. *(Continued)*

indicates that I frames numbered 0 to 3 were received correctly. Because the P bit is not set to 1, this frame does not grant permission to send. It only acknowledges I frames received. The primary station finds that it has an information frame destined to C. Hence, it sends I00 to C with the poll bit set to 1. C responds by sending the receive ready frame with $N(\mathrm{R})$ equal to 1.

Example 4.2 Supervisory Frames (RR and RNR), Unnumbered Frames, and Broadcast Addressing. This example, shown in Figure 4.18*b*, assumes that the secondary stations B and C have been setup and the stations are in the data transfer mode. The primary station sends three frames to B and sets the P bit to 1 in the third frame indicating the poll function. B responds with three frames, identifying the final I frame with an F bit set to 1. The receiving sequence number in each of the three I frames is set to 3, thus acknowledging the correct reception of frames 0, 1, and 2 from A. The primary station returns frame B, RNR3, acknowledging the I frames received, but also indicating that the primary station is not ready to receive additional I frames from B.

At this point, let us assume that the primary station has three information frames to be sent to all of the secondary stations on the link, but the information is not so important to the operation of each secondary station to warrant individual delivery with individual acknowledgments. The nature of the information is such that should it be lost in transit, it will not cause a serious problem at any of the secondary stations. Examples of such information might include (1) periodic time checks or weather reports or (2) some sort of updated hourly production report (missing one out of a series of such updated reports may not pose a problem). The use of the UI plus the broadcast address G (all ones) allows the information to be sent to all secondary stations at the same time without impacting the send or receive sequence variables at any of the stations.

Following the UI frames, the primary station sends one I frame to C, with the P bit set to 1. The secondary station C acknowledges the receipt of frame I00 by using frame C, RR1. The 1 indicates that I frame numbered 0 was received correctly and the next I frame expected from A should have a send sequence number of 1.

Example 4.3 Error Recovery (SREJ, go-back-N, timeout). This example, shown in Figure 4.18*c*, demonstrates three major techniques used in error recovery; selective-reject, go-back-N, and timeout procedures. Again, assume that the secondary stations B and C have been set up. The primary station polls B. B transmits the I frames and sets the F bit to 1. As indicated, I frame B, I00 sent by B was received in error and was discarded. The two following I frames were received free of error, and so only B, I00 is needed by the primary station to complete the reception of the three I frames. Thus, I frames B, I10, and B, I20 are held in the primary station's buffer, awaiting correct reception of the first I frame B, I00. The primary station sends a selective-reject frame (B, SREJ0, P) with P bit set to 1, telling B that I frame numbered 0 will have to be retransmitted when permission to transmit is granted. B responds by retransmitting frame B, I00, but does not retransmit I frames numbered 1 and 2 that were transmitted originally. Since its last transmission, B has obtained some additional I frames for transfer to the primary station. The SREJ command provided B with the opportunity to transmit I frames up to the point where it would have *W* unacknowledged I frames outstanding, where *W* is the window size. Consequently, B follows the I00 frame with I frames numbered 3, 4, and 5. The I frame numbered 5 has the F bit set to 1 to identify it as the final frame in the transmission. As shown, frames numbered 3 and 4 are subjected to transmission errors.

Upon correct receipt of frame B, I00, the primary station passes it plus I frames B, I10 and B, I20 from the original transmission up to a higher level. Frames B, I30 and B, I40 are in error. In this instance, there are two frames requiring retransmission, and the primary station will have to decide which type of recovery procedure to take.

If the SREJ function is used, it will have to be used twice. First, it would be used to have B, I30 retransmitted. Then after B, I30 is correctly received, it would be used to acknowledge B, I30 and to have B, I40 retransmitted. After the successful transmission of frame B, I40, I frame B, I50, F would not have to be retransmitted, because it was correctly received in the original transmission.

On the other hand, if the SREJ function is not used, the primary station can utilize the piggy-backed acknowledgment mechanism included in the I-frame, or the seperate acknowledgment mechanism by the use of the supervisory frames RR or RNR. With this approach, only a single control exchange is required, but any already successfully transmitted I frames (such as B, I50, F) will be retransmitted.

As shown in Figure 4.18c, the primary station detects that I frames 3 and 4 are in error, chooses not to use the SREJ recovery action, and has an I frame of its own to deliver to B. The primary station sends frame B, I03,P to B, acknowledging receipt of I frames numbered 0 to 2 from B. B responds by sending all of the I frames starting from frame 3 and in the mean time acknowledging I frame 0 sent by the primary station. The primary station sends I frame to C, with poll bit set to 1, C responds back with RR1, F. But the response suffers transmission errors, causing it to be lost or to fail the FCS check at the primary station. To determine when sufficient time has elapsed waiting for an expected response, the primary station would probably activate a timeout counter when it sends the C, I00, P frame. When this timeout counter runs out, the primary station may initiate an appropriate recovery action. In the case shown, the recovery action is to reissue the C, I00, P and activate the timeout function again. It is also possible for the primary station to send a supervisory command with the P bit set to 1 to find out the number of the I frame that C expects to receive.

4.7 PERFORMANCE EVALUATION OF DATA-LINK-CONTROL PROTOCOLS

First, we present the performance evaluation of the stop-and-wait protocol followed by that of the sliding-window protocol. For both protocols, we study the two cases of error-free transmission and the case with errors.

4.7.1 Stop-and-Wait Protocol

We start first with operation under ideal conditions (i.e. an error-free link). Figure 4.19*a* shows a typical timing diagram for this case. We use the same notation in Section 4.5.1.2. Hence, the efficiency of the stop-and-wait proto-

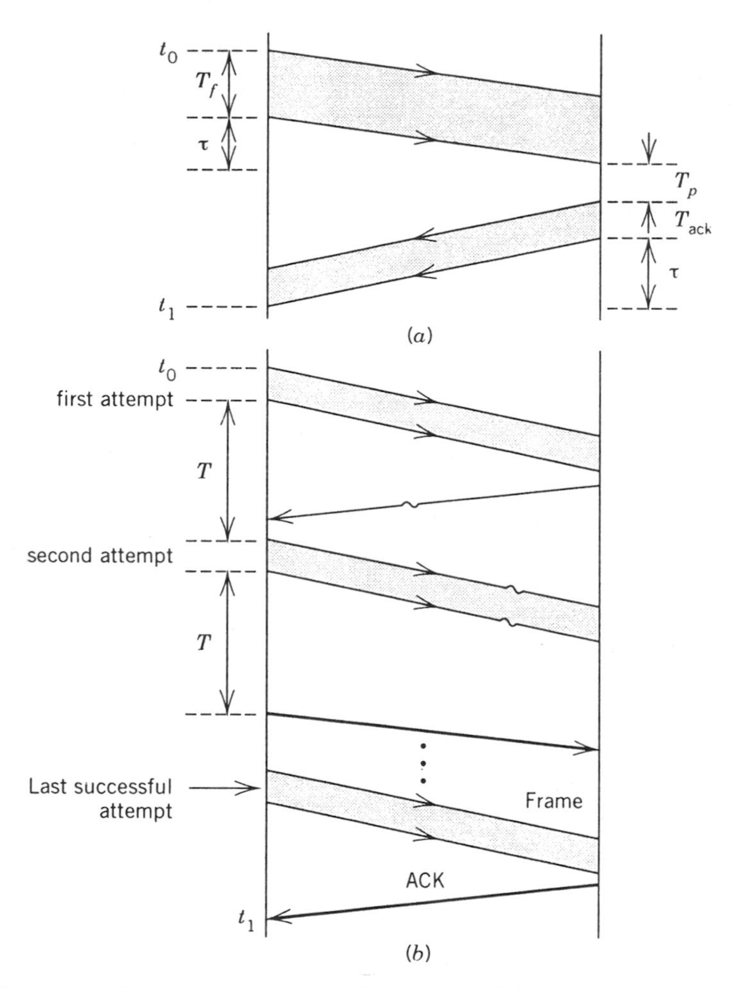

Figure 4.19. Stop-and-wait protocols. (*a*) No errors. (*b*) With errors, ~ = error.

col η is given by Eq. 4.1. Note that the utilization also indicates the maximum rate or throughput at which new frames can be transmitted under the error-free assumption.

Now, let us consider the effect of noise (Fig. 4.19*b*). A damaged or lost information frame or acknowledgment frame will cause the sender to retransmit the frame after a timeout period T. The value of T should be $> 2\tau + T_{\mathrm{ACK}} + T_p$. Thus, every unsuccessful transmission wastes $(T + T_F)$ second of the channel time. To determine the average number of transmissions a frame goes through, let P be the probability of an unsuccessful transmission of a data or acknowledgment frame, and assume that the errors on different frames are independent. Then the probability of i transmissions per frame, a_i, equals the probability of $(i - 1)$ unsuccessful transmission times the probability of the ith successful transmission. Thus,

$$a_i = P^{i-1}(1 - P)$$

and the average number of transmissions per frame N_f is given by

$$N_f = \sum_{i=1}^{\infty} ia_i = \sum_{i=1}^{\infty} iP^{i-1}(1 - P)$$

$$= \frac{1}{1 - P} \tag{4.2}$$

Since the first $N_f - 1$ transmissions waste $(T + T_f)(N_f - 1)$ seconds of the channel time and the last successful transmission uses $(T_f + 2\tau + T_p + T_{ACK})$ seconds, the total time T_t needed for successful frame transmission is given by

$$T_t = t_1 - t_0$$

$$= (T + T_f)(N_f - 1) + (T_f + 2\tau + T_p + T_{ACK})$$

$$= \frac{(T + T_f)P}{(1 - P)} + (T_f + 2\tau + T_p + T_{ACK})$$

and the throughput of the stop-and-wait protocol (with error) η is given by

$$\eta = \frac{T_f}{(T + T_f)P/(1 - P) + (T_f + 2\tau + T_p + T_{ACK})} \tag{4.3}$$

Effect of the Propagation Delay and Transmission Time. To see the impact of the propagation and transmission times on the protocol throughput, notice that in many cases T_p is very small and can be neglected. The acknowledgment frame is also very small and thus T_{ACK} can be neglected. Equation 4.1 can be written as

$$\eta \approx \frac{1}{(1 + 2\tau/T_f)}$$

$$= \frac{1}{(1 + 2a)} \tag{4.4}$$

where a is the propagation time (one way) divided by the frame transmission time. Since the propagation time τ is the distance of the link d divided by the propagation velocity V and the frame transmission time T_f is the frame length L divided by the transmission rate R, then a is given by

$$a = \frac{\tau}{T_f}$$

$$= \frac{(d/v)}{(L/R)} \tag{4.5}$$

For systems with a large a, typical of a satellite link, the stop-and-wait protocol has a very low efficiency η and hence is not recommended, whereas for systems with a small a, typical of a local environment, the stop-and-wait protocol can be used because it provides reasonable efficiency.

4.7.2 The Sliding Window Protocol*

To simplify the analysis here, assume that the acknowledgment frame transmission time is small and can be neglected. Also, assume that the processing time at the receiver T_p is negligible.

The maximum throughput of the sliding window protocol will depend on the window size W as well as the frame transmission time T_f and the propagation time τ. First, we assume an error-free link. If the transmission time for all W frames WT_f is larger than the round trip delay 2τ, then the transmitter can continue transmitting frames at will. Thus the sender is fully utilizing the channel, and the protocol efficiency is unity (see Fig. 4.20a).

On the other hand, if the transmission time for all W frames is smaller than 2τ, then the transmitter will stop sending frames after it has exhausted its window size, and it will resume transmission after it receives the acknowledgment for the first frame. In this case, the protocol efficiency η is $WT_f/(T_f + 2\tau)$. Hence, the efficiency of the sliding-window protocol for an error-free link is given by

$$\eta = \begin{cases} 1, & \text{if } WT_f \succ 2\tau + T_f \\ \dfrac{WT_f}{(T_f + 2\tau)}, & \text{if } WT_f \prec 2\tau + T_f \end{cases}$$

Using Eq. 4.5,

$$\eta = \begin{cases} 1, & \text{if } W \succ 2a + 1 \\ \dfrac{W}{(1 + 2a)}, & \text{if } W \prec 2a + 1 \end{cases} \tag{4.6}$$

*The reader should refer to Section 4.5.3 for a description of the sliding-window protocol

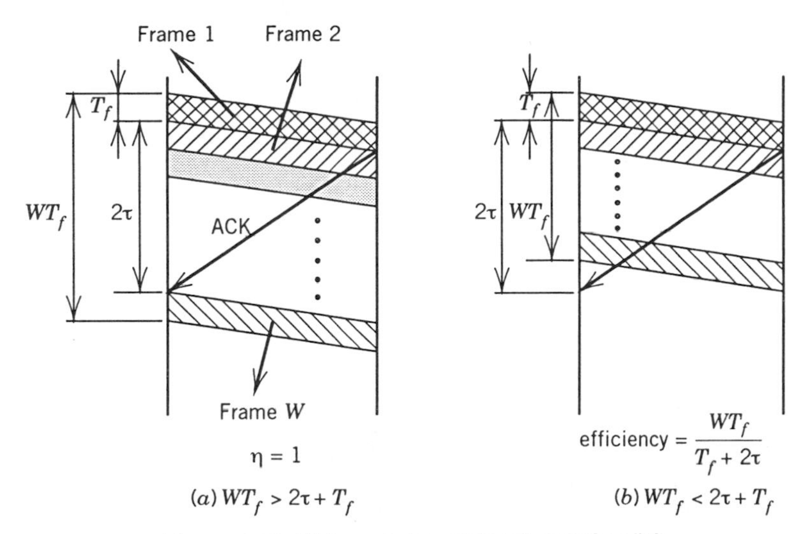

Figure 4.20. Sliding-window protocol, error-free link.

Equation 4.6 assumes error-free operation. Now we investigate the effect of error on sliding-window protocols. We need to address the two schemes; selective repeat ARQ, and go-back-N ARQ cases. When N_f is the average number of transmissions needed by one frame to be successful, the protocol throughput η is then given by

$$\eta = \begin{cases} \dfrac{1}{N_f}, & \text{if } W \succ 2a + 1 \\ \dfrac{W/N_f}{(1 + 2a)}, & \text{if } W \prec 2a + 1 \end{cases} \tag{4.7}$$

Now we need to determine N_f for both selective-repeat and go-back-N ARQ schemes.

For the selective-repeat ARQ, every frame is transmitted on the average $1/(1 - \mathrm{P})$ times, as given in Eq. 4.2; and W frames are transmitted on the average $W/(1 - P)$ times. Hence, Eq. 4.7 becomes

$$\eta = \begin{cases} (1 - P), & \text{if } W \succ 2a + 1 \\ \dfrac{W(1 - P)}{(1 + 2a)} & \text{if } W \prec 2a + 1 \end{cases} \tag{4.8}$$

For go-back-N, a corrupted frame requires a retransmission of N frames, $0 \preccurlyeq N \preccurlyeq W$. Thus, a given frame requires one transmission with probability $(1 - P)$, $(N + 1)$ transmissions with probability $P(1 - P)$, or $(2N + 1)$ transmissions with probability $P^2(1 - P), \ldots,$ or $(iN + 1)$ transmissions with probability $P^i(1 - P)$, and so on.

Thus, the number of transmissions needed for one successful frame N_f is given by

$$\begin{aligned} N_f &= 1 \cdot (1 - P) + (N + 1)P(1 - P) + (2N + 1)P^2(1 - P) \\ &\quad + \cdots + (iN + 1)P^i(1 - P) + \cdots \\ &= \sum_{i=0}^{\infty} (iN + 1)P^i(1 - P) \\ &= 1 + \frac{NP}{1 - P} \end{aligned}$$

Now we proceed to determine N. When $WT_f \succ 2\tau + T_f$, after 2τ seconds the sender receives an ACK or NAK, thus $NT_f \simeq T_f + 2\tau$, leading to $N = 1 + 2a$ and thus $N_f = (1 + 2aP)/(1 - P)$. When $WT_f \prec 2\tau + T_f$, $N = W$ (this is the worst case); thus for go-back-N and using Eq. 4.7, we get

$$\eta = \begin{cases} \dfrac{(1 - P)}{(1 + 2aP)}, & \text{if } W \succ 2a + 1 \\[2ex] \dfrac{W(1 - P)}{(2a + 1)(1 - P + WP)}, & \text{if } W \prec 2a + 1 \end{cases} \tag{4.9}$$

4.8 X.25

Public packet-switching networks around the world use CCITT recommendation X.25, which is the standard interface between packet network equipment, called a DCE (Data Circuit-terminating Equipment) by CCITT, and user devices, called DTEs (Data Terminal Equipment). CCITT recommendation X.25 was first approved in March, 1976 and was revised in 1980, 1984, and 1988.

The X.25 interface (Fig. 4.21) between the DTE and the DCE specifies the first three levels of the OSI model, namely, the physical level (layer 1), the frame level (layer 2), and the packet level (layer 3).

The physical layer specifies the use of a duplex, point-to-point synchronous circuit, thus providing physical transmission path between the DTE

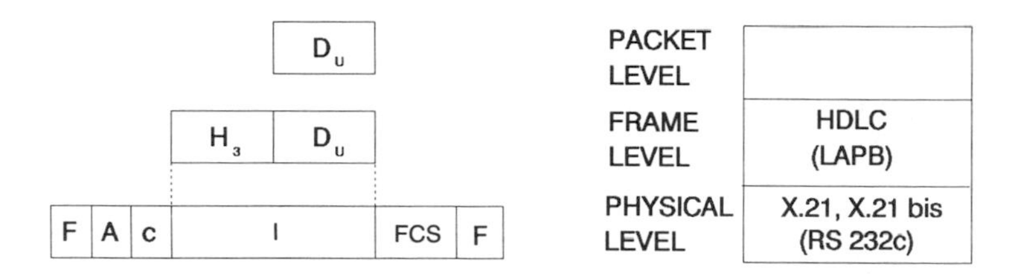

D_U : Data Unit
H_3 : Layer 3 Header (X.25 Packet Header)
F : flag A : address C : Control FCS : Frame Check Sequence

Figure 4.21. Structure of X.25 interface.

and the network. It specifies the use of X.21 which is used mostly in Europe. It also specifies the use of the V.24 physical interface (i.e., the EIA RS232-D standard).

The frame level (data link layer) specifies the use of the balanced link access procedure (LAP-B), which is a subset of HDLC. The packet level (network layer) is the highest level of the X.25 interface and specifies the manner in which control information and user data are structured into packets. The control information, including addressing information, is contained in the packet header field and allows the network to identify the DTE for which the packet is destined. It also allows a single physical circuit to support communications to numerous other DTEs concurrently.

4.8.1 Network Services Available to X.25 DTEs and Logical Channels

The X.25 interface recommendation provides access to the following services that may be provided on public data network.

1. Virtual call (VC), also called switched virtual call (SVC).
2. Permanent virtual circuit (PVC).
3. Fast select call.

A *virtual circuit* is a bidirectional transparent, flow-controlled path between a pair of logical or physical ports. A permanent virtual circuit is a permanent association existing between two DTEs, which is analogous to a point-to-point private line. Thus, it requires no call setup or call clearing action by the DTE. A switched virtual circuit is a temporary association between two DTEs and is initiated by a DTE signaling a call request to the network. The fast-select call provides for the exchange of up to 128 bytes of data while the call is established and clearing procedures for virtual call service. Fast-select effectively extends the capability of virtual call service to satisfy more transaction-oriented applications where at least one inquiry and one response is needed for a communication. When the fast-select facility is activated for a virtual call, the call request packet, to be discussed below, contains a facility request indicating fast-select with one of two possible parameters.

4.8.2 X.25 Packet Formats

Figure 4.22 shows the general X.25 packet format. The minimum packet header is 3 bytes. There are two main types of packets: data packets and control packets. The first bit in the third byte (octet) of the packet header distinguishes between a data packet (D = 0) and a control packet (C = 1).

4.8.2.1 Data Packets. Data packets contain the user's information. Figure 4.23 shows the data packet format. The data packet header consists of three bytes described as follows;

General Format Identifier (GFI). The GFI consists of four bits (bits 4, 5, 6, and 7 in Byte 1) described as follows;

The qualifier Bit (Q-bit, bit 7). The Q-bit distinguishes between a packet containing qualified data (i.e. user information, $Q = 0$) and one containing control information ($Q = 1$). It is particularly useful when the DTE is connected to the network via packet assembler/disassembler.

The delivery confirmation bit (D-bit, bit 6). When the D-bit is set to zero, then flow control and delivery confirmation information are conveyed locally

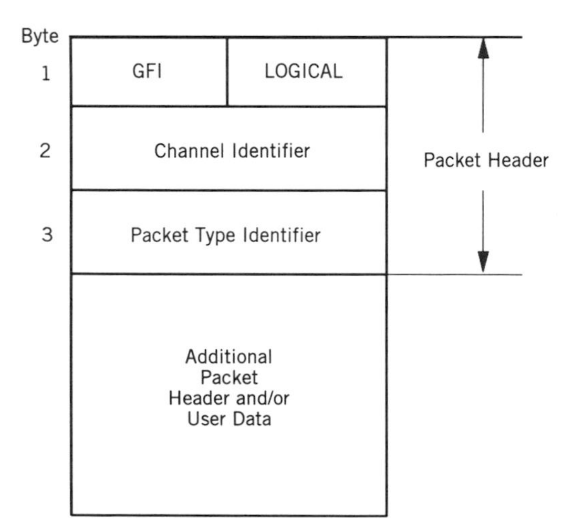

Figure 4.22. General X.25 packet format.

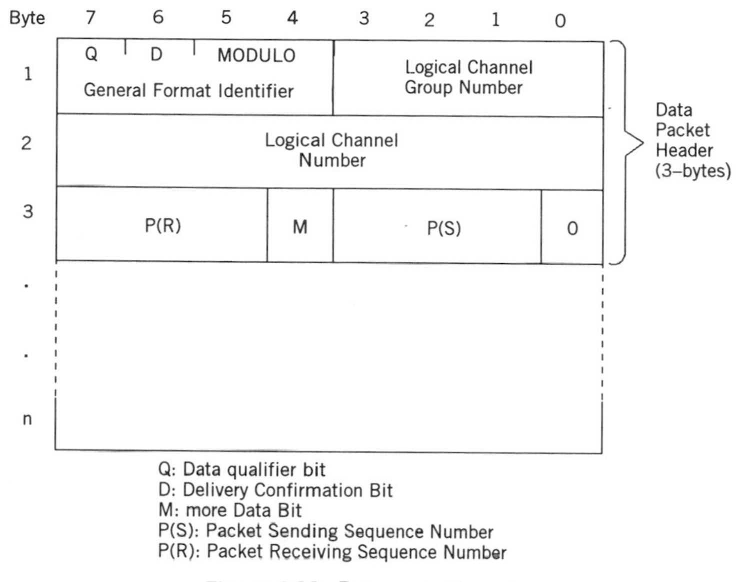

Figure 4.23. Data packet format.

(i.e., between DTE and the local DCE). When the *D*-bit is set to 1, flow control and delivery information are conveyed end-to-end (i.e., between DTE and DTE).

Modulo bits (bits 4 and 5). The modulo bits provide information about the packet sequence numbering. The sequence numbers are modulo 8 (i.e., the packet count will vary between 0 and 7) if bits 5 and 4 are set to binary 01. The packet sequence numbers are modulo 128 if bits 4 and 5 are set to binary 10. Most public switched data networks (PSNs) support a modulo 8 packet sequence.

Logical channel numbers (LCNs). X.25 uses LCN, or a logical channel identifier, to identify the DTE connections in the network. Thus, each packet contains a logical channel number that identifies the packet with a switched or permanent virtual circuit for both directions of transmission.

The LCN consists of a logical channel group number, (4 bits), and an LCN of 8 bits. This gives a total of 12 bits with a possible maximum of 4095 logical channels on the same physical interface line. The range of LCNs that can be used by a customer for virtual circuits is assigned at subscription time by the network administration.

Logical channel numbers for VCs are dynamically assigned (within the allocated range) during call establishment and identify all packets (i.e., control and data) associated with the VC. The LCNs are only significant at a

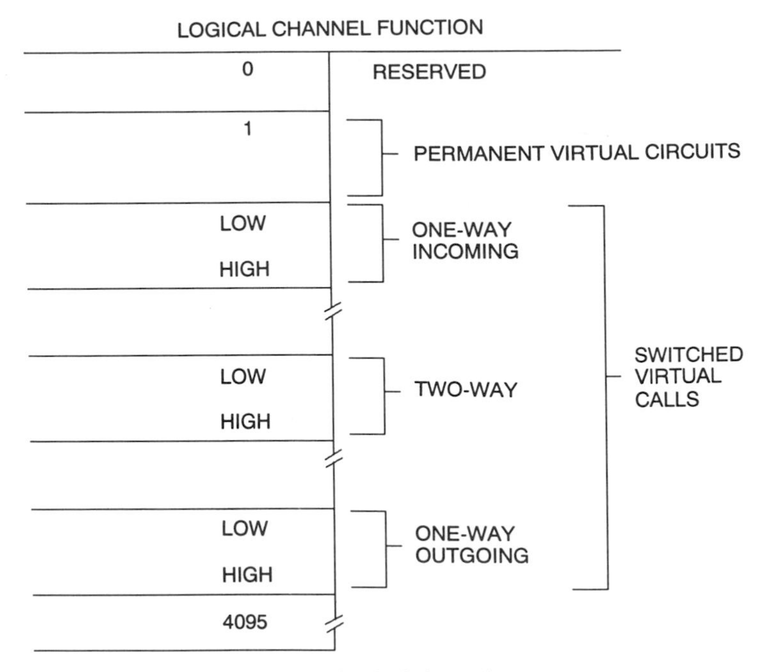

Figure 4.24. Logical channel ranges.

particular DTE/DCE interface. Figure 4.24 shows the logical channel ranges. The number zero is reserved and permanent virtual circuits (PVCs) are allocated LCNs starting from LCN 1, followed by VC assignment. Virtual calls can be divided into three categories; one-way incoming, two ways, and one-way outgoing.

Thus, X.25 allows a DTE to establish concurrent virtual circuits with a number of DTEs over a single physical access circuit. In effect, the X.25 packet level acts as a packet-interleaved statistical multiplexer (see Fig. 4.25).

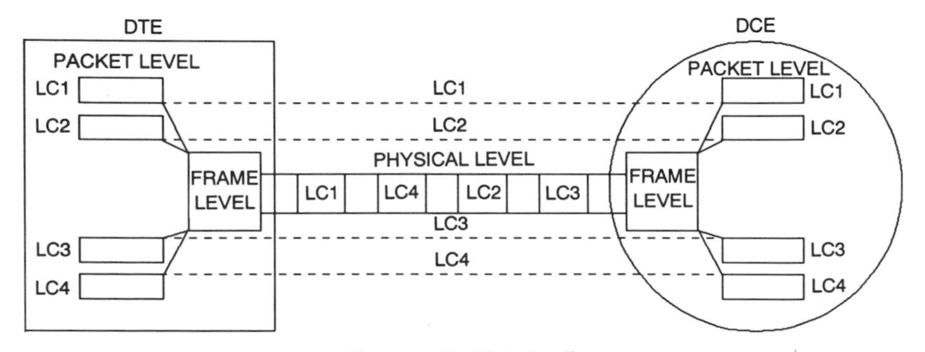

Figure 4.25. Virtual calls.

Sending and receiving sequence numbers. The third byte in the data packet header contains the sending and receiving sequence numbers $P(S)$ and $P(R)$. Figure 4.23 shows $P(R)$ and $P(S)$ fields for modulo 8 packets.

The more bit (M-bit). The M-bit identifies contiguous data transmission across data packets. Public switched networks have a maximum limit on the data packet size. Long messages are broken into a number of packets. Each packet except the last one would have the more bit on ($M = 1$).

The user data field length is variable according to the data it contains. Its maximum size is agreed between the subscriber and the network at subscription time, typically 128 bytes.

4.8.2.2 Control Packets. There are many types of control packets. The packet-type identifier distinguishes one control packet from another. Control packets can be classified into six groups; call setup, flow control, supervisory, confirmation, diagnostic, and interrupt.

Call Setup Packets. This group includes four types of call setup packets; CALL REQUEST, INCOMING CALL, CALL ACCEPTED, and CALL CONNECTED. These packets are used during the switched virtual circuit call setup phase. Figure 4.26*a* shows the call setup packet format. For example, the CALL REQUEST packet includes the LCN chosen by the DTE to be used to identify all packets associated with the call. It also includes the calling DTE and the called DTE address.

Some useful functions are available to the end user to enhance the call's quality of service. These functions are called facilities and are indicated by the user in the facility field of the call setup packets. The user requests the facilities at subscription time and network administration allocates the required network resources when needed for a certain user's call. The optional user facilities include the following:

1. The closed-user-group facility allows *a* group of users to communicate with each other, but precludes communication with all other users for security reasons. A user can belong to more than one closed user group.
2. Flow control parameter selection allows negotiation on a per-call basis for the packet size and the window size for each direction of transmission. The maximum allowable packet sizes are 16, 32, 64, 128, 256, 512, and 1024 octets. The window size can be 1, 2, 3, 4, 5, 6, or 7. The default values are 128 bytes for the packet size and 2 for the window size. For PVCs, the flow control parameters are established at subscription time.

3. The throughput class-negotiation facility permits the negotiation on a per call basis of the throughput classes. Because of the statistical sharing of transmissions and switching resources, the throughput class cannot be guaranteed 100 percent of the time. The throughput class may vary from 75 bps to 48 kbps.
4. The logical channel ranges facility allows the use of a range of logical channels to one-way PVCs, outgoing calls, one-way incoming calls, or two-way calls as shown in Figure 4.24.
5. The reverse charging facility allows acceptance and generation of reverse charge calls. The user data may follow the facility field and may contain up to a maximum of 16 octets.

BITS 8 7 6 5	4 3 2 1	
GFI	LOGICAL	OCTETS 1
CHANNEL IDENTIFIER		OCTETS 2
PACKET TYPE IDENTIFIER		OCTETS 3
CALLED DTE ADDRESS LENGTH	CALLED DTE ADDRESS LENGTH	OCTETS 4
CALLED DTE ADDRESS		
CALLING DTE ADDRESS		
00	FACILITY LENGTH	
FACILITY FIELDS		
USER CALL DATA (16 OCTETS MAXIMUM) ONLY IN CALL REQUEST AND INCOMING CALL PACKETS		

(*a*) CALL SET-UP PACKETS

GFI	LOGICAL	OCTETS 1
CHANNEL IDENTIFIER		OCTETS 2
P(R)	PACKET TYPE IDENTIFIER	OCTETS 3

(*b*) FLOW CONTROL PACKETS

GFI	LOGICAL	OCTETS 1
CHANNEL IDENTIFIER		OCTETS 2
PACKET TYPE IDENTIFIER		OCTETS 3
CAUSE		OCTETS 4
DIAGNOSTIC CODE		OCTETS 5

(*c*) SUPERVISORY PACKETS

Figure 4.26. X.25 control packets.

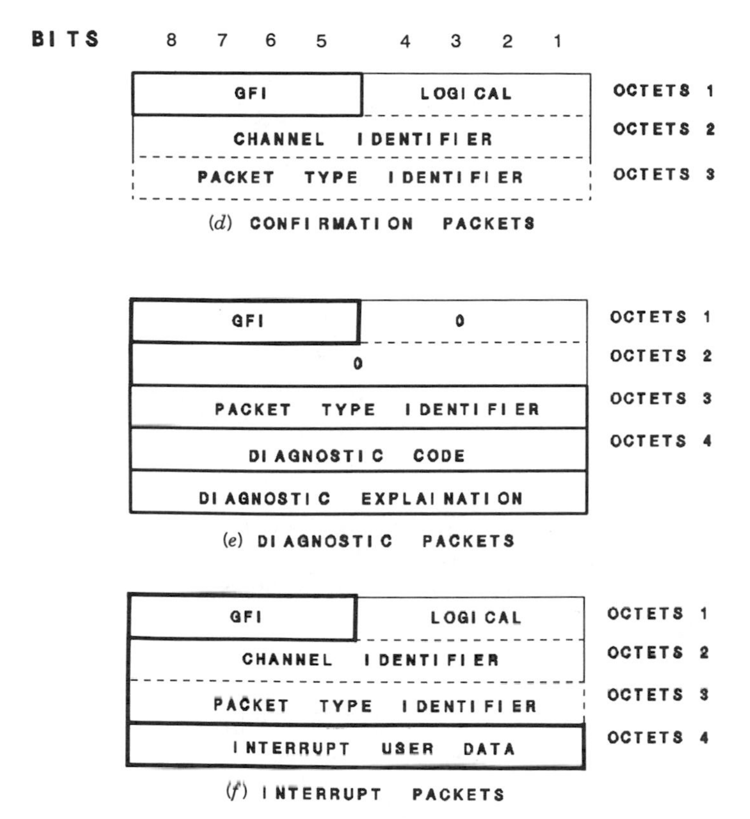

Figure 4.26. (*Continued*)

Flow Control Packets. This group includes three types of flow control packets: RECEIVE READY (RR), RECEIVE NOT READY (RNR), and REJECT (REJ) (Fig. 4.26*b*). These packets are used only during the data transfer phase. Each of these packets carries a packet receive sequence number, $P(R)$. The RR packet indicates readiness to receive data packets with a send sequence number, $P(S)$, equal to the value of $P(R)$ encoded in the RR packet. Thus, the RR packet provides on a given logical channel in the reverse direction, separate acknowledgment when there is no data flow to piggyback onto. The RNR packet is used by the DTE or DCE to indicate that it is temporarily unable to receive data packets with a higher send sequence number than that encoded in its $P(R)$ field. The RR packet can then be used to inform the other side to continue transmission.

The REJ packet permits a DTE to request retransmission of the data packet containing the sequence number encoded in its $P(R)$ field. Some public networks may not support the REJ packet.

Supervisory packets. This group includes the RESTART, REQUEST / INDICATION, CLEAR REQUEST / INDICATION and RESET REQUEST / INDICATION packets (Fig. 4.26*c*). The RESTART REQUEST packet is used in a disaster situation, such as a host crash, to clear all the SVCs and reset all the PVCs currently held by the DTE issuing this packet. The CLEAR REQUEST packet disconnects the virtual circuit identified by the packet's LCN. The CLEAR REQUEST packet is not used in a permanent virtual circuit.

Octet 4 of the supervisory header format is the cause field. For example, the cause field in the CLEAR INDICATION packet may indicate "number busy" or "reverse charging acceptance not subscribed". A RESET REQUEST packet is used to reset a particular send P and receive P sequence numbers to zero in the data transfer mode. The virtual circuit associated with that reset is identified in the LCN field. The diagnostic code in octet 5 is generated by the network, and it provides error information for the DTE. The issuance of a RESTART REQUEST packet is equivalent to sending a CLEAR REQUEST on all logical channels for SVCs and RESET REQUEST on all logical channels for PVCs.

Confirmation packets. This group includes four types of packets: RESTART CONFIRMATION, CLEAR CONFIRMATION, RESET CONFIRMATION, and INTERRUPT CONFIRMATION (Fig. 4.26*d*). They are used to acknowledge the execution of a previously requested action.

Diagnostic packet. This special packet is generated by the network for fault diagnosis (Fig. 4.26*e*). The diagnostic field in octet 4 indicates the reason why the packet is refused. The diagnostic explanation fields include the header of the refused packet.

Interrupt packet. The INTERRUPT packet is transmitted in the data transfer phase, and it does not contain either the sending sequence number nor the receiving sequence number (Fig. 4.26*f*). That is, the INTERRUPT packet is not subject to flow control. Only one unconfirmed INTERRUPT packet may be outstanding at a given time.

4.8.3 Establishing and Clearing a Virtual Circuit

Three phases are needed for exchanging packets (see Fig. 4.27). These phases are the call-setup phase, the data-transfer phase, and the call-clearing phase. In the *call-setup phase*, the calling DTE sends a CALL REQUEST packet. The header of this packet contains the address of the remote (called) DTE. On the other side of the network, the called DTE receives the call request in the form of an INCOMING CALL packet. If the called DTE accepts the call it transmits a CALL ACCEPTED packet, which causes the calling DTE to receive a CALL CONNECTED packet.

In the *data transfer phase*, the two DTEs on a virtual circuit can simultaneously exchange data packets. The two numbers shown in Figure 4.27 indicate the sending and receiving sequence numbers, $P(S)$ and $P(R)$, associated with each data packet.

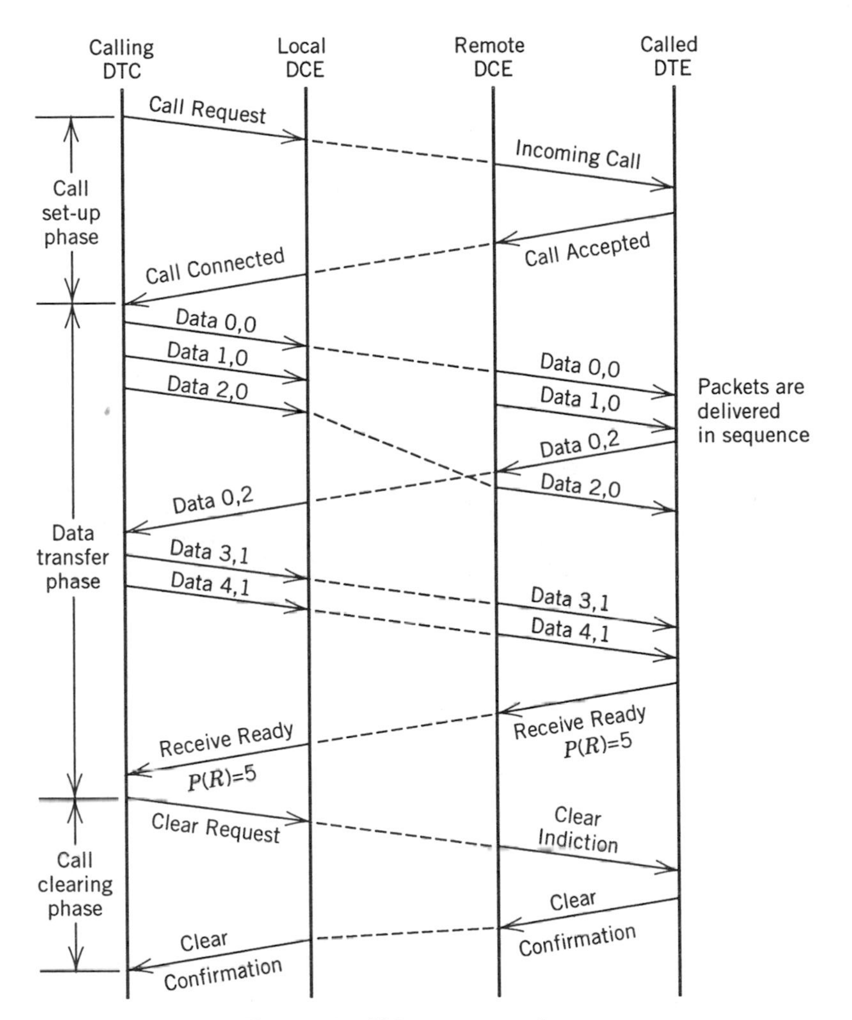

Figure 4.27. X.25 sequence of events.

The DTE (or the network in case of failure) may issue a CLEAR REQUEST to start the *call clearing phase*. The DTE that receives a CLEAR INDICATION must reply with a CLEAR CONFIRMATION. The logical channel number can be used again for another call when the clearing procedure is completed by the transmission of a CLEAR CONFIRMATION.

4.8.4 Packet Assembler and Disassembler

Remote asynchronous devices, such as terminals, printers, and plotters, do not have the capabilities to implement the three levels of X.25. Thus, standards were developed to provide protocol conversion and packet assembly and disassembly (PAD) functions for these asynchronous devices: These devices are referred to as DTE-Cs, the C signifying that they transmit and

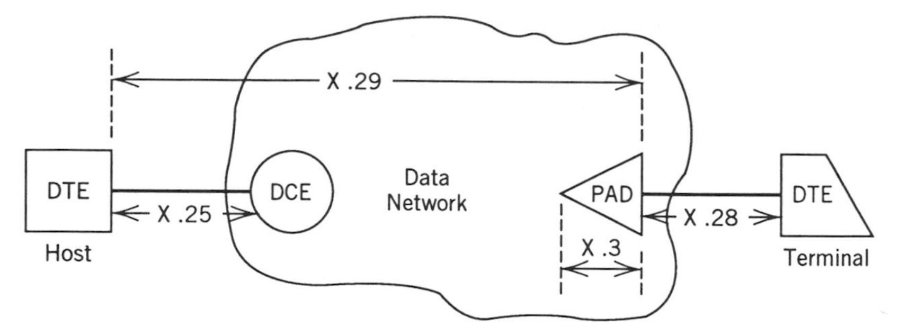

Figure 4.28. Terminal access to X.25 network.

receive characters as opposed to the packets of a standard DTE. A DTE-C connects to a data network via a translation device called a PAD. The operation of the DTE-C to PAD interface, the services offered by the PAD, and the interaction between the PAD and the host system are defined in three CCITT recommendations: X.28, X.3, and X.29, respectively (Fig. 4.28).

SUGGESTED READINGS

For an extensive presentation of bit-oriented data link protocols, see [CARL80]. For character-oriented link protocols, see [CONA80]. [LIN84] provides a comprehensive treatment of ARQ error-control schemes. [BLAC89] provides a detailed study of the data link layer, whereas [HALS92] discusses the window mechanism in details. [RYBC80] is an excellent presentation of X.25. [PROC83] is a good reference for the ISO model and X.25. [HEWL88] presents a detailed explanation of the three layers of X.25. The work in [GOPA84] describes and analyzes the use of ARQ schemes for point-to-multipoint conversations. The work in [AMMA92] describes an improvement to multipoint ARQ schemes.

PROBLEMS

4.1 Assuming a stop-and-wait protocol, describe a scenario to show the need for sequence numbers.

4.2 In go-back-N, what is the maximum size of the transmit window W_t and the receive window W_r? Assume a sequence number field n bits long. The transmit window size represents the maximum number of frames the sender is allowed to transmit before receiving an acknowledgement. Similarly, the receive window size is the number of frame sequence numbers a receiver must remember in order to detect duplicates.

4.3 Repeat Problem 4.2 for selective ARQ. Obtain a relationship between W_t, W_r, and n.

4.4 A finite-state machine is an important tool in modeling protocols. For the stop-and-wait protocol, define the status of the system as SRC, where S is the sender [$S = 1$ when the sender sends frame 1 and $S = 0$ when the sender sends frame 0]. Similarly, R is the receiver [$R = 1$ when the receiver expects to receive frame 1 and $R = 0$ when the receiver expects to receive frame 0]. C is the channel, which can be in one of four states: frame 1 is on the channel ($C = 1$), frame 0 is on the channel ($C = 0$), the acknowledgement is on the reverse channel ($C = A$), or the channel is empty ($C =$ —). Assume a half-duplex communications link connecting two stations (A and B) with A only sending traffic to B and B only sending back acknowledgment traffic.

a. Draw the finite state diagram; assume an error-free link.

b. Repeat (a) assuming that the communications link may have errors.

4.5 Figure 4.29 shows a simple protocol in which each interacting process (node) is modeled by a finite-state graph. Each process has four states (0 to 3). The initial state is identified by the state labeled 0. The message exchanger between the processes are represented by integers. Message transmission is represented by the negative value of the corresponding integer, and the message reception by its positive value. The protocol shown exhibits some possible errors. Identify these errors.

4.6 In Figure 4.30a, the primary station A communicates with the secondary stations B, C, and D using a standard bit-oriented data link protocol (HDLC, for example) on a two-way half-duplex line. Figure 4.30b, shows a typical sequence of data transmission between A and B, C, and D with the shorthand used in Section 4.6.2.

a. In Figure 4.30b, fill in $A, YN(S)N(R), P/F$.

b. What is the size in bits of the sending (or receiving) sequence number field? Why?

c. What is the window size? Why?

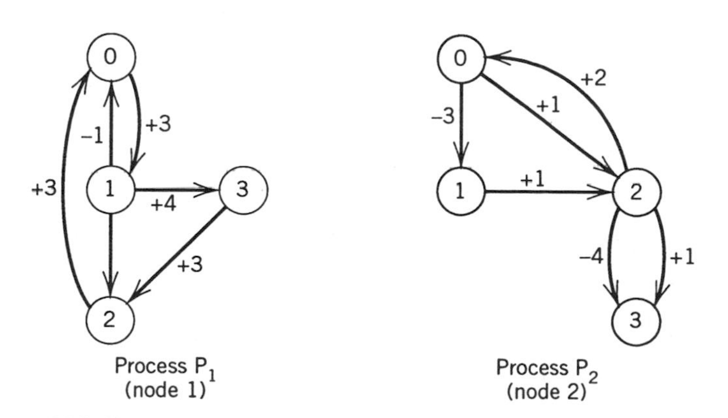

Figure 4.29. Two process-interaction protocols containing various design errors.

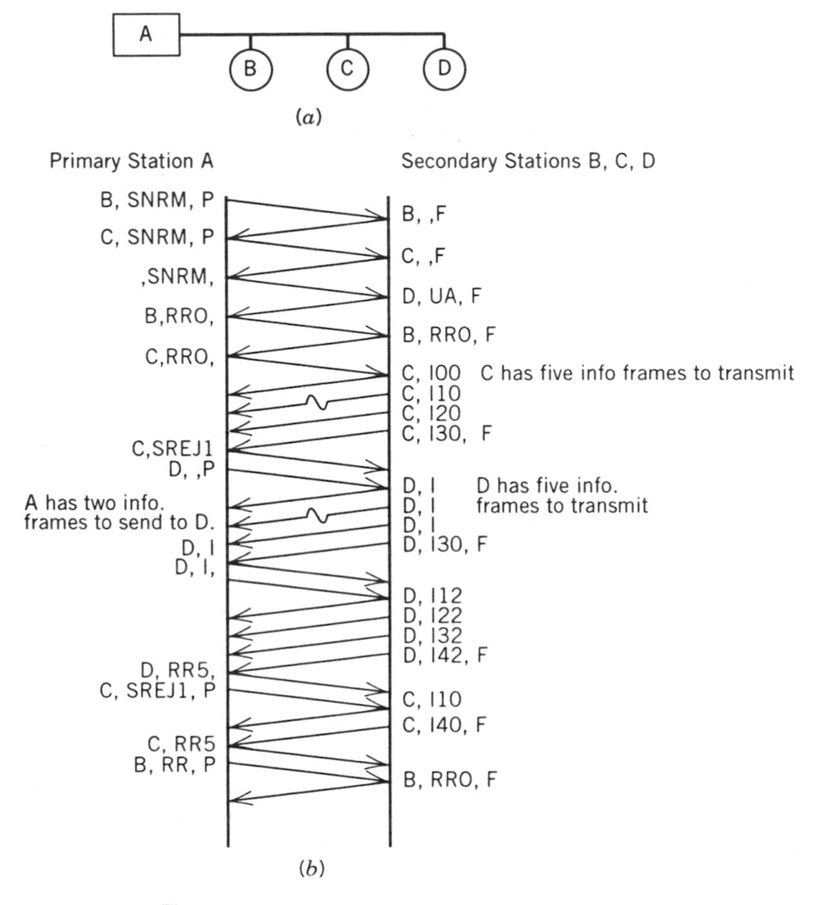

Figure 4.30. Problem 4.6, ~ represents frame error.

4.7 The following events occur between the primary station A and the two secondary stations B and C on a multidrop error-free half-duplex line using HDLC protocol: *Events* e_1, e_2, e_3, e_4, where

e_1 = A activates the link with B and C using a normal response mode

e_2 = A polls B for traffic, B responds by sending four I frames, then A only acknowledges B without granting B additional further transmission rights

e_3 = A polls C for traffic and C only acknowledges A

e_4 = A sends three frames to B and grants B the right to transmit. B responds by sending five additional frames and A acknowledges

a. Show the frames exchanged between the primary station A and the two secondary stations (B and C).

b. Now assume that the noise corrupted the transmission of the first frame out of the five frames sent out by B to A in event e_4. Show two possible procedures for error recovery. Also, assume that the window size is seven.

Note: Use the abbreviations $A, YN(s)N(R), P/F$ to describe a frame where A is the address field, Y is the type of frame, P/F is the poll/final bit, $N(s)$ is the sending sequence number, and $N(R)$ is the receiving sequence number.

4.8 **a.** Apply the bit stuffing rules to the following:

0110111110011111101011111111110

b. Suppose the following string of bits is received:

0111111011111011011111001111100000111110010111111 0

Remove the stuffed bits and show where the actual flags are.

4.9 Consider a 1-Mbps satellite channel. Frame length is 1000 bits and the probability of an unsuccessful frame transmission is p. Assume that the errors on different frames are independent. Neglect acknowledgment transmission time and receiver processing time and assume a sliding window protocol, using the selective-repeat ARQ method of error recovery.

a. What is the average number of transmission per frame N_f? Assume $p = 0.01$.

b. What is the protocol efficiency η? Assume a window size $W = 4$.

4.10 Draw the X.25 packet format for the following packets using modulo 128:

a. Data packets

b. Flow control packets

4.11 A user message is 310 bytes. The PSN can support the X.25 data packet size with a maximum user data field of 128 bytes. The PSN is modulo 8 at the network layer. Draw the X.25 data packets for this message. Show the modulo field, M-bit, $P(s)$, and the data user field values.

4.12 Explain why only one unconfirmed INTERRUPT packet may be outstanding at a given time.

4.13 Two users (U_1 and U_2) have established a virtual circuit connection to an X.25 network through their host H. The timing diagram shows the packets' arrival to the network layer, where P_{ij} stands for the jth packet from the ith user ($i = 1, 2$). The network layer inserts, among other parameters, the logical channel number (LCN) and the sending sequence number $P(s)$ in the network layer header. Assume that for U_1 its LCN is 5 and for U_2 its LCN is 17. Then all packets are multiplexed and passed to the data link layer which, in turn, inserts, among other parameters, the sending sequence numbers $N(s)$ in the frame header.

Draw a timing diagram showing the packets traveling on the interface line between the host H and the network. Show the values for $N(s)$, LCN, $P(s)$, in that order, for each packet.

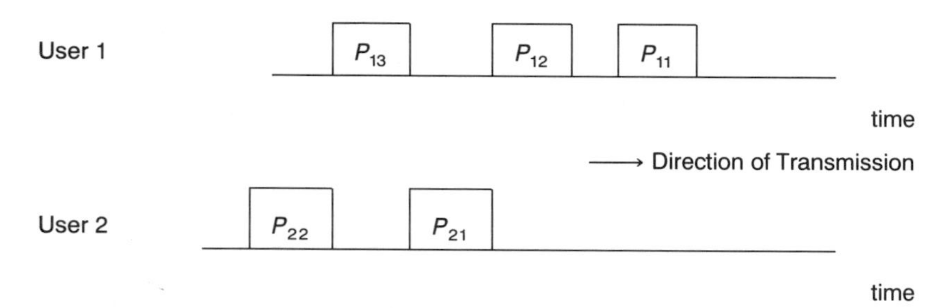

Timing diagram for the packets' arrival to the network layer.

5

ROUTING IN PACKET SWITCHED NETWORKS

5.1 INTRODUCTION

The routing of packets to their intended destinations is perhaps the most basic service provided by a packet switched network (PSN). The routing functions described in this chapter belong to the network layer. Before describing the mechanisms used to determine routes in a PSN, we first give an overview of the fundamental principles underlying the operation of routing procedures.

5.1.1 Virtual Circuit and Datagram Routing

Packets entering a PSN are first presented to an entry node by an attached host. Typically, such a packet needs to be delivered to another host attached to an exit node (see Fig. 5.1). The source–destination host pair are said to use the services of the subnetwork (i.e., the collection of switching nodes) to reach their goal of end-to-end communication. Two types of services may be provided by the subnetwork:

1. *Connection-oriented service.* A connection needs to be established between the source and destination hosts before data can be transferred. The connection also needs to be explicitly torn down at the end of data transfer. All data appears to be carried over an end-to-end logical pipe. In particular, sequenced delivery of packets to the destination host is guaranteed.
2. *Connectionless service.* No connection setup is required. No guarantee of sequenced or reliable delivery is made.

It should be evident to the reader that reliable and sequenced delivery (a connection-oriented service) is a requirement at some level for almost all

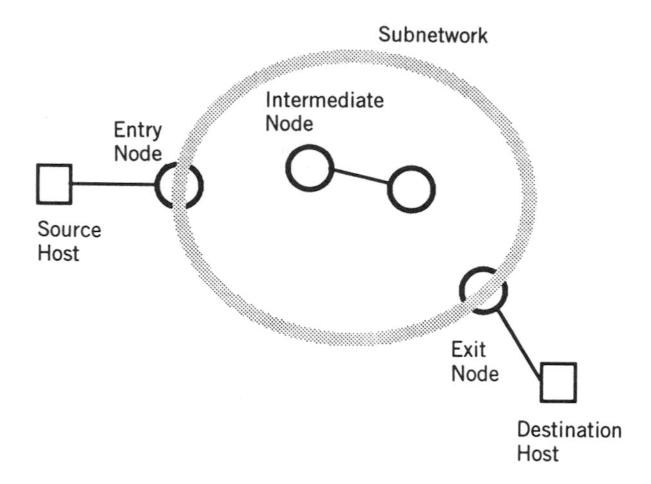

Figure 5.1. Nodes and hosts in a packet switched network.

data communication (except perhaps the most elementary) between any two entities. In cases where the subnetwork does not provide such a service, the source–destination hosts are burdened with the task of building it. This is typically done at the transport layer. If the subnetwork already provides a connection-oriented service, then the communication tasks at the end hosts become simpler.

As in most communication tasks, the details of a service are independent of how that service is provided. A subnetwork may provide its connection-oriented or connectionless service by either using virtual circuits or datagrams internally to route packets between entry and exit nodes. The terms *external* virtual circuit and *external* datagram are also sometimes used to describe connection-oriented and connectionless service, respectively.

When datagram routing is used, packets are routed from entry to exit node independently of each other. Consequently, each packet may follow a different path through the network and packets may arrive at the exit node out of sequence. Also, packets transmitted by an entry node may never reach the exit node. With virtual circuit routing, a connection between the entry and exit nodes is set up. As part of the connection setup procedure, a path through the network is selected. This path is followed by all packets traveling from the entry node to the exit node and belonging to the virtual circuit. Since all packets follow the same path, reliable and sequenced arrival of packets to the exit node may be guaranteed when virtual circuit routing is used.

A subnetwork that uses datagrams internally naturally provides connectionless service to its attached hosts. It can, however, also provide connection-oriented service if the exit node is capable of resequencing out-of-order packets and requesting any missing packets from the entry node. A special entry-to-exit node protocol is required for such a purpose.

When virtual circuit routing is used internally, the subnetwork can provide connection-oriented service in a straightforward manner. It does not make much sense to have a network that uses virtual circuit routing provide a connectionless service; all the work done by the subnetwork to achieve virtual circuit routing is wasted in such a situation.

5.1.2 A Typical Switching Node

Figure 5.2 shows a typical switching node in a PSN. Packets enter and leave the node via a set of incoming and outgoing links as shown. For clarity, we logically separate incoming and outgoing links, although they will typically coexist on one physical connection.

As packets enter the node, they are examined by the node's central processing unit (CPU), which performs checks on the packet. (Sometimes some of the work is done by a peripheral line processor.) Included in such processing is an examination of the packet's network layer destination address. Based on this, the packet is scheduled for transmission on the appropriate outgoing link by placing it in the corresponding queue. The process by which a node selects the appropriate outgoing link is known as the *routing function*.

A packet's delay through a switch has three major components: queueing time in the CPU and outgoing link queues, CPU processing time, and packet transmission time. This latter component is a function of the packet length and the data rate of the outgoing link. When the link data rates are low (e.g., 56 kbps), the transmission time of the packet dominates the required CPU processing time, which can be considered negligible. In modern packet switching networks, fiber optic links capable of fast data transmission (100s of Mbps) are used. This reduces a packet's transmission time by orders of

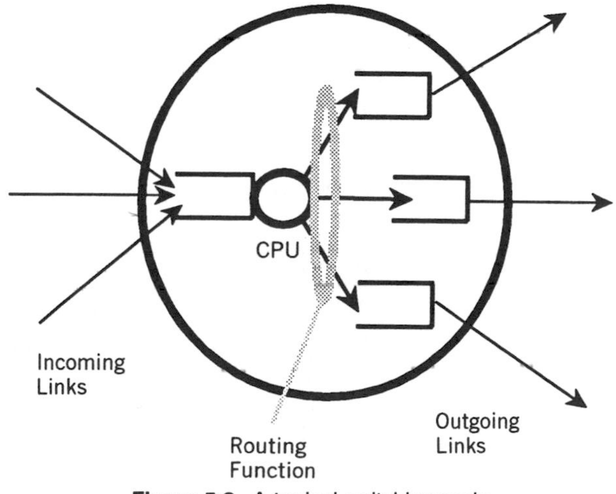

Figure 5.2. A typical switching node.

magnitude, making the CPU processing time (which is not affected by the increase in data rate) a more significant component of a packet's delay. Thus routing in these higher speed networks attempts to minimize the CPU processing required.

Two basic approaches are used to implement the routing function:

1. *Table-driven routing*. This is the most popular approach and requires each node to store and maintain a routing table, which contains an association between a packet's identification (ID) and an outgoing link. The packet's ID can be the packet's destination address, a combination of the packet's source and destination or an indication of the virtual circuit to which the packet belongs. Determination of the appropriate outgoing link involves the examination of a packet's header to extract the packet ID followed by a search of the routing table to determine the outgoing link. Routing tables need to be initialized and updated and may require extensive storage capacity in large networks.
2. *Table-free routing*. This approach is used when maintaining routing tables is not desirable. For example, the use of high-speed links requires that the CPU processing required for each packet be minimal; thus looking up and maintaining long tables is not feasible. Routing decisions for each packet are made without consulting a table. Different techniques are possible, some of which are described later in this chapter. With these techniques, the overhead for maintaining and storing the tables is avoided.

5.1.3 Determining the Best Path

Generally, it is desirable to route packets over the best possible path available. An important aspect of the design and implementation of routing procedures is the criteria used for path selection. A straightforward approach is always to select a path that will carry a packet to its destination in the least amount of time. Although minimizing packet delay is a noble goal, it is easier said than done.

A packet's delay is comprised of two main components:

1. *Packet transmission time* is the sum of the time required to transmit the packet (including propagation delays), which is a function of the packet's length, the data rates, and the length and type of the links on which the packet travels.
2. *Queueing and processing time* includes the time spent awaiting transmission on an outgoing link and the time spent processing and queueing at the switch's CPU. (This latter component is normally negligible in a traditional low-speed long haul network.)

As a packet starts its journey on a known path, the only predictable component of its delay is its transmission time. Queueing and processing times can only be estimated, because they are strongly dependent on traffic conditions in the network and tend to vary over time.

Most routing algorithms attempt to route a packet over the best guess of the shortest path from the source to the destination. This is done by assigning fixed or variable costs to links in the network and performing a shortest-path calculation. The results of this calculation are reflected in the routing tables at each node.

Different approaches exist to determining a link's cost. For example, each link may be assigned:

a unit cost in which case packets are routed over a minimum hop path

a cost that is inversely proportional to the link data rate, which is an easy way to account for differences in transmission time due to varying link data rates

a cost that is equal to the average delay experienced by packets traveling on the link (including queueing delays); this average may be taken over some predetermined time interval

two costs, that is, a low cost when the queue at the node is below a certain threshold and a high cost if the queue grows beyond that level

When virtual circuit routing is used inside a network, a path is chosen at circuit setup time and used throughout the connection lifetime. Although the path may provide minimum delay when chosen, there is no guarantee that packets using the path will continue to experience little delay throughout. When datagram routing is used, a routing decision may be made for each packet, and thus the ideal of achieving minimum delay for each packet is closer to fulfillment.

5.1.4 Classification of Routing Procedures

Static and Dynamic. As mentioned previously, a shortest-path calculation is typically performed using the costs assigned per link and producing the shortest paths reflected in the routing tables. If the shortest path calculation is performed often (e.g., 10 times an hour) and is based on some real-time measurement of network conditions, the routing procedure is said to be *dynamic*. Otherwise, it is known as a *static* procedure. It should be emphasized that routing tables change even when a static procedure is used. This, however, happens less frequently (e.g., once a week) and typically is based on long-term averages of network conditions.

Centralized and Distributed. In a *centralized* routing procedure, a central site is in charge of computing the shortest paths in the network. If the procedure is dynamic, each node needs to report periodically the status of its links to the central site, which, in turn, needs to provide periodically new routing tables to all the nodes. Centralized routing procedures are extremely vulnerable to the central site's failure, and thus its capabilities need to be replicated for reliability purposes.

In a *distributed* routing procedure, all network nodes are involved in shortest-path calculations. As a node typically possesses direct knowledge of only its local links, a distributed procedure needs to somehow pool the information available at each node to perform the distributed computation. In the next section, we present two examples of routing algorithms suitable for a centralized (either dynamic or static) implementation. This is followed by a description of two dynamic distributed procedures.

5.2 SHORTEST-PATH ROUTING PROCEDURES

As mentioned above, a routing procedure is used by a PSN to determine the shortest path. We input the link costs and obtain routing tables that show the shortest paths from every source node to every destination node. We start our discussion with two algorithms that can be used by a centralized routing procedure to determine shortest paths. Next we describe two distributed routing procedures that are based on the centralized algorithms.

5.2.1 Centralized Procedures

A centralized routing procedure is depicted in Figure 5.3. The link costs are provided to the central site as input to its shortest-path calculation, which results in routing tables provided to the network nodes. The link costs and routing tables are typically transmitted as ordinary packets to the central site, which is a host connected to the network. How often the shortest-path

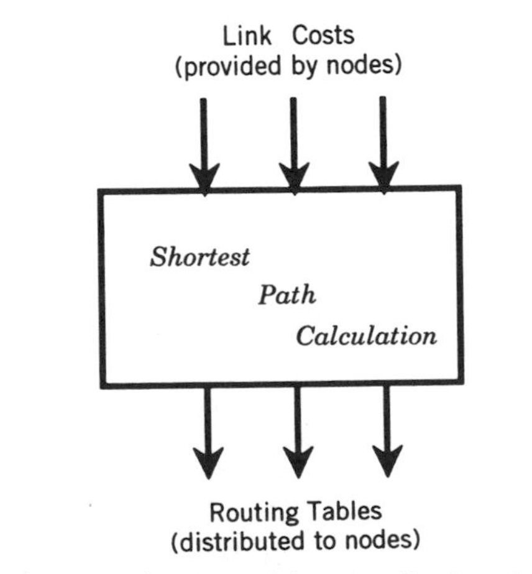

Figure 5.3. Inputs and outputs of the centralized routing procedure.

$\mathcal{N} := \{s\}$
For all nodes $v \neq s$
begin
    $\mathcal{C}_{sv} := c_{sv}$
    If $\mathcal{C}_{sv} < \infty$ then $\mathcal{P}_{sv} := (s, v)$
End.
Do while ($\mathcal{N}$ does not contain all nodes)
    Find $w \notin \mathcal{N}$ for which $\mathcal{C}_{sw} = \min_v \mathcal{C}_{sv}$
    $\mathcal{N} := \mathcal{N} \cup \{w\}$
    For all $v \notin \mathcal{N}$
    Begin
        temp $:= \mathcal{C}_{sv}$
        $\mathcal{C}_{sv} := \min[\mathcal{C}_{sv}, \mathcal{C}_{sw} + c_{wv}]$
        If $\mathcal{C}_{sv} <$ temp then
            $\mathcal{P}_{sv} := \mathcal{P}_{sw} \| v$
    End
End

Figure 5.4. Algorithm for finding the shortest forward path tree.

calculation is performed determines whether the procedure is static or dynamic.

We describe two algorithms that can be used to perform the shortest-path calculation. In the description to follow we use the following notation.

c_{vw} is the cost of the v-to-w link ($c_{vw} = \infty$, if no link exists from node v to node w)

$\mathcal{C}_{vw}$ is the cost of the best-known path from node v to node w ($\mathcal{C}_{vw} = \infty$, if no path is currently known)

$\mathcal{P}_{vw}$ is an ordered list of nodes describing the currently known shortest path from v to w. For example, $\mathcal{P}_{vw} = (v, n_1, n_2, w)$ means that the shortest path goes from v to w via n_1 and n_2 in that order.

5.2.1.1 Shortest Forward Path Tree. The shortest-path algorithm described here is generally attributed to E. Dijkstra. It determines the shortest paths from a source node to all other nodes in the network. We call the collection of such paths (which forms a tree) the *shortest forward path tree*.

The algorithm's formal description is given in Figure 5.4. It maintains a set $\mathcal{N}$ of nodes to which it has determined shortest paths. The set is initialized to $\{s\}$, the source node. Nodes are added to the set until it contains the complete list of nodes. At any point, the algorithm has determined the shortest paths for nodes that are included in the set $\mathcal{N}$. For all other nodes, the algorithm determines shortest paths that can be constructed by concatenating one hop to a shortest path that has already been determined. The node, not in $\mathcal{N}$, with the best such path is then added to $\mathcal{N}$. The process continues until no more nodes remain. Note that the central site needs to run this computation once for every node to determine the shortest paths from every node.

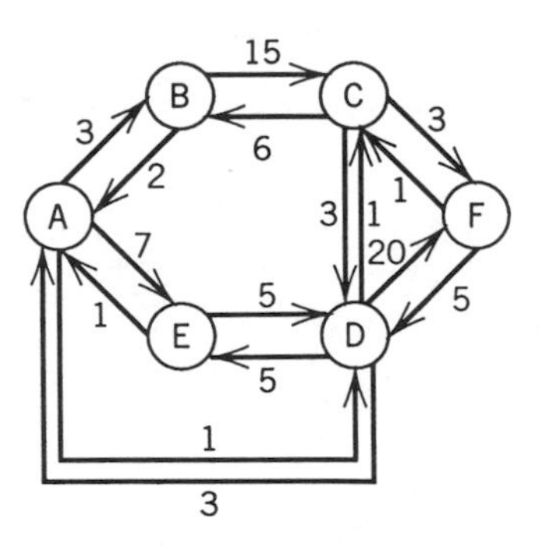

Figure 5.5. Example of a network.

TABLE 5.1 Illustration of Shortest Forward Path Tree Calculation

Iter.	$\mathscr{N}$	$\mathscr{P}_{AB}$	$\mathscr{C}_{AB}$	$\mathscr{P}_{AC}$	$\mathscr{C}_{AC}$	$\mathscr{P}_{AD}$	$\mathscr{C}_{AD}$	$\mathscr{P}_{AE}$	$\mathscr{C}_{AE}$	$\mathscr{P}_{AF}$	$\mathscr{C}_{AF}$
0	{A}	(A, B)	3	—	∞	(A, D)	1	(A, E)	7	—	∞
1	{A, D}	(A, B)	3	(A, D, C)	2	(A, D)	1	(A, D, E)	6	(A, D, F)	21
2	{A, D, C}	(A, B)	3	(A, D, C)	2	(A, D)	1	(A, D, E)	6	(A, D, C, F)	5
3	{A, D, C, B}	(A, B)	3	(A, D, C)	2	(A, D)	1	(A, D, E)	6	(A, D, C, F)	5
4	{A, D, C, B, F}	(A, B)	3	(A, D, C)	2	(A, D)	1	(A, D, E)	6	(A, D, C, F)	5
5	{A, D, C, B, F, E}	(A, B)	3	(A, D, C)	2	(A, D)	1	(A, D, E)	6	(A, D, C, F)	5

As an example, consider the application of the algorithm to the network shown in Figure 5.5. The number on each link indicates the cost of the link in the indicated direction. As more nodes are explored in subsequent iterations, lower cost paths are discovered for some destinations. Table 5.1 shows the running of the algorithm that terminates in five iterations. (The algorithm has actually discovered all shortest paths by the second iteration.) For example in the second iteration, nodes A, D, and C have been explored as possible next nodes on the path to all destinations. The indicated costs and paths reflect this fact. The resulting shortest forward path tree for node A is shown in Figure 5.6.

5.2.1.2 Shortest Backward Path Tree. We now describe an algorithm that can be used to determine the shortest paths from all sources to a single destination. As in the previous section, the collection of these paths forms a

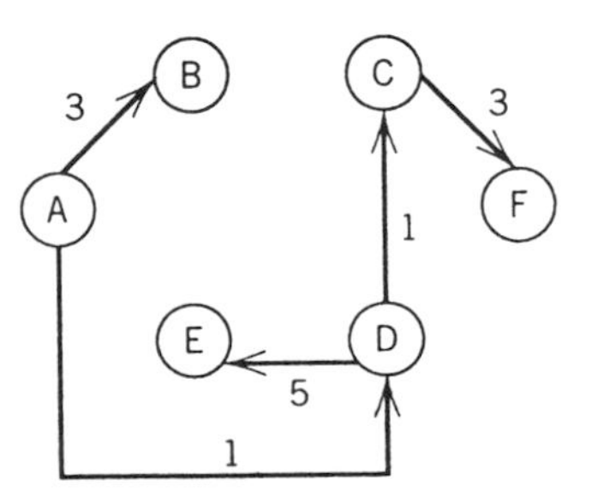

Figure 5.6. Example of a shortest forward path tree from node A.

tree, which we call the *Shortest backward path tree*. The algorithm described here is suitable for a centralized implementation. For such an implementation, however, Dijkstra's algorithm, described above, requires less time to compute the shortest paths and is thus the preferred algorithm to use. The reason we describe this algorithm is that it forms the basis for one of the distributed algorithms described later.

The algorithm operates by attaching a label to each node. We let node d be the destination to which we desire to establish shortest paths from all other nodes in the network. With respect to d, a node v is labeled

$$\mathscr{L}_d(v) = (n_{vd}, \mathscr{C}_{vd})$$

where $\mathscr{C}_{vd}$ is, as before, the cost of the shortest known path from node v to node d and n_{vd} is the next node along the shortest path from v to d.

If no path from v to d is known, then $\mathscr{C}_{vd} = \infty$ and n_{vd} is set to a null value, that is $\mathscr{L}_d(v) = (\bullet, \infty)$. Note that when a shortest path is known, then $\mathscr{C}_{vd} < \infty$ and n_{vd} is one of node v's neighbors. (A node's neighbor is a node that is directly connected to it.)

The algorithm is shown in Figure 5.7. Initially no shortest paths are known, and thus all nodes are labeled $(\bullet, \infty)$. The destination node d is labeled $(\bullet, 0)$ reflecting that the cost of going from node d to itself is zero, and that no next node is required. As part of the initialization process, an ordering of the nodes in the network (excluding d) is established.

The algorithm operates in cycles. In each cycle, all the nodes' labels are considered for update. If all labels remain unchanged from the previous cycle, the algorithm terminates with node labels indicating the shortest backward path tree. A node's label is updated by first updating the cost of the shortest path to the destination. For each neighbor w of node v, the cost of going from node v to the destination d via w ($c_{v,w}$) is determined. This is done by adding the cost of the link from v to its neighbor w to the cost of the best currently known path from w to d. This latter quantity is obtained from

$\mathscr{L}_d(d) := (\bullet, 0)$
For all nodes $v \neq d$
  $\mathscr{L}_d(v) := (\bullet, \infty)$
$\mathscr{L}(d) :=$ the set of all node labels
Choose node ordering $(v_1, v_2, \ldots, v_N)$
Repeat
  $\mathscr{L}^p(d) := \mathscr{L}(d)$
  For $i = 1$ to N
  Begin
    $\mathscr{B}(v_i) :=$ set of v_i's neighbors
    $\mathscr{C}_{v_i d} := \min_{w \in \mathscr{B}(v_i)} [\mathscr{C}_{wd} + c_{v_i w}]$
    $n_{n_i d} := w$ that minimizes $\mathscr{C}_{v_i d}$
  End
Until $(\mathscr{L}(d) = \mathscr{L}^p(d))$

Figure 5.7. Algorithm for finding the shortest backward path tree.

TABLE 5.2 Illustration of Shortest Backward Path Tree Calculation

Cycle	A	B	C	D	E	F
0	(•, 0)	(•, ∞)	(•, ∞)	(•, ∞)	(•, ∞)	(•, ∞)
1	(•, 0)	(A, 2)	(D, 8)	(A, 3)	(A, 1)	(D, 8)
2	(•, 0)	(A, 2)	(D, 6)	(A, 3)	(A, 1)	(C, 7)
3	(•, 0)	(A, 2)	(D, 6)	(A, 3)	(A, 1)	(C, 7)

node w's label. The neighbor that provides the shortest path from v to d is then selected as the next neighbor (this is determined from the equation $\mathscr{C}_{vd} = \min_w[c_{vw} + \mathscr{C}_{wd}], w \in \beta(v)$ as given in Figure 5.7), and the values in node v's label are updated accordingly.

We now apply the algorithm to the network shown in Figure 5.5. We are interested in determining the shortest paths to A and updating the node labels in the order B, C, D, E, F. The operation of the algorithm is shown in Table 5.2 and the resulting shortest backward path tree is shown in Figure 5.8. Note that this is different from the shortest forward path tree from node A (see Figure 5.6). In general the two trees will be the same only if the link costs are *symmetric*, that is, if $c_{vw} = c_{wv}$ for all nodes.

Note that the number of cycles the algorithm requires may depend on the ordering used to update the labels. As an exercise, the reader should try running the algorithm with a different ordering of the nodes (say B, E, C, D, F).

5.2.2 Distributed Procedures

In a network using a distributed routing procedure, network nodes cooperate in determining the shortest paths in the network. Here we describe two procedures that employ different basic approaches. The first is a distributed version of the shortest backward path tree algorithm described earlier. This approach was popularized through its use in ARPANET. Although ARPANET eventually abandoned that approach, variations of it have been used successfully by other networks such as Datapac, the Canadian public data network. The second approach is one that the ARPANET designers adopted in the late

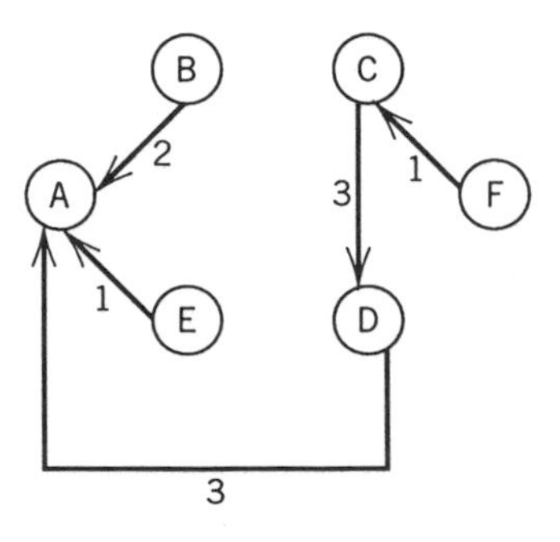

Figure 5.8. Example of the shortest backward path tree to node A.

1970s. Although the ARPANET is no longer in operation, many of the distributed routing techniques described here continue to be used in other networks such as the Internet.

As mentioned previously, distributed routing algorithms rely on the continuous exchange of information among network nodes. Generally, the information exchanged is in one of two forms:

1. In the form of *internodal distances*, with each node estimating the total distance (i.e., cost) to other nodes based on its knowledge of local link conditions and the information available from other nodes. These distances are then communicated to other nodes.
2. In the form of *link state information*, with each node communicating only the condition of its local links to other nodes.

Distributed routing procedures have the objective of determining the best path to take between any two nodes. This can be done on a per packet basis in datagram networks or once per connection in a virtual circuit network. Some distributed routing procedures may cause packet looping to occur. Although these loops are eventually removed from the paths, they make such procedures unsuitable for providing internal virtual circuit service.

The distributed routing procedures we describe below rely on the ability of each node to determine the state and evaluate the performance of each of its outgoing links. An outgoing link may or may not be operational. If it is operational, a measure of its performance is required. Several methods for determining this performance can be used:

1. An instantaneous sample of queue length plus a constant (to avoid zero cost links) was used in the first ARPANET procedure.
2. Delays of packets carried over a link averaged over a 10-second period were used as the basis of link performance in the second ARPANET procedure.
3. The speed of the link is used as the basis of link cost in the Datapac network.

5.2.2.1 A Distributed Internodal Distance Exchange Procedure. In this procedure, each node maintains a distance/routing (D/R) table. An example of a D/R table at node v for a network of N nodes is shown in Figure 5.9. Each row corresponds to a network destination and thus each node in the network will have a corresponding row. Each column corresponds to one of node v's neighbors. (Note that a node's neighbors may also be network destinations and thus the same node may appear as a row and a column index.) For each network destination, the table entries indicate the cost of reaching that destination via the corresponding neighbor. The routing

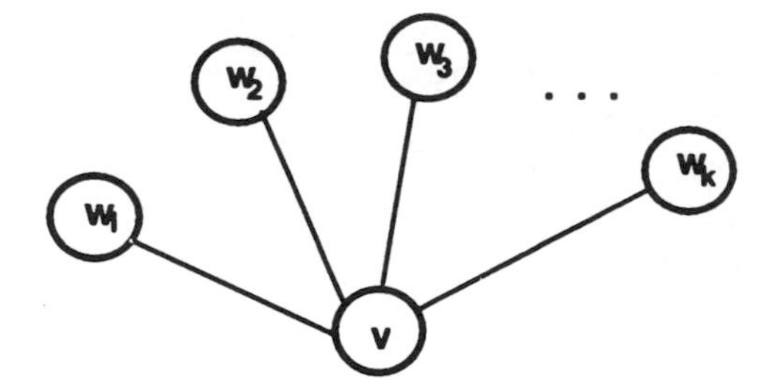

Destination	cost via			
	w_1	w_2	...	w_k
d_1	5	2	...	12
d_2	17	10	...	22
$\vdots$				
d_N	10	3	...	14

Figure 5.9. The distance routing table at node *v*.

of a particular packet is achieved by examining the row corresponding to the packet's destination. The packet is routed via the neighbor with the minimum cost entry.

The distributed routing procedure is based on the update and exchange of entries in the D/R tables. It can be viewed as two parallel procedures (see Figure 5.10):

1. The *inside updating* procedure is responsible for updating the node's D/R table and generating update messages that need to be communicated to the node's neighbors. This procedure is initiated by the occurrence of events that are described later. The basic product of the inside updating procedure is the accumulation of an update list. Updates added to the update list are generated whenever a node determines that the cost of the shortest path from the node to a destination has changed (increased or decreased). This may happen every time an entry in the D/R table is changed. Updates are of the form $v, d, \mathscr{C}_{vd}$, where v is the node at which the update is generated, d is the destination node, and $\mathscr{C}_{vd}$ is the new lowest cost from v to d.
2. The *outside updating* procedure, once triggered, causes the node to communicate the update list accumulated to all the node's neighbors. The outside updating process could be triggered periodically (say every $\frac{2}{3}$ of a second) or it could be tied to the inside updating procedure in some manner (for example, every ten additions to the update list may trigger an outside update).

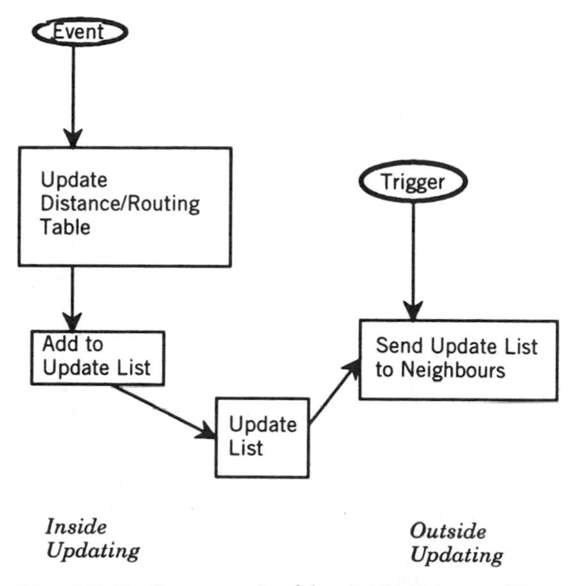

Figure 5.10. Components of the distributed procedure.

The inside updating procedure may be initiated by any one of four events. These events and the resulting changes to the D/R table are described below.

1. **A change in cost of the v to w link is detected by node v.** Let the cost change (positive or negative) detected be Δ_{vw}. This change in cost implies that all costs in the paths that go from node v via neighbor w have increased by Δ_{vw}. Thus the D/R table is updated by adding this increase in cost to all the costs in the column corresponding to neighbor w, as illustrated in Figure 5.11.

The minimum cost to all destinations from node v is then computed from the changed D/R table. For each destination, d, the minimum of the values that appear in the row corresponding to that destination is determined. If

Destination	cost via		
	...	w	...
d_1		$\smile + \Delta_{vw}$	
d_2		$\smile + \Delta_{vw}$	
$\vdots$		$\vdots$	
d_N		$\smile + \Delta_{vw}$	

Figure 5.11. The effect of a change in link cost on the D/R table.

this minimum is different from the previously computed minimum (i.e., the minimum before the table was changed), then node v has determined a new cost for reaching destination d. The update $(v, d, \mathscr{C}_{vd})$ is added to the update list. Recall that this means that the new shortest path from v to d now has cost $\mathscr{C}_{vd}$.

Since the link cost change affects all rows in the D/R table, an update may need to be generated for each destination. Also, note that we add to the update list if the shortest path cost to a destination changes, even if the neighbor that node v uses to reach the destination remains the same.

2. **A failure of the v to w link is detected by node v.** A node detects link failure by observing that, over a period of time, none of the packets it tries to send over the link have been acknowledged. In addition to actual link failure, this may be observed because the neighboring node on the other side of the link has failed or has been removed from the network. In any case, the node detecting the failure will set its cost to infinity (in practice this is some very large number). The procedure is then the same as for event 1 above when the link cost change was detected with $\Delta_{vw} = \infty$.

3. **A repair of the v to w link is detected by node v.** As in the discussion for event 2 above, this event will be detected if a link that was down gets repaired, a neighbor that had failed gets repaired, or if a new neighbor is installed. (In case a new neighbor is added, a new row and column needs to be added to the D/R tables at all the new node's neighbors.)

When this event occurs, the just-repaired link's cost, c_{vw} is determined. This cost is entered as the cost of the path from node v to destination w via neighbor w (the entry in row w and column w, see Fig. 5.12). If c_{vw} is less than the previously established minimum cost from node v to node w, then an update (v, w, c_{vw}) is added to the update list.

To make node w, which could be a new node, aware of the costs it can achieve by routing through node v, a message is sent to node w (only) informing it of the cost of the shortest paths from node v to all destinations. This is essentially an outside update directed to neighbor w and not to all neighbors.

4. **An update $(w, d, \mathscr{C}_{wd})$ is received by node v from neighbor w.** This update indicates to node v that neighbor w has established a new cost for its best path to destination d. Node v can thus determine that its path to destination d via neighbor w now has cost $c_{vw} + \mathscr{C}_{wd}$. This new cost is

Destination	cost via		
	...	w	...
⋮			
w		c_{vw}	
⋮			

Figure 5.12. The effect of link repair on the D/R table.

Destination	cost via		
	...	w	...
⋮			
d		$c_{vw} + C_{wd}$	
⋮			

Figure 5.13. The effect of an update receipt on the D/R table.

entered in the entry in the row corresponding to d and column corresponding to w (see Fig. 5.13). As in the description of event 3, if this new cost changes the previously established cost from v to d then an update $(v, d, \mathscr{C}_{vd})$ is generated and added to the update list. The reader should compare this step with the minimization step of the shortest backward path tree algorithm shown in Figure 5.7. Also observe how the reception of an update message may result in the generation of another update message. This is of concern because, if it always happens, the algorithm will never stop (or converge). (In the example that follows, we see that the exchange of messages eventually terminates.) This termination can be proved, but such a proof is outside the scope of our discussion.

To illustrate the operation of the algorithm, we consider its use in the network shown in Figure 5.5. To simplify our presentation, we only consider the computation of the shortest paths to node A. Initially, we assume that node A was not operational. Thus, in all the nodes, the D/R table entries in the rows corresponding to destination A contain entries set to infinity. The initial network state is shown in Figure 5.14, where we also show pictorially

Node	B		C			D				E		F	
Neighbor	A	C	B	D	F	A	C	E	F	A	D	C	D
Costs To Destination A Via Neighbor	∞	∞	∞	∞	∞	∞	∞	∞	∞	∞	∞	∞	∞
Paths to A	Ⓑ Ⓒ Ⓐ Ⓕ Ⓔ Ⓓ												

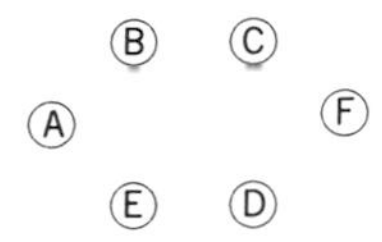

Figure 5.14. Distributed routing example — initial state.

Node	B		C			D				E		F	
Neighbor	A	C	B	D	F	A	C	E	F	A	D	C	D
Costs To Destination A Via Neighbor	2	∞ *	∞	∞	∞	3	∞	∞	∞ *	1	∞ *	∞	∞
Paths to A													

Figure 5.15. Distributed routing example — Nodes determine direct costs to A.

that no node has determined a path to A. For each node, we show only one row from its D/R table: the row for destination A. The tables at node A are not shown as it does not need to compute a path to itself. We assume that all the update lists at all nodes are initially empty. Also, for the duration of this example, we assume that no other changes are occurring in the network.

At this point node A becomes operational. This is determined by the nodes directly connected to A, that is nodes B, D, and E. These nodes can also determine the cost of the links that connect them directly to A and this is reflected in their D/R tables, shown in Figure 5.15. (An underlined entry represents the minimum cost.) As nodes B, D, and E have just discovered a new cost for reaching A, each generates an update that is added to their update list. The fact that the update lists are now nonempty is indicated by the asterisk (*) in each column. The arrows to node A in the network picture indicate the paths currently established to A.

Node B is now triggered to perform an outside update. The message (B, A, 2) is sent to C. Note that as a neighbor of B, node A is also sent a copy of the message. We ignore such transmissions, because A is not a participant in the computation of this example. The update indicates that the new cost of the B to A path is now 2. Node C updates its D/R table by adding the cost of the C to the B link (= 6) to the cost reported by B. This results in the network state shown in Figure 5.16.

Node C is next triggered to perform an outside update sending the message (C, A, 8) to B, D, and F. These nodes update their tables resulting in the state shown in Figure 5.17. Note that at this point all nodes have established paths to destination A. These paths, however, are not the shortest possible and the procedure continues as shown in Figure 5.17.

Next, both nodes D and E are triggered simultaneously to perform outside updates. As a result they send the messages (D, A, 3) to C, E, and F and

Node	B		C			D				E		F	
Neighbor	A	C	B	D	F	A	C	E	F	A	D	C	D
Costs To Destination A Via Neighbor	2	∞	8	∞	∞ *	3	∞	∞	∞ *	1	∞ *	∞	∞
Paths to A													

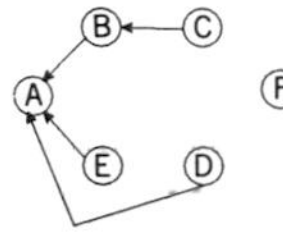

Figure 5.16. Distributed routing example — Node B performs an outside update.

(E, A, 1) to D. The recipients of these messages update their tables as shown in Figure 5.18. Observe that node F has determined a different cost for reaching A than in the previous state and will thus add an update to the update list of the form (F, A, 8); previous to this, the update list contained the update (F, A, 9), which was generated in the previous iteration but never communicated to F's neighbors. The new update is allowed to overwrite the old one, because they both indicate paths to the same destination.

Next, node C is triggered to perform an outside update, sending (C, A, 6) to B, D, and F. This is node C's second outside update for this computation. The resulting network state is shown in Figure 5.19. Note that the nodes have determined the shortest paths to A (cf. Fig. 5.8). The procedure is not

Node	B		C			D				E		F	
Neighbor	A	C	B	D	F	A	C	E	F	A	D	C	D
Costs To Destination A Via Neighbor	2	23	8	∞	∞	3	9	∞	∞ *	1	∞ *	9	∞ *
Paths to A													

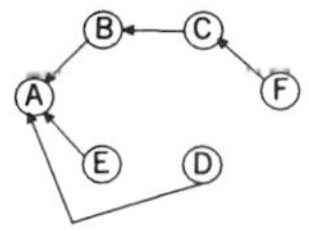

Figure 5.17. Distributed routing example — Node C performs an outside update.

Node	B		C			D				E		F	
Neighbor	A	C	B	D	F	A	C	E	F	A	D	C	D
Costs To Destination A Via Neighbor	2	23	8	6	∞ *	3	9	6	∞	1	8	9	8 *
Paths to A													

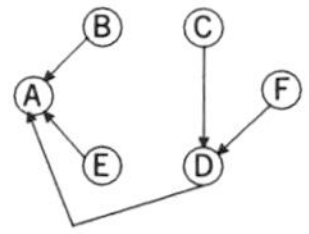

Figure 5.18. Distributed routing example — Nodes D and E perform outside updates.

complete, because F still contains an update it has not communicated to its neighbors.

When node F performs its outside update, sending (F, A, 7) to C and D, the network state shown in Figure 5.20 ensues. All the update lists are now empty and no more messages relating to this computation will be exchanged. Before proceeding with the example, let us examine Figure 5.20 for D. The costs to A from D via A, C, E, and F are 3, 7, 6, and 27, respectively. The cost via C is a concatenation of D's cost to get to C (= 1) and C's cost to get to A (= 6). Node C's cost actually corresponds to the path C–D–A. Thus, if D were to ever send a packet with destination A to C, node C would send the packet back to D and so on. This looping of packets does not actually occur

Node	B		C			D				E		F	
Neighbor	A	C	B	D	F	A	C	E	F	A	D	C	D
Costs To Destination A Via Neighbor	2	21	8	6	∞	3	7	6	∞	1	8	7	8 *
Paths to A													

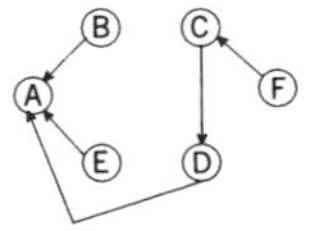

Figure 5.19. Distributed routing example — Node C performs an outside update.

Node	B		C			D				E		F	
Neighbor	A	C	B	D	F	A	C	E	F	A	D	C	D
Costs To Destination A Via Neighbor	2	21	8	6	10	3	7	6	27	1	8	7	8
Paths to A													

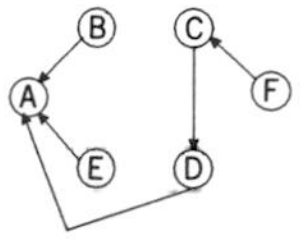

Figure 5.20. Distributed routing example — Node F performs an outside update.

as D does not send packets to A via C. However, the presence of this cost entry (i.e., 7 in D's table) constitutes a "time bomb" waiting to explode.

Our example continues in Figure 5.21 with the cost of the link from D to A changing from 3 to 10 and with node D detecting this change. Node D updates its D/R table and changes its best route to A to the one via neighbor E. Node D also adds an update to its update list to indicate this change.

Next, the cost of the D to E link changes from 5 to 7. The detection of this event by node D causes the network state shown in Figure 5.22. Node D has now decided to route packets destined for A through node C and the "time bomb" has exploded. While the tables are in this state, packets from C, D,

Node	B		C			D				E		F	
Neighbor	A	C	B	D	F	A	C	E	F	A	D	C	D
Costs To Destination A Via Neighbor	2	21	8	6	10	10	7	6	27 *	1	8	7	8
Paths to A													

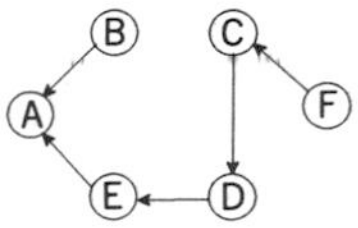

Figure 5.21. Distributed routing example — D to A link cost increases.

Node	B		C			D				E		F	
Neighbor	A	C	B	D	F	A	C	E	F	A	D	C	D
Costs To Destination A Via Neighbor	2	21	8	6	10	10	7	8	27 *	1	8	7	8
Paths to A													

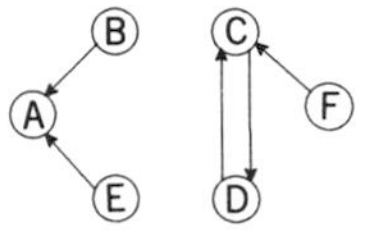

Figure 5.22. Distributed routing example — D to E link cost increases.

and F destined to A will loop between C and D, which is what makes this routing procedure unsuitable for use when internal virtual circuits are desired; a virtual circuit from C, D, or F to A cannot be set up while the network is in this state. Datagrams, however, are not affected as this loop is temporary.

An outside update is triggered at node D resulting in the transmission of the message (D, A, 7) to C, E, and F. The ensuing network state is shown in Figure 5.23. The loop in the network is removed. The paths to A are not, however, the shortest possible.

An outside update triggered at node C causes the message (C, A, 8) to be sent to B, D, and F resulting in the network state shown in Figure 5.24.

Node	B		C			D				E		F	
Neighbor	A	C	B	D	F	A	C	E	F	A	D	C	D
Costs To Destination A Via Neighbor	2	21	8	10	10 *	10	7	8	27	1	12	7	12
Paths to A													

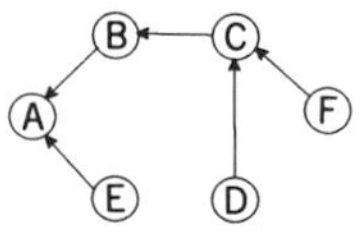

Figure 5.23. Distributed routing example — D performs an outside update.

Node	B		C			D				E		F	
Neighbor	A	C	B	D	F	A	C	E	F	A	D	C	D
Costs To Destination A Via Neighbor	2	23	8	10	10	10	9	8	27 *	1	12	9	12 *
Paths to A													

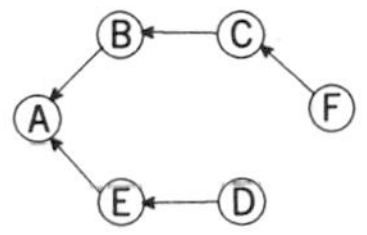

Figure 5.24. Distributed routing example — Node C performs an outside update.

Node	B		C			D				E		F	
Neighbor	A	C	B	D	F	A	C	E	F	A	D	C	D
Costs To Destination A Via Neighbor	2	23	8	11	12	10	9	8	29	1	13	9	13
Paths to A													

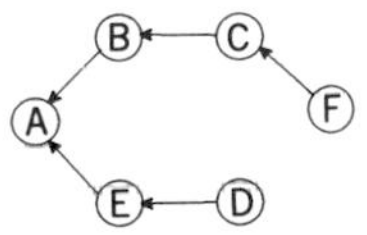

Figure 5.25. Distributed routing example — Nodes D and F perform an outside update.

Simultaneous outside updates are triggered at D and F with the message (D, A, 8) being sent to C, E, and F and the message (F, A, 9) being sent to C and D. This causes the network state shown in Figure 5.25. The network has now converged to using the shortest paths to A with no more pending updates.

The example above should help to clarify the basic properties of the distributed procedure we described. These properties are summarized below.

1. The procedure eventually converges to the use of the shortest paths. The amount of time required for convergence will be a function of the policy used to trigger outside updates as well as the time required to detect changes and transmit and process update messages.

2. While the procedure is running, non-shortest paths, as well as loops may develop in the network.
3. A certain overhead in the form of the generation, exchange, and processing of update messages is required to achieve the distributed nature of the shortest-path computation.

5.2.2.2 A Distributed Link-State Exchange Procedure. ARPANET's implementation of the distributed procedure described in the previous section had several shortcomings, including the high cost incurred by the frequent exchange of update messages and the inconsistency of routes used by different nodes. Link delays were estimated by considering the number of packets awaiting transmission on the link. This turned out to be a bad predictor of delay for various reasons, including the fact that it did not take into account the link data rate and the packet size.

To overcome these shortcomings, the ARPANET procedure was redesigned with a different philosophy. In the new procedure, each node maintains a complete view of the network—its nodes, links, and link costs. (In the old procedure, a node was only aware of its own outgoing link costs and only the total cost for reaching a destination.) When a node detects a sufficiently large cost change in one of its outgoing links, it informs all the other nodes in the network (not just its neighbors as in the old procedure). All nodes update their local view of the network based on the updates received. Each node now computes the shortest paths to reflect updates that have been received. This computation is carried out using the node's own local database that contains the link costs for the entire network. The computation uses a variant of Dijkstra's shortest forward path tree algorithm described earlier.

We discuss two aspects of this link state exchange procedure in more detail: delay measurement and update policy.

Delay Measurement. A link's cost is determined by first obtaining a raw value for the delay experienced by each packet forwarded on the link. When the packet arrives at the node, the time of arrival, say t_1, is noted. The time that the packet is transmitted successfully, say t_2, is also noted. A packet is considered successfully transmitted if an acknowledgement is received within the timeout period. A packet's delay is computed as

$$t_2 - t_1 + \text{link propagation delay} + \text{packet transmission time}$$

The packet's transmission time can be determined from its length and the link's data rate.

Packet delays are averaged over a 10-second measurement period. A transformation is then applied to the average packet delay to obtain the link's cost. This transformation is designed to help with the heavy-traffic behavior of the routing procedure. (In its first incarnation, the new ARPANET procedure used the average packet delay "as is" to determine link cost. This was later modified.)

At the end of a measurement period, the computed link cost is reported to the rest of the network (using the update procedure, which is described later) only if

$$|\text{last reported link cost} - \text{recently computed link cost}| > \tau$$

This ensures that only significant changes to the link cost are reported. The threshold τ varies over time in the following manner:

τ is set to 64 mseconds when an update is reported.

If the next computed link cost is not reported (because it does not satisfy the criterion above) then τ is reduced by 12.8 mseconds.

This process is repeated until an update is reported, at which point τ is reset to 64 mseconds, or until τ reaches 0, in which case the criterion for reporting an update is automatically satisfied and an update is reported.

The reporting procedure above is illustrated through the example shown in Figure 5.26.

This link reporting method has the following desirable properties:

1. Small changes in link cost are not reported unless they are long lasting.
2. Large changes in link cost are reported as soon as they are detected.
3. An update is reported at least once every minute.

An exception to the reporting procedure described here takes place when a topological change occurs; for example, a link fails or gets repaired. In such a situation, the update is reported immediately.

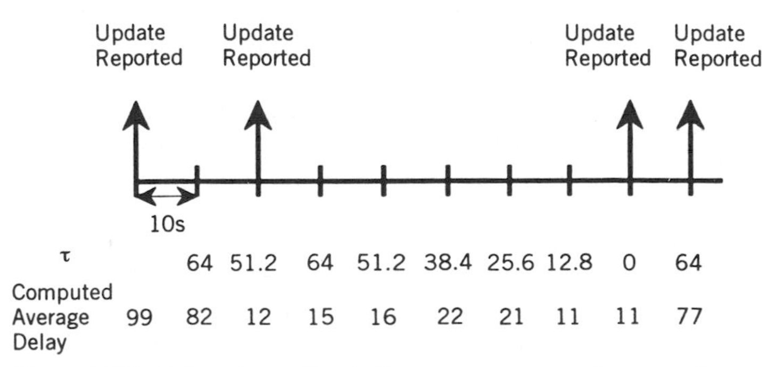

Figure 5.26. Link cost reporting in the new ARPANET routing procedure.

Update Policy. We now examine the mechanisms by which a link's cost is reported to the network nodes. For efficiency reasons, a node groups information about costs for all its outgoing links into one message. This update message is then broadcast to the entire network; that is, it is transmitted in such a way that each network node receives at least one copy of the update message. As is seen later, different methods can be used to broadcast a packet in a PSN. The new ARPANET algorithm uses a technique based on the flooding of the packet.

In its purest form, *flooding* works as follows:

1. The originator of the packet transmits a copy of the packet on each of its outgoing links
2. All nodes that receive the packet transmit a copy on each outgoing link except the one on which the packet was received

This approach has the obvious property that each node in the network will eventually receive a copy of the packet and our broadcast objective is met.

It should be clear that if this pure form of flooding is used, the network will quickly collapse from the very fast growth in the number of packets being forwarded. A mechanism is thus needed to control the forwarding of the packets while still achieving the broadcast objective. To achieve this, a sequence number is placed on each update message as it is generated. The combination of the sequence number and the source address identifies a message uniquely. Each node maintains a list of the identifiers of the recently flooded messages. A message that has been seen before will not be flooded again. Using this approach, each node will flood each message once and the number of packets is limited while the broadcast objective is achieved.

The basic controlled flooding scheme is enhanced to ensure reliable delivery of the broadcast messages. In the basic scheme, an update is forwarded on all outgoing links except the link on which the update was received. In the new ARPANET procedure, an update is also forwarded (or echoed) on the link on which it was received. This echo is used by the node transmitting an update as an acknowledgment that the update has been received. If an echo is not heard within a timeout period after an update is forwarded, the update is retransmitted. All retransmitted updates are marked as such to force an echo from the neighboring node even if the update had been received successfully before. If a newer update message is received from the same source node while an update message is being forwarded, the older message is discarded.

For the correct and efficient operation of the update scheme, it is necessary for a node to determine the relative age of two updates that originated at the same node. This is required so that the contents of an old update do not overwrite the information in the database produced by a newer update. Note that sequence numbers cannot be used to determine the age of an update since it is not possible for a node to continue labeling messages it originates uniquely forever. At some point the sequence numbers will "wrap around" and two distinct messages m_1 and m_2 originating at the same node with sequence numbers s_1 and s_2, respectively, may be such that m_1 is older than m_2 whereas $s_1 > s_2$. Thus sequence numbers cannot be used to determine the relative age of different updates. In the new ARPANET procedure, each update message contains an age field which is initially set to some value. As each node processes and attempts to forward an update, the age field is decremented by an amount proportional to how long the update spends at the node. If an update's age reaches zero, it is discarded and no more attempts are made to forward it.

To insure that all nodes maintain a consistent view of the network, a link that has been repaired is not used for the forwarding of data packets for 1 minute. Update messages, however, can be forwarded using the link during this time. Since each node generates an update at least every minute, this guarantees that the link is not used until an update from every node has traversed the link, which ensures that nodes on both sides of the link have a consistent view of the network.

5.3 HIERARCHICAL ROUTING

As seen above, the routing function is typically achieved by maintaining a routing table at each node. The table is consulted to determine which outgoing link is to be used for any particular destination address or some other packet identification. The use of the routing table requires switching node memory resources for the storage of the routing information, as well as processing resources for looking up the required table. The expenditure of these two types of resources can be significant, particularly if the network is large.

The *hierarchical routing* technique was designed to reduce the size of the routing table maintained at each node and thus require less of a node's resources. In this scheme, the set of nodes in the network are divided into groups known as *first level clusters*. Such clusters are in turn grouped into *second level clusters* and so on until the $m - 1$ level clusters are formed. The mth level cluster is the union of all the $m - 1$ level clusters and includes all the nodes in the network. This clustering needs to have the property that nodes in the same first level cluster are reachable from each other. An example of the clustering is shown in Figure 5.27 where a set of 17 nodes are subdivided into five first level clusters. Clusters 2 and 3 are formed to incorporate two first level clusters each and cluster 1 contains one 1st level cluster. In this example, $m = 3$. With hierarchical routing, each node is identified by a number $i_{m-1}, i_{m-2}, \ldots i_0$, where m is number of the cluster that contains all nodes and i_j is the identifier of the jth level cluster

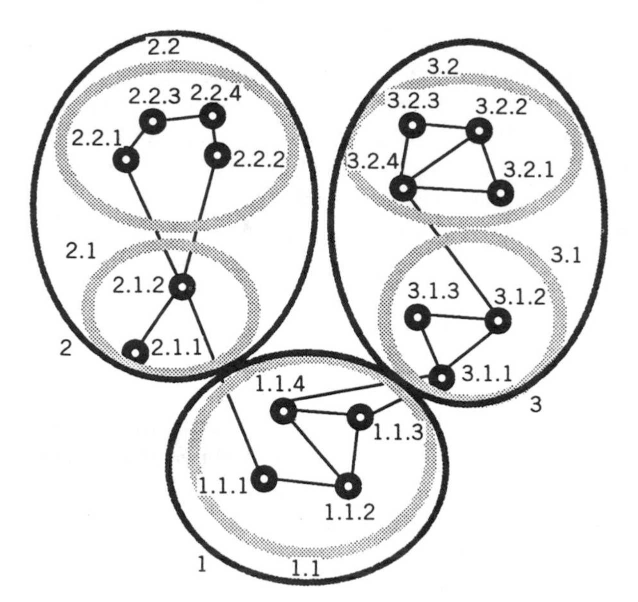

Figure 5.27. An example of the clustering of nodes for hierarchical routing.

containing the node within the $(j + 1)$th level cluster. (Individual nodes are defined as 0th level cluster.)

A node's routing table in the hierarchical routing scheme contains an entry for each destination in the node's own first level cluster and an entry for each of the jth level clusters (except the one containing the node) in the same $j + 1$th level cluster. Figure 5.28 shows the routing table for node 2.2.1 in the example of Figure 5.27.

To illustrate how routing is accomplished in the hierarchical routing scheme, consider a packet, for the example in Figure 5.27, originating at node 1.1.1 with destination node 3.2.2. Node 1.1.1's routing table does not contain an entry for node 3.2.2; it does, however, contain an entry for cluster 3. This entry indicates where to send packets destined to that cluster, say

Destination	Next Node
2.2.2	2.2.3
2.2.3	2.2.3
2.2.4	2.2.3
2.1	2.1.2
1	2.1.2
3	2.1.2

Figure 5.28. Hierarchical routing table for node 2.2.1.

node 1.1.2. The packet will be forwarded until it enters cluster 3 at node 3.1.1, which will have a routing table entry for cluster 3.2 indicating a next node, say 3.1.2. The packet will then be forwarded until it enters cluster 3.2 at node 3.2.4, where a routing table entry for the destination node will be found. The packet will then be forwarded within cluster 3.2 until it reaches its destination.

It is clear that the use of hierarchical routing will result in smaller routing tables, thus reducing the memory processing requirements at each node. However, this reduction is generally achieved at the expense of longer paths through the network. To illustrate this, let us assume that for our example (Figure 5.27) node 2.1.2's routing table indicates that all packets heading to cluster 2.2 should be routed via node 2.2.2. Then packets from node 2.1.2 will require four hops to reach node 2.2.1, instead of a single hop, which would be the case if the direct link to node 2.2.1 were used. This latter option would be available if clusters 2.1 and 2.2 are merged, which obviously would require longer routing tables. Thus there is a tradeoff between the size of the routing table and the length of the paths through the network. Studies into the hierarchical routing scheme have shown that the clustering can be chosen to require approximately $\log_2 N$ table entries (where N is the number of nodes in the network). For very large networks, this significant reduction in table size is achieved with essentially no increase in network path lengths.

5.4 TABLE-FREE ROUTING

Whereas hierarchical routing attempts to reduce the size of the routing table, there are occasions where it is desirable to eliminate the routing tables entirely. We describe several examples of table-free routing here.

5.4.1 Random Routing

Perhaps the simplest method for table-free routing is to disregard the destination address or other packet identification in an incoming packet and select the outgoing link over which to send a packet randomly. This technique obviously sends packets on circuitous and long routes to their destinations. It is, however, extremely robust in that if a destination is connected to a source, the packet will always reach it.

One way to augment this random-routing scheme is to have a node maintain a routing table for neighboring (directly connected) nodes only. Packets with destinations in the table are routed on the appropriate outgoing link whereas other packets are routed randomly. This scheme is actually a *reduced table* scheme rather than a completely table-free scheme. The scheme can be generalized by maintaining tables only for nodes that are less than k hops away.

5.4.2 Source Routing

Another approach to table-free routing is to have the packet's source host rather than the network nodes determine the packet's path through the network. This can be done by having the path a packet should take specified in the packet header. The path specification is of the form $H_{src}, X_1, X_2, \ldots, X_k, H_{dst}$, where H_{src} and H_{dst} are the addresses of the source and destination hosts, respectively. The identifiers X_i represent nodes in the network. This node list should represent a path in the network; that is, consecutive nodes in the list should be neighbors. Furthermore, nodes X_1 and X_k are the entry and exit nodes, respectively. A node that receives the packet forwards the packet to the node or host whose identification code follows that of the receiving node.

For source routing to work, source nodes need to be able to determine a path through the network to each destination. One straightforward method to achieve this path discovery task is to provide the network with a *path server* that continuously receives information about the state of network links and computes shortest paths between nodes. Hosts desiring to send packets request path information from the server. This approach has the problems that any centralized scheme would have; namely, vulnerability to failure and requirement of all nodes to maintain constant communication with the path server.

A distributed method to determine paths to a destination proceeds as follows. First, the source sends a path discovery packet using a controlled flooding technique (see Section 5.2.2.2). The path discovery packet initially contains the destination address and an empty path list. Before each node forwards the packet, it adds its address to the end of the path list. The node to which the destination host, identified by the discovery packet's destination address, if attached forwards the packet only to that host. The flooding of the path discovery packet is controlled by having a node that finds its address already in the path list discard the packet. The destination host receives several copies of the path discovery packet, each containing a different path list. The destination then chooses one of these paths and sends a response packet containing this path back to the source. The source now uses this path in all of its packets. Figure 5.29 shows an example of how the path discovery packet is forwarded in a network.

When source routing is used, two design decisions need to be made:

1. *How does the destination choose from among the paths it receives?* One approach is to have the destination return the path in the first discovery packet it receives. The reasoning here is that the first packet to arrive has most likely taken the least delay path and thus indicates a good source-to-destination path. Alternatively, the destination can wait until a certain number of path discovery packets have been received (or until a timeout expires after the reception of the first one) and return the one from among the ones received that it deems best (e.g., has the minimum number of hops).

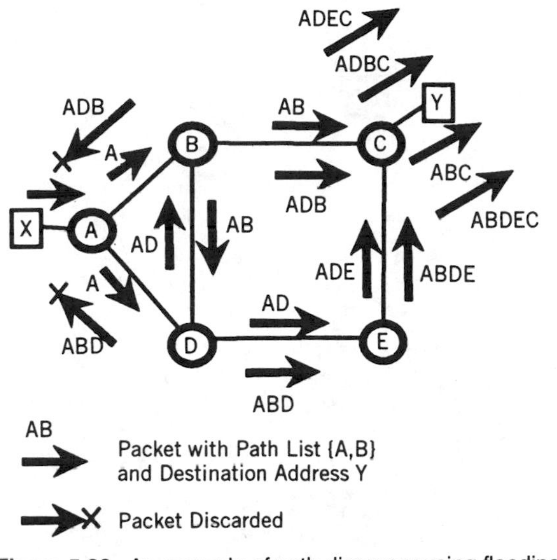

Figure 5.29. An example of path discovery using flooding.

2. *How often does a source need to go through the path discovery process?* Note that once a good path is known it should be used by the source to transmit several packets. However, the least-cost path from a source to a destination may change over time as network conditions change. It is thus desirable that a source rediscover a path to a destination periodically. In a connection-oriented environment, the connection establishment time is a natural time to do path discovery. At connection tear-down time the path information is discarded and rediscovered when a new connection is established. In a connectionless environment, there is no natural time to perform path discovery. Rediscovering a path too frequently (e.g., once every two or three packet transmissions) can lead to a significant amount of overhead. On the other hand, using a path for a long period of time defeats one of the advantages of the connectionless mode of operation, namely the ability to select paths in a network based on current network conditions.

5.4.3 Computed Routing

In some situations, a switching node can determine the appropriate outgoing link over which to forward a packet by computing the identity of the desired outgoing link rather than looking it up in a table. This computation takes as input the packet identification (e.g., the source and destination addresses in the packet) and the address of the node making the decision. Typically, this kind of routing requires that the network have a regular topology, such as a ring, a grid, or a mesh.

We demonstrate the computed-routing approach by considering one example of a regular topology known as *ShuffleNet*, which is shown in

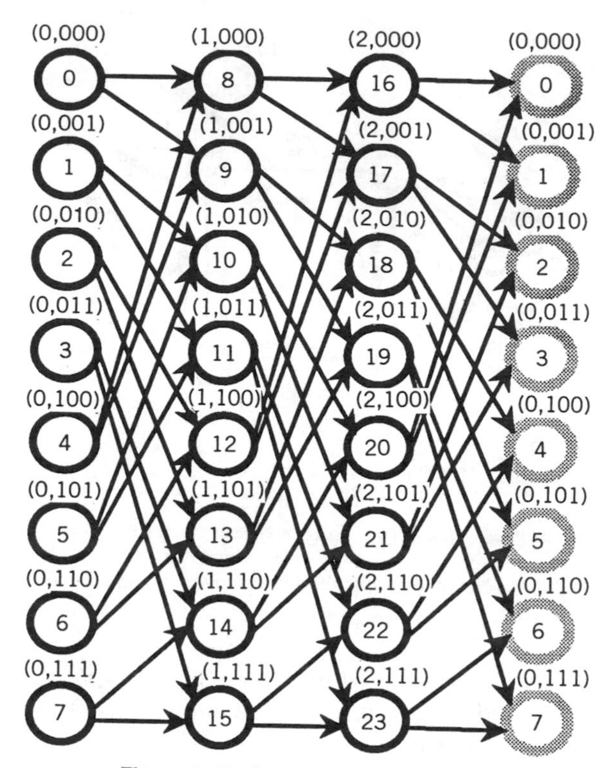

Figure 5.30. An example of ShuffleNet.

Figure 5.30. In this 24-node network, all links are unidirectional and a node can send packets to two other nodes and receive packets from two nodes. Note that the nodes in the last column are the same as those in the first column, so that all nodes are reachable from all other nodes. In this network, each node can be identified by the label $(c, \mathbf{r})$ where c identifies the column to which the node belongs (0, 1, or 2) and $\mathbf{r} = (r_2 r_1 r_0)$ is the binary representation of the row number to which the node belongs (000 to 111).

A node with label $(c, r_2 r_1 r_0)$ which receives a message for destination $(c^d, r_2^d r_1^d r_0^d)$ computes the next node label as follows:
(1) Compute X according to

$$X = \begin{cases} (3 + c^d - c) \bmod 3, & \text{if } c^d \neq c \\ 3, & \text{if } c^d = c \end{cases}$$

(2) The next node address is then computed as

$$\left[(c+1) \bmod 3, r_1 r_0 r_{X-1}^d\right]$$

Note that no tables are maintained at any node in the above computation.

TABLE 5.3 Path Followed from Node 9 to Node 1 [Label = (0, 001)]

Node	$(c, r_2r_1r_0)$	X	r^d_{X-1}
9	(1,001)	2	0
18	(2,010)	1	1
5	(0,101)	3	0
10	(1,010)	2	0
20	(2,100)	1	1
1	(0,001)	Destination	

The above algorithm is guaranteed to find the minimum hop path from any source to any destination.

Table 5.3 shows an example of a packet being routed from node 9 with address (1, 001) to node 1 with address (0, 001).

5.5 MULTIDESTINATION ROUTING

The type of routing we have been dealing with thus far can be called single destination routing, since it is concerned with determining a path to a single destination. Our objective in this section is to discuss how the delivery of a packet to multiple destinations, or *multidestination routing*, may be accomplished. The need for multidestination communication arises in several distributed computing applications such as maintaining the consistency of replicated data and determining the location of a resource known only by name or property. The most general form of multidestination routing, known as *multicast*, occurs when a packet is to be delivered to a subset of the nodes in the network. *Broadcast* routing is a special case in which the packet is to be delivered to a set that includes all the network nodes. We focus here on protocols to achieve broadcast communication. The reader will be able to readily determine that some of the techniques we discuss are also suitable for multicast communication.

The following are some techniques by which broadcast routing can be accomplished.

5.5.1 Flooding

In section 5.2.2, we saw how flooding can be used as a technique for broadcast communication. Flooding can be wasteful of network bandwidth as the number of packets transmitted far exceeds the minimum required. It can however be very robust, because delivery to all destinations will be achieved even after some link or node failures.

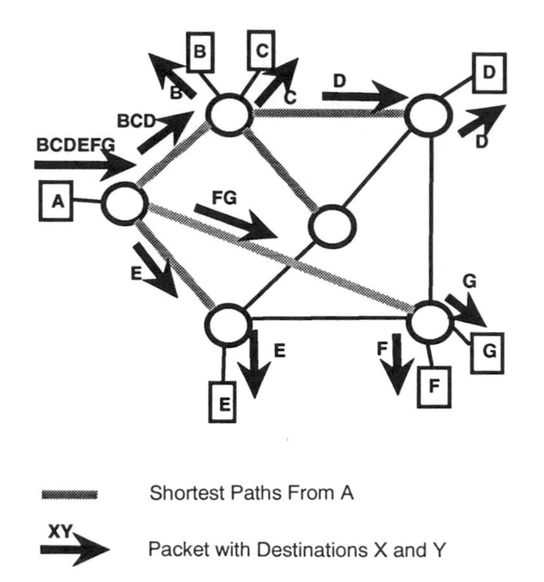

Figure 5.31. An example of the multidestination addressing scheme.

5.5.2 Separately Addressed Packets

In this straightforward technique, a node desiring to send a packet to multiple destinations sends a copy to each destination. The already available single-destination routing procedures are used to deliver each packet. The scheme can consume a significant amount of bandwidth. On the other hand, it does not require any changes in the way the network operates, that is, how routing is done, the type of information maintained at the packet switches, or the format of each packet.

5.5.3 Multidestination Addressing

Another approach to broadcasting packets is to modify the packet format so that each packet can hold a set of destination addresses, not just one. A node delivers a copy of an incoming packet to its attached hosts and deletes their addresses from the list of destination addresses. The node then subdivides the remainder of the list of destination addresses in the packet into groups. Two addresses are put in the same group if the routing table in the switching node indicates that the same outgoing link is to be used for both destinations. For each group of addresses, a new packet is formulated with the addresses in the group used to compile the destination address list.

An example of the operation of this technique is shown in Figure 5.31. The multidestination addressing scheme is very efficient in terms of bandwidth consumption, because the number of copies of the packet generated is minimal. In addition, it does not require any information to be maintained at the switching node for the purposes of carrying broadcast packets. (Note that the routing table used in the multidestination addressing scheme is required for the single-destination routing of packets.) However, this scheme requires a modification to the packet format to allow it to carry multiple destination addresses.

5.5.4 Spanning Tree Forwarding

In this scheme a *spanning tree* of the network is defined. The spanning tree is the smallest set of links that provide complete connectivity among all switching nodes and attached hosts (see Fig. 5.32). Each node maintains information about which of its outgoing links belongs to the spanning tree. Broadcast packets in this scheme are distinguished by a special broadcast destination address (e.g., all 1s). When a node receives a broadcast packet, it forwards a copy on each of its outgoing links that belong to the spanning tree, except the incoming link. In addition, the node delivers a copy of the packet to all of its attached hosts.

An example of spanning tree forwarding is shown in Figure 5.32. This scheme attempts to emulate the bandwidth efficiency of the multidestination addressing scheme while not requiring packets with multiple destination addresses. This is accomplished at the expense of maintaining at each switching node additional information, which is required solely for broadcast routing.

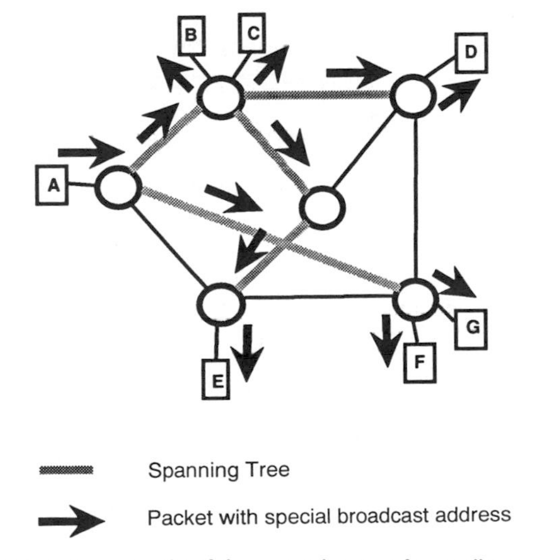

Figure 5.32. An example of the spanning tree forwarding scheme.

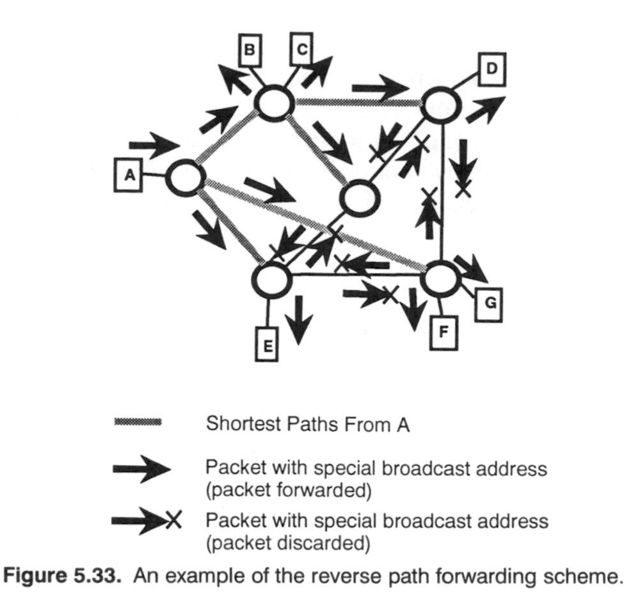

Figure 5.33. An example of the reverse path forwarding scheme.

5.5.5 Reverse Path Forwarding

Here broadcast packets carry the special broadcast destination address. When a node receives a copy of a broadcast packet, it first determines, using its routing table, the outgoing link that it normally would use to reach the source of the broadcast packet. It then checks to see if the link over which the broadcast packet arrived is the same as the link just determined. If there is no match, the packet is simply discarded. If there is a match, a copy of the packet is sent on all of the node's outgoing links (including links to any attached hosts) except the incoming link.

The operation of this scheme is illustrated in Figure 5.33. This scheme attempts to achieve lower bandwidth consumption than is incurred with flooding or separately addressed packets, although somewhat higher than multidestination addressing or spanning tree forwarding. The scheme, however, does not require changes to the packet format, nor does it require nodes to maintain any additional information for the purposes of packet broadcasting.

SUGGESTED READINGS

Some of the earliest work on routing in packet switched networks can be found in [PROS62a], [PROS62b], and [BARA64]. Survey articles on routing include [SCHW80] and [MAXE90]. The discussion in [SCHW87] contains a good survey of how routing is accomplished in some real-life networks. The old ARPANET routing procedure is described in [MCQU78] along with some of the problems that were encountered in the implementation. A description of the use of variations of the distributed ARPANET routing procedure can be found in [SPRO81] and [CEGR75]. A proof of the convergence of the procedure is contained in [TAJI77]. The new ARPANET routing procedure is described in [MCQU80], and the update policy is developed in detail in [ROSE80]. The revision of the delay metric for the new ARPANET algorithm is explained in [KHAN89]. Hierarchical routing was investigated in [KLEI77], [KAMO79], and [KLEI80a]. Our discussion of broadcast routing algorithms is based on the work in [DALA78]. Random routing was first discussed in [PROS62a] and source routing is discussed in [SUNS77]. The routing procedure described for ShuffleNet is presented in [ACAM89].

Routing continues to be the subject of further research, particularly in the context of high-speed networks. Examples of such studies include those found in [DZIO90], [BERN90], [YEE93], [MAXE93], and [AMMA93].

PROBLEMS

5.1 For the network shown in Figure 5.34:

a. Use Dijkstra's algorithm to obtain the shortest paths from node A to all other nodes.

b. Use the shortest backward path tree algorithm to obtain the shortest paths from all nodes to node A.

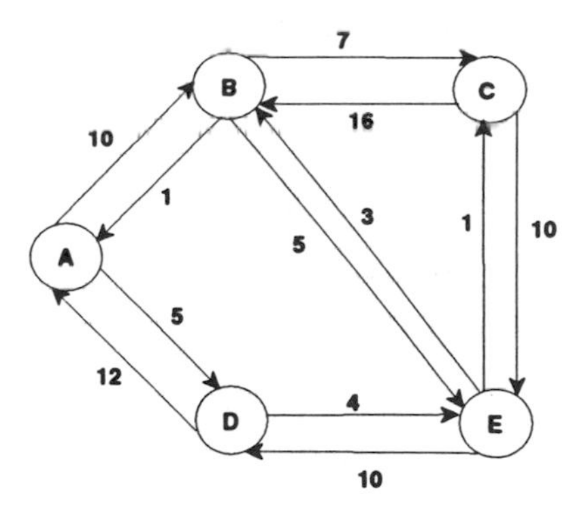

Figure 5.34. Problem 5.1.

5.2 Assume that the network in Problem 5.1 above has been using the internodal distance exchange distributed routing procedure for a long time and that the costs have not changed during that time.

a. Show the distance–routing table entries at all nodes (except A) that indicate paths to destination A.

b. Assume that the A to B and the B to A links fail simultaneously. Show the operation of the algorithm including all the messages transmitted.

5.3 It is said that in the internodal exchange distributed routing procedure "good news travels fast and bad news travels very slowly." Confirm this by considering the network shown in Figure 5.35 with symmetric link costs.

a. Assume that the A–B link is initially down and show all the destination A rows of the D/R table in all nodes as they would be initially.

b. Now assume that the A–B link is repaired. Assuming one outside update per iteration, how many iterations are required before all nodes discover that link A–B has been repaired?

c. Now assume that the A–B link fails again (cost goes to infinity). How many iterations are required before all nodes discover this fact?

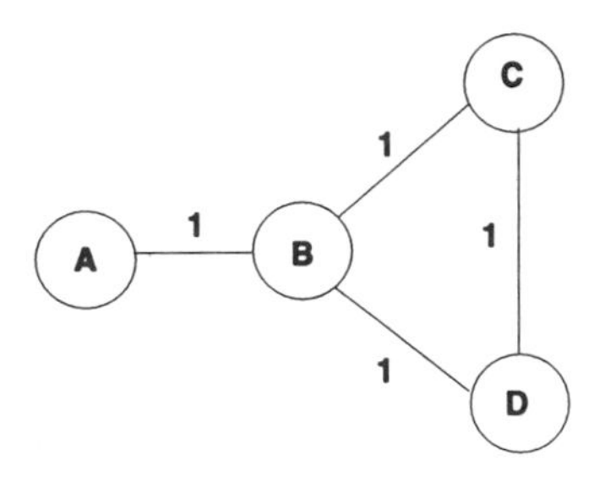

Figure 5.35. Problem 5.3.

5.4 A modification of the internodal exchange distributed routing procedure is known as the *predecessor* algorithm. This algorithm operates in the same manner as the procedure described in this chapter with the following changes:

- Whenever a node v discovers a new path to a destination d, it sends a $P(v, d)$ message to the neighbor w that it would use as the next node. A node receiving the $P(v, d)$ message sets the cost of going to d via v in the D/R table to infinity. Node v will inform all neighbors except w of the new cost it discovered.
- If a cost update message is received, the D/R table is updated. If a new cost to a destination is discovered, then a cost update message is sent to all neighbors except the one that would be used as the next node for the destination. If the next node to the destination has changed, then a $P(\cdot)$ message is sent to that neighbor.

a. For the network shown in Figure 5.36, show the rows of the D/R tables for destination A for both the original algorithm and the predecessor algorithm.

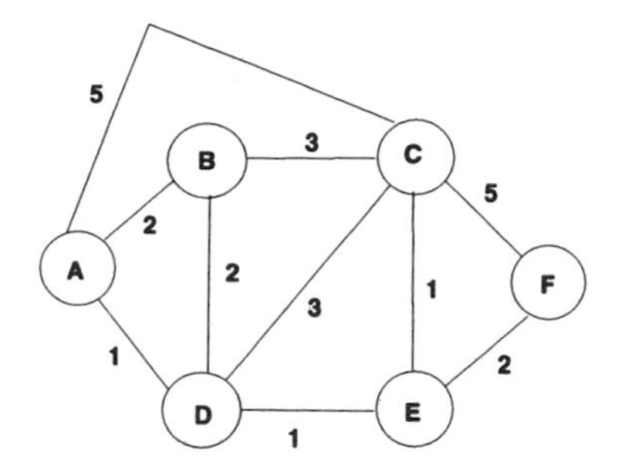

Figure 5.36. Problem 5.4.

b. Now assume that link A–D fails. Show the operation of the original and the predecessor algorithm and compare in terms of the number of iterations required until convergence, the number of updates generated (do not count updates sent to A), and whether or not packet looping is possible before convergence takes place.

5.5 Another modification of the internodal distance exchange procedure uses the *split horizon* technique to propagate updates (in the split horizon procedure, if a new cost to a destination is discovered via a neighbor w, all nodes are informed of the cost change except w). Repeat Problem 4 using this modification.

5.6 In this problem, we evaluate the bandwidth cost of the various broadcast routing techniques presented in Section 6.5. We measure this cost by counting *packet hops*. One packet hop of cost is incurred for every packet transmitted on any link in the network, including links from a host to an entry node. Consider the network shown in Figure 5.37, where we assume that there is exactly one host attached to each node. Assume that the routing tables (used for single-destination routing) indicate the paths shown from node A.

a. Regardless of the broadcast technique used, what is the minimum number of packet hops required to broadcast a packet in this network? Generalize your answer to a network with N nodes and exactly one host attached to each node.

b. Determine the number of packet hops required for broadcasts starting at the host attached to node A for the following schemes:

i. Controlled flooding. Assume that the following control technique is implemented: A broadcast packet starts with a hop counter that is set to 4. Each node decrements the hop counter by one as it receives the packet. If this results in a counter value of 0 the packet is only delivered to the attached host. Otherwise, a packet with the reduced hop counter is forwarded to all neighbors and delivered to the attached host, except the one from which the packet was received.

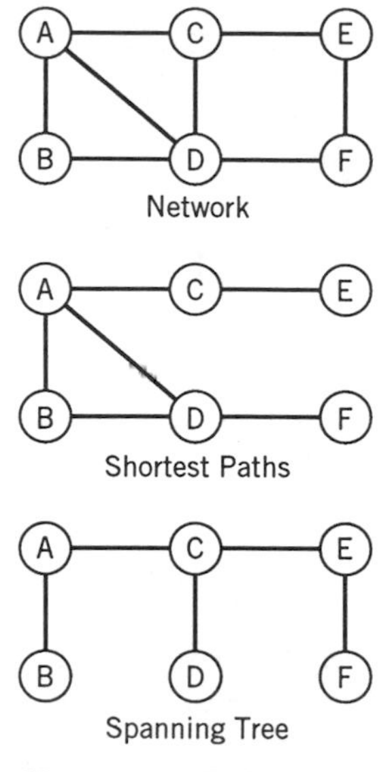

Figure 5.37. Problem 5.6.

ii. Separately addressed packets. Separately addressed packets will be routed over shortest paths.

iii. Spanning tree forwarding with the indicated tree.

iv. Reverse path forwarding.

5.7 We would like to broadcast from one of the sources in the network shown in Figure 5.38 to all the destinations. The nature of the broadcast is that any of two sources (A or C) can undertake the broadcast. The network already uses routing tables based on minimum-hop routing for single-destination traffic. The network managers will not allow you to maintain any information in the switching nodes specifically for the

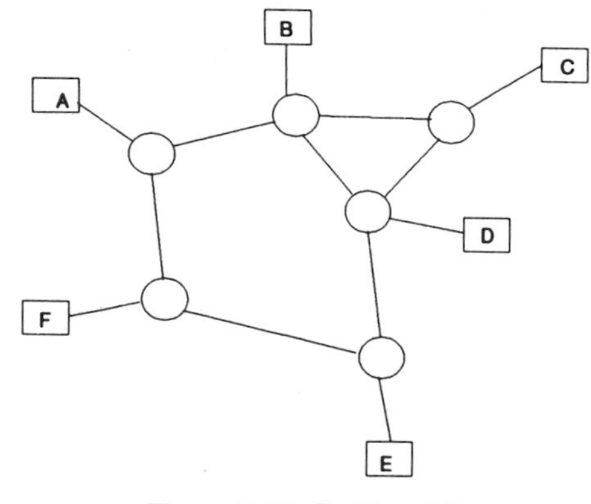

Figure 5.38. Problem 5.7.

broadcast. You may however install your own software. You are also allowed to change the packet format if need be. Answer the following questions:

a. What are the possible broadcast approaches.

b. The network management tells you that the total broadcast cost is given by

$$\text{Cost} = 1 * \#\text{of packets\_hops} + 2 * \text{Max} - \text{Delay} + 5 * \delta$$

where δ is 1 if you change the packet formats, 0 otherwise. Determine the best (minimum total cost) broadcast routing technique and the best source to use for the broadcast.

5.8 Consider the three-way 18-node ShuffleNet shown in Figure 5.39. Devise a route computation procedure, similar to the one in Section 6.4, for this network. (*Hint*: Number rows using the base 3 number system, e.g., $5_{10} = 12_3$.)

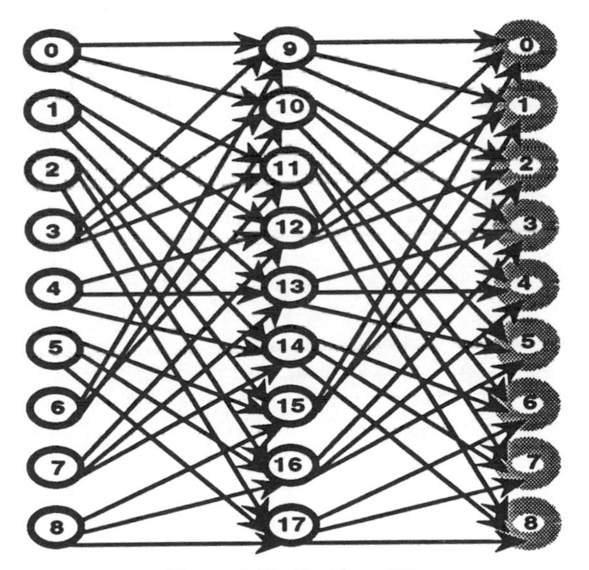

Figure 5.39. Problem 5.8.

6

FLOW AND CONGESTION CONTROL

6.1 INTRODUCTION

Packet switched networks (PSNs) derive their efficiency from the sharing of network resources by the various network users. For example, a link in the network may be used for traffic between different source–destination pairs. Node buffering and central processing unit (CPU) capacity are also shared and may be called upon to accommodate the communication requirements of different source-destination pairs. It is precisely this sharing of resources, an intrinsic and desirable feature of packet switched networks, that can lead to problems of congestion in the network.

Simply stated, congestion occurs when the total demand for a resource exceeds its capacity. We have all experienced rush-hour traffic when the highway resources are oversubscribed; for the few rush hours, the demand for highway real estate is greater than what is available. In most situations, the solution of adding more capacity cannot be economically justified as the highway is typically severely underutilized during nonrush hours. If only we could spread out the rush hour traffic over the entire day, there would be no congestion problem. In general, for a highway system as well as for a computer network, this *global scheduling* of the distributed demands is not feasible. We, therefore, have to be able to cope with the potential for congestion and attempt to minimize its effects.

Dealing with congestion requires that the network be designed and operated in a manner that will make the probability of congestion low. Since the goal of a congestion-free network is, for the most part, unattainable due to the distributed and random nature of the demand, procedures that allow for the system to recover from a congested state are also required. This chapter describes the flow and congestion control policies and procedures in a PSN that are required to avoid, detect, and recover from congestion situations.

6.1.1 Behavior of Uncontrolled Networks

Before considering flow and congestion control procedures, we motivate the discussion by investigating the behavior of networks that do not implement

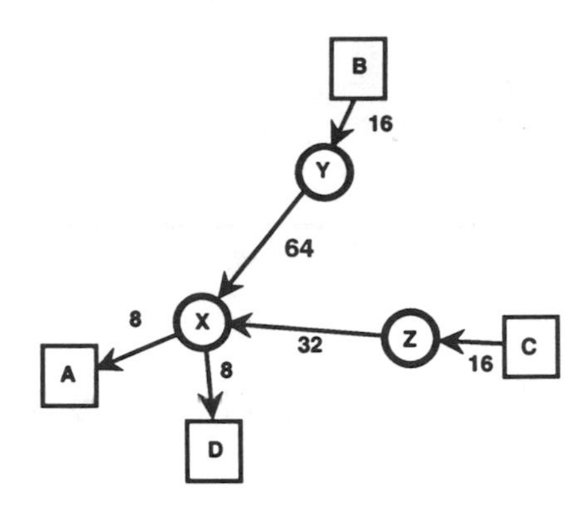

Figure 6.1. Example of a congested network.

such procedures. There are basically two types of problems that can arise in an uncontrolled network: degradation of throughput and unfairness.

To illustrate these problems, we consider the network shown in Figure 6.1. The numbers on the links represent their respective capacities in kilobits per second. We assume that the following traffic requirements exist in the network: a host B to host A requirement of λ_{BA} kbps and a host C to host D requirement of λ_{CD} kbps. The traffic from B to A follows the path B → Y → X → A and the C to D traffic follows the path C → Z → X → D.

We assume that the network is *uncontrolled*; that is, all traffic can access all network resources and no limits are placed on the traffic rate originating at any host. Particularly, the packet buffers available (i.e., the memory available in the switching node to store incoming packets) in nodes X, Y, and Z can be occupied by all packets. We assume (for simplicity) that the network is error-free.* Packets, however, may still be lost, as they will be discarded if they arrive at a node with no available packet buffers. To recover from these lost packets, the network operates with the usual hop-by-hop protocol in which a packet is retransmitted by a node or host if no acknowledgment is received within a timeout period. A node or host retains a copy of a packet until this packet is acknowledged by the next node or host along the path.

To see how the lack of control can cause degradation of throughput and unfairness, we consider the following scenarios:

Scenario 1. $\lambda_{BA} = 7$ kbps and $\lambda_{CD} = 0$. This does not represent a congestion situation, as the B-to-A traffic requirement can be handled by the existing network capacity. The rate at which packets are delivered to host A is the same as the rate at which they are offered to the network by host B. Links B → Y, Y → X, and X → A each carry 7 kbps.

Scenario 2. $\lambda_{BA} = 8 + \delta$ kbps and $\lambda_{CD} = 0$ $(\delta > 0)$. Here packets are offered to the network for delivery to A at a rate higher than the link X → A can handle. Because of this, node X's buffers will eventually fill up, causing

*Actually the presence of errors can only exacerbate any congestion problems.

packets received from node Y to be discarded and not acknowledged. Since node Y retains copies of unacknowledged packets for retransmission, this eventually causes node Y's buffers to fill up. Another interesting phenomenon develops. Since node X can deliver 8 kbps and it is initially offered $8 + \delta$ kbps, it will at first reject δ kbps. The Y → X link will now carry $8 + 2\delta$ kbps to account for the retransmissions of δ kbps of discarded packets. But node X can only deliver 8 kbps, causing it to discard 2δ kbps, which will have to be retransmitted and now the Y → X link will have to carry $8 + 3\delta$. Because of the retransmissions, the traffic on the Y → X link will continue to increase up to the capacity of the link, which is 56 kbps. For the same reason the B → Y link will be carrying 16 kbps worth of new and retransmitted packets. Here we observe that if the network is asked to deliver packets at a higher rate than its capability, a significant amount of the network resources can be consumed by this excess demand.

The congestion problem in this scenario can be addressed in one of two ways:

1. Provision the network with enough capacity in the X → A link so as to accommodate the maximum possible traffic requirement from B, or
2. Restrict B to a maximum of 8 kbps worth of traffic.

The solution in 1 is the right one if the maximum traffic requirement is known and makes sense economically only if traffic is offered by B at the maximum level frequently and for prolonged periods of time. The solution in 2 is reasonable if most of the time the B-to-A traffic requirement is low (say 2 kbps) but occasionally it peaks to above 8 kbps. Restricting the instantaneous flow rate of B to a maximum of 8 kbps is feasible as any excess traffic can be delayed until this overload condition subsides. We note a critical difference between these two solutions: The first is a design consideration and cannot be implemented in real time. The second is a policy for which network protocol mechanisms have to exist to enforce the policy in real time and when the need arises.

Scenario 3. $\lambda_{BA} = 7$ kbps and $\lambda_{CD} = 7$ kbps. As in scenario 1, this does not represent a congestion situation. Data is delivered to A and D at a total rate of 14 kbps with each getting 7 kbps. Each network link will carry 7 kbps.

Scenario 4. $\lambda_{BA} = 8 + \delta$ kbps and $\lambda_{CD} = 7$ kbps ($\delta > 0$). Note that in this scenario the C-to-D path has enough capacity to accommodate the traffic requirements; however, the B-to-A traffic requirement exceeds the capacity available on that path. The problem is that, in this uncontrolled network, the B-to-A and the C-to-D packets have to share node X's buffer capacity. As was seen in scenario 2, the B-to-A traffic requirement causes node X's buffers to fill up. This in turn causes packets from host C as well as host B to be discarded frequently as they arrive at node X. This happens to all packets despite the fact that it is host B's traffic that is causing the problem. As was seen in scenario 2, the ultimate result of the full buffers at node X is that node Y's and node Z's buffers also fill up, and all links carry traffic at their capacity.

Whenever node X delivers a packet to either A or D, it accepts and acknowledges an incoming packet. Because it receives packets from Y at twice the rate it receives from Z (the Y $\rightarrow$ X capacity is twice the Z $\rightarrow$ X capacity), node X is twice as likely to replace a delivered packet with a packet destined for A than for B. Thus the ratio of packets in node X's buffers destined for A to those destined for B is 2 to 1. Packets to A are delivered at the maximum rate of the X $\rightarrow$ A link (8 kbps), and hence packets to D are delivered at half that rate, i.e., 4 kbps.

If we compare this scenario to scenario 3, we find that by increasing λ_{BA} from 7 to $8 + \delta$ kbps

1. Total throughput was decreased. The total traffic delivered by the network decreased from 14 kbps to 12 kbps.
2. Host C's traffic was treated unfairly. The rate at which traffic is delivered to D from C decreased from 7 kbps to 4 kbps, and thus host C was penalized more than host B, although host B's traffic is the cause of the problem.

To address the congestion problem in this scenario, the same two solution approaches discussed for scenario 2 are possible. A third solution is also possible: reserve some number of buffers in node X for traffic destined to D. This will insure that host C's packets will receive guaranteed access to node X's buffers and will thus be treated fairly regardless of whether node B is overloading the network or not. Of course reserving resources runs contrary to the resource sharing ideal that made packet switching desirable in the first place. It seems that sacrificing some of the resource sharing benefits is a reasonable price to pay to insure fairness.

Figure 6.2 demonstrates the general behavior of *throughput*, that is, the rate at which the network delivers packets as a function of the *offered load* or the rate at which packets (new and retransmitted) are offered to the network. Ideally, the network should deliver all offered packets as long as the offered load is below the network's capacity. When the offered load exceeds the network's capacity, the network should (again ideally) continue to deliver packets at its capacity. The curve labeled *ideal* reflects this expectation.

In reality, if the network is uncontrolled, the network will deliver all the offered load (matching the ideal situation) only for values of the offered load below a certain level. As the offered load increases beyond that point, the network's actual throughput (although still increasing as a function of offered

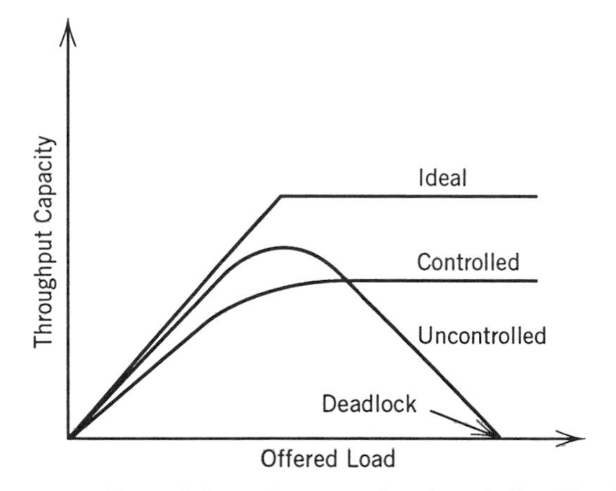

Figure 6.2. Network throughput as a function of offered load.

load) will begin to deviate from the ideal. As the offered load increases further in an uncontrolled network the throughput will start to decrease (see Fig. 6.2). The more traffic the network is offered, the less it will deliver. In some situations a high enough offered load can lead to *deadlock* with no or very few packets being delivered.

Flow and congestion control procedures are normally introduced in a computer network to insure that even in overload situations, the network continues to do useful work. These procedures, however, require a certain amount of overhead, for example, the exchange of control messages or resource reservations. In a controlled network, the throughput continues to increase (up to a maximum) as the offered load increases, as shown in Figure 6.2. This maximum throughput is typically less than the ideal network capacity. Also for certain values of the offered load, the controlled network throughput can be less than the uncontrolled network throughput due to the overhead of the control procedures.

6.2 CLASSIFICATION AND ELEMENTS OF CONTROL PROCEDURES

6.2.1 Flow and Congestion Control

Flow control procedures are typically used to insure that a source does not inundate a destination with more traffic than it can handle. In a computer network, two types of flow control can be present: end to end and hop by hop. *End-to-end* control attempts to insure that a source host does not send packets to a destination host at a rate higher than the destination can consume (process or deliver to a user). The *hop-by-hop* form of flow control is normally exercised between two consecutive nodes on a path from a source host to a destination host.

Congestion control procedures are those used to insure that the network as a whole is not asked to carry more packets than it can handle. Here also, two types of procedures are generally used: network access and buffer allocation. The *network access* procedures are used to control the amount of traffic a host can inject into the subnetwork. The *buffer allocation* procedures are typically used to protect a switching node's buffer capacity to insure fairness and liveness (i.e., freedom from deadlock).

Although the objectives of flow and congestion control are well defined, often flow control procedures can be used to help with congestion control and vice versa.

6.2.2 Levels of Control

Flow and congestion control protocols can and do appear in all protocol layers. They are, however, most prominent in the data link, network, and transport layers. Hop-by-hop flow control is a data link layer function, whereas end-to-end flow control is a transport layer function. Congestion control is predominantly a network layer function.

6.2.3 Mechanism vs Policy

To achieve the objectives of flow and congestion control the network needs to have control mechanisms and control policies. Control *mechanisms* are low-level protocol and network features that determine how some desired actions may be performed. *Policies* are higher level specifications of what needs to be done. Typically, to enforce a policy, a set of mechanisms needs to already be available. Also typically, different policies may be implemented using the same set of mechanisms.

6.2.4 Reactive and Preventive Control

Reactive flow and congestion control policies generally rely on observation or feedback of network status and detection of certain network conditions that indicate or forecast the onset of congestion. The policies then define what actions need to be taken as a result of particular observations. Note that, in general, the information contained in a feedback message requires a certain amount of time to reach the point at which the control actions are to be applied, that is, the *control point*. Typically, a reactive congestion control scheme will not work if it attempts to address congestion that lasts for a duration shorter than the time required to carry the feedback message indicating its occurrence to the control point. As the duration of congestion in a network can vary over a wide range, a network is typically equipped with different procedures designed to operate at different time scales.

Preventive congestion control procedures attempt to ensure that congestion will not take place. If network delays are not significant, feedback can be used in preventive procedures. Otherwise, congestion prevention can be achieved only through open loop procedures that anticipate traffic demand and allocate network resources accordingly.

6.2.5 Our Classification System

In the next three sections, we discuss a set of flow and congestion control policies and mechanisms. We group the procedures into three categories: buffer allocation schemes, window schemes, and network access schemes. It is convenient to use this classification; however, while going through the remainder of this chapter the reader should attempt to relate our description of the schemes below with the discussion in this section.

6.3 BUFFER ALLOCATION SCHEMES

Packet-switching nodes typically store received packets in local *buffers* while these packets are awaiting routing decisions or awaiting the availability of an outgoing link. These buffers represent one of the most important shared resources in a PSN. As demonstrated by our example in Section 6.1, allowing uncontrolled access to switching node buffers can result in severe degradation of throughput or unfairness. It can also result in deadlock as shown in the following example.

Consider two switching nodes (see Fig. 6.3), each with limited packet buffer capacity. Assume that this buffer capacity is available for all packets, that is, uncontrolled buffer access is provided. We can, therefore, enter a state where both nodes' buffers are full with each node's buffers containing packets destined for the other. In networks where reliable hop-by-hop transmissions are required, a node does not release the space for a transmitted packet until it receives a positive acknowledgment for the packet. In this situation, if node A attempts to transmit a packet to node B, the packet will be discarded by node B because it does not have buffer space to store the incoming packet. Therefore, node B cannot generate an acknowledgment for the packet and node A cannot remove the packet from its buffers. In this

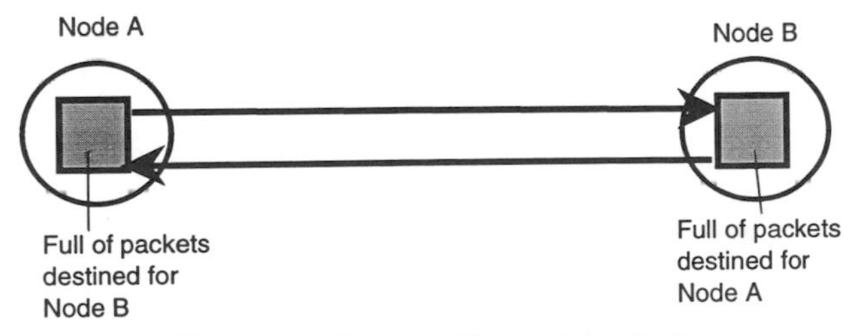

Figure 6.3. Store-and-forward deadlock.

case, no further progress can be made. The above is known as *direct store-and-forward deadlock*. Note that this deadlock is a direct consequence of the requirement of reliable hop-by-hop delivery. Such deadlock cannot occur in networks that do not require reliable hop by hop delivery, but such networks have to implement some form of end-to-end loss detection and retransmission strategies.

A solution to this deadlock situation is to impose a limit (less than the total available capacity) on the number of buffers that can be used to store packets awaiting transmission on a link. As a result, there will always be buffers available to store incoming packets and the deadlock above will not occur. This is known as *channel queue limiting* (CQL) and is a general technique by which the number of packets stored in a switching node and awaiting transmission on a particular outgoing link is not allowed to exceed a certain limit. This limit is normally calculated based on the number of outgoing links, the total buffer capacity available, and the traffic load.

Channel queue limiting can be applied to other situations as well. For example, in scenario 4 described in Section 6.1, if part of node X's buffers are allocated to the B-to-A traffic and the other part to the C-to-D traffic, both the degradation of throughput and the unfairness effects will be controlled.

Another form of buffer allocation is exemplified by the *input buffer limit* (IBL) schemes. Input buffer limits address the contention for buffer resources at a node between input traffic (arriving from directly attached hosts) and transit traffic (arriving from other switching nodes). When such contention occurs, it makes sense to favor transit traffic since the network has already invested its resources to allow this traffic to reach the node. One version of IBL allows input traffic to occupy no more than a certain percentage of the total available buffers at a node. It was found that there exists an optimal value for this percentage that will maximize heavy load throughput of the network.

6.4 WINDOW SCHEMES

Window schemes operate between source and destination pairs connected by an intermediate subsystem (see Fig. 6.4.) This subsystem can be as simple as a wire or microwave link that is dedicated to the transmission of data between the source and destination, or it can be an elaborate internetwork that is shared by numerous source and destination pairs.

Figure 6.4. Environment in which a window scheme operates.

Packets are sent from the source, traverse the intermediate subsystem, and, if everything goes well, reach the destination. The destination generates acknowledgments that are sent to the source to indicate the acceptance of packets received. Packets may never reach the destination if they are discarded by the intermediate subsystem. They may not be accepted by the destination if they are received in error or if they arrive to find the destination buffers full. To recover from such situations, the source always maintains copies of packets that have been transmitted but not acknowledged. A timeout mechanism is used to cause retransmission of unacknowledged packets.

The main function of window schemes is to limit, when necessary, the flow of packets from the source. This has the dual effect of limiting the flow of packets arriving at the destination as well as limiting the rate at which packets are offered to the intermediate subsystem. Window schemes can thus be used to achieve both flow and congestion control objectives. Central to all window schemes is a variable called the *window size*, which indicates the maximum number of packets that the source may send to the destination before it receives an indication that it may transmit more. Such an indication may be received from the destination or from the intermediate subsystem. The source may also receive instructions that may force it to decrease or increase its window size.

A window scheme consists of various components, some of which are not required in certain environments. The following is a list of these components that may be present in various parts of the system shown in Figure 6.4.

Functions at the source:

- How to determine the state of the intermediate subsystem.
- How to react to indications to increase or decrease window size.

Functions in the intermediate subsystem:

- What status measurements to take and how and when to report changes.

Functions at the destination

- When and how to acknowledge receipt of packets.
- When to signal the source for an increase or decrease of window size.

6.4.1 The Window Mechanism as a Flow Control Scheme

We illustrate the use of the window mechanism to achieve flow control objectives by considering the system shown in Figure 6.5. To concentrate on the flow control properties of the window mechanism, we assume that the system is error free. The source has a number of packets to transmit to the destination, and the direct connection between the source and destination is not shared by any other source–destination pairs. The destination processes

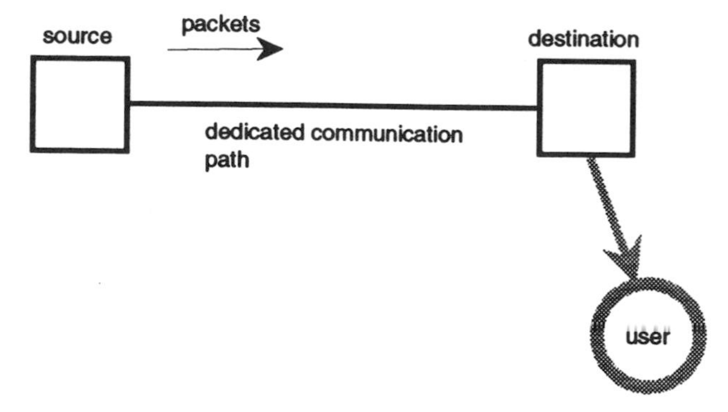

Figure 6.5. Example illustrating the use of the window mechanism for flow control.

and then delivers packets it receives to a user (which could be the next switching node or an attached host). The number of packets that the destination can store is limited to K packets; any packet it receives while its buffers are full is discarded by the destination.

One of the objectives in the operation of the system is to provide a match between the rate at which the user who is attached to the destination consumes packets and the rate at which these packets are delivered to the destination. A guarantee that no packets will be discarded at the destination from lack of buffers can be achieved by the following procedure:

1. Set the window size to K, the number of available buffers at the destination.
2. Generate an acknowledgment at the destination and return it to the source only after the destination has delivered a packet to the user.
3. The source maintains a window counter which is initially set to K. Upon receiving an acknowledgment, the source increments its window counter by 1.
4. Upon the transmission of a packet, the source decrements its window counter by 1. A window counter value of 0 indicates that the source cannot transmit any more packets before receiving an acknowledgment.

Figure 6.6 shows how the above works in a system with $K = 3$. Since the acknowledgments are returned only after a packet is delivered by the destination, a packet is guaranteed to find available buffer space at the destination. This conservative approach provides, in addition, a close match between the rate at which packets are sent by the transmitter and the rate at which the destination can deliver packets to the user. There is, however, a problem with this approach. Because the acknowledgment is only returned after a packet is delivered to the user, it can be delayed for a long time. Meanwhile, the source may think the packet never reached its destination,

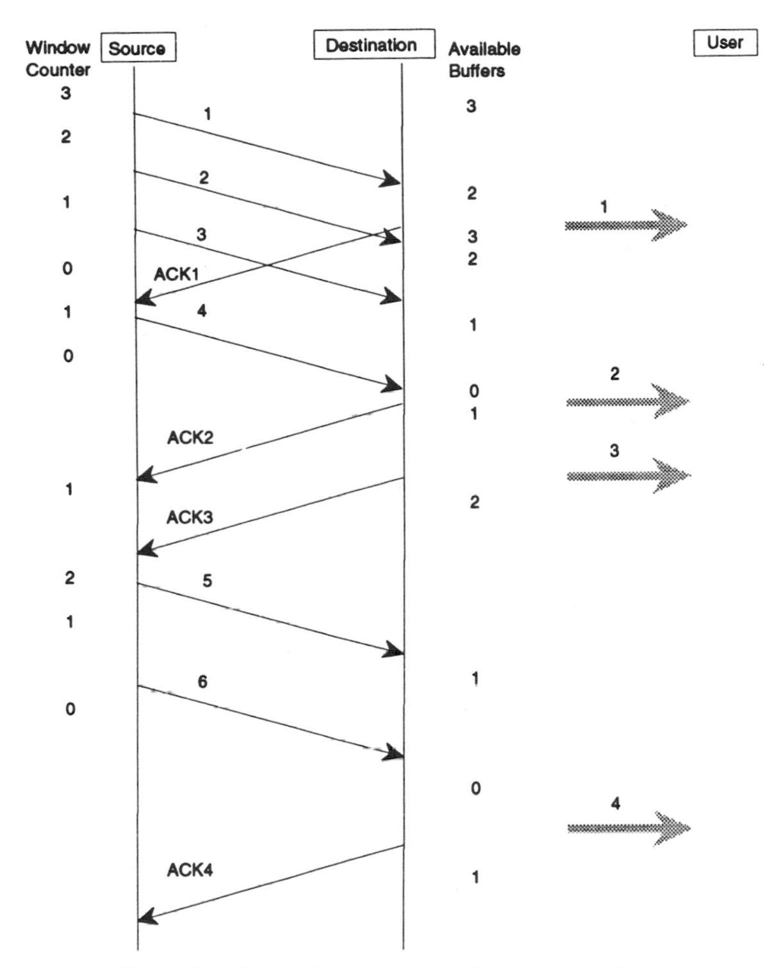

Figure 6.6. Acknowledgment upon delivery by destination.

have its timeout timer expire, and (unnecessarily) retransmit the packet. Also since the source has to maintain a copy of the packet during this time, this approach may require excessive buffering capacity at the source.

Another alternative, is requiring that the acknowledgment be returned as soon as a packet is received at the destination. As illustrated in Figure 6.7, this may cause packets to be discarded at the destination because of a lack of buffering space. A possible compromise consists of setting a timer at the time the packet is received and transmitting an acknowledgment upon the expiry of the timer or the delivery of the packet to the user, whichever comes first.

An even better approach is to separate the two functions that an acknowledgment performs in the above scenarios, namely, the indication of packet reception and the indication of the willingness to accept another packet. The

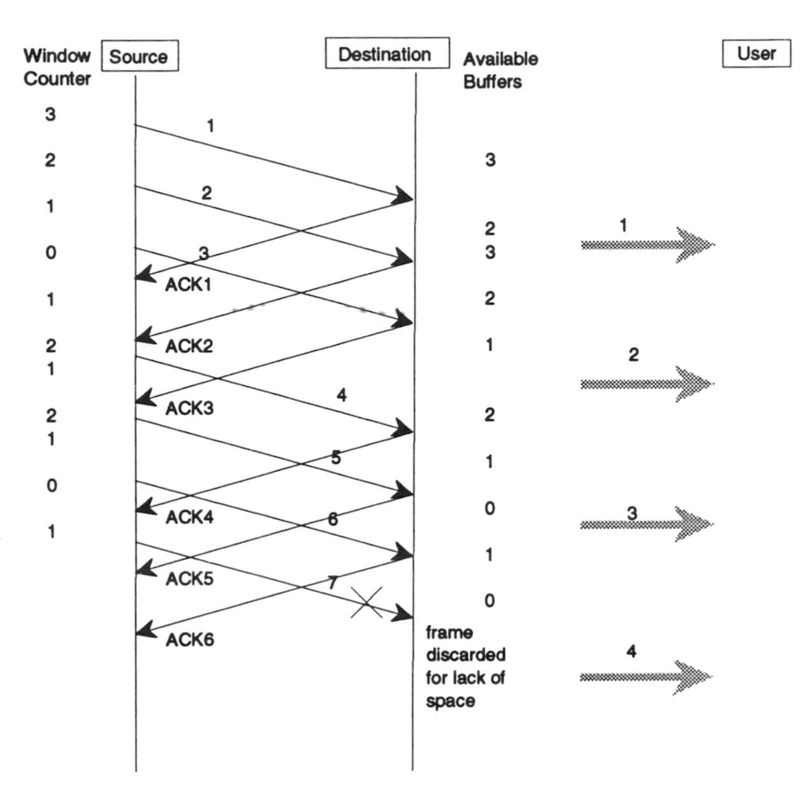

Figure 6.7. Acknowledgment upon reception by destination.

simplest form of this is the use of the receive not ready (RNR) supervisory frame in high-level data link control (HDLC), where the destination can indicate the reception of source packets without allowing the source to transmit any more packets.

A more general approach is through the use of the acknowledgment–credit scheme (Fig. 6.8). An acknowledgment is transmitted whenever a packet is received and accepted, a credit packet is used to indicate the ability to accept more packets. A more elaborate acknowledgement–credit scheme is also possible and will be discussed in our coverage of transport protocols in Section 10.5.2.

6.4.2 The Window Mechanism as a Congestion Control Scheme

We now examine situations where the intermediate subsystem in Figure 6.4 is a shared network that carries traffic among several sources and destinations including the pair under consideration. In such a situation, it is possible to use the window mechanism to control congestion inside the intermediate

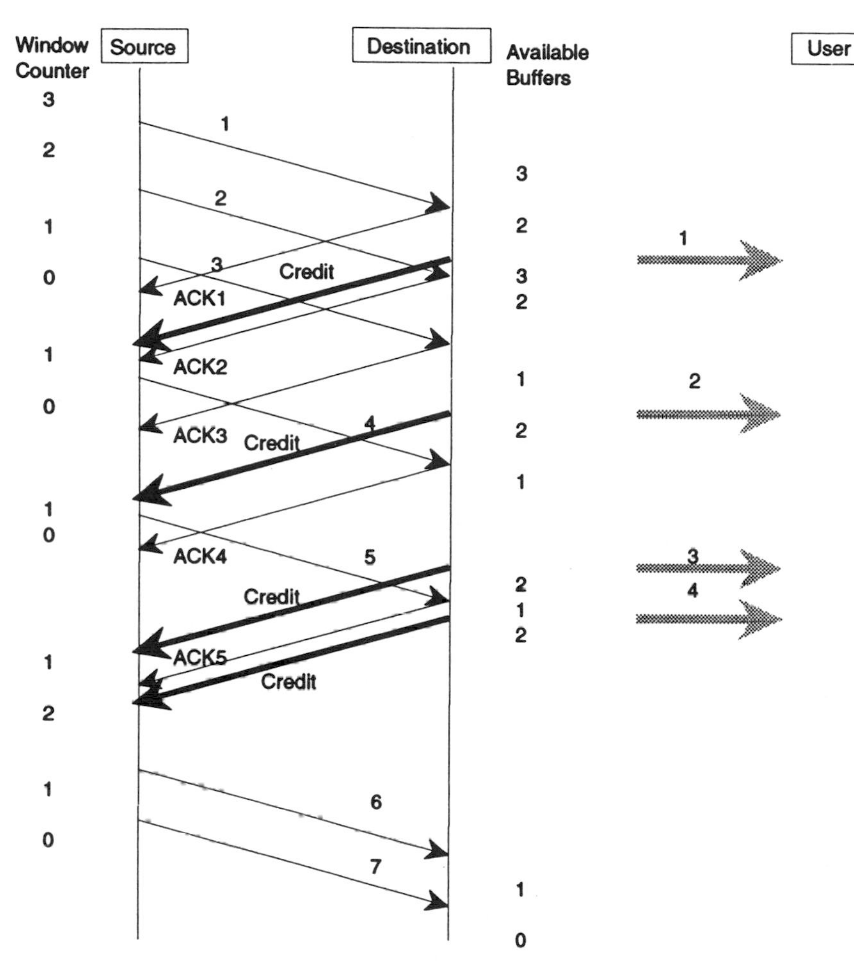

Figure 6.8. The acknowledgment – credit scheme.

subsystem. To understand how this might be done, we first observe that the window size is an important factor in determining the packet input rate (into the network) that a source can have. The smaller the window size, the less the input rate can be. Figure 6.9 illustrates the effect of a change of window size on the performance of a network. The figure assumes that all source-to-destination conversations use either a small or a large window size. For systems with both large and small window sizes, an increase in the activity of the sources (the offered load) will initially cause a corresponding increase in throughput. With a small window size, however, the throughput does not go up as fast because of the limiting effects of the small window size. Further increases in the offered load will cause a degradation of the throughput when a large window size is used since a large window size cannot effectively limit

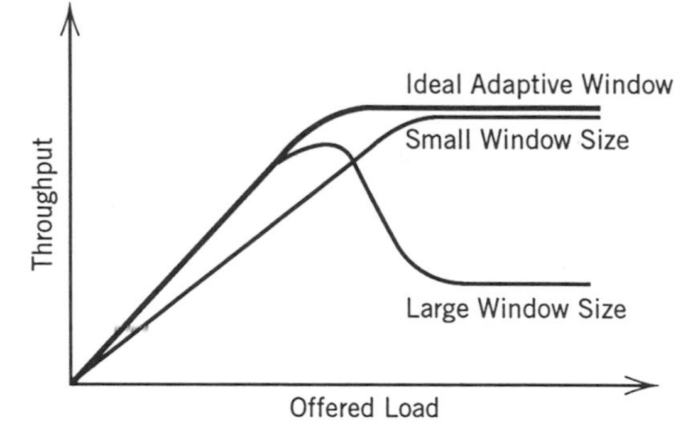

Figure 6.9. Throughput as a function of network offered load for small and large window sizes.

the input rate. With a small window size in use, however, effective control of the input rate is achieved and there is no degradation of throughput.

If several active sources are routing packets through the same switching node, we should make sure that their collective input rates does not exceed the capabilities of the resources present at the node. This will avoid the degradation of the throughput effect shown in Figure 6.9. If we do this conservatively, we have to assume that all sources are active at the same time and determine the window size accordingly. This tends to allow only small window sizes and thus restricts the input rate of a source even if other sources are not active. Therefore, as shown in Figure 6.9, the total throughput at light offered loads might suffer. In this situation it is better to allow the active source a large window size until another source with which it shares switching resources becomes active, in which case the window size is reduced.

It is thus desirable to have the window size adapt to the state of the network. If congestion is experienced, the window size is reduced and increases only if the congestion condition disappears. In an ideal situation, we can switch sources between large and small window sizes in such a way that the throughput-vs-offered load behavior would follow the ideal curve in Figure 6.9.

For a source to adjust its window size, the source must be able to tell the level of offered load in the network. Alternatively, it must be able to tell whether the network (or intermediate subsystem) is in a congested state or is about to enter such a state. It also needs to know when the danger of congestion has passed.

We distinguish between two types of intermediate subsystems:

1. *Active Subsystems.* Ones that are smart enough to determine when congestion has occurred or is about to occur and can in some way inform the source.

2. *Passive Subsystems.* These cannot detect congestion and thus the burden is placed on the source to guess at the start and end of a congestion situation.

Below we describe schemes that can be used to adapt the window size for both types of subsystems.

6.4.2.1 *Active Intermediate Subsystem.* A technique that uses the window mechanism to control active intermediate subsystem congestion operates as follows:

1. Switching nodes detect the onset of congestion (e.g., buffer occupancy has exceeded a certain threshold).
2. Once congestion is detected, the switching node relays that information to all or some sources transmitting packets through the node.
3. A host reacts to congestion information it receives from the intermediate subsystem by reducing its window size.
4. A host is allowed to increase its window size once it determines that the congestion situation, of which it had been informed earlier, has abated.

Note that a host continues to reduce its window counter because of new transmissions and continues to increment its window counter after receiving acknowledgments or credits from destinations as before. The window size should not be increased beyond some maximum value determined by the destination's buffering capability.

There are different methods by which the onset of congestion in a switching node can be determined. All involve the consideration of the queueing systems at work inside the node (see Figure 5.2). A simple *threshold* policy identifies a particular queue occupancy value. If the number in the queue exceeds that value, the queue is said to be congested. Typically average queue occupancy over a certain time interval is considered rather than instantaneous queue sizes. This avoids declaring a congested state based on a momentary and short-lived increase in the queue length. Another approach is to look at switch CPU or link utilizations.* If that exceeds a certain threshold (say 0.5) then the queue is said to be congested. Note that different queues may be experiencing different traffic loads over the same period of time. For example a particular outgoing link and its queue may be congested, whereas other outgoing links at the same node are not. Also note that the processor queue may also be congested independently of the node's outgoing links.

**Utilization* is the average percentage of time that a server is being used.

While a queue is in a congested state, each packet arriving at the queue generates a congestion indication to be transmitted to the packet source. The packet is still forwarded. The most straightforward way to relay this indication is via the transmission of a *choke packet* from the node to the packet's source. Such packets, however, can further aggravate a congested situation. Another approach operates by having a *congestion bit* in each packet, initially set to 0. This bit is set to 1 if the packet passes through a congested queue. Acknowledgments contain a congestion-experienced bit, which is set to 1 if the packet being acknowledged was received with the congestion bit set to 1. The congestion-experienced bit is set to 0 otherwise.

A choke packet, or an acknowledgment received with the congestion-experienced bit set to 1, indicates to the source that congestion is being experienced somewhere along the path used to reach the destination. The source needs to use these indications to determine both the beginning and end of a congestion situation. The source should not react to every choke packet or congestion-experienced bit received, as this might cause a premature reaction thus causing unnecessary degradation of throughput. A better way is to count the number of congestion indications received during a period of time. If the number exceeds a certain value, then the source can declare that a congestion situation has started. If the number is below a certain threshold, then the source can assume that the congestion state has ended.

Another issue of concern is the extent of window size decrease or increase once this is indicated by the detection of the beginning or the end of a congestion situation. The two options are additive and multiplicative.

1. *Additive* Increase/Decrease. In this case $W_{new} = W_{old} + I$ where W_{new} and W_{old} are the new and old window sizes respectively and I is some integer with absolute value greater than or equal to 1.
2. *Multiplicative* Increase/Decrease. In this case $W_{new} = \alpha * W_{old}$ and α is a factor that is greater or less than 1 depending on whether an increase or a decrease is desired.

In both options, if the procedure yields a noninteger window size, the nearest integer is used. The value of the window size, however, should be maintained as a real value to feel the effect of cumulative increases or decreases.

Studies have shown that a good combination is additive increase and multiplicative decrease. This allows a source to respond to the onset of congestion by proportionately reducing its input rate. The end of a congested state causes an input rate increase that is independent of the current one.

Let us illustrate the above concepts with an example. We consider a source–destination pair communicating over an active, error-free, intermediate subsystem that returns congestion-experienced bits in acknowledgments if the packets that are being acknowledged have passed through a congested

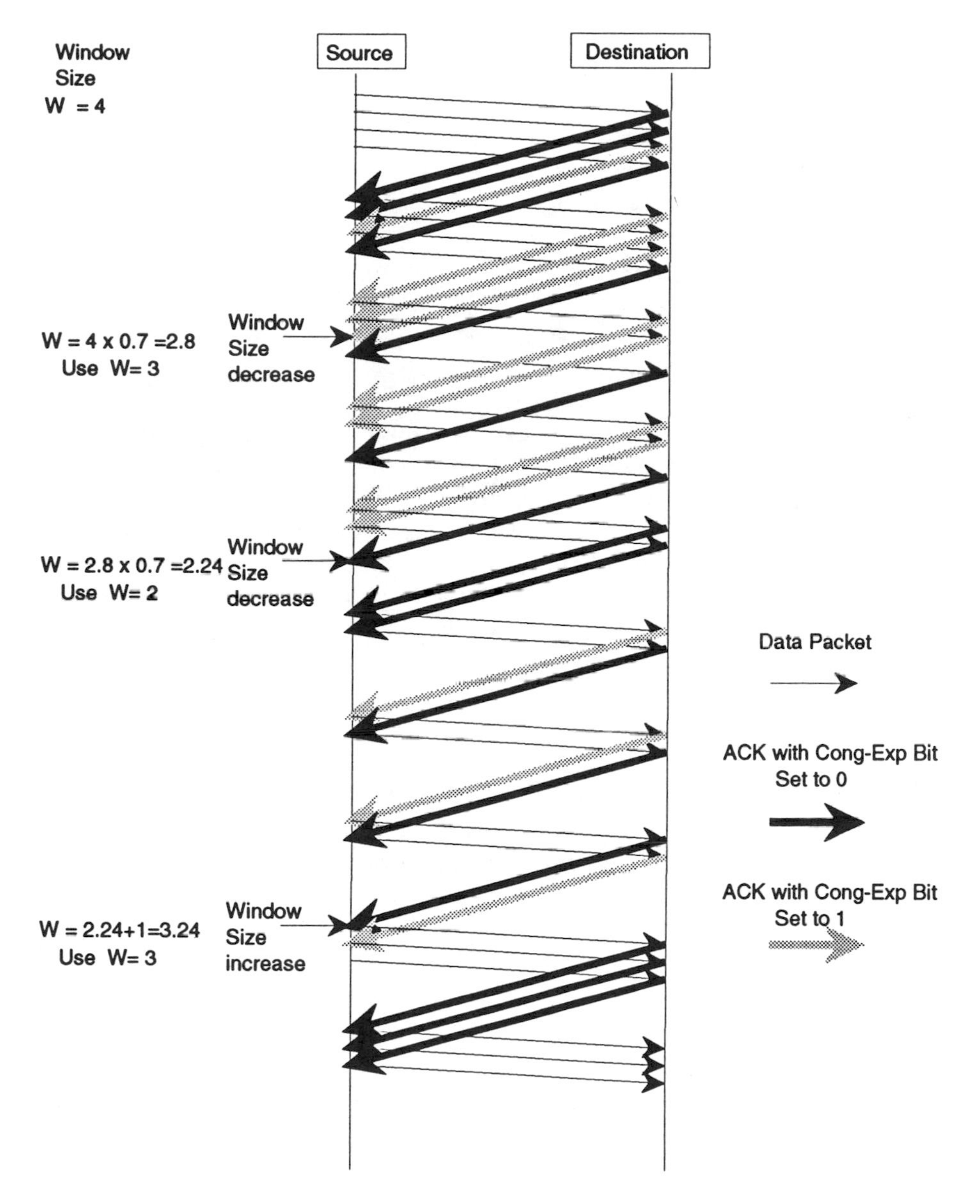

Figure 6.10. The use of congestion-experienced bits to increase or decrease window size.

node. The initial window size used is 4. The source uses a typical filtering scheme to determine the onset or end of congestion. A decision to increase or decrease the window size is made once for every seven acknowledgments it receives. If the number of congestion-experienced bits in the last seven acknowledgments is greater than or equal to four, then the window size is reduced. Otherwise, the window size is increased, if its current value is less than 4. The window size is never reduced below 1. We choose multiplicative decrease and additive increase with $\alpha = 0.7$ and $I = 1$. The operation of the system is illustrated in Figure 6.10.

6.4.2.2 Passive Intermediate Subsystem. We now consider the case where the intermediate subsystem is unable to determine congestion situations. This can be the case if it is desired to have simple switching systems that are not equipped to monitor queue sizes or utilizations and thus cannot determine whether they are congested.*

In this kind of environment, a source cannot rely on the intermediate subsystem to return congestion indications. Therefore, the burden is placed on the source to guess when congestion has started or ended. A simple approach is to use the fact that packets and/or acknowledgments may be excessively delayed or altogether lost when the network is in a congested state. Typically, a delayed acknowledgment will cause a source to timeout and retransmit unacknowledged frames. This can serve as a good indication of intermediate subsystem congestion.

The above arguments lead us to the following timeout-based, adaptive window congestion control scheme:

1. A source is initially set up with a maximum window size, W_{max}, that is, $W = W_{max}$.
2. Whenever a source's timeout timer expires and the source needs to retransmit a packet, it sets $W = 1$.
3. The window size is incremented by 1 for every n packets the source receives acknowledgments for. The window size is never incremented beyond W_{max}.

By reducing the window size to 1 upon any indication of congestion, this scheme can swiftly throttle input traffic. Its reaction to improvements in the congestion situation can be made more cautious by increasing the value of n. We also observe that a timeout should be viewed as a severe congestion event caused by the loss of a packet. (This assumes that error rates are low and the timeout threshold value has been chosen to reflect typical delay values.) Therefore, a significant reduction of input rate is warranted. This should be

*In some cases fast switching and forwarding is desired and, therefore, there is no time to check congestion situations as packets pass through switching nodes.

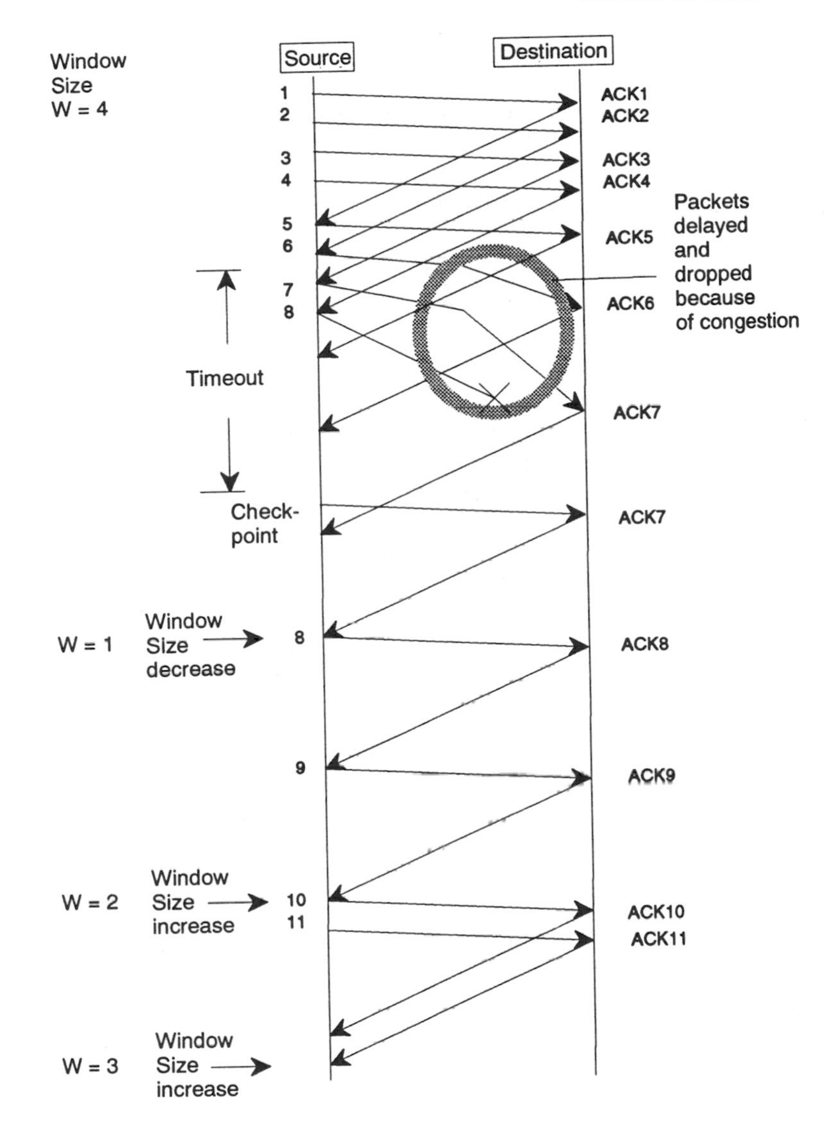

Figure 6.11. The use of timeout and acknowledgments to decrease or increase window size.

contrasted with the relatively milder reaction to the reception of congestion indications as described previously.

An example of the operation of this scheme is shown in Figure 6.11. In this example, the maximum window size is 4 and the value of n is 2. We assume that the network (or intermediate subsystem) may delay or drop packets and/or acknowledgments if they pass through congested switching nodes. This will cause the transmitter to time out and use a checkpointing procedure (see Chapter 2) to determine the state of the receiver and what it needs to retransmit. Upon retransmission, the window size is set to 1. It is increased by 1 (up to 4) for every two acknowledgments the source receives.

6.5 NETWORK ACCESS SCHEMES

Network access schemes operate at the entry point of the network to control and prevent congestion. We have already discussed two techniques that can be viewed as network access techniques: the window mechanism and input buffer limit schemes. Both schemes operate at network access points and limit the traffic that a host can inject into a network.

Other network access schemes have been proposed. The simplest is known as the *isarithmic* scheme and has the goal of imposing a limit on the total number of packets in a PSN. In the scheme, a fixed number, L, of permits circulates in the network. A packet must capture and destroy a permit before it is allowed to enter the network. The permit is regenerated when the packet leaves the network. Therefore, the total number of packets is always less than or equal to L.

Permit management is an important aspect of the isarithmic scheme. It is important to have a fair system for initially allocating and redistributing permits. One problem is that a permit is typically regenerated at a node different from the node where it was captured and destroyed. There is, therefore, the danger that permits will accumulate in some nodes, whereas others will be starved, which is obviously unfair. One distribution scheme sets a limit on the number of permits allowed to accumulate at a node. If a node receives more permits than this limit, it reassigns its excess permits to other nodes.

Limiting the total number of packets in a network does not guard against congestion. After all, all L packets can be in the same location! Imposing this limit, however, does work well in some network situations, particularly those without other means of congestion control and those with uniform traffic requirements.

Another form of network access scheme is possible in connection-oriented networks. Such networks provide a clear event (connection establishment) that can be used to make congestion control decisions. Connection requests are indications to the network that the source making the request desires to establish a communication path to a destination. In some instances, no bandwidth or traffic demands are made as part of this request. In this case, the network, as part of the connection establishment procedure, sets limits on the traffic demands allowable for that connection to insure congestion-free access to network resources. In other situations, connection requests carry a traffic requirement that could be as simple as a throughput class designation or as elaborate as a specification of data rate, utilization, and burstiness requirements. The network, then, has the option of rejecting (or blocking)

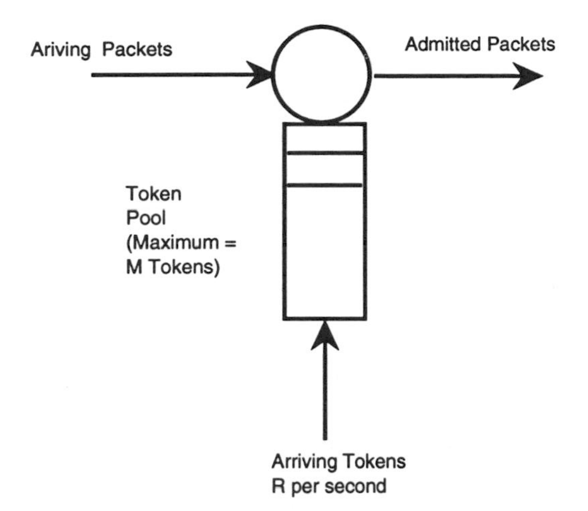

Figure 6.12. The leaky bucket scheme.

the connection request if admission implies an oversubscription of the network resources that will lead to congestion.

In both situations above, end-to-end, hop-by-hop, or other mechanisms are used to enforce the agreed upon traffic characteristics of a connection. These mechanisms perform what is called a *policing* function. For example, if it is agreed that a connection will be limited to a maximum of M packets in the network at any point in time, an entry-to-exit window mechanism can be used to enforce this limit.

An interesting method for policing a connection's input traffic is known as the *leaky bucket* scheme. The scheme operates at a connection's input point. The version we describe is designed to enforce two characteristics of the input traffic: (1) The average packet rate should not exceed R packets/second and (2) the maximum *burst length* is M. That is, the input traffic will not have a busy period (with back-to-back packets) with more that M packets. The scheme operates as follows (see Fig. 6.12):

> The input point is equipped with a token pool that can store a maximum of M tokens. Tokens are generated at a constant rate of R tokens per second and are added to the token pool. Tokens arriving to a full token pool are discarded. Arriving packets that find a nonempty token pool are admitted into the network and capture and destroy a token. Packets arriving into an empty token pool are discarded.

It is clear that the long-term average packet input rate will not exceed R, the token generation rate, since each packet is required to capture a token before it is admitted. Furthermore, the maximum number of back-to-back packets that can be transmitted is M (assuming that the time to transmit M

packets is $\ll 1/R$). The leaky bucket scheme is an *open loop* input rate control scheme, and requires no feedback from the network or the destination. It is thus suitable for use in high-speed networks where the feedback might arrive too late to control the input traffic.

SUGGESTED READINGS

Flow and congestion control have been the subject of extensive investigation for some time. Some of the early work on the topic can be found in [DAVI72], and [KAHN72]. An excellent survey of the topic that includes techniques and ideas up to 1980 can be found in [GERL80]. The advent of high-speed networking has required a rethinking of some of the established techniques, a discussion of these issues can be found in [JAIN90], [JAIN91], [HONG91], and [MAXE90].

Input buffer limiting schemes are discussed in [LAM79, KAMO81]. Dynamic window adjustment approaches can be found in [JAIN86], [JACO88], [BUX85], [RAMA88], [MITR91a], and [HAHN91]. A discussion and analysis of window increase and decrease alternatives can be found in [CHIU89]. The issues involved in the connection admission and policing have been the subject of a flurry of activities. Examples of this work are [BERG91], [RATH91], [BUTT91], [HABI91], [SOHR91], [YAZI92], and [TARR94].

PROBLEMS

6.1 In Section 6.1, we assume the so-called *retransmission model*. That is, packets discarded because of buffer overflow are retransmitted by the source. Another possible model is the *loss model*, in which discarded packets are totally lost. Discuss the behavior of an uncontrolled network using the four scenarios in Section 6.1 and using the loss model.

6.2 A switch with finite buffer capacity has input and output links as shown in Figure 6.13. Traffic coming in on link X is at a constant rate of 0.8 packets per unit time. 50% of this traffic leaves over link A and the other 50% leaves over link B. Traffic coming in over link Y is at a variable rate L and is divided 25% and 75% for output links B and C. Assume that the buffers in the switch are strictly split into two partitions; one for outgoing link B and another partition shared between outgoing links A and C.

a. Using the loss model, sketch the following (i.e., show the general shape of the curve and the parameter values):

i. The total throughput and X's throughput on link B as a function of L.

ii. The total throughput as a function of L on links A and C.

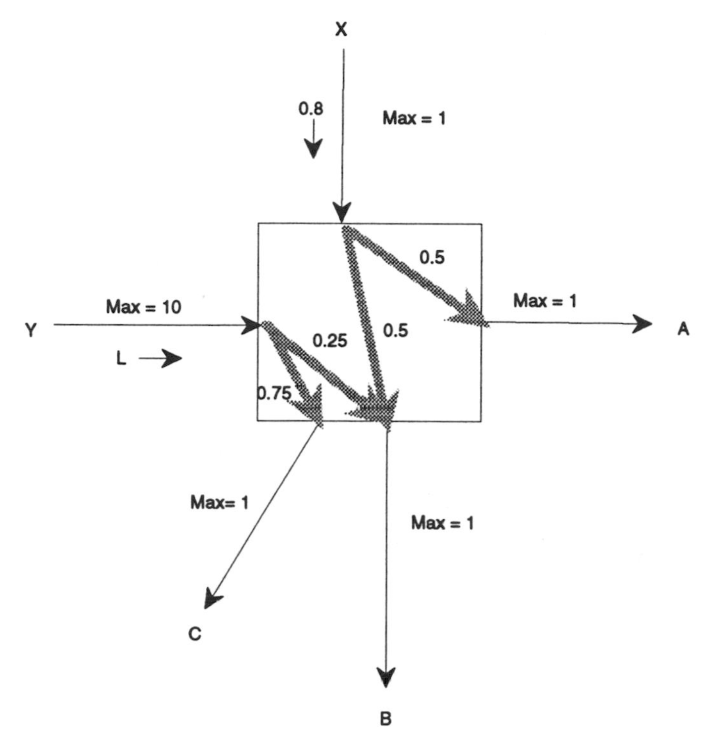

Figure 6.13. Problem 6.2.

b. What are the problems with these throughput curves? Devise a solution (by further partitioning the buffers) to alleviate the problems.

6.3 Draw a diagram showing packet transmissions from a source to a destination using the congestion bit feedback scheme. Assume the following:

a. The source always has packets to transmit.

b. The propagation delay is equal to three packet transmission times.

c. The initial and maximum window size is 5.

d. The source filters congestion indications by making a window reduction/increase decision once for every nine ACKs. If five of the nine have congestion indications, then the window is reduced and it is increased only if none of the frames have a congestion indication. The window is left the same otherwise.

e. Multiplicative decrease and additive increase are used with factors 0.8 and 1, respectively.

f. The network is congested during the first 12 transmissions from the source and returns to the uncongested state afterward.

Show the diagram from the starting point until the window size returns to its initial value.

6.4 In this question we consider the use of a window scheme for broadcast communication.

a. For *flow control* purposes what would be the criteria by which one can determine the window size used by the source. Are there any unfairness problems you see in your scheme?

b. For *congestion control* purposes how would an adaptive window scheme operate. (You may assume a form of broadcast routing that uses a spanning tree, e.g. spanning tree forwarding.) Again in this context do you see any unfairness problems arising?

c. Can you propose any solution to the unfairness problems in the above two sections? Are there any tradeoffs that your proposed solution must address?

6.5 A network has been designed to guarantee that no congestion will occur unless the total input traffic increases above 50 packets/second. Assume that the users of the network can give you one of two guarantees:

a. The input traffic into the network will be a steady 45 packets/second. That is for any period of time, no more than 45 packets/second will enter the network.

b. The input traffic averaged over any 2-day period will not exceed 45 packets/second.

Would you need to implement congestion control procedures in the network if you are given the first guarantee, why or why not? What about if you are given the second guarantee?

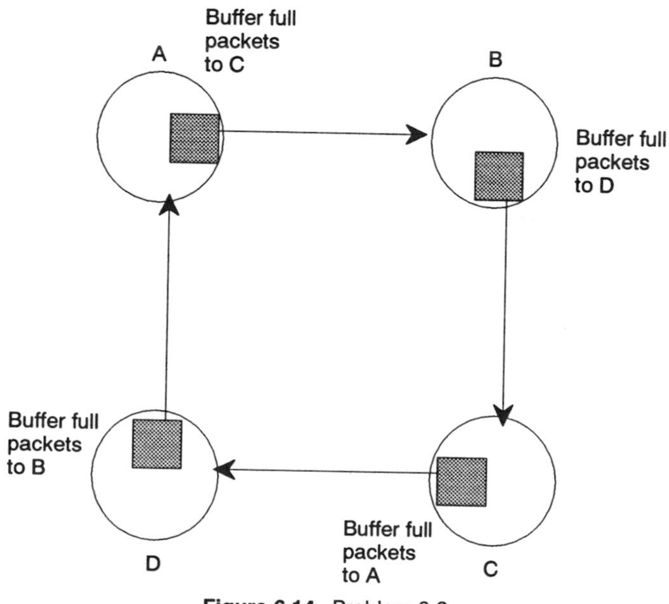

Figure 6.14. Problem 6.6.

6.6 Figure 6.14 illustrates a condition known as *indirect store-and-forward deadlock*. The traffic in the illustrated ring network is such that each outgoing queue is filled and that packets queued at each link are destined to a node two or more hops away. The packets cannot make forward progress because all the buffers are full. Devise a scheme that can be employed in a network to avoid this type of deadlock.

6.7 Another form of deadlock is *reassembly deadlock*, which arises when a message (e.g., a file) transmitted to a destination is fragmented into several packets with each packet transmitted independently in the network. The packets have to be reassembled at the destination before they can exit the network. Figure 6.15 illustrates a deadlock situation where two messages, A and B, are each fragmented into four packets each. The destination can only store four packets at a time and currently has three packets from message A and one packet from message B. It cannot accept any more packets but cannot assemble any message either and the system is deadlocked. Devise a scheme by which this type of deadlock can be avoided.

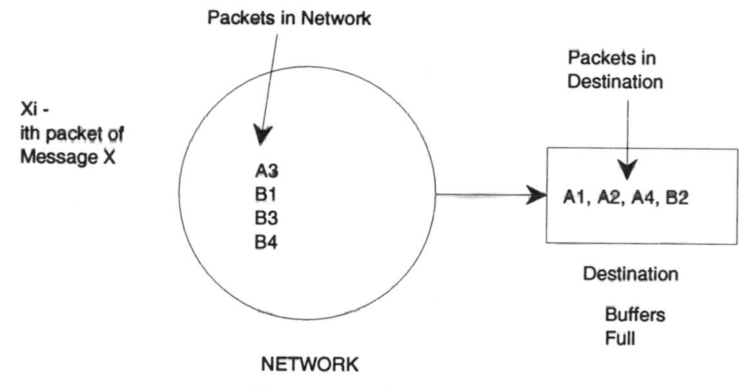

Figure 6.15. Problem 6.7.

7

MULTIPLE ACCESS COMMUNICATION PROTOCOLS

7.1 INTRODUCTION

A multiple access communication protocol is a set of rules to control access to a shared communication channel capacity among the various contending users. Access protocols use different multiple access media. Examples of such multiple access media are satellite channels, radio networks, and optical fiber and copper wires used in local area networks (LANs) and metropolitan area networks (MANs). Multiple access communication protocols are sometimes referred to, especially in the standards arena, as medium access control (MAC) procedures. In relation to the open-systems-interconnection (OSI) reference model, the data link layer of the network is divided into the medium access control (MAC) lower sublayer and the logical link control (LLC) upper sublayer. In this chapter, we discuss the different multiple access communication protocols. The following chapter discusses some of the protocol standards regarding the MAC lower sublayer and the LLC upper sublayer.

It is important to distinguish between multiplexing (which was discussed in chapter 2) and multiple access. Both refer to the sharing of a communication resource (channel); however, there is a difference. With multiplexing, the system controller (which may be an algorithm, or a human, and may be centralized or distributed) has instantaneous knowledge of all user's requirements or plans for sharing the communication channel. There is no overhead needed to organize the resource allocation, and it is typically implemented at a local site. On the other hand, multiple access usually involves the remote accessing of a resource, and there may be a finite amount of time required for the controller to become aware of each user's channel requirements.

Most of the access protocols take advantage of the fact that all these multiple access media are *broadcast channels*. A broadcast channel is a communication channel in which a signal generated by one transmitter can be received by many receivers. Each multiple access medium has different characteristics, which may influence the design of the multiple access protocol. For example, in packet radio, the radio frequency (RF) spectrum is a scarce commodity, and hence the multiple-access protocols must be devised to allow the dynamic allocation of the spectrum to a large population of bursty mobile users.

On the other hand, as pointed out in Chapter 1, satellite channels are characterized by a long propagation delay of approximately 0.25 seconds. This delay is long compared to the transmission time of a packet and thus affects the design of a satellite multiple access protocol, its procedure for allocating bandwidth, and the error and flow control protocols.

Optical fiber and copper wires when used in LANs and MANs are other examples of multiple access channels. Local and metropolitan area networks are characterized by a large and often variable number of devices requiring interconnection. Thus, we face the situation in which a high-bandwidth channel with short or medium propagation delays is to be shared by independent users.

7.2 PERFORMANCE EVALUATION AND CLASSIFICATION OF MULTIPLE ACCESS PROTOCOLS

Multiple-access protocols are evaluated according to various criteria. The performance parameters are average throughput, average packet delay, stability, and miscellaneous factors. These parameters are discussed below:

7.2.1 Average throughput

High bandwidth utilization is a major objective of access schemes. Average throughput provides a measure for the percentage of capacity used in accessing the channel. More precisely, the average throughput is defined as the ratio of the number of packets that are successfully transmitted in a very long interval to the maximum number of packets that could have been transmitted with continuous transmission on the channel.

7.2.2 Average Packet Delay

The time it takes for a packet to access the channel is the packet delay (or response) time. The delay is the time from the instant that a packet arrives at a source to the instant that it is successfully received. Thus, the average packet delay is the ratio of the total delay of the packets in a very long interval to the number of packets in the interval.

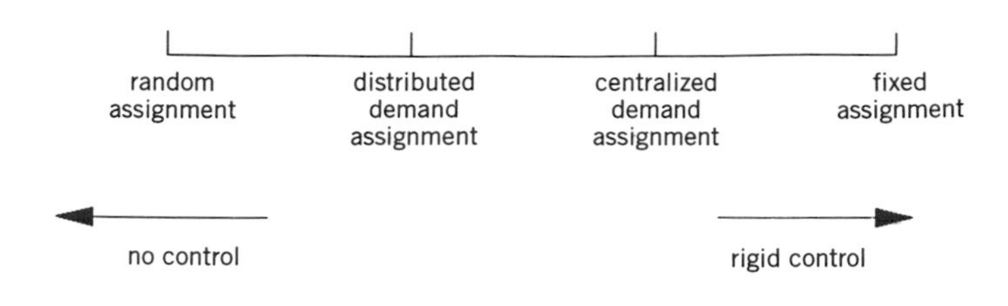

Figure 7.1. Multiple access protocols spectrum.

7.2.3 Stability

Although for some multiple access schemes, their delay and throughput properties might be satisfactory in the short term, they are quite poor when observed over a long interval of time. These schemes are unstable (eventually they reach a situation where the number of stations having packets ready for transmission becomes large and the throughput tends to zero). Some form of control must be applied to these schemes to achieve stability.

7.2.4 Miscellaneous Factors

Multiple access schemes may include other features. Of particular concern in some situations is the ability for an access protocol to simultaneously support traffic of different types, different priorities, with variable message lengths, and differing delay constraints.

Multiple access schemes may be classified according to the bandwidth allocation scheme, which may be static or dynamic. Also, the bandwidth allocation scheme may be implemented using a centralized controller or in a distributed manner. In general, multiple access schemes can be classified into the following five main categories (see Fig. 7.1):

1. Fixed assignment techniques
2. Random access
3. Demand assignment with distributed control
4. Demand assignment with centralized control
5. Hybrid modes

7.3 FIXED ASSIGNMENT SCHEMES

Fixed assignment protocols partition the channel into subchannels in the time, frequency, or code domain. These subchannels are assigned to the users in a fixed manner independently of their traffic statistics. That is, each station has exclusive ownership of its subchannel whether or not it is actively generating traffic. The three well-known fixed-assignment schemes are frequency division multiple access (FDMA), time division multiple access (TDMA), and code division multiple access (CDMA).

7.3.1 Frequency Division Multiple Access

In FDMA systems, each station is permanently allocated a portion of the frequency spectrum (that is, a frequency band). This frequency band is dedicated to a given station all the time and the transmitted signal spectral component must be confined to the allocated frequency band, otherwise it will cause interference with adjacent channels.

7.3.2 Time Division Multiple Access

In TDMA systems, each station is allocated the entire available bandwidth of the transmission media, but only for a limited portion of the time, that is a time "slot."

Both FDMA and TDMA schemes work well under heavy traffic and with a small and fixed number of stations. Under these conditions, we guarantee full utilization of the channel bandwidth. Basically every station is continuously using its allocated channel capacity. But if the number of stations is large and continuously varying (stations are added and deleted), and if the traffic is bursty (typical of interactive applications), dividing the available bandwidth is inefficient. The basic problem is that the bandwidth allocated to the station remains unused even when they have nothing to send, thereby wasting the channel bandwidth.

7.3.3 Code Division Multiple Access

The following example (Figure 7.2) aids in better understanding CDMA. Assume a communication channel of bandwidth W. M users are using the channel. W is divided into b frequency bins (i.e., subchannels) and each user is assigned a frequency hopping pattern (or code). The pattern specifies the code (or sequence) of the frequency bins in which the user is permitted to transmit during a time slot.

If the codes assigned to the different users do not overlap in frequency at any time (i.e., they are orthogonal codes), then CDMA is equivalent to TDMA or FDMA. Orthogonal codes are limited and it is reasonable to assume that codes may partially overlap causing collisions. The advantage of a CDMA system is that collisions are not destructive; each of the signals involved in a collision would be received with only a slight increase in error rate. The use of long codes and extensive error control procedures make it possible to allow several, nonorthogonal, simultaneously transmitted signals without serious performance degradation. Thus, the effect of interference is

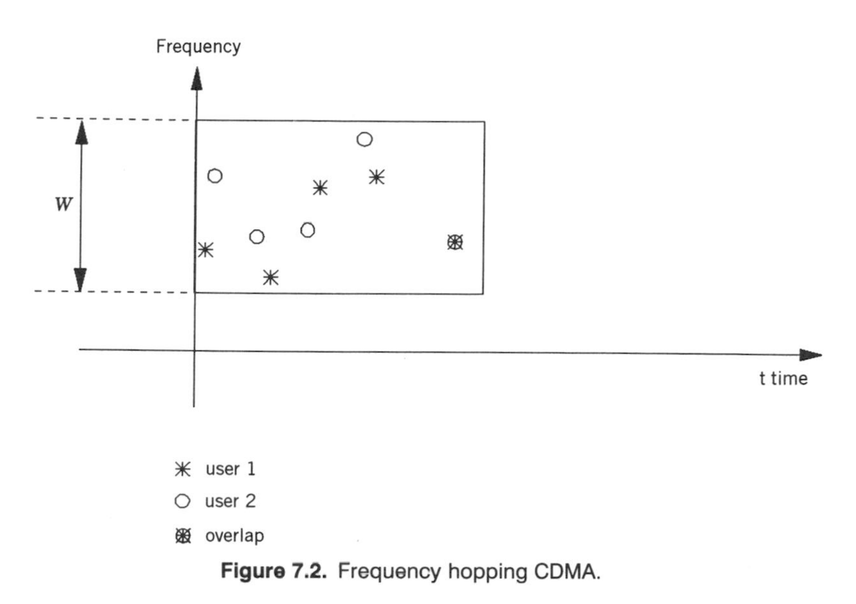

Figure 7.2. Frequency hopping CDMA.

minimized by the ability of the receiver to lock on one packet (signal) while all other overlapping packets appear as noise (this capability is sometimes referred to as the *capture effect*).

The number of simultaneous transmissions that a CDMA system can tolerate depends upon the modulation and coding scheme used, the receiver implementation, the propagation characteristics of the communication medium, and the required acceptable performance (such as bit error rate).

7.4 RANDOM ACCESS TECHNIQUES

Random access protocols (sometimes called contention protocols) are characterized by the lack of strict ordering of the stations contending for access to the channel. In a random access technique, a station is free to broadcast its messages at a time determined locally without any coordination with the other stations. That is, each station transmits packet bursts at random. Systems, in which multiple stations share a common channel in a manner that can lead to conflicts, are widely known as contention schemes. The simplest protocol of this type is the well-known ALOHA protocol, wherein stations completely ignore the channel state and transmit whenever they have data to send. Collisions may occur due to the overlap of two or more packets. Colliding packets will be destroyed and will have to be retransmitted again after a random amount of time.

A variation of the ALOHA protocol that can reduce the potential for collision is called the carrier sense multiple access (CSMA) protocol. It incorporates a listen-before-transmitting feature, whereby, a station listens for a transmission (senses a carrier) and decides according to the state of the channel (the channel could be busy or idle) to either transmit or wait. In spite of the added listen-before-transmit feature, collisions may still occur because of nonzero propagation delays and the nonexistence of any coordination procedures between stations. Yet, CSMA provides better performance than the ALOHA protocol.

To reduce the bandwidth wasted by collisions, the additional feature of listening while transmitting is added to the CSMA protocol. This allows the colliding stations to cut off their transmission as soon as they detect collisions. This protocol, known as CSMA with collision detection (CSMA/CD), is used by Ethernet and is the basis of the IEEE 802.3 standards. Still, like CSMA, CSMA/CD cannot completely eliminate collisions. In this section, we will study the following random access protocols; ALOHA, ALOHA with capture, CSMA, and CSMA/CD.

7.4.1 ALOHA

The ALOHA system was first implemented at the University of Hawaii using the terrestrial radio medium. The ALOHA protocol can be used by any broadcast medium (including a satellite channel, coaxial cable, etc.). There are two versions of the ALOHA protocol: pure-ALOHA (P-ALOHA) and slotted-ALOHA (S-ALOHA).

In P-ALOHA, stations transmit their packets any time they desire. Since the channel is a broadcast channel, if the station hears only its own transmission, it assumes that no conflict occurred (i.e., a positive acknowledgment) and the packet is considered a successfully transmitted packet. On the other hand, if the station hears something other than what it has transmitted, it assumes that its packet has overlapped with one or more other stations' packets (i.e., it has collided with other packets on the channel) and must retransmit (i.e., the station assumes a negative acknowledgment).

If all collided stations (stations that have their packets collided with other stations' packets during transmission) retransmit immediately upon hearing a conflict (remember there is no cooperation among the stations) then they are sure to overlap again, and so some procedures must be followed to randomly and independently reschedule the retransmission of the collided packets.

Figure 7.3 shows a typical example of the ALOHA channel. In this example, three users; U_1, U_2, and U_3 share an ALOHA channel. We assume, for simplicity, that the propagation delay is zero (i.e., the three stations are very close to each other).

In case of P-ALOHA, Figure 7.3*a*, U_1 transmits a packet and just before the end of packet transmission, U_2 starts transmitting a packet; as a result both packets partially overlap, as shown on the channel-timing diagram. Thus both U_1 and U_2 must retransmit after a random time. U_3 transmits a packet and its tail overlaps with U_2's retransmission and hence both transmissions are

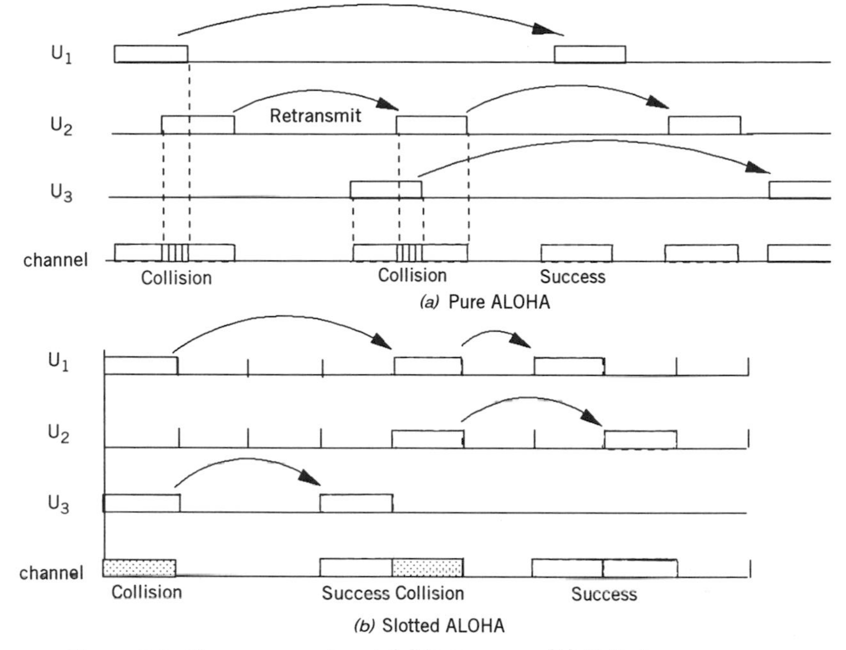

Figure 7.3. The ALOHA protocol. (*a*) Pure ALOHA. (*b*) Slotted ALOHA.

destroyed, U_1 retransmits successfully (the only packet on the channel). U_2 retransmits for the second time and succeeds, and then U_3 retransmits and succeeds. At this moment, all three users have successfully transmitted their packets (three successful packets). Hence, the output of the channel is three packets during the observed time. Note that, in this example, the channel actually carried seven packets (four packets in collision and three successful packets) although the output of the channel, or throughput of the channel, is only three packets during the observed time. Later we define channel throughput and show that for P-ALOHA only a maximum of approximately 18 percent of the channel time is used in transmitting successful packets, whereas the other 82 percent of the channel time is used in transmitting collision packets and idle time.

In S-ALOHA, the channel time is slotted. The packet transmission time T is exactly equal to the slot time (i.e., all packets must have the same length). All stations must begin transmitting their packets only at the beginning of a slot, and as a result, S-ALOHA has an improved channel utilization compared with P-ALOHA.

As shown in Figure 7.3*b*, U_1 and U_3 must begin transmitting their packets at the beginning of the slot (even though their packets may have been generated earlier than the beginning of the slot). As a result, their packets completely overlap and are destroyed; U_3 successfully retransmits the packet, U_1's retransmitted packet collides with a newly transmitted packet from U_2.

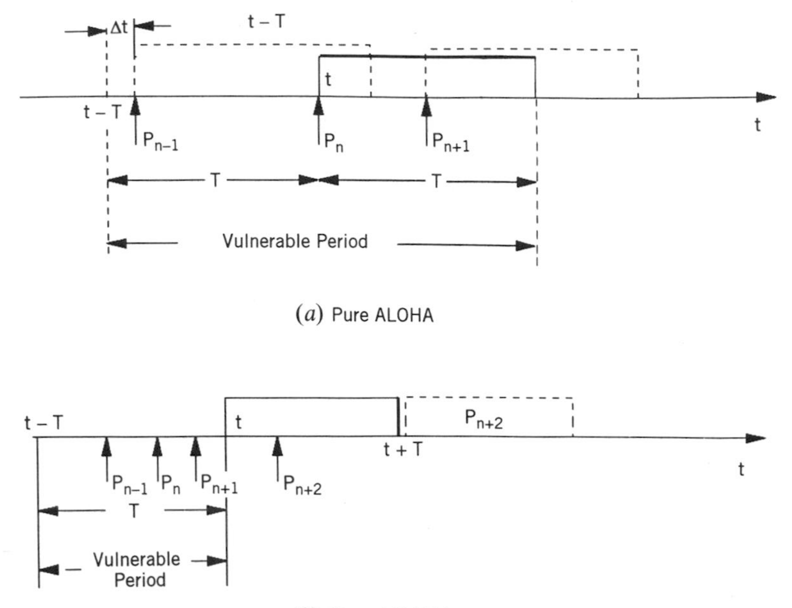

Figure 7.4. Vulnerable period for ALOHA channel. (*a*) Pure ALOHA. (*b*) Slotted ALOHA.

U_1 retransmits again and the packet succeeds, and similarly U_2 retransmits and the packet succeeds.

To better understand the behavior of P-ALOHA and S-ALOHA, let us define the *vulnerable period* of a transmitted packet P as the period during which if another packet $\bar{P}$ begins its transmission, P and $\bar{P}$ will collide. As shown in Figure 7.4, let P_n be the nth packet transmitted on the channel, P_{n-1} is the transmitted packet before P_n, and P_{n+1} is the one after P_n.

Let the packet transmission time be T seconds, assume fixed length packets, and assume P_n is transmitted at time t. In the case of P-ALOHA (Fig. 7.4*a*), if P_{n-1} is exactly transmitted at time $(t - T)$ then P_n will not collide with P_{n-1} and both will be successful. Now if P_{n-1} is transmitted at time $t - (T - \Delta t)$, then P_n and P_{n-1} will overlap for Δt and hence a collision will occur. Therefore, if P_{n-1} is transmitted at any time of T seconds or less before the transmission time of P_n a collision will occur.

Similarly, P_{n+1} will collide with P_n if P_{n+1} transmission begins at any time less than T seconds after the beginning of the transmission of P_n. If any station begins its transmission during the period of $2T$ seconds (i.e., from time $t - T$ to $t + T$) it will definitely collide with packet P_n. Hence the vulnerable period for P-ALOHA is $2T$.

As mentioned earlier, in S-ALOHA, whenever a station generates a packet, it must delay its transmission until the beginning of the next possible channel slot. As shown in Figure 7.4*b*, all packets generated just before time t

(packets; P_{n-1}, P_n, and P_{n+1}) will begin their transmission at time t, whereas P_{n+2} is generated just after t and thus its transmission is delayed until $t + T$. As a result the vulnerable period for slotted ALOHA is only T seconds. Now we proceed to analyze the throughput for the ALOHA protocols.

7.4.1.1 Throughput Analysis. The ALOHA channel (whether P-ALOHA or S-ALOHA) can be modeled as a feedback system (see Fig. 7.5). Let the average number of successful transmissions per transmission period T (i.e., channel throughput) be S, the average number of packet transmissions attempted per T seconds (i.e., the average channel traffic) be G, and the probability of successful transmission be γ.

Note that, because the channel cannot store and destroy packets, the user input traffic to the channel must be the output from the channel. G represents the average channel traffic, a certain percentage of G will succeed and hence the channel throughput S equals that percentage, that is,

$$\begin{aligned} S &= \text{channel throughput} \\ &= \text{Channel traffic} \times \text{Probability (no additional packet is transmitted)} \\ &= G \times \gamma \qquad (7.1) \end{aligned}$$

Also, note that the collided percentage of G [i.e., the collided packets $(1-\gamma)G$] is retransmitted (i.e., fed back to the channel), hence the channel traffic G is the sum of the newly input traffic S plus the feedback traffic $(1-\gamma)G$.

Here we will assume that every station can hold a maximum of one packet in the station's buffer including any previously collided packets. It is also assumed that the channel traffic follows a Poisson process with an average rate of G/T packets attempting transmission per second. Hence,

$$\text{Probability \{that } K \text{ packets arrive at the channel in time } t\text{\}}$$

$$= \frac{(Gt/T)^k}{K!} e^{-(G/T)t} \qquad (7.2)$$

and γ is the probability that no additional packets are generated during the

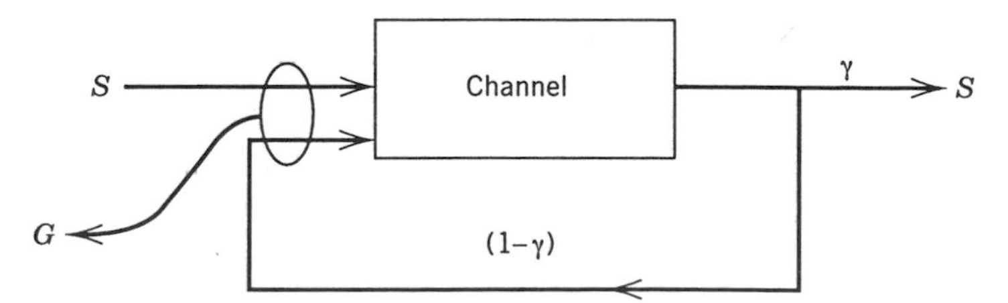

Figure 7.5. Modeling the ALOHA channel.

vulnerable period $2T$ for P-ALOHA or

$$\gamma = e^{-(G/T)2T} = e^{-2G}$$

In Eq. 7.1, we get

$$S = Ge^{-2G} \tag{7.3}$$

The maximum throughput for P-ALOHA S_{max} occurs at $G = 1/2$ and is given by

$$S_{max} = \frac{1}{2e} \simeq 0.184$$

For S-ALOHA, the vulnerable period is T and hence γ is the probability that no additional packets are generated during the vulnerable period T for S-ALOHA or

$$\gamma = e^{-(G/T)T} = e^{-G} \tag{7.4}$$

From Eq. 7.1, we obtain

$$S = Ge^{-G} \tag{7.5}$$

The maximum throughput for S-ALOHA, S_{max}, occurs at $G = 1$ and is given by

$$S_{max} = \frac{1}{e} \simeq 0.368$$

which is twice that of a pure-ALOHA system.

Equations 7.3 and 7.5 give the throughput equation for the case of an infinite population of users whose aggregate rate of new packet generations is S. In practical systems, stations are finite in population, and they have enough buffer to hold more than one packet. In this case, simulation results and approximate analysis have shown higher throughput values (see Problems 7.3 and 7.5).

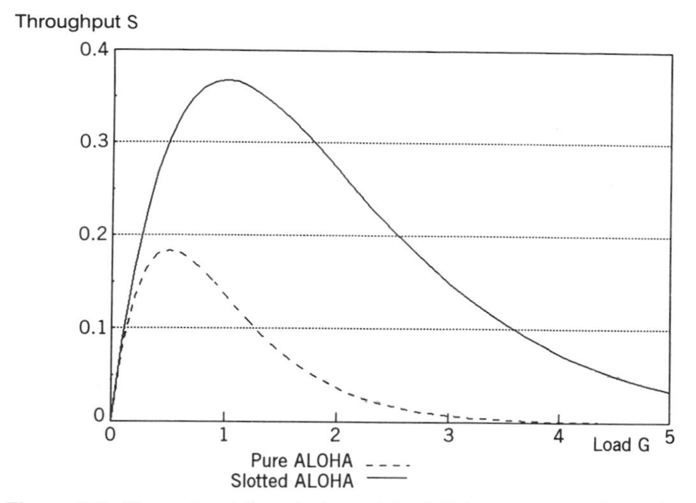

Figure 7.6. Throughput S and channel load G for an ALOHA channel.

In Figure 7.6, we plot S against G. Note that $G > 1$ is possible, because more than one packet/slot may attempt a transmission. However, in this case, the collision rate increases, and as a result the channel throughput decreases, eventually reaching a very low value.

7.4.1.2 Delay Analysis. We now proceed to analyze the delay for the ALOHA protocols. Again, we assume a large number of stations, each generating new packet arrivals from a Poisson process with mean arrival rate S packets every T seconds, or $\lambda = S/T$ packets/second, where T is the packet transmission time. The channel traffic follows a Poisson process with mean arrival rate G packets every T second. Each station has at most one packet ready for transmission, including any previously collided packets.

The end-to-end propagation delay is τ seconds. For a satellite channel (see Fig. 7.7), τ is approximately 0.25 seconds. A station hears its own transmission, and decides whether or not transmission is successful after τ seconds or equivalently after R slots where R is the smallest integer greater than or equal to τ/T. As mentioned earlier, after unsuccessful transmission, all collided stations reschedule their transmission randomly to smooth the retransmitted traffic and reduce the possibility of a second collision. One strategy is to retransmit during the next K slots, with each slot being chosen at random with probability $1/K$. By analogy, imagine having a roulette with numbers 1 to K; the ball will settle on one of the numbers (the slot number to transmit at) with probability $1/K$. Thus, the average number of slots a station has to wait after detection of a collision before transmission $\bar{i}$ is given

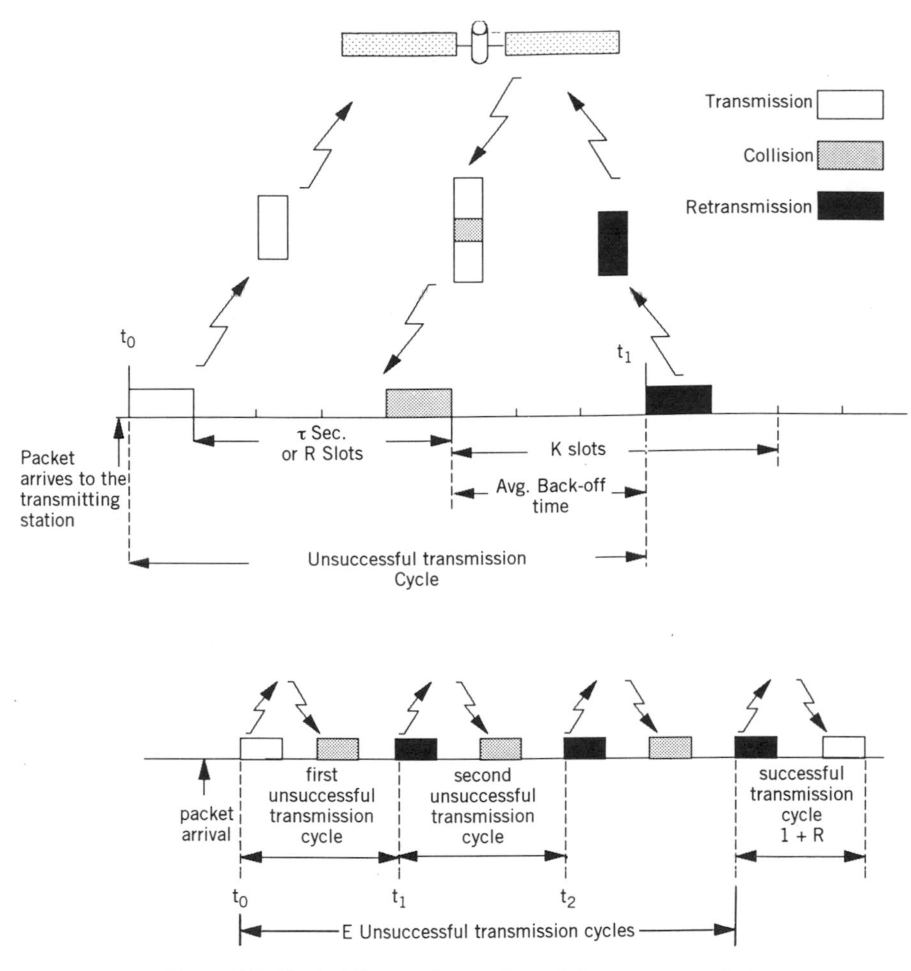

Figure 7.7. Typical timing diagram for a slotted-ALOHA packet.

by

$$\bar{i} = \sum_{i=0}^{K-1} i \times \text{Probability of transmitting during } i\text{th slot}$$

$$= \sum_{i=0}^{K-1} i \frac{1}{K} = K \frac{K-1}{2} \cdot \frac{1}{K}$$

$$= \frac{K-1}{2}$$

We will refer to the time $\bar{i}T$ seconds as the *average backoff time*.

Define the unsuccessful transmission cycle (Fig. 7.7) as the time between the beginning of the packet transmission time t_0 and the beginning of the packet's first retransmission time t_1. Thus, the average unsuccessful cycle time T_u equals the packet transmission time T plus the end-to-end propagation delay τ plus the backoff time $\bar{i}T$, that is, the average unsuccessful transmission cycle time

$$T_u = 1 + R + \frac{K-1}{2} \text{ slots}$$

$$= R + \frac{K+1}{2} \text{ slots}$$

$$= T\left[R + \frac{(K+1)}{2}\right] \text{ seconds}$$

Because a packet may collide more than once, we define E as the average number of unsuccessful transmissions. Note that the successful transmission cycle time (last retransmission attempt) is equal to the packet transmission time plus the propagation time. In slotted-ALOHA, the transmission of any packet that arrives at the station during a slot time has to be delayed to the beginning of the following slot. The arrival process is assumed to be Poisson; therefore, a packet may arrive at any point during a slot time with equal probability. The average waiting time between arrival and the beginning of the following slot is thus half the slot time (i.e., a packet, on the average, arrives in the middle of the slot). Now we can write the equation for the average time D a packet experiences from the moment it arrives at the transmitting station until it is successfully received as follows (refer to Fig. 7.7):

$$D = \frac{1}{2} + \frac{T_u}{T} \times E + [1 + R]$$

$$= \frac{1}{2} + \left[R + \frac{K+1}{2}\right] \times E + [1 + R] \text{ slots} \tag{7.6}$$

Now, we need to determine the average number of unsuccessful attempts, E. A simplified analysis is as follows: From Eq. 7.4, the probability that no additional packets are generated during a time slot is given by e^{-G} and the probability that one or more packets are generated by other users is given by $(1 - e^{-G})$. For a packet to go through n transmissions until it succeeds, it will collide for $(n - 1)$ transmissions and succeeds in the nth transmission. Thus, the probability that a packet is transmitted exactly n times, P_n, equals the probability that it collides $(n - 1)$ times [i.e., $(1 - e^{-G})^{n-1}$] multiplied by the probability that it succeeds in the nth attempt [i.e., e^{-G}]. Hence,

$$P_n = (1 - e^{-G})^{n-1} e^{-G}$$

The average number of transmissions $\bar{n}$ is given by

$$\begin{aligned}\bar{n} &= \sum_{n=1}^{\infty} nP_n \\ &= \sum_{n=1}^{\infty} n(1 - e^{-G})^{n-1} e^{-G} \\ &= e^G\end{aligned}$$

Thus,

$$\begin{aligned}E &= e^G - 1 \\ &= G/S - 1\end{aligned} \tag{7.7}$$

Using Eqs. 7.5, 7.6, and 7.7, we obtain the average packet delay D as a function of the channel throughput for slotted ALOHA.

Similarly, we can write the delay equations for pure-ALOHA. In this case, the first term in Eq. 7.6 disappears, because a newly arriving packet is immediately transmitted. Thus, the average packet delay D for pure-ALOHA is given by

$$D = \left(R + \frac{K+1}{2}\right) \times E + [1 + R] \text{ slots} \tag{7.8}$$

Using Eqs. 7.3, 7.7, and 7.8, we obtain the average packet delay D as a function of the channel throughput for pure-ALOHA. Figure 7.8 shows the delay–throughput characteristics for the ALOHA channel.

7.4.2 ALOHA with Capture

The utilization of the ALOHA channel can be improved by dividing users into two groups: one transmits at high power and the other group at low power. For example, assume that there are three simultaneous packet transmissions; one packet from the high-power group and two packets from the low-power group; the packet from the high-power group emerges from the packet collision successfully and is *captured* by the intended receiver. The other two packets from the low-power group are unsuccessful and will attempt retransmission at a later time. Obviously, if the three packets all come from users in the high-power group, all three packets are unsuccessful. In other words, high-power packets are not affected by low-power packets, whereas the opposite is not true. Below we show that the maximum utilization of the

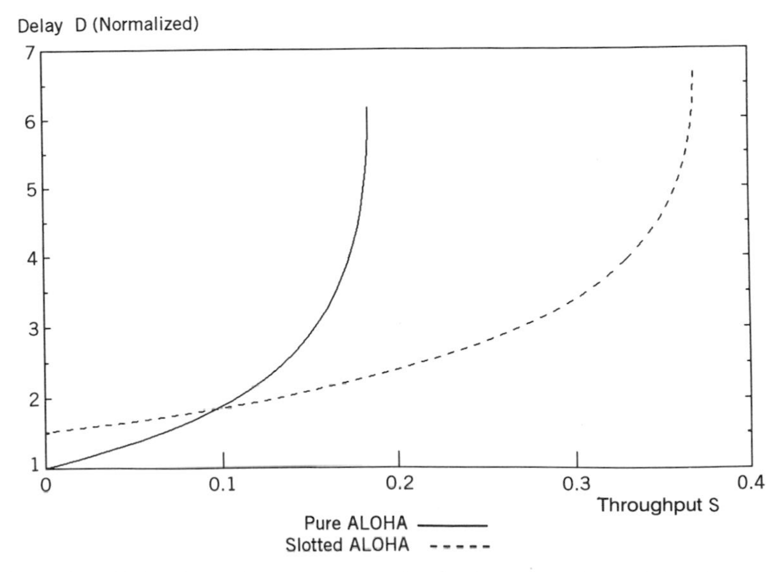

Figure 7.8. Delay-Throughput characteristics for the ALOHA channel, $R = 0$, $k = 5$.

slotted ALOHA channel for the infinite population model used above can be increased from 0.368 to approximately 0.53 by using the capture capability.

Let S_H be the average number of successful high-power packet transmissions per time slot (i.e., high-power packets throughput), and let G_H be the average number of packet transmissions attempted per time slot (i.e., newly generated plus retransmitted high-power packets). Let S_L and G_L be the corresponding parameters for the low-power packets. With similar assumptions of a large population of users and Poisson arrivals, we can apply Eq. 7.5; thus

$$S_H = G_H e^{-G_H} \tag{7.9}$$

Note that for the high-power group transmissions, the lower power signals have no effect, because it cannot destroy the high-power packets.

Successful low-power packet transmission requires two events: no other low-power packet in the same time slot, which happens with probability e^{-G_L}, and no high-power packet in the slot, which happens with probability e^{-G_H}. Assuming the two events are independent, the probability of successful transmission equals $e^{-(G_L+G_H)}$.

Since G_L is the newly generated low-power packets plus the retransmitted low-power packets

$$G_L = S_L + G_L \times \text{Probability of retransmission}$$

$$G_L = S_L + G_L[1 - e^{-(G_L+G_H)}] \tag{7.10}$$

Using Eq. 7.9 in Eq. 7.10,

$$= S_L + G_L\left[1 - \frac{S_H}{G_H}e^{-G_L}\right]$$

Hence,

$$S_L = \frac{S_H}{G_H}G_L e^{-G_L} \tag{7.11}$$

For a given S_H (and hence G_H), S_L is maximum (the point of saturation for low-power signals) when $G_L = 1$, hence

$$S_{L\,\max} = \frac{S_H}{G_H}\left(\frac{1}{e}\right) \tag{7.12a}$$

$$= e^{-G_H}\left(\frac{1}{e}\right) \tag{7.12b}$$

As seen the value of $S_{L\,\max}$ depends on S_H (or G_H). $S_{L\,\max}$ has its largest value when the high-power packets do not exist, that is, when $G_H = 0$. $S_{L\,\max}$ decreases as S_H increases. Defining the total average utilization S as the sum of $S_{L\,\max} + S_H$, we get

$$S = S_{L\,\max} + S_H$$

$$= \frac{1}{e} \cdot e^{-G_H} + G_H e^{-G_H}$$

$S_{\max}$ occurs at $G_H = 1 - 1/e$; therefore,

$$S_{\max} = e^{-(1-1/e)} = 0.53$$

Thus, the maximum utilization of the ALOHA channel can be increased to 0.53 by the use of the capture capability. Note that this increase is obtained at the expense of a somewhat unfair treatment of low-power transmitters.

7.4.3 Carrier Sense Multiple Access (CSMA) Schemes

In CSMA protocols, like in ALOHA, there is no central controller managing access to the channel, and there is no preallocation of time slots or frequency bands. A station that has data to transmit to another station first senses the channel (that is, uses carrier sensing) and decides to either transmit or wait according to the state of the channel. If a carrier is sensed on the channel (i.e., another station is sending data), the station defers transmission of its packets until the channel becomes idle (i.e., no other transmissions taking place). If the channel is sensed idle, the station may transmit. Thus, carrier sensing adds the requirement that stations listen to the common channel and only attempt to transmit if this channel is found idle. This added feature of listening-before-transmitting reduces the probability of collision between packets when compared with the ALOHA protocol. Collisions may still occur from nonzero propagation delays (i.e., a busy channel may be sensed idle while another station is transmitting, because the signal has not yet reached the sensing station) and from the lack of coordination between the stations.

Three basic variations of CSMA exist: nonpersistent, p-persistent, and 1-persistent. A *nonpersistent* CSMA station upon sensing the channel busy does not persist in listening to the media in order to transmit; rather, it schedules transmission for some future time, according to a certain retransmission delay distribution. At the scheduled time, if the channel is found idle, the station transmits its packet; otherwise it repeats the nonpersistent algorithm. In the *p-persistent* station, the station sensing the channel idle transmits with probability p and delays its transmission with probability $(1 - p)$. The *1-persistent* method is a special case of p-persistent for which the probability $p = 1$. Note that all three variations of CSMA differ mainly in the action that a terminal takes after sensing the channel.

In the slotted version of these CSMA protocols, the time axis is slotted, and the slot size equals the maximum propagation delay τ of all source-destination pairs. Normally, these slots are called *minislots*. Here, the packet transmission time T is much larger than τ. Hence, the packet transmission time is equivalent to many minislots. All stations start their transmission at the beginning of the minislot. As in slotted ALOHA, when a packet's arrival occurs in a minislot, the station waits until the next minislot and proceeds according to the protocols described above.

One of the most successful implementations of a variant of the CSMA protocol is the Ethernet LAN. The Ethernet is a multiaccess, packet-switched LAN. The shared communications channel is a single coaxial cable. In addition to sensing the carrier, a collision detection mechanism is used. So a transmitter listens to the cable and compares the data bits transmitted with those present on the cable. Any difference indicates a collision, upon which transmission must stop. This feature of listening-while-talking (that is collision detection) has proven to improve the performance of CSMA protocols, because collided packets can be aborted as soon as they are detected, thus reducing the time wasted from collisions. This scheme is called CSMA with collision detection (CSMA/CD). Below, we will provide the throughput analysis for both the nonpersistent CSMA and CSMA/CD.

7.4.3.1 Analysis of the Nonpersistent CSMA. As discussed earlier, nonpersistent CSMA works as follows; a station with a ready packet (ready station) senses the channel and follows this procedure: If the channel is sensed idle, it transmits the packet. In case of collision, the station reschedules another test for the channel, choosing the next test time according to a retransmission distribution and repeats the algorithm. If, however, the channel is sensed busy, the station reschedules the transmission choosing a time from the retransmission distribution and repeats the algorithm.

We assume an infinite user population sharing the channel (say a coaxial cable). The total generated new traffic is S packets every T second. All packets have the same transmission time, T seconds. As mentioned earlier, S is also the channel utilization, or throughput. G is again defined as the average channel traffic (new and retransmitted packets) and follows a Poisson process with an arrival rate of G/T packets per second. Each terminal has a buffer size of exactly one packet size. The propagation time is assumed to be τ seconds between all source-destination pairs, where τ is the maximum one-way propagation delay on the bus. The one-way normalized delay a is defined as,

$$a = \tau/T \tag{7.13}$$

Again the channel is a broadcast channel (that is, all terminals can hear each other), and it is assumed to be noiseless so that corrupted transmission is due to collision only.

Now consider a nonpersistent CSMA bus network, as shown in Figure 7.9. Station k transmits its packet (labeled 0) at time t (the channel is idle at time t). The packet propagates onto the bus and takes τ seconds to reach all other stations. Therefore, during τ seconds, stations j, and m each have a packet ready for transmission but do not sense packet 0; as a result, they transmit

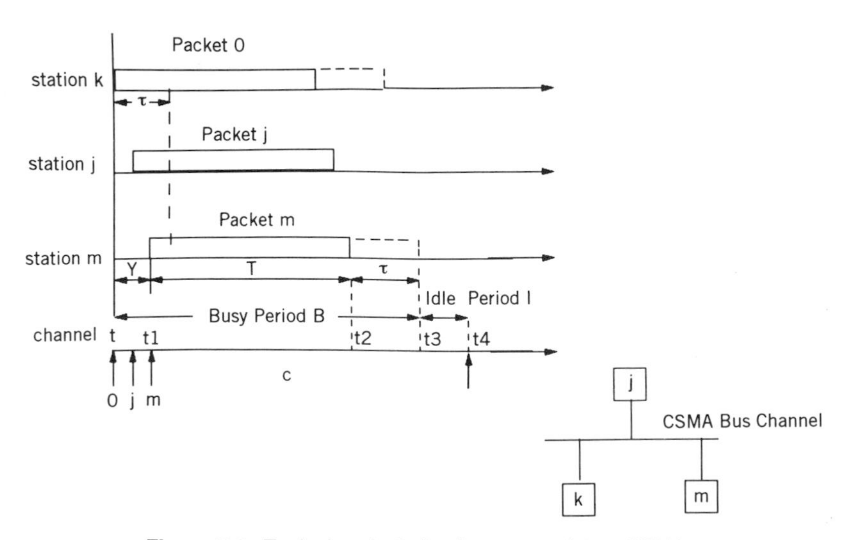

Figure 7.9. Typical cycle timing for nonpersistent CSMA.

causing overlapping. Packet m is the last packet transmitted at time $t_1 = t + Y$, where $0 < Y < \tau$. Note that any station having a packet that arrives after time $t + \tau$ will find the channel busy and according to the protocol will defer its transmission and repeat the algorithm. On the other hand, all stations will sense the end of packet m transmission by time $t_3 = t + Y + T + \tau$ and new transmission may start at time t_4. The time interval $t_3 - t$ is the busy period B, the time interval $t_4 - t_3$ is the idle period I, and $t_4 - t$ is the cycle time C. In Figure 7.9, the busy period contains collision. Other times, the busy period may contain successful transmission. In this case, the duration of the busy period is $Y + T + \tau$.

It is clear by now that the vulnerable period for the nonpersistent CSMA is τ seconds. Normally, τ is much smaller than the packet transmission time T (τ is typically of the order of 0.1 T or less). Recalling that for pure-ALOHA, the vulnerable period is $2T$, whereas that for slotted-ALOHA is T, we then expect the CSMA protocols to perform better than the ALOHA protocols.

Now we have

$$C = B + I$$

where C, B, and I are all random variables. Taking the average of both sides of the previous equation, we get

$$\bar{C} = \bar{B} + \bar{I}$$

Then, the channel utilization, or throughput S, is given by

$$S = \frac{\text{average duration of successful transmission}}{\bar{C}}$$

$$= \frac{\text{average duration of successful transmission}}{\bar{B} + \bar{I}} \tag{7.14}$$

The average duration of successful transmission equals the packet transmission time multiplied by the probability that no other packet arrives during the vulnerable period τ. Thus,

$$\text{average duration of successful transmission} = Te^{-(G/T)\tau}$$

$$= Te^{-aG} \tag{7.15}$$

where a is the one-way normalized propagation time as given in Eq. 7.13.

The average duration of an idle period equals the average interarrival time between packets arriving at the channel

$$\bar{I} = \frac{1}{G/T} \tag{7.16}$$

The average busy period $\overline{B}$ is given by

$$\overline{B} = \overline{Y} + T + \tau \tag{7.17}$$

To determine $\overline{Y}$, we need first to determine the probability distribution function of Y, $F_Y(y)$

$$\begin{aligned}
F_Y(y) &= P\{Y < y\} \\
&= \text{Probability that a zero packet is generated in the interval } (y, \tau) \\
&= e^{-(G/T)(\tau - y)}
\end{aligned}$$

The probability density function of Y, $f_Y(y)$, is given by

$$f_Y(y) = \frac{d}{dy} F_Y(y) = \frac{G}{T} e^{-G(a - y/T)}$$

Hence

$$\begin{aligned}
\overline{Y} &= \int_0^\tau y f_Y(y)\, dy \\
&= \tau - \frac{T}{G}(1 - e^{-aG})
\end{aligned}$$

Substituting into Eq. 7.17, we get

$$\overline{B} = 2\tau + T - \frac{T}{G}(1 - e^{-aG}) \tag{7.18}$$

and substituting Eqs. 7.15, 7.16, and 7.18 into Eq. 7.14, we get

$$\begin{aligned}
S &= \frac{Te^{-aG}}{2\tau + T - \dfrac{T}{G}(1 - e^{-aG}) + \dfrac{T}{G}} \\
&= \frac{Ge^{-aG}}{G(1 + 2a) + e^{-aG}}
\end{aligned} \tag{7.19}$$

Figure 7.10 shows the relationship between S and G for several values of a. Note that as a decreases, the vulnerable period decreases, and, as a result, the maximum throughput increases up to unity. This is also evident from

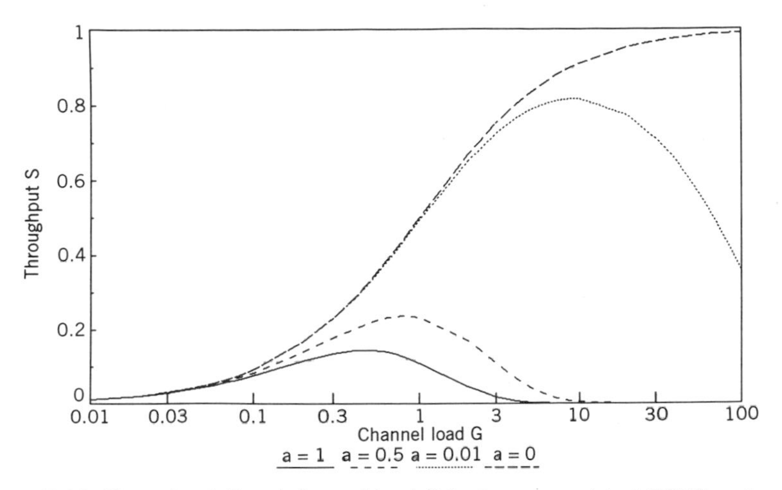

Figure 7.10. Throughput S and channel load G for the nonpersistent CSMA protocol.

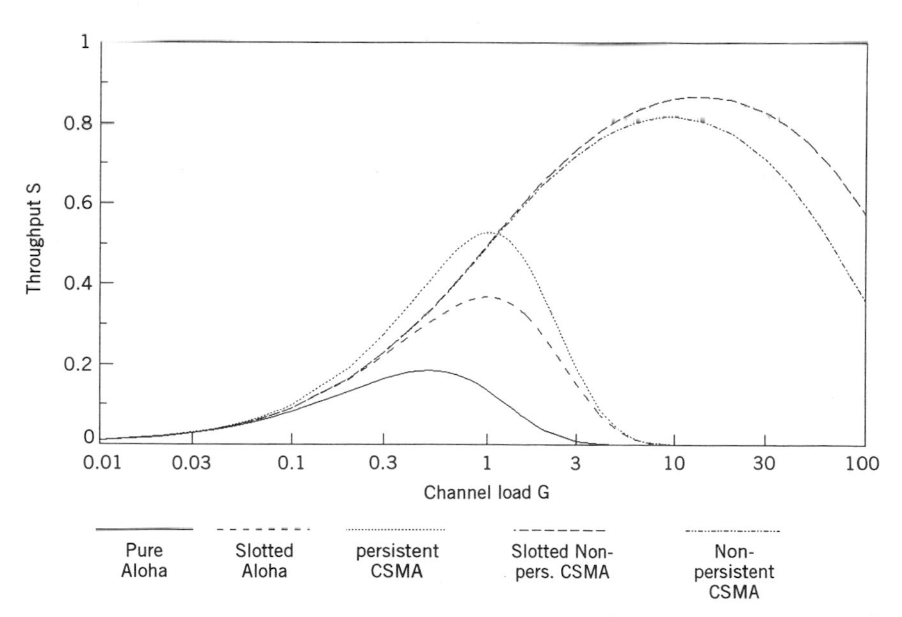

Figure 7.11. Throughput for various random access modes (propagation delay $a = 0.01$)

TABLE 7.1 Maximum Throughput for Different Access Protocols ($a = 0.01$)

Protocol	Maximum Throughput
Pure ALOHA	0.184
Slotted ALOHA	0.368
1-Persistent CSMA	0.529
Slotted 1-Persistent CSMA	0.531
0.1-Persistent CSMA	0.791
Nonpersistent CSMA	0.815
0.03-Persistent CSMA	0.827
Slotted nonpersistent CSMA	0.857

Eq. 7.19, when $a = 0$, S is given by

$$S_{a \to 0} = \frac{G}{1 + G} \tag{7.20}$$

Similar throughput equations can be found for the other CSMA protocols; however, the analysis is more complex. Figure 7.11 shows the relationship of S and G for various random-access schemes with $a = 0.01$. Note that the throughput of the ALOHA channel is not a function of the propagation delay, whereas that of the CSMA channel depends on the propagation delay. For $a = 0.01$, any CSMA protocol is superior to ALOHA protocols. The CSMA channel capacity (that is, the maximum throughput) in some cases may be as high as 90 percent of the available bandwidth. Table 7.1 shows the maximum attainable throughput (i.e., capacity) for various access protocols when $a = 0.01$.

7.4.3.2 Throughput Analysis of CSMA / CD. In CSMA/CD, the maximum time it takes to detect a collision is equal to the end-to-end round-trip delay. To understand why this time equals 2τ, let us refer to Figure 7.9. Assume station k starts transmitting at t. At time $t + (\tau - \epsilon)$, station m senses the channel idle and transmits its packet. Obviously, station m detects the collision at time $t + \tau$ and truncates its transmission. The first bit in station m's packet will arrive at station k at time $t + (\tau - \epsilon) + \tau$, and station k detects collision at time $t + 2\tau - \epsilon$. Thus, in the worst case a station can be sure of successful transmission only after transmitting its packet for 2τ and hearing no collision.

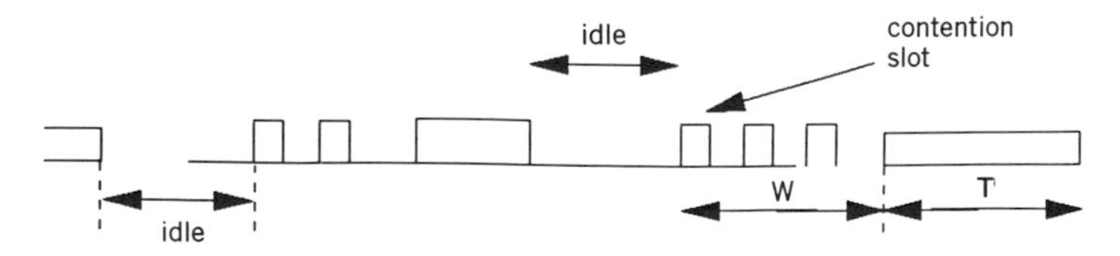

Figure 7.12. CSMA / CD channel.

In the following we present a simple analysis for CSMA/CD under the assumption of heavy traffic (i.e., all k stations on the network have packets ready for transmission). We also assume that channel time is divided into slots each of length 2τ (the maximum collision detection time). Notice that the channel time is composed of two alternating intervals. The first, called a success interval, occurs when a station succeeds in transmitting its packet. The second, called a contention interval, composed of collision slots of duration 2τ, during which more than one packet attempted transmission. Based on the assumption of a heavy traffic situation, there is no idle time on the channel and the channel time alternates between success and contention intervals.

We assume that a ready station attempts to transmit in the current slot with probability p, or delays with probability $(1 - p)$. The main idea of the analysis is to obtain the duration of the collision interval W (Fig. 7.12). The duration of the success interval equals the packet transmission time, T. Since, the channel time is divided between success intervals and contention intervals, the efficiency of the network (or the throughput) is the fraction of time the channel carries successful packets. That is

$$\text{Channel throughput} = \frac{T}{T + W} \tag{7.21}$$

To find W, we define A as the probability that exactly one station attempts a transmission in a slot and therefore succeeds in transmission. Since there are k ways in which one station can choose to transmit (with probability p) while $k - 1$ stations choose to wait (with probability $1 - p$), A is given by

$$A = kp(1 - p)^{k-1} \tag{7.22}$$

Let B_j be the probability that the contention interval has exactly j slots (each slot equals 2τ).

Notice that the contention interval is exactly one slot when more than one station transmits in the first slot [with probability $(1 - A)$] followed by the transmission of exactly one station in the following slot [with probability A]. Similarly, the contention interval is exactly two slots when more than one transmission occur in each of the first two slots followed by exactly one transmission in the third slot. Hence, B_j equals the probability of having more than one transmission in each of the first j slots followed by exactly one transmission in the $(j + 1)$ slot. Specifically,

$$B_j = (1 - A)^j A$$

and the average number of slots in the contention period $\overline{B}$ is given by

$$\overline{B} = \sum_{j=0}^{\infty} jB_j$$

$$= \frac{(1 - A)}{A} \qquad (7.23)$$

Since each slot is of duration 2τ seconds, W equals $2\tau\overline{B}$ seconds. Substituting into Eq. 7.21, we get

$$S = \frac{T}{T + 2\tau\overline{B}} \qquad (7.24)$$

In general, it is necessary to optimize the probability of successful transmission. In Eq. 7.22, A is maximum when p equals $1/k$. Substituting into Eq. 7.22, we get

$$A_{\max} = (1 - 1/k)^{k-1}$$

which approaches e^{-1} as k goes to infinity. Hence, $\overline{B}$ in Eq. 7.23 equals $(e - 1)$ and $w = 2\tau(e - 1) = 3.4\tau$. The channel efficiency S from Eq. 7.24, becomes

$$S = \frac{T}{T + 3.4\tau} \qquad (7.25a)$$

$$= \frac{1}{1 + 3.4a} \qquad (7.25b)$$

with $a = \tau/T$, as before.

Again, we see here the importance of the parameter a. As a approaches zero the channel efficiency approaches unity, and as a goes to infinity the channel efficiency approaches zero. In practice $a < 0.1$ is desired. Notice also that for a given cable length (i.e., for a given τ) and for a given channel speed, the ratio a becomes inversely proportional to the packet length L. Increasing L results in decreasing a and increasing channel throughput, whereas decreasing L results in increasing a and decreasing channel throughput.

Experimental results also verify this conclusion. Figure 7.13 plots the channel efficiency against the number of stations on the network for varying packet lengths. Increasing the packet length in the Ethernet results in an increase in the channel efficiency. That is why terminal-type traffic (typically small length packets) results in degradation of Ethernet performance.

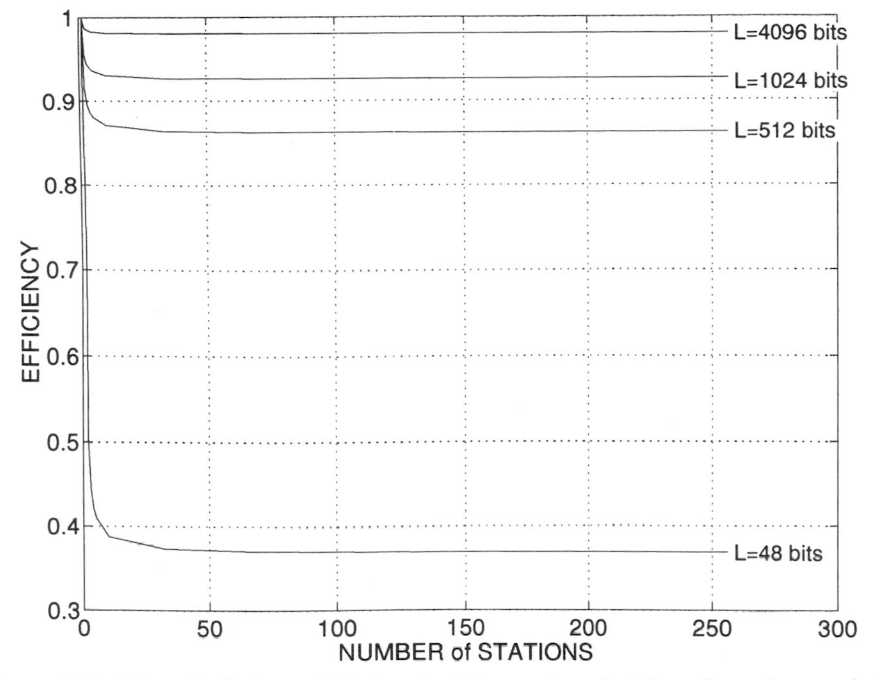

Figure 7.13. Ethernet efficiency plotted against the number of stations for various packet lengths, 3 Mbps and 1 Km cable.

The CSMA/CD protocol has been implemented for both baseband and broadband bus-type local area networks, see Chapter 8. In a broadband LAN, the channel is divided into subchannels, and stations can access a given subchannel using the CSMA/CD protocol. Modems are used at the user's stations to allow modulating information signals into the right subchannel. All transmitted packets are sent on an upstream subchannel to the head-end of the cable, where it is frequency converted to the downstream subchannel. Thus, a packet travels the cable twice to reach the destination. In a baseband LAN, there is only one channel and hence only one packet can be transmitted at a given time. A major difference between baseband and broadband CSMA/CD is the slot size, which is chosen to be at least equal to the maximum time it takes for all stations to detect a collision. In baseband CSMA/CD, we have shown that the slot size is 2τ. In broadband CSMA/CD, because a transmitted packet must go first to the cable head-end, the slot size is 4τ (see Problem 7.11).

7.5 CENTRALIZED DEMAND ASSIGNMENT SCHEMES

In demand assignment access protocols, stations are required to provide explicit or implicit information regarding their needs for communication bandwidth. As discussed before, these schemes are classified into two main groups: centralized demand assignment schemes, in which a central unit allocates the requested bandwidth, and distributed demand assignment schemes, in which a distributed algorithm, executed by all users, controls the channel access times.

Unlike fixed assignment, demand assignment schemes minimize wasted channel bandwidth by avoiding assignment of the channel to idle stations. Also, unlike random assignment, they eliminate channel bandwidth wasted by collision, since collision cannot occur in these protocols. Demand assignment schemes perform well for moderate to high channel loads, particularly for applications that require predictable channel access time and bounded packet delay. However, the overhead in some of these schemes may be large, which leads to inefficiency. In this section, we discuss polling and adaptive polling as examples of multiple-access centralized demand assignment schemes. Section 7.6 examines some examples of distributed demand assignment schemes.

7.5.1 Polling

Polling is appropriate for multidrop lines, for example, a number of terminals trying to access a mainframe computer. Polling is widely used in many data networks. There are various polling techniques. We emphasize here roll-call polling and hub polling.

In roll-call polling, a central controller sends polling messages to the terminals, one-by-one. In essence, the controller is asking the polled terminals to transmit. If a polled terminal has data to transmit, it transmits the data as soon as it receives the poll message. If the polled terminal has nothing to send, it sends back a negative reply to the controller. The controller then polls the next terminal in sequence. Polling requires this cyclic exchange of control messages between the controller and the terminals. The polling message has a terminal address field to identify the terminal being addressed. In this multidrop environment, all terminals listen to the polling message and only the terminal that identifies its own address in the polling message responds.

In hub polling, the controller polls only the terminal furthest away from it, and the terminal sends back data messages or a negative reply. The terminal then sends the poll message to its neighbor, which assumes control. If a station assuming control has no messages, it immediately transfers control to the next station upstream. The algorithm repeats until the cycle is completed, with all terminals polled, the controller then regains control.

In hub polling, the time required to transfer control to a station upstream is much less than the time required in roll-call polling to transmit a polling message and await a reply.

We can immediately see another variation of the polling system. We can eliminate the controller and have the stations pass a polling message, or token, among themselves resulting in a distributed scheme. Such a variation is used in LANs and is referred to as the token scheme, which is discussed in Section 7.6.2.

Generally, polling is efficient when (1) the overhead from the polling messages is low, (2) the round-trip propagation delay is small, and (3) the number of terminals is not large. Note that the overhead in a cycle is proportional to the number of terminals in the system. Adaptive polling (or probing), discussed below, is introduced to deal with cases in which the user population is large.

7.5.2 Probing and Adaptive Polling

In polling, when the system is lightly loaded, for example when few terminals are active, we still need to poll each terminal, which results in high overhead. The heart of adaptive polling, or probing, is based on the idea of isolating and determining which of the terminals have data messages. This function of isolating the active terminal is implemented by splitting the terminal population into subsets.

Consider the centralized network shown in Figure 7.14. Here, we assume that the controller can broadcast signals to all terminals. The controller begins the cycle by sending a signal, referred to as probe message. The probe message is a specially coded message to which only stations in a certain user subset may respond. The probe message is basically asking those stations if they have messages. All stations having messages respond by putting a signal on the line. This signal basically means "yes, I have messages." Overlapping positive responses are also interpreted as a positive response. Following a positive response, the controller breaks down the terminal population into subsets and repeats the question to each of the subsets. This process is repeated until the individual terminals having messages are isolated. Now, when the isolated terminal is interrogated, it transmits the message that it has been holding in its buffer.

To see the improvement of this probing technique over the conventional polling, let us assume that the number of terminals, M, is a power of 2, that is $M = 2^n$. Thus each station is assigned an n-bit binary address. The controller can probe a group of stations by transmitting the probe message with the address of that group. The group address is chosen to be the common prefix of the addresses of the stations in that group. For example, if a group of terminals is composed of the two terminals whose addresses are 010 and 011, the group address is 01 (two bits long). If a group of terminals is composed of the four terminals whose addresses are 100, 101, 110, 111, the group address is only 1 (one bit long).

Now let us explain the protocol through the following example. Suppose that there are eight stations ($U_0, U_1, \ldots, U_7$) on a multidrop line (Fig. 7.14) whose binary addresses are 3 bits long. Assume that station U_2 (with address

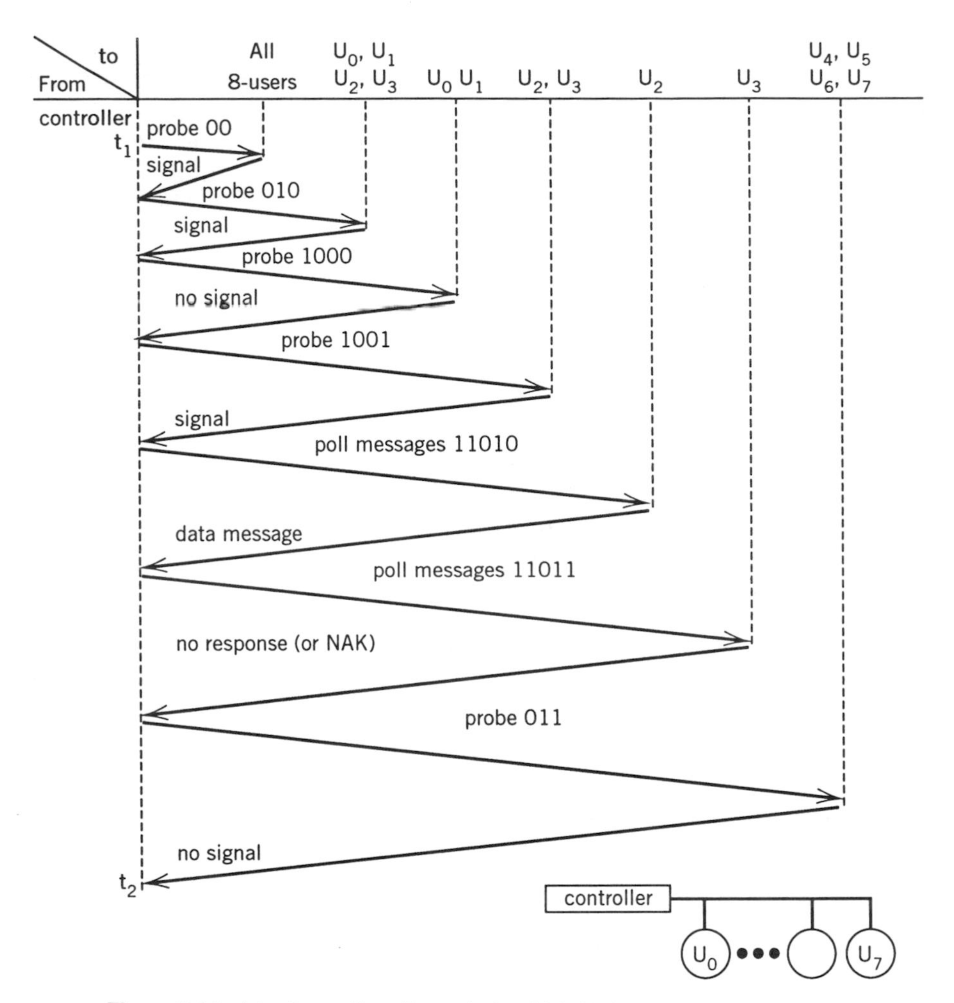

Figure 7.14. Adaptive polling: Example in which U_2 is the only active station.

010) is the only active station (that is, U_2 has a data message ready for transmission to the controller). Figure 7.14 shows the sequence of probing and polling messages needed to isolate U_2. The probe message has a group address field that is variable in length. The first two bits indicate the length of the group address (which is the common prefix of the station's address). The first probe message sent has an address length of 00 indicating that no common prefix, hence all stations are probed. In this example, we assume that U_2 is the only active station and subsequently, U_2 sends a signal on the line. The controller splits the eight users into two groups and sends a probe message to the first group (U_0 to U_3). The first two bits 01 of the probe message indicate the length of the address field, and the common prefix of

that group is 0. A signal is returned. The controller breaks down that group into two subgroups and probes the first subgroup (U_0 and U_1) and gets no response. It then probes the other subgroup (U_2 and U_3) and gets a response signal. It further divides that subgroup and probes (or polls in this case) U_2, and U_2 sends back its data messages. Polling U_3 results in no response. Now, the controller probes all users with the common prefix of 1 (that is users U_4, U_5, U_6, and U_7) and receives no signal. Here the time $(t_2 - t_1)$ represents a probing (or polling) cycle, which is defined as the time required for the polling and transmission of all data messages that were generated in the preceding cycle.

Note that in the previous example, the controller initiated seven inquiring (probe or poll) messages, whereas in conventional polling it would have initiated eight polling messages. In general, if there are 2^n terminals and there is only one ready terminal, then the number of inquiries per cycle for probing (adaptive polling) is $2n + 1$, and the number of inquiries per cycle for conventional polling is 2^n.

For example, for 256 terminals and for the case of one active terminal, probing requires 17 as compared to 256 inquiries for conventional polling. It is now obvious that there is a big saving in the number of inquiring messages in probing when the traffic load is light.

On the other hand, probing has a penalty in heavy loads. For example, if all 2^n terminals have messages, probing requires $2^{n+1} - 1$ inquiries as

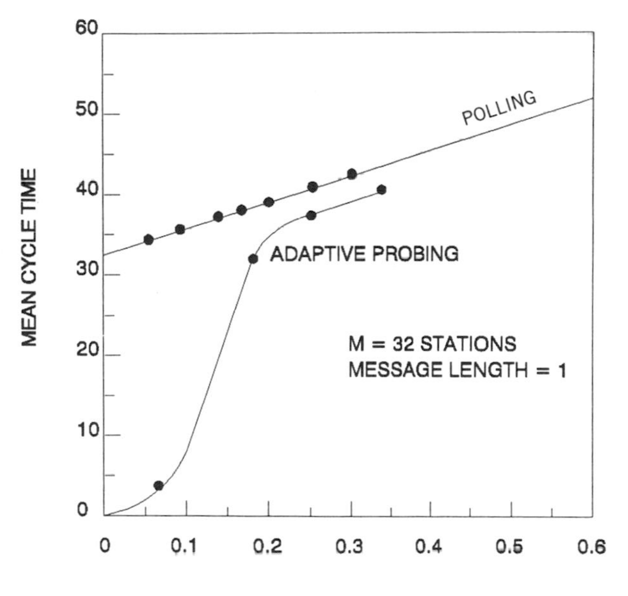

Figure 7.15. Polling and adaptive probing mean cycle time plotted against message arrival rate (simulation results for 32 stations). From [HAYES 78], © 1978 IEEE.

opposed to 2^n for conventional polling. To improve the performance of probing under heavy load situations, the controller starts a cycle by probing smaller groups. The size of these groups can be determined by the duration of the immediately preceding polling cycle. For the case of eight users, if the previous cycle has, say six active terminals indicating a heavy load situation, in the present cycle the controller may break the users into four groups. The size of each group is two terminals. The first probe message is addressed to the U_0 and U_1 group and the algorithm repeats as described before.

Simulation of the adaptive probing technique has shown that this scheme is always superior to polling in that its mean cycle time is always smaller than that of polling (see Fig. 7.15).

7.6 DISTRIBUTED DEMAND ASSIGNMENT SCHEMES

Distributed demand assignment schemes have the two main advantages of reliability and performance. Reliability in distributed control is higher than centralized control, since the system operation is not dependent on a central scheduler. Performance, in terms of delay and channel utilization, is, in general, better than centralized schemes. The basic idea in distributed algorithms is the need to exchange control information among the users, either explicitly or implicitly. With this information, all users then independently execute the same algorithm resulting in some coordination in their actions. We will discuss here two schemes: the tree algorithms and the ring protocols.

7.6.1 Tree Algorithms

These are basically random-access schemes based on the idea of resolving the collision between the collided users. These algorithms are based on the idea that contention among several active stations (an active station is one that has a packet ready for transmission) is resolved if, and only if, all the sources are somehow subdivided into groups such that each group contains at most one active station. Let us explain the algorithm through the use of an example as shown in Figure 7.16. In the example, the communication channel is assumed to be a slotted broadcast channel. The length of each slot equals that of a packet and each station transmits the packet at the beginning of the slot. As in all random-access schemes, the station is capable of determining after transmission of a packet whether there have been zero (i.e., idle channel), one (i.e., success), or multiple packets (i.e., collision). For convenience purposes, the propagation time is assumed negligible.

Example 7.1. Assume eight stations (with users $U_1, U_2, \ldots, U_8$) are using the slotted channel. Each user is assigned a leaf on the tree (i.e., assign addresses

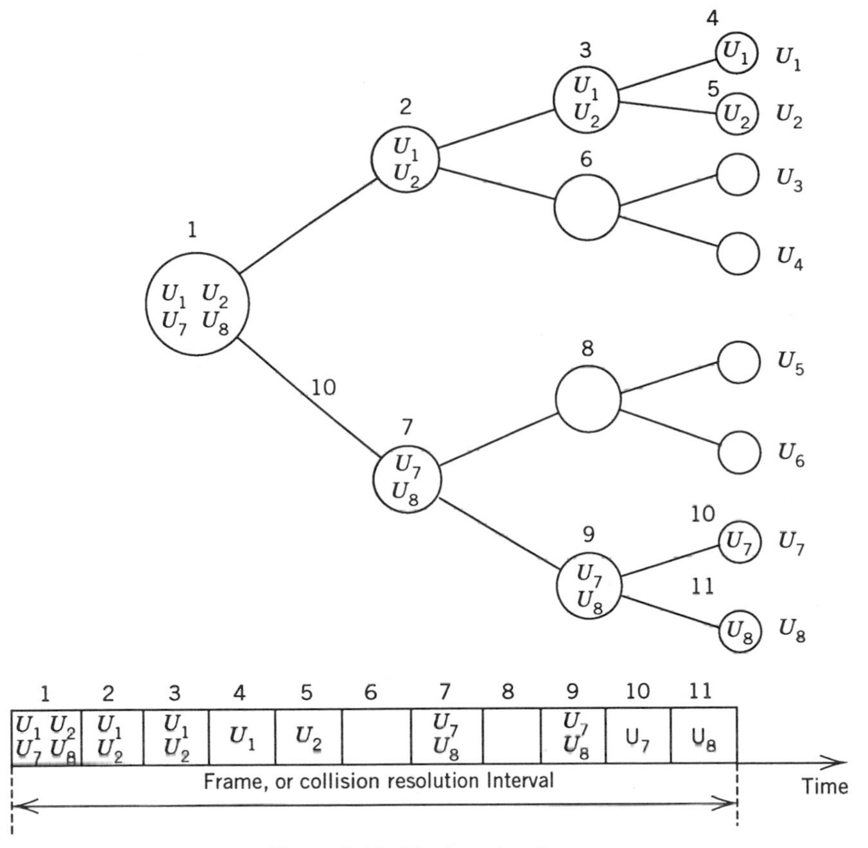

Figure 7.16. The tree algorithm.

to all stations). U_1, U_2, U_7, and U_8 transmit their packets in the first slot and all users detect a collision. In Figure 7.16, the nodes of the tree represent time slots. The number on top of the node represents the slot number, and the users transmitting in a slot are shown inside the node. The lower part of Figure 7.16 shows the channel time, the slot number, and the users who are transmitting in the slot.

The algorithm states that once a collision has occurred (as in slot 1, the root node, in Fig. 7.16), new packets are not allowed to enter the channel until all packets involved in this collision have been able to succeed in transmission. Now, those who are involved in the collision (in our example U_1, U_2, U_7, and U_8) split into two subgroups. The first subgroup is composed of all users who own a leaf in the upper half of the tree (basically, all users whose addresses are from 1 to 4, that is U_1 to U_4). The second subgroup is composed of all users who own a leaf in the lower half of the tree (that is U_5 to U_8).

In the second slot, only the collided users who belong to the first subgroup retransmit while all users belonging to the second subgroup wait until those in the first subgroup transmit successfully. As a result, U_1 and U_2 retransmit in slot number 2 and collide, while U_7 and U_8 wait until U_1 and U_2 resolve their collision and successfully transmit. Now, the algorithm repeats. With collision occurring in slot 2, collided users further split into two subgroups. The first subgroup consists of all users in the upper half of the subtree, that is U_1 and U_2, and the second subgroup is all users in the lower half of the subtree, that is U_3 and U_4. Hence, in slot 3, U_1 and U_2 retransmit again and collide splitting again; the first subgroup is now only composed of U_1, and the second subgroup is only composed of U_2. In slot 4, U_1 retransmits and succeeds followed by U_2 in slot 5.

In slot 2, all collided users split into U_1 and U_2 as the first subgroup and U_3 and U_4 (if they were involved in the collision) as the second subgroup. As a result, slot 6 is empty.

At this moment U_7 and U_8 know for sure that the first subgroup of users (U_1 through U_4) has resolved their collision. In slot 7, U_7 and U_8 retransmit and collide. With splitting, slot 8 is empty (there is a chance that U_5 or U_6 are involved in the collision) and then U_7 and U_8 retransmit and again collide in slot 9. With splitting, U_7 succeeds in slot 10 and U_8 follows in slot 11. At this point, the collision that occurred in slot 1 has been completely resolved and all packets involved in the slot 1 collision have transmitted successfully. This period (from slot 1 to slot 11 in this example) is referred to as a collision resolution period (CRP) or, simply a frame. Note that the frame length is variable and it depends on the number of users involved in the transmission in the first slot of the frame and also on their addresses (location on the tree). Problem 7.18 emphasizes this point.

Since the channel is a broadcast channel, all users observe the events in the channel. Any new packets generated during the frame are delayed until the first slot of the following frame. The tree algorithm is sometimes referred to as the *collision resolution algorithm*.

7.6.2 Ring Protocols

A variety of schemes have been devised for ring networks. The ring (loop) topology connects the stations in a closed network and circulates all messages in one direction. Often messages are amplified and repeated at each station as they pass through. Ring networks utilize a set of point-to-point (unidirectional) links to achieve connectivity. Packets are forwarded through the ring nodes acting as repeaters. These nodes also act as insertion points for messages originating locally. Multiple-access techniques are required in the ring network to insure that inserted messages do not interfere with messages being forwarded at a given node. Ring topologies are naturally suited to demand assignment schemes. We discuss three main ring protocols: token passing, register insertion, and the slotted ring protocol.

7.6.2.1 Token-Passing Ring Protocol. In the token-passing protocol, a control token is explicitly passed from node to node (Fig. 7.17). The control token is a special bit pattern that indicates the state of the channel (free or busy). A free token represents an access permission to the channel. A station having data in its buffer waits for a free token, changes it to a busy token and puts its queued data onto the ring. The station continues transmission until a specified token holding time elapses or until the queue becomes empty, whichever event occurs first. Stations in the ring regenerate bits received from the incoming link, repeating them into the outgoing link. The sending station is responsible for removing its own packet from the ring. At the end of the station's transmission, it passes the access permission to the next station by generating a free token. The token-passing protocol allows only one station at a time to transmit a message of variable length. The IEEE 802 standards committee has adopted the token-passing ring protocol as one of its standards for LANs (i.e., the IEEE 802.5 standard). Also, the American National Standard Institute (ANSI) has accepted the token passing as the MAC layer for its fiber distributed data interface (FDDI) standard. The token-passing protocol can also be applied to the bus topology network, which has been adopted (IEEE 802.4) for the token bus standard LAN. A detailed discussion of these standards is presented in the next chapter.

Now, we use a simplified model to analyze the performance of the token ring. Assume that a total of N stations with infinite queues are served by the ring (see Fig. 7.18a). Messages arrive at station i according to a Poisson

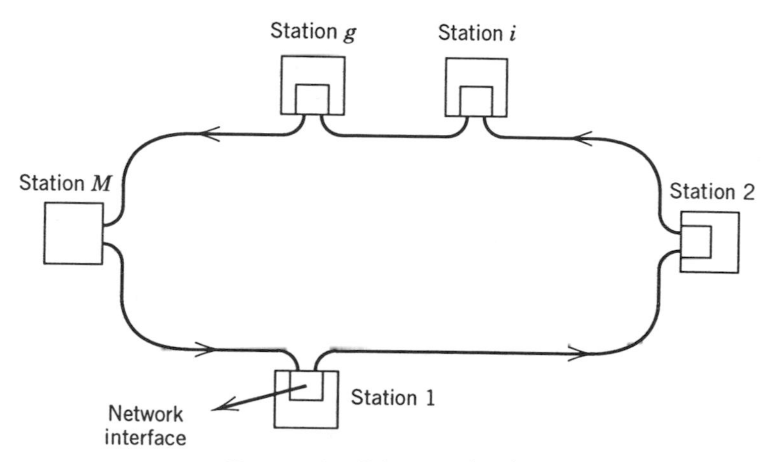

Figure 7.17. Token-passing ring.

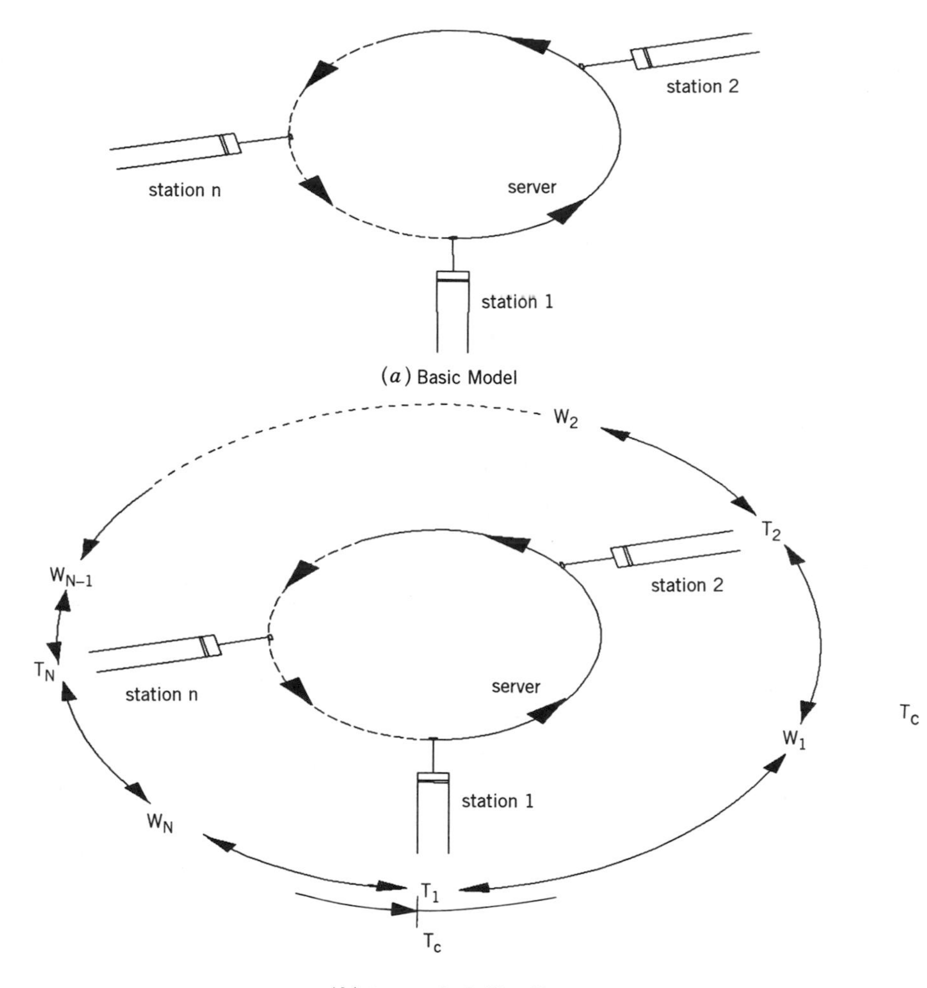

Figure 7.18. Token ring model. (*a*) Basic model. (*b*) Average cycle time T_c.

process with rate λ_i. The total message arrival rate is $\lambda = \sum_{i=1}^{N} \lambda_i$. The average message length is t_i seconds governed by a general distribution. The offered traffic from station i is ρ_i, and the total offered traffic is ρ. The time it takes a bit to travel from station i to the next station $i + 1$ on the ring is called the walk time, and it is a random variable with mean, W_i (W_N is the average walk time from station N to station 1; see Fig. 7.25*b*). The walk time per cycle (i.e., the time for a bit to go all the way around an idle ring) is

$$W = \sum_{i=1}^{N} W_i$$

Let T_i be the average dwell time per visit of the server at station i (basically it is the time needed by station i to transmit all its messages in the buffer). Thus, the cycle time of the system, which is the mean interval between token arrivals at a given station and is sometimes referred to as the scan time, consists of the dwell times at each of the stations plus the sum of the walk time. The average cycle time T_c is given by

$$T_c = \sum_{i=1}^{N} T_i + \sum_{i=1}^{N} W_i \tag{7.26}$$

The average number of messages arriving at station i during a cycle is $\lambda_i T_c$. Thus, the average dwell time T_i is given by

$$T_i = \lambda_i T_c t_i = \rho_i T_c, \quad i = 1, 2, \ldots, N$$

Substituting into 7.26, we get

$$T_c = \sum_{i=1}^{N} \rho_i T_c + \sum_{i=1}^{N} W_i$$

$$= T_c \rho + W$$

Solving for the average cycle time T_c, we have

$$T_c = \frac{W}{1 - \rho} \tag{7.27}$$

For a stable system, the average number of messages transmitted by station i during a cycle, m_i, must equal the average number of messages that arrive at each station during a cycle T_c. Therefore, we have

$$m_i = \lambda_i T_c = \lambda_i \frac{W}{1 - \rho}$$

and the mean dwell time at station i

$$T_i = \rho_i T_c = \rho_i \frac{W}{1 - \rho}$$

7.6.2.2 Slotted Ring. In a slotted ring, the ring is divided into a constant number of fixed size slots that continuously circulate around the ring. Figure 7.19 shows the structure of a Cambridge ring, as an example of a slotted ring.

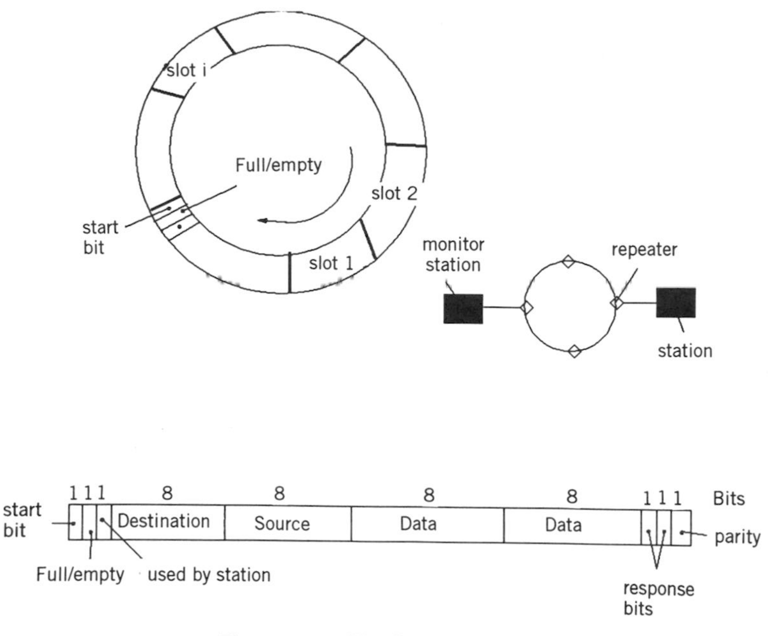

Figure 7.19. The Cambridge ring.

The *Cambridge ring*, developed at the University of Cambridge in England, consists of a set of repeaters connected by cables. A monitor station has the responsibility for synchronizing and maintaining the network.

Stations access the ring through the repeater and access interface. The slot size in the Cambridge ring is 38 bits, which are divided into 8 bits for source address, 8 bits for destination address, 16 bits for data, and the remaining 6 bits for control information. Thus, each slot contains 16 bits of data out of 38 total bits. The control field has one start bit indicating the beginning of the slot, and a status bit indicating whether the slot is empty or full.

To transmit, a station must wait until an empty slot arrives, marks it full by setting the status bit, and fills it with its data. When the sender receives back its own transmission, it removes data from the slot and marks it empty by resetting the status bit. It also inspects the two response bits, which are located after the data bits, to see whether the data was accepted, rejected, or marked busy by the destination address. For reception, each station looks at the destination addresses of all full slots circulating the ring and recognizes its own address. If a match occurs, it copies the source address and data field and sets the response bits. The Cambridge ring repeater introduces a delay of 3 bits. Because slots do not fit exactly into the physical length of the ring, a small part of the available capacity is wasted. In this case, padding bits are added.

For example, assume a 10-Mbps slotted ring is supporting 30 nodes. Average spacing between stations is 30 m. The propagation speed is 2×10^8 m/second. Thus, one bit propagates in

$$\frac{30 \text{ m}}{2 \times 10^8 \text{ m/sec}}$$

The bit length between two stations is

$$10^7 \text{ bps} \cdot \frac{30}{2 \times 10^8} = 1.5 \text{ bits}$$

with a 3-bit delay at each station, the total bit length is (3 + 1.5) bits × 30 stations, or 135 bits. Thus, the number of slots circulating in the ring is

$$\frac{135}{38 \text{ bits per slot}} = 3 + \frac{21}{38}$$

That is, 3 slots circulating with the remaining 21 bits are padding bits.

The slotted ring, unlike the token-passing ring, allows more than one node to simultaneously transmit messages, with each transmission fitting within the fixed-length slot. Thus, a long packet may have to be transmitted within several slots.

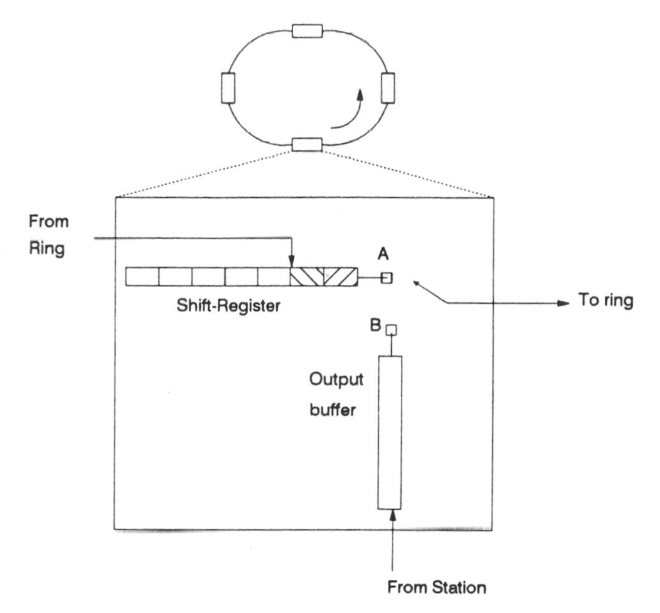

Figure 7.20. Register-insertion ring.

7.6.2.3 Register-Insertion Ring. A register-insertion ring, developed at Ohio State University, derives its name from the fact that a shift register resides at each station interface (Fig. 7.20). Frames received from the ring are shifted into the shift register, whereas frames to be transmitted are queued into the output register. When the channel becomes idle, the frame is inserted into the loop by shifting it from the output register onto the ring, and any incoming frame at that time is shifted into the shift register.

One advantage of the register-insertion ring is the possibility that variable-length frames are simultaneously transmitted by more than one node. Another advantage is that the register-insertion ring provides a fair allocation of the channel bandwidth to the nodes. The register-insertion concept has also been implemented in the SILK network developed in Germany.

7.7 HYBRID SCHEMES

Hybrid schemes attempt to incorporate some of the properties or advantages seen in the previous schemes. Reservation-type schemes are good examples of this category.

7.7.1 Reservation Schemes

A number of schemes have been designed with explicit reservation capabilities. The major motivation for these schemes is the satellite system, which consists of a set of earth stations and a satellite in synchronous orbit. In these explicit reservation schemes, a short control packet (reservation packet) is first sent by stations to request transmission at scheduled times.

The first explicit reservation scheme, referred to as reservation ALOHA and developed by L. Roberts, uses slotted ALOHA for the transmission of reservation packets. Many other variations of reservation ALOHA-type schemes have been suggested. The Defense Advanced Research Project Agency (DARPA) has included many features of these schemes and has sponsored the development of a satellite reservation scheme called PODA (priority-oriented demand assignment), which extended the reservation concept to include integrated packet voice and data traffic and priority allocation of the satellite capacity. The PODA scheme was first implemented on SATNET, an experimental satellite that used a 64-Kbps channel on an INTELSAT satellite and later evolved to a version accommodating channel rates of several megabits per second. Another application for PODA is on the FLTSAT satellite using a 19.2 Kbps channel. FLTSAT allows for ship-to-ship and ship-to-shore communications. Below we describe reservation-ALOHA.

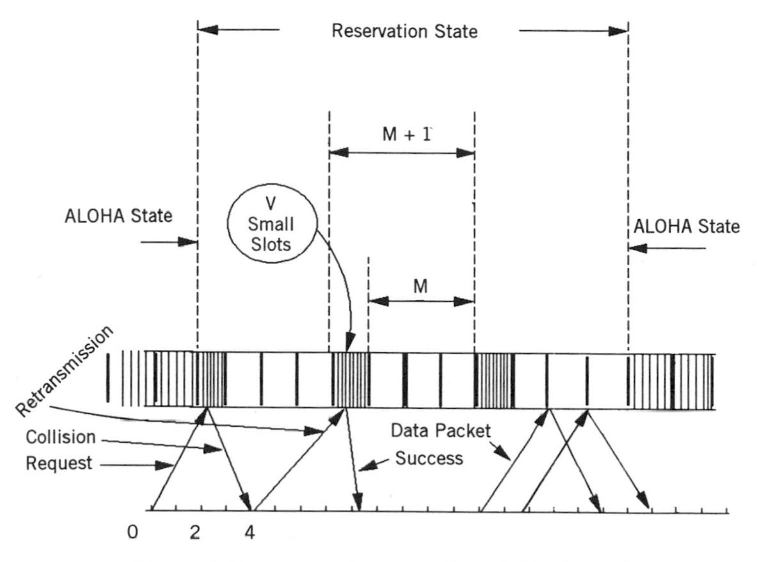

Figure 7.21. Reservation ALOHA for satellite channel.

7.7.2 Reservation ALOHA

In reservation ALOHA, the satellite channel is divided into time slots. After every M slots, one slot is subdivided into V small slots. Thus, a frame is $(1 + M)$ slots. The small slots are for reservations and acknowledgment. Requests (or reservations) are transmitted using the slotted-ALOHA protocol. The size of a reservation packet equals the size of a small slot. The remaining M slots (large slots) are used for actual data packet transmission. Each station has a counter J; thus, there is one common queue for all stations and by broadcasting reservations they can claim space on the queue.

Figure 7.21 shows typical operation of reservation-ALOHA. Initially, a reservation packet for two slots is transmitted. In its first attempt, a collision has occurred. In the second attempt, the station hears its own reservation and knows where it is supposed to transmit its two packets (based on the counter J). Note that whenever the reservation queue goes to zero (i.e., when there are no messages in the system waiting to be transmitted) the system goes to the ALOHA state with all $(M + 1)$ slots per frame available for reservation requests until the next valid reservation is received.

7.7.3 Channel Utilization Analysis

Define Z as the channel rate in large slots (data packets) per second. We assume that there are K stations (users) in the system and on the average a user requests transmission of B packets per request (i.e., block size or the message size is an average of B packets).

Each user generates λ requests per second. Thus each user's arrival rate is $B\lambda$ packets/second resulting in a total data traffic rate of $KB\lambda$ data packets/second and the total requesting traffic rate is $K\lambda$ requesting packets/second.

When the system is idle (i.e., when no packets are waiting for transmission), the channel is in the ALOHA state. Then all time slots are small slots, and the channel rate available for the request packets is ZV small slots/second. Thus the small slot channel utilization in the ALOHA state, S_1, is given by

$$S_1 = \frac{K\lambda}{ZV} \tag{7.28}$$

When the channel is in the reserved state, only one slot every $(1 + M)$ slots is dedicated to the reservation packets. Thus the small slot channel rate in the reserved state is $ZV/(1 + M)$ small slots/second. Thus, the small slot channel utilization in the reserved state, S_2, is given by

$$S_2 = \frac{K\lambda}{ZV/(1 + M)} \tag{7.29}$$

Since one slot every $(1 + M)$ slots is used for reservation packets and the M slots are used for actual data packet transmission, the effective channel capacity is $Z \times M/(1 + M)$ slots/second. The data channel utilization, S_3, is given by

$$S_3 = \frac{KB\lambda}{Z[M/(1 + M)]} \tag{7.30}$$

Note that the actual utilization of the satellite channel, ρ, is the ratio of the data traffic rate to the satellite channel rate

$$\rho = \frac{BK\lambda}{Z} \tag{7.31}$$

An upper limit on M can be obtained from Eqs. 7.28, 7.29, and 7.30. Notice that both S_1 and S_2 must be less than $1/e = 0.368$ and $S_3 < 1$. In Eq. 7.29, we have

$$S_2 = \frac{S_3 M}{VB} < \frac{1}{e}$$

However, a much lower value of M minimizes the time delay (see problem 7.22 for delay analysis) and occurs when S_3 is close to 1, therefore,

$$M < \frac{BV}{e} \tag{7.32}$$

For example, assume that the block size is uniformly distributed in length from one to eight packets. Hence, $B = 4.5$ packets per block size. Let the number of small slots per frame V equal six. Hence, M is less than or equal to nine slots.

7.8 SUMMARY

We have classified multiple access schemes into five major categories and discussed examples of developed protocols for each category (see Fig. 7.22). Although these schemes were developed for low-to-moderate speed multiuser channels (on the order of kilobits per second to a few megabits per second), some of these are also extendable to high-speed multiuser channels (in hundreds of megabits per second).

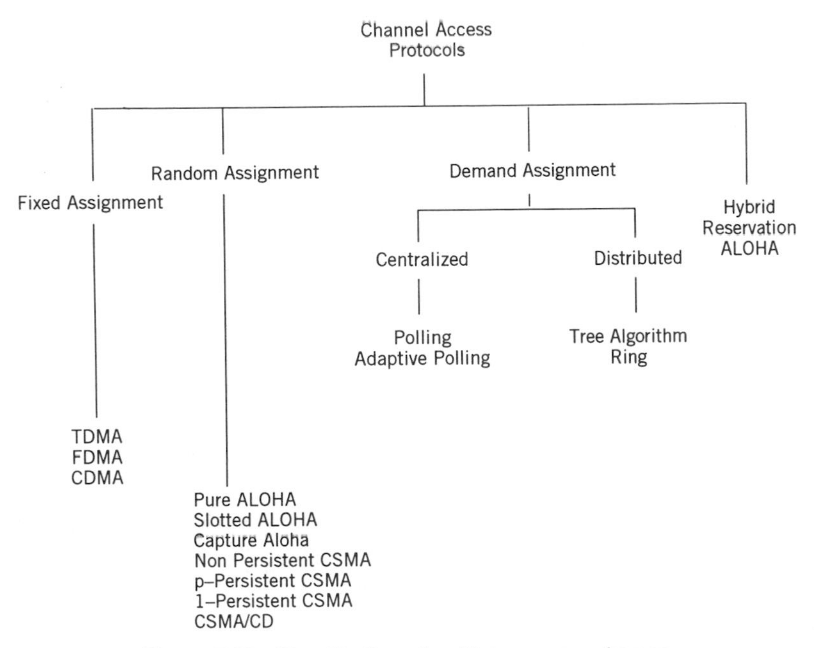

Figure 7.22. Classification of multiple access schemes.

SUGGESTED READINGS

[TOBA 80] provides an excellent survey of multiple access-scheme classification, and [SACH 88] updates such classification. [RUBI 90] and [KAMA 94] present survey papers on access schemes for high-speed networks.

The original work on ALOHA is found in [ABRA 70] and further work in [KLEI 75], [ABRA 77], and [ABRA 85]. [SAAD 80] discusses the analysis of ALOHA with the finite number of buffered users. [METZ 76], [DAVI 80] and [GOOD 87] provide detailed studies of ALOHA with capture.

[HAMM 86], [KLEI 75b], and [KLEI 75c] study in detail the performance of CSMA protocols. [TOBA 80b] and [LAM 80] present detailed analysis of CSMA/CD. [METC 76] and [SHOC 82] describe the implementation of CSMA/CD in the Ethernet. [HAYE 78] proposed adaptive polling. [CAPE 79] and [BERT 92] are excellent references for tree-type algorithms.

[HOPP 87] is a good reference on the Cambridge ring, and [LIU 82] presents the register-insertion ring. [HUBE 87] discusses the implementation of the register insertion concept in the SILK network. [ROBE 78] and [TOBA 84] provide discussions of an ALOHA-type reservation scheme with applications to packet radio and satellite networks. [MEHM 88] analyzes a star LAN access method.

[PICK 82] is a good tutorial on spread spectrum, and [KAVE 84] and [PURS 87] are examples of civilian applications for spread spectrum communications. [EPHR 87], [MAHM 89], [STEE 92], [CHUA 93], and [ELHA 94a and b] present applications of spread spectrum techniques to packet and mobile radio networks and personal communications networks.

PROBLEMS

7.1 A group of N very-small-aperature-terminal (VSAT) stations share a 56-kbps pure-ALOHA satellite channel. Each station outputs a 500-bit packet on an average of one every 10 seconds, even if the previous one has not yet been sent (e.g., the stations are buffered). What is the value of N for which maximum throughput is achieved? Repeat for slotted-ALOHA and for capture-ALOHA?

7.2 A packet radio network with a data rate of 10 Mbps has two stations at a distance of 1.5 km from each other. Let the packet length be 1000 bits and the propagation speed be 3×10^8 m/second. Assume that each station generates packets at an average rate of 1000 packets/second and that the Poisson assumption is valid. If one station begins transmission, what is the probability of collision?

a. For pure-ALOHA?

b. For slotted-ALOHA?

7.3 Consider the following ALOHA channel with a finite population. S_i represents user i throughput and G_i represents the total transmission probability for user i (including new packets plus retransmitted). N is the total number of users in the system. Let

$$S \doteq \sum_{i=1}^{N} S_i \quad \text{and} \quad G \doteq \sum_{i=1}^{N} G_i$$

a. Prove that

$$S_i = G_i \prod_{j \neq i} (1 - G_j)$$

b. If the N users are identical with each having the same throughput and the same total transmission rate, prove that

$$S = G[1 - G/N]^{N-1}$$

c. Assume two groups of users: N_1 is the number of users in group 1 and N_2 is the number of users in group 2, with throughput S_1 and S_2 (per user), respectively. Using the results in (a), obtain equations relating S_1, S_2, G_1, and G_2.

d. Apply the results in (c) for $N_1 = N_2 = 1$ and obtain S_1 and S_2 in terms of G_1. Plot curves for S_1, S_2, and S versus G_1.

7.4 Consider a satellite channel of 56 kbps using slotted-ALOHA as the access scheme. Packet size is 1000 bits. If the total users' traffic (i.e., throughput S) is 0.1, what is the average packet delay in slots? What is the average channel traffic, G?

Assume the satellite is at a distance of 36,000 km and the speed of light is 3×10^5 km/second, and assume K (maximum back-off time) is five slots.

7.5 Assume the following slotted-ALOHA system. Stations transmit their packets (newly generated or collided ones) in any given slot with probability P. There are N stations in the system. Write an expression for the channel throughput. What is the value of P that achieves maximum throughput? Obtain the maximum throughput for $N = 2$, 4, 8, and ∞. Plot the maximum throughput versus N.

7.6 In a two-user slotted-ALOHA system, let σ be the probability that a user generates a new packet in a given slot, let α be the probability that a collided user retransmits a packet in a given slot, and let P_n be the probability that n packets are collided, where $n = 0, 1, 2$. Using a Markov chain model, obtain expressions for P_0, P_1, P_2 and solve for P_0, P_1, P_2, when $\sigma = 0.1$ and $\alpha = 0.5$.

7.7 An approximate model for the ALOHA channel is an $M/G/1$ system in which a packet collides and immediately joins the queue again with probability p, or succeeds and departs forever with probability $(1 - p)$. Service is first-come first-serve (FCFS), and the service time for a colliding packet is independent of its previous service times (this is not really true in ALOHA). Let $X(s)$ be the Laplace transform for the service time (x) probability density function (PDF) [i.e., $X(s) = E\{e^{-sx}\}$] and let $Y(s)$ be the Laplace transform for the packet's total service time PDF [i.e., $Y(s) = E\{e^{-sy}\}$].

a. Find $Y(s)$ in terms of $X(s)$ and p.

b. Show that

$$\bar{y} = \frac{\bar{x}}{1 - p}$$

[*Hint*: Let X_i be the packet service time per cycle i and y be the packet's total service time.]

7.8 For the network shown in Figure 7.23, external traffic is generated by users who are connected to the branches. Each user generates traffic at

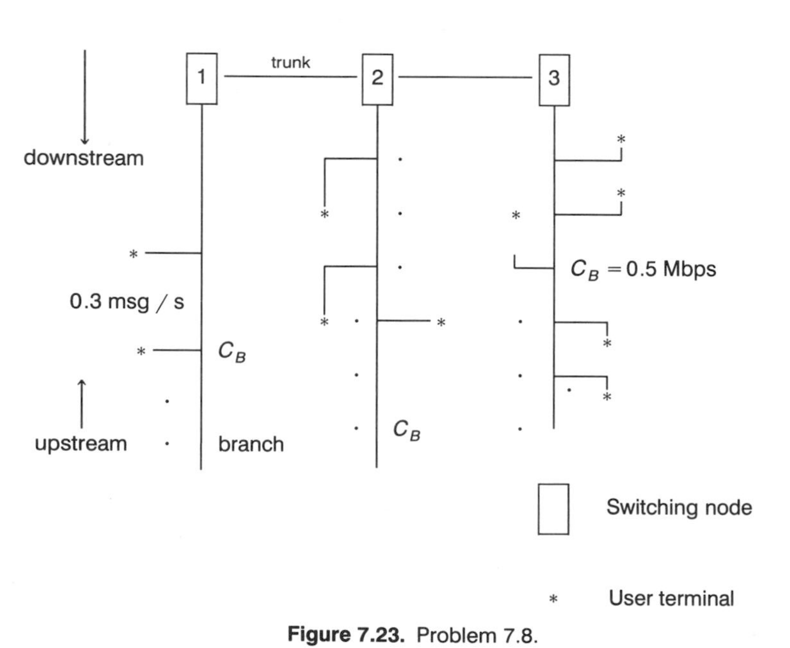

Figure 7.23. Problem 7.8.

a rate of 0.3 messages per second, which is sent to the nearest switch. The switch routes the traffic to the destination switch, which broadcasts the message on the destination branch. There are 1000 users per branch. The branch channel C_B is 0.5 Mbps, whereas the trunk channel C_T is 1 Mbps. The probability that a message originated at a given branch and is being destined to any branch equals $\frac{1}{3}$ (i.e., all users are equally likely to communicate with each other).

a. Determine the following: downstream traffic λ_b on each of the branches; upstream traffic λ_i on trunks (1–2) and (2–3); downstream traffic Λ_i on trunks (2–1) and (3–2); total external traffic γ (msg/seconds).

b. How many output queues exist in each node? Write an equation for the average message delay T over the network. What is the value of T? Assume an $M/M/1$ queueing system and the average message length is 1000 bits.

c. Is it possible for the users to use the slotted-ALOHA protocol to access the branch channel? Explain why?

7.9 The total number of stations N in an asynchronous CDMA packet radio network equals four. Users' communication with each other is equally likely. Packets arrive according to a Poisson process with rate λ packets/unit time. The packet length (and as a result, the service time) follows an exponential distribution with mean $1/\mu$ time units. A user may not transmit or receive simultaneously. Assume perfect acknowledgment, zero propagation time, and perfect capture, whereby correct reception is guaranteed once the packet is locked onto. If multiple users (packets) are destined to the same receiver, one packet is captured; the others are considered to be collided.

Let i represent the number of successfully communicating pairs, $i = 0, 1, 2$ and j represent the number of collided single terminals in the network, $j = 0, 1, \ldots, 4$. Then, the channel state can be represented by (i, j).

a. Draw the steady-state transition diagram representing the system states. Show the transition probabilities.

b. Obtain $P(i, j)$ for all i and j as a function of λ and μ.

c. Obtain the average network throughput.

7.10 **a.** Can CSMA/CD protocol be used as the access scheme for a satellite channel? Why?

b. Explain why Ethernet performs poorly with terminal type traffic?

7.11 Assume two CSMA/CD networks as follows:

Network A is an LAN baseband Ethernet with transmission speed of 5 Mbps, a 1-km cable, and 1000 bits/packet.

Network B is an MAN broadband Ethernet with a 50-km cable, and 1000 bits/packet.

What transmission speed is required for network B to maintain the same channel throughput as network A?

7.12 K stations share a 0.5 Mbps bus cable using CSMA/CD as the access scheme (i.e., an Ethernet LAN). The length of the bus is 500 m. The packet length is L bits. Assume that the K stations on the network are always having traffic and are ready to transmit (case of heavy load). Let p be the probability that a station transmits during a slot, let $k = 10$, and let the propagation speed be 3×10^8 m/second. Obtain the average number of slots for the contention period, the average duration of the contention period in microseconds, and the channel utilization for the two cases of

a. $L = 100$ bits.

b. $L = 1000$ bits.

7.13 A 1-Mbps, 0.5-km Ethernet has a propagation speed of 200 m/μsecond. Assume that data packets are 500 bits long, including 64 bits of header. The first slot after a successful transmission is reserved for the receiver to capture the channel to send a 64-bit acknowledgment packet. What is the effective data rate, excluding header, assuming that there are no collisions.

7.14 Assume two stations on an Ethernet and assume that both always have a frame to send. Assume also that the back-off scheme is the IEEE 802 binary exponential back-off algorithm described as follows: The number of slot times delayed before the nth retransmission attempt is chosen as a uniformly distributed random integer r in the range $0 < r < 2^K - 1$, where $K = \min(n, 10)$. The delay is an integral multiple of time slots, and the slot time approximately equals twice the round-trip propagation delay.

After a collision, what is the mean number of retransmission attempts before one station successfully transmits? Obtain an approximate solution.

7.15 N stations are connected on an Ethernet. One of the stations had a hardware problem and continuously transmits its packets whether the channel is busy or idle. Suggest a trouble-shooting method to allocate and isolate the faulty station.

7.16 In Figure 7.24, the controller polls each of the 2^n terminals attached to the multidrop line using the probing (adaptive polling) technique. Let K be the number of active terminals at the beginning of a probing cycle. Obtain the average number of probe messages sent out by the controller for the following cases and compare with conventional

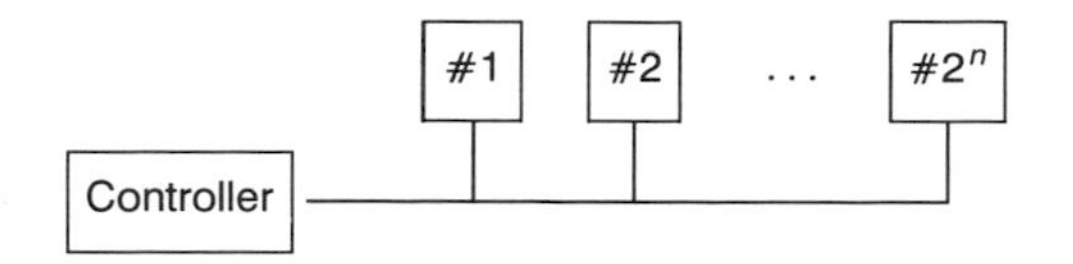

Figure 7.24. Problem 7.16.

polling;

a. $n = 2 \quad K = 1$

b. $n = 2 \quad K = 2$

c. $n = 3 \quad K = 1$

d. $n = 3 \quad K = 2$

7.17 Two users, U_1 and U_2, use the tree algorithm as the access method for a communication channel.

a. Draw the tree diagram and the channel timing diagram showing situations when the frame (i.e., collision resolution period) is maximum and minimum?

b. Now, assume U_1 and U_2 transmit new packets in a slot with probability P_1 and P_2, respectively. Obtain expressions (in terms of P_1 and P_2) for

average frame length N

average delay for U_1 packets D_1

average delay for U_2 packets D_2

channel throughput S

c. Repeat (a) when $P_1 = P_2 = P$ and plot N, D_1, D_2, and S versus P.

d. What is the maximum throughput S_{max} and what is the value of P that achieves S_{max}.

7.18 Eight users $(U_1, U_2, \ldots, U_8)$ implement the tree algorithm (collision resolution algorithm) as the channel access scheme. In a given slot, users U_1, U_5, and U_8 transmit and as a result collide.

a. Show how the users can resolve their collision by drawing the tree diagram and the timing diagram. Indicate on both diagrams the slot number and the users who transmit in every slot.

b. What is the frame length in slots?

c. What is the maximum frame length and under what conditions it might occur?

7.19 Consider a 1-Mbps token ring having N stations (assume $N = 50$), each with infinite buffer size. Each station generates 10 messages/second. The average message length is 1000 bit. The stations are equally spaced on the ring and the ring is of length 1 km and has a propagation speed of 200 m/μsecond. Determine the average cycle time T_c (i.e., scan time), the approximate average channel access time, and the average dwell time per visit of the server at any station.

7.20 A 0.5-km-long, 1-Mbps token ring has a propagation speed of 200 m/μseconds. Twenty-five stations are uniformly spaced around the ring. Assume data packets are 500 bits, including 64 bits of header. Acknowledgments are piggybacked onto the data packets and are thus effectively free. The token is 3 bytes long. Compare the maximum effective data rate with that of the Ethernet of Problem 7.13.

7.21 A 1-Mbps slotted ring is supporting 25 stations uniformly spaced around the 750-m-long ring. Each station introduces a 3-bit delay. Each slot contains 2 bytes for addressing, 6 bits for control, and 2 bytes for data. How many slots are on the ring?

7.22 In reservation-ALOHA, the average packet delay is the delay incurred in sending requests, that is, the delay to access the slotted-ALOHA channel D_S, plus the $M/G/1$ queueing delay incurred by packets waiting for their reserved time slot to arrive D_q plus the round trip delay R. Determine the average packet delay in terms of the traffic intensity S_3. Plot the average packet delay versus S_3 assuming $M = 2$ and compare with slotted-ALOHA.

[*Hint*: D_S is the delay experienced by the request packets in either the ALOHA state, with probability $(1 - S_3)$ or the reserved state with probability S_3].

8

LOCAL AREA NETWORK PROTOCOLS AND STANDARDS

8.1 INTRODUCTION

Local area network (LAN) technology provides for the interconnection of computing devices that are closely located within a single building or a collection of buildings such as a college campus. Such LANs are characterized by four main features:

1. Limited geographical coverage, typically in the range of a few meters to a few kilometers*
2. Ownership by a single organization
3. Low bit-error rates
4. Simple network topologies, such as a unidirectional ring or a bus

As LAN technology gained in popularity, it became clear that adoption of standards is necessary to allow for the coexistence of equipment from multiple vendors in a single LAN. In this chapter we discuss some of the more prominent LAN standards: the IEEE LAN standards and the Fiber Distributed Data Interface (FDDI) standard promulgated by American National Standards Institute (ANSI). The intent of this discussion is to present some of the principles underlying LAN technology; in particular, how that relates to the theory of multiple-access communication as discussed in Chapter 7.

Before proceeding, it is important to note that the term metropolitan area network (MAN) is sometimes used to refer to some of the standards we are about to discuss. As the name implies, MANs are designed to operate over larger distances on the order of 10 to 100 km and also require enhanced reliability characteristics. Of the protocols we discuss here, the IEEE 802.6

*In Chapter 9, we consider how individual LANs may be interconnected to increase their geographical coverage.

and the FDDI standards are suitable for MAN implementation. It should be emphasized, however, that they can also be used to provide LAN coverage.

8.1.1 The IEEE 802 Family of Standards

The IEEE has developed a set of LAN standards. These standards deal with the physical and data link layers as defined by the International Standards Organization (ISO) Open-System Interconnection Reference Model. The standards were a result of contributions from the members of the IEEE Project 802 Working Group and are sometimes known as the *802 LAN Standards*.

In the 802 standards, the data link layer is subdivided into two sublayers: the logical link control (LLC) and the medium access control (MAC). (This is perfectly legal as far as the OSI reference model is concerned.) The LLC sublayer deals with error and flow control issues on a point-to-point connection, similar to the high-level data link control (HDLC). It does not worry about the idiosyncrasies of the particular medium used. It is the job of the MAC sublayer to make sure that the frame given to it by the local LLC is delivered to the remote LLC. In this way, the same LLC protocol can be used to send frames over different network topologies, for example, bus or ring (see Fig. 8.1).

At last count, there were a total of eight working groups charged with developing standards and two technology advisory groups charged with developing recommended guidelines and serve as liaisons with other standards bodies. Figure 8.1 shows the IEEE 802 working groups and their already approved or in-progress standards. The two technical advisory groups

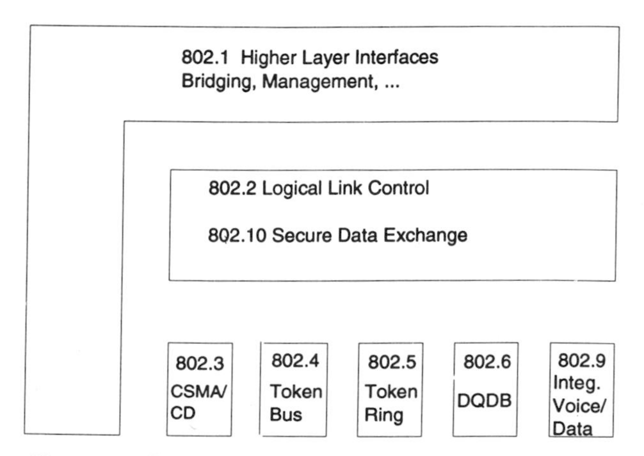

Figure 8.1. The architecture of the IEEE 802 LAN standards.

(not shown) are 802.7 on broadband technology and 802.8 on fiber optic technology.

The higher layer standard (802.1) is primarily concerned with network management and bridging (including MAC layer bridging discussed in Chapter 9). In this section, we discuss some of the details of IEEE 802 MAC layer standards for CSMA/CD (802.3), token bus (802.4), token ring (802.5), and the distributed queue dual bus protocol (802.6). Our intention is to illustrate how some of the multiple-access concepts discussed in Chapter 7 translate into working protocol standards.

8.2 CARRIER SENSE MULTIPLE ACCESS WITH COLLISION DETECTION

The IEEE 802.3 standard describes the operation of a LAN bus utilizing carrier sense multiple access with collision detection (CSMA/CD). The basic protocol was described in Chapter 7. The IEEE 802.3 standard is largely based on Xerox's Ethernet network product. The two standards however, are, not compatible. There are two basic configurations allowed in a CSMA/CD network depending on whether baseband or broadband signaling is used. With baseband signaling, a signal can propagate in both directions, and thus a single bus configuration is possible (Fig. 8.2). Broadband signaling dictates unidirectional propagation and thus a folded bus topology (Fig. 8.3) is required to achieve the desired broadcast connectivity. Alternatively, a broadband bus can employ frequency division multiplexing to provide two channels on a single cable. A frequency translating head end is then employed at one end of the bus (Fig. 8.4). Transmissions on one channel travel toward the head end and are then frequency translated by the head end and retransmitted on the other channel.

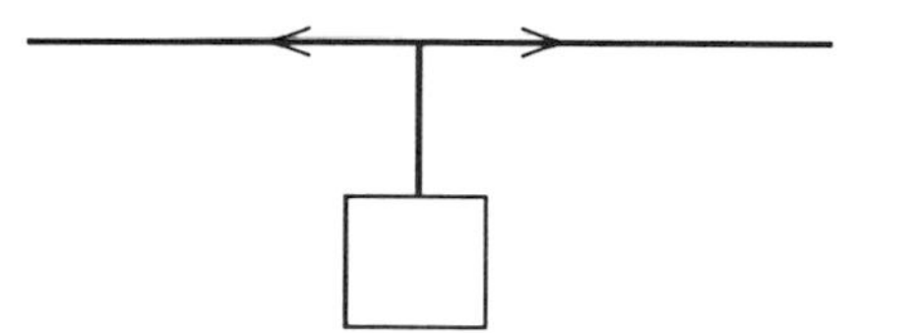

Figure 8.2. A baseband bus.

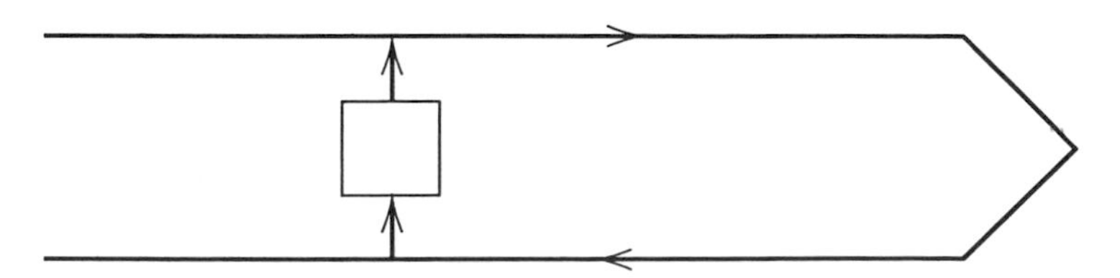

Figure 8.3. A broadband bus-folded bus.

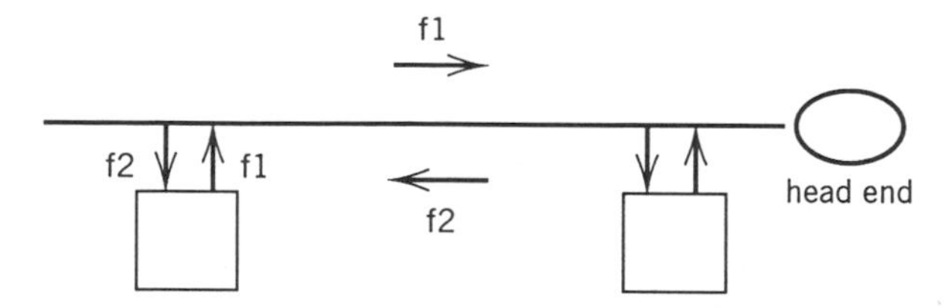

Figure 8.4. A broadband bus-frequency translating head end.

8.2.1 MAC Layer Procedures

All stations implement a 1-persistent CSMA/CD protocol with the so-called truncated binary exponential back-off algorithm. The protocol operates as follows:

1. Sense medium and transmit if idle, otherwise wait until medium becomes idle and then transmit immediately. This is the 1-persistent feature of the protocol.
2. For the first δ time units of the transmission, listen to the bus to determine if a collision has taken place.
3. If no collision is detected in the first δ time units, then *bus acquisition* has taken place; stop listening and continue transmission to conclusion.
4. If a collision is detected then abort the transmission and transmit a jamming signal. Wait for a random amount of time and repeat the process starting at step 1 above.

We now address three issues that arise in the description above.

8.2.1.1 Bus Acquisition Time. The value of δ (in step 2 above) has to be such that if no collision is detected during the first δ time units of a packet's transmission, then no collision will subsequently occur. If we let τ be the maximum propagation delay between the transmitter of any station and the receiver of another station (see Chapter 7), then $\delta = 2\tau$. In other words, if after transmitting a packet with no collision detected for 2τ time units, a station is guaranteed to have acquired the bus and no other station will interfere with this current transmission.

8.2.1.2 Jamming Signal. The jamming signal sent after a collision is detected is used for *consensus enforcement*. Basically, the system is trying to avoid a situation where one station detects a collision while the other station involved in the collision does not and continues to transmit. This can happen because of the difference in the relative powers of the received signals at different stations or because of the difference in the time required by a station to detect collisions. The jamming signal is a way for a station to inform others when it detects a collision.

8.2.1.3 *Back-off Algorithm.* After a transmission is aborted (because of collision detection), a station waits for a random amount of time before it attempts a retransmission. The waiting time is random in order to reduce the likelihood of further collisions between the same packets. This process of waiting a random time and retransmitting is known as the *back-off process*. The method of computing this random time is known as the back-off algorithm. The IEEE 802.3 specifies the use of the truncated binary exponential back-off algorithm which operates as follows:

1. Pick a random integer, r, uniformly between 0 and $2^{\min(k,10)}$, where k is the number of transmissions of the frame so far.
2. The transmitter waits $r \times 2\tau$ time units, where τ is the maximum transmitter-to-receiver delay in the network.

In the standard, a maximum of 16 transmission attempts is allowed, after which the transmitter gives up on the frame and reports a problem to its higher layer.

8.2.2 Frame Format

The MAC layer frame format for the CSMA/CD standard is shown in Figure 8.5. The frame contains the following fields:

Preamble. This is a 7-octet field that is used by the receiver to achieve synchronization with the received signal.

Start Delimiter. The bit sequence 10101011 is used to indicate the start of the frame.

Source/Destination Addresses. These are either 2 or 6 octets long. All addresses on a particular network will be the same size.

Length. A two-octet field whose value indicates the length (in octets) of the data field.

Pad. This is a number of octets that are added if necessary to insure that the CSMA/CD frame is no shorter than a minimum allowable length. This minimum is typically defined as the number of bits that can be transmitted during 2τ time units. This insures that a transmitting

Preamble	Start delimeter	Source Address	Destination Address	Length	Data	Pad	Frame Check Sequence
7 octets	1 octets	2 or 6 octets	2 or 6 octets	2 octets	variable	variable	4 octets

Figure 8.5. Frame format for the IEEE 802.3 CSMA/CD protocol.

station can tell whether its frame has been involved in a collision before completing its transmission. This is important to allow retransmission of collided frames at the MAC layer rather than at a higher layer which might cause unnecessary delays.

Frame Check Sequence. A four octet cyclic redundancy check (CRC) calculated based on the entire frame excluding the preamble and the start delimiter.

8.3 TOKEN RING

The IEEE 802.5 token ring standard describes a protocol that has come to be viewed as the main rival to the CSMA/CD protocol. In general, the main opposition to the CSMA/CD protocol is due to the nondeterminism and potentially unbounded delays that can result from the back-off mechanism involved. Token passing is viewed as a mechanism to insure bounded delays and to impose priority access, both of which are not possible in CSMA/CD.

8.3.1 Medium Access Control and Logical Link Control Frames

The basic service provided by the MAC layer is the delivery of frames given to it by the local LLC layer to a remote LLC layer. This is accomplished by encapsulating the data passed down with a MAC header and a MAC trailer and following the MAC procedures. Such frames passed between MAC layers are known as *LLC data frames*.

Also in the operation of the MAC protocol some frames are exchanged between communicating MAC layers. These frames do not originate in a higher LLC layer and are known as *MAC frames*. These frames are normally exchanged without the knowledge and intervention of the LLC layer. An example of such a frame is the token exchanged to transmit access rights (see Fig. 8.6). Medium access control frames are not necessary in the operation of the CSMA/CD protocol, so we will first encounter them in the discussion to follow on the token ring protocol.

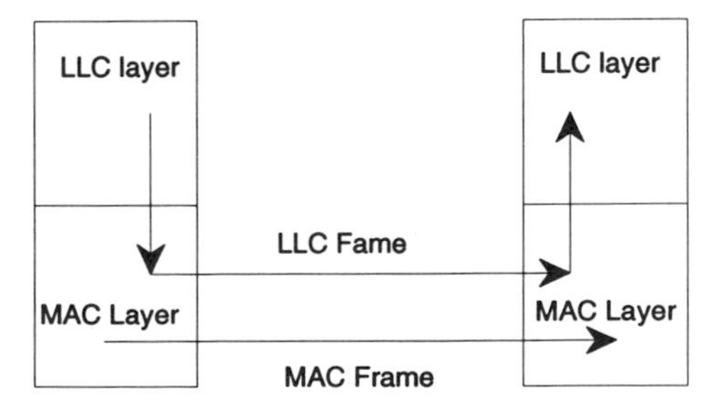

Figure 8.6. MAC and LLC frames.

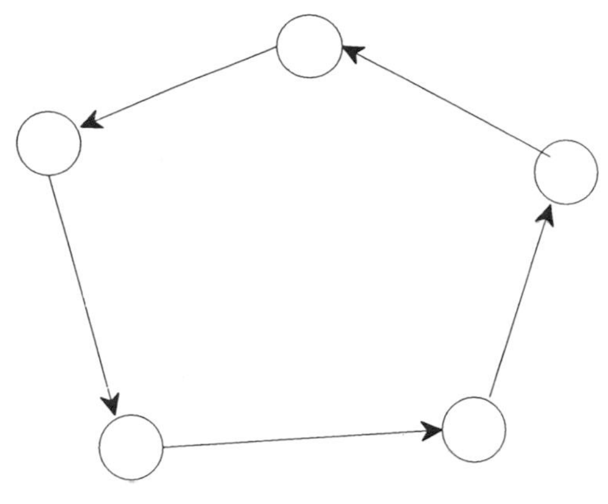

Figure 8.7. Ring architecture.

8.3.2 Basic Operation

The main idea behind the token ring protocol has been described in Chapter 7. Here we describe some details of the specific IEEE 802 version of this protocol. A token ring network consists of a number of stations connected by point-to-point links to form a unidirectional ring (see Fig. 8.7). The stations can be viewed conceptually as being in one of two configurations:

Repeating and Copying. All data arriving on the incoming link is repeated on the outgoing link after being stored inside the station for some brief period of time, typically one bit transmission time. The station can also make a copy of the received data for further processing. A station in this configuration (see Fig. 8.8) is allowed to modify the data slightly as it is repeating it (e.g., by flipping a bit).

Removing and/or Inserting. All incoming data is removed from the ring and new data can simultaneously be inserted on the outgoing link (see Fig. 8.9).

To describe the operation of a token ring protocol consider Figure 8.10, which describes a state of the ring network where all stations are in the

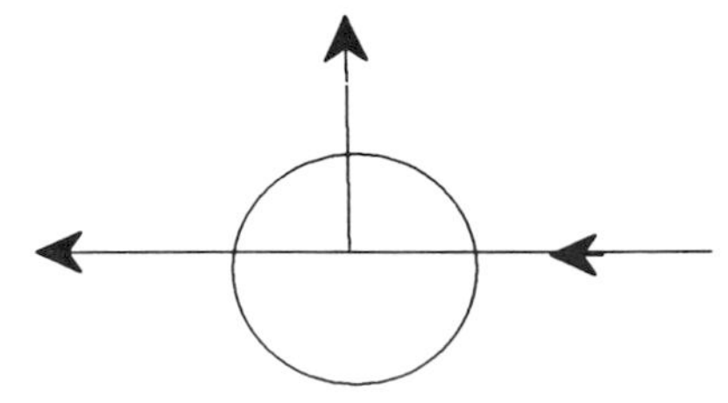

Figure 8.8. The repeating and copying configuration.

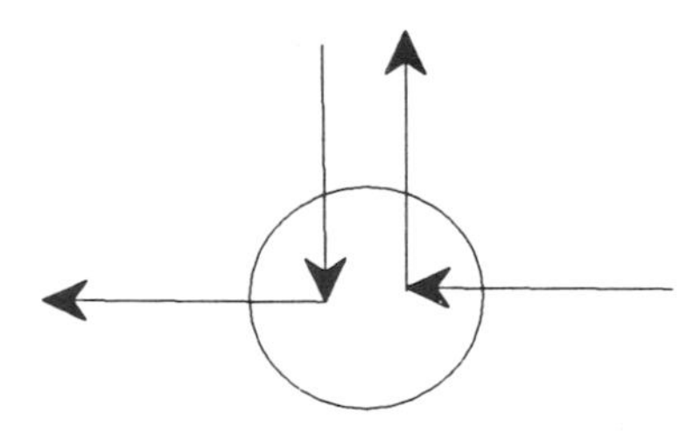

Figure 8.9. The removing / inserting configuration.

repeating and copying configuration with a token progressing from station B to station A. Assuming that node A is ready to transmit, upon reception of the token, it will switch to the removing and/or inserting configuration and transmit its frame (see Fig. 8.11). The station remains in this configuration until it completes transmission of its frame or until it receives the header of its own transmission (whichever occurs last). A new token is issued at that time. After completely removing its own frame and after issuing a new token, a station reverts to the repeating and copying configuration.

There are two points to observe about this process:

1. The IEEE 802.5 protocol specifies *source removal* of transmitted packets. This allows for a receiver to piggyback indications to the transmitter that the frame has been successfully received. This also insures that the ring network operates as a broadcast network, where each transmission is heard by all stations.
2. Whether a station finishes a transmission before it begins to receive the header or not depends on the *bit length* of the ring, which is defined as

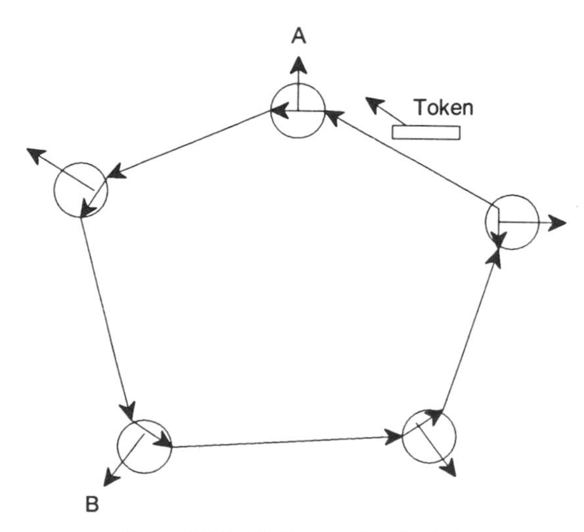

Figure 8.10. Station waiting for token.

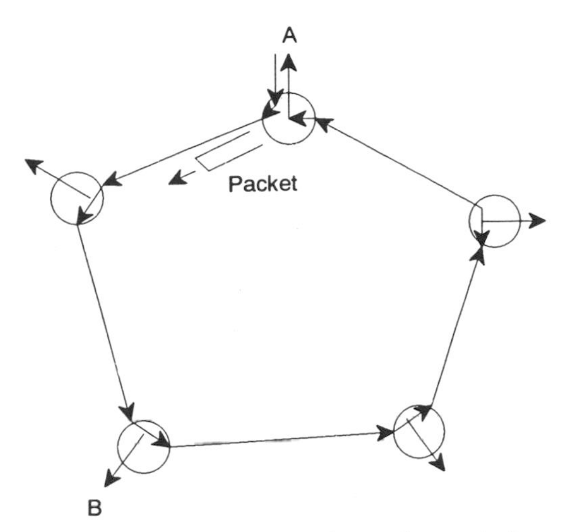

Figure 8.11. Station captures token and transmits frame.

the number of bits a station can transmit before it begins to hear its transmission. This length is a function of the physical length of the ring as well as the delay incurred by each repeating station and the data rate.

8.3.3 Frame Format

There are two basic kinds of data units transmitted on an IEEE 802.5 token ring: a token and a frame.

8.3.3.1 Token. A token is shown in Figure 8.12 and consists of the following fields:

Start delimiter consists of an octet containing the pattern JK0JK000 where J and K are nondata symbols. A discussion of the use of nondata symbols can be found in the next section.

Access control is an octet that contains four subfields as indicated in Figure 8.12. The use of the monitor (M) bit will be explained later. It is important to note that the T bit is always set to 0 to indicate that this is a token. The P bits are used to indicate the priority of the token, and the R bits are used by stations to request a token of a certain priority.*

End delimiter consists of an octet containing the pattern JK1JK1IE. The I bit indicates whether or not this is an intermediate frame in a transmis-

*The priority mechanism is somewhat complex, and we do not discuss it further here. Suffice it to say that the token ring standard allows frames to be of different priorities and frames of higher priority will normally be transmitted before frames of a lower priority.

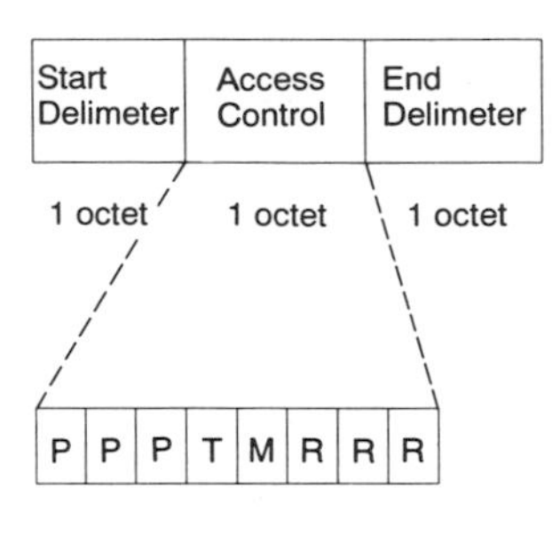

Figure 8.12. Token format.

sion sequence or not. The E bit is set by any station that detects an error in a frame.

Note that there is no error checking capability in a token. If errors in the token do occur they can be detected and recovered from using the monitor mechanism to be described later.

8.3.3.2 Frame. A frame consists of the following fields (see Fig. 8.13).

Start/end delimiters are the same as in the token.

Access control is basically the same as in the token except that the T bit is set to 1. Note that from a receiving stations viewpoint a token and a frame are indistinguishable until the T bit is received.

Frame control is an octet containing two subfields:

Type (2 bits) indicates whether this is a MAC or LLC frame.

Control (6 bits) indicates the type of MAC frame. We discuss some types of MAC frames in Section 8.3.4.

Destination and source address

Information field

Frame check sequence is a 4-octet CRC covering the frame control, destination and source address, information, and frame check sequence fields.

Frame status contains one octet as shown in Figure 8.13. The field contains two pieces of information regarding the reception of a frame by its intended destination:

A-bit indicates that the destination address in the frame was recognized by a station on the network.

C-bit indicates that the frame has been copied by the intended destination.

Start Delimeter	Access Control	Frame Control	Destination Address	Source Address	Data	FCS	End Delimeter	Frame Status
1 octet	1 octet	1 octet	2 or 6 octets	2 or 6 octets	0 or more octets	4 octets	1 octet	1 octet

Frame Control

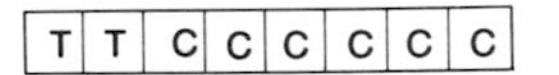

TT: Frame Type (MAC or LLC)
CCCCCC: MAC frame type

Frame Status

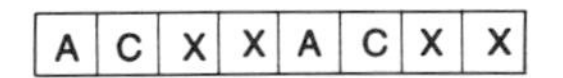

A,A: Duplicate bits indicating address recognized
C,C: Duplicate bits indicating frame copied
X : Reserved bits (not currently used)

Figure 8.13. Frame format.

Both bits are initially set to zero by the source station. Note that each bit is repeated twice in the frame status field. This provides for error detection since this field is outside the scope of the coverage of the frame check sequence.

8.3.4 Token Ring Operation Revisited

Armed with the description of frame formats we now briefly revisit the token ring procedures.

8.3.4.1 Token Passing. First we observe that the first two fields of a token and a frame are the same, the distinguishing feature being the value of the T bit. When a station recognizes a start delimiter, it cannot tell whether it is receiving a token or a frame until it receives the T bit. If the T bit is 1, it continues to repeat the frame, checks the destination address, and makes a copy of the frame if it is indeed the intended destination of the frame. If the T bit is 0 and the station has nothing to transmit, it continues to repeat the token. Otherwise, it overwrites the T bit with a 1 before sending it and continues to repeat the incoming access control field. It then appends its own frame (starting with the frame control field), while at the same time absorbing the tail of the incoming token.

8.3.4.2 Frame Reception. A station that recognizes its address in the destination address field of an incoming frame makes a copy of the frame and continues to repeat it. The station then sets the two A and the two C bits in the frame status field to 1, indicating that it has recognized its address and successfully made a copy of the frame. This is a mechanism to indicate to the source that the frame has been properly received. When a frame is removed by its transmitter, the frame status field can be examined to determine what happened to the frame at its intended receiver.

8.3.4.3 Monitor Functions. In the token ring standard, one of the stations on the ring is assigned to be a monitor. The monitor is in charge of detection and recovery of eternally circulating frames and lost token conditions. A frame can circulate indefinitely, if its source station malfunctions after completing its transmission but before removing the frame from the ring. To recover from this, the monitor sets the M bit in a frame to 1. If the monitor receives a frame with the M bit set to 1, this means that the frame has already made more than one full circle around the ring; an error condition. The monitor will thus remove the frame from the ring.

Tokens can be lost if the token holder malfunctions while holding the token. This results in the absence of activity on the ring. The monitor is equipped with a timer that is reset whenever a valid frame or token is heard. If this timer expires, the monitor then issues a new token.

The monitor in a token ring network performs vital functions, and, therefore, the network is extremely vulnerable to the loss of the monitor. In the IEEE 802.5 standard, all stations can act as monitors and are known as *standby monitors*. The station assigned to be the monitor is called the *active monitor*. An active monitor periodically issues a broadcast active-monitor-present MAC frame to reassure the rest of the network. Each station maintains a timer that is reset when the station hears the active-monitor-present frame. Absence of an active monitor causes the expiry of the timer, which causes the stations on the network to go through an election process that uses the claim-token MAC frame. The process results in identifying the highest address station, which then becomes the active monitor.

8.4 TOKEN BUS

The token bus protocol was conceived as a compromise between the token ring and CSMA/CD protocols. The intent was to develop a protocol that had the advantages of each and none of the disadvantages (real or perceived) of the other two. In particular, it shares the token passing technique with the token ring protocol, which makes network access (and therefore delay and fairness) more controllable than in the random access CSMA/CD network. On the other hand, the bus architecture is viewed as more reliable than the

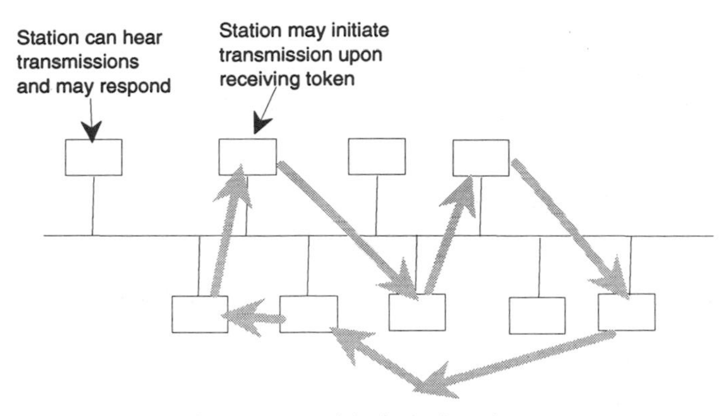

Figure 8.14. A logical token ring.

ring architecture.* As will be seen in the discussion that follows, this compromise is achieved at the price of added complexity to the protocol procedures.

8.4.1 Basic Operation

In a network using the token bus access method, all stations are interconnected by a broadcast bus. This shared medium is such that a transmission by one station can be heard by all operational stations connected to the bus. There are two groups of stations: those that are allowed to initiate transmissions and those that can only hear and are allowed to only respond to transmissions. We describe later the procedures involved if a station desires to move from one group to the other. Stations that may initiate transmissions are arranged in a *logical ring*, which bears no relationship to the physical arrangement of the stations on the bus (see Fig. 8.14).

In the logical ring, a token, representing the right to transmit, is passed sequentially from one station to the next. A station that possesses the token may transmit and passes the token after concluding its transmission. As we describe in Section 8.4.3, procedures are defined to handle lost tokens and other error situations.

8.4.2 Frame Format

In this section, we describe the contents of the MAC frames, that is, the frames exchanged between MAC layer entities. Figure 8.15 shows the various fields in a MAC frame. These are described next. A MAC frame can be a maximum of 8191 bytes (or octets) long not counting the start and end delimiters.

*This is perhaps more of a perception than a reality.

Preamble	Start Delimeter	Frame Control	Destination Address	Source Address	Data	Frame Check Sequence	End Delimeter
1 or more octets	1 octet	1 octet	2 or 6 octets	2 or 6 octets	0 or more octets	4 octets	1 octet

Figure 8.15. IEEE 802.4 frame format.

Preamble. This field precedes every transmitted frame and serves two purposes: (1) It allows the receiving station to synchronize with the transmitter and to acquire the signal level, that is, to turn the receiver's volume up or down so it can hear the frame comfortably. (2) It guarantees a minimum separation between consecutive frames so that a receiver will have time to process a frame before it receives the next one.

Start and End Delimiters. The start delimiter (SD) denotes the beginning of a frame and an end delimiter (ED) denotes the end of a frame (similar to flags in HDLC). To insure that no SD or ED appears in the transmitted data, these delimiters contain nondata symbols (as explained below). The SD is coded as NN0NN000 and the ED is coded as NN1NN1IE where N is a nondata symbol, 1 and 0 are the usual bits.

How a nondata symbol is implemented depends on the details of the physical layer signaling procedures. For example, if frequency shift keying (FSK) is used, a transmitter can change the signal on the bus from some low frequency (L) to some higher frequency (H) or vice versa. In this case a 1 may be coded as a transition from L to H in one bit time and a 0 may be coded as a transition from H to L within a bit time. A nondata symbol can thus be encoded by not having a transition within a bit time.

The I bit in the ED is set to 0 to indicate that this is the last frame transmission from this station. The E bit is set to 1 by a repeater and indicates that the frame immediately preceding the ED is in error. This helps to isolate the segment of the bus causing the error. A receiving station may treat a frame with the E bit set to 1 as an invalid frame.

Frame Control. The first two bits of this field indicate the type of the frame as follows:

00 MAC control
01 LLC data
10 Station management data
11 Special purpose data

The first two types are discussed below.

In a MAC control frame, the last six bits indicate the type of MAC frame. There are seven types defined in the standard (the use of these frames is discussed later).

1. Claim_Token
2. Solicit_Successor_1
3. Solicit_Successor_2
4. Who_Follows
5. Resolve_Contention
6. Token
7. Set_Successor

In an LLC data frame the last six bits are subdivided into three bits that indicate the priority of the data frame and three bits that indicate the required MAC layer action (if any) in response to this frame.

Source / Destination Addresses. The standard allows for either 16-bit or 48-bit address fields. The all ones address is designated for broadcast to all stations on the bus.

Data-Unit Field. The contents of this field depend on the class of frame as defined in the frame control field. In an LLC data frame, this field contains the LLC frame passed down from the local LLC layer. In a MAC control frame, if needed, information specific to the type of control frame being transmitted is put in the data-unit field.

Frame Check Sequence. This is a 32-bit cyclic redundancy check covering all fields except the preamble and start/end delimiters.

8.4.3 MAC Layer Procedures

In this section, we describe the procedures required for the token bus protocol to function properly in the presence of transmission errors and losses. These procedures are implemented by exchanging MAC control frames between MAC entities. The complexity of some of the procedures derives from the fact that in the token bus protocol a logical ring is maintained. In the token ring, activities such as bypassing a failed station and adding new stations have to be done in hardware and are thus external to the MAC protocol. In the token bus, such activities are performed in software and are thus part of the MAC protocol.

In what follows, the term *slot-time* is used to refer to the maximum time that a station needs to wait for a response to a MAC control frame. This time is measured in units of *octet-times*, the time required to transmit 8 bits. The standard defines the slot time as

$$\text{Slot-time} = \text{INTEGER}((1/8) * \{[2 * (\text{MaxPropDelay} + \text{StationProcDelay}) + \text{SafetyMargin}] * \text{DataRate} + 7\})$$

Note that the above definition of slot time is basically a determination of the number of octets that would be transmitted in twice the maximum propagation delay plus the station processing time. A safety margin is added to this time.

8.4.3.1 Token Passing. The token passing procedure is shown in Figure 8.16. The logical ring is organized so that each station (with one exception) passes the token to a lower address station. To close the ring, one station passes the token to a higher address station. Each station maintains its own address in the register TS (this station), the address of the station to which it passes the token in the register NS (next station), and the address of the station it receives the token from in the register PS (previous station).

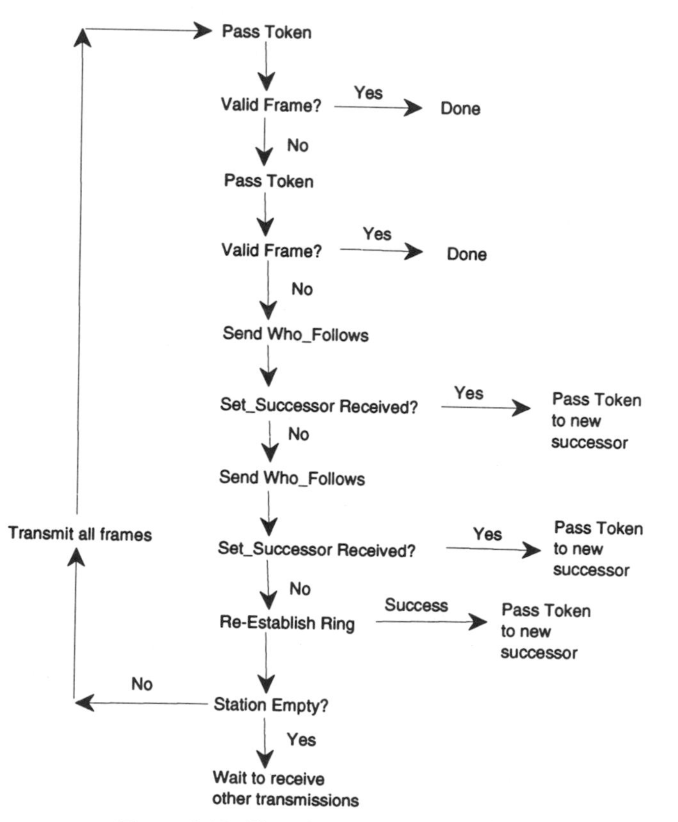

Figure 8.16. The token-passing procedure.

The token is a MAC control frame with a null data field; SA is the sending station's address and DA is the sending station's NS. Note that the token is actually broadcast on the bus and is heard by everybody (as are all frames). It is, however, only addressed to the station whose address appears in the DA field and other stations hearing the token will ignore it.

In the token bus protocol, the station passing the token is responsible for insuring that it reaches its destination. If, within a timeout period after passing the token, the sending station hears a valid frame (i.e., a frame that is error and collision free) on the bus, it assumes that its successor in the logical ring has received the token and sent a data frame or is passing the token. If a valid frame is not heard, the station assumes that the token was lost and attempts to pass the token a second time. If nothing is heard within a timeout period, the station passing the token assumes that its successor has failed. Note that this is just an assumption; the station passing the token has no way of finding out for sure whether the next station has failed or whether it was unlucky and the token it passed was corrupted twice in a row.

The station holding the token now attempts to bypass the failed station. This is done by sending a *who_follows* MAC control frame. This frame has a broadcast DA (all 1s), sending station's address SA, and a data field containing the sending station's NS. All stations on the logical ring receive this frame and compare their predecessor's address (in the PS registers) with the address in the data field of the who_follows frame. The station that matches these two addresses (there should be only one) responds with a *set_successor* MAC control frame. In this frame, DA is the address of the sender of the who_follows frame and the data field contains the value of the sending station's TS or NS addresses.

If no response is heard to the who_follows frame within three slot-times, the token holder sends another who_follows frame. If still no response is heard, the token holder assumes the ring is dead and it is the only station holding a token; the token holder therefore attempts to reestablish the ring using the procedure described in Section 8.4.3.5.

If reestablishing the ring fails, one of several catastrophic events could have taken place such as failure of all stations, breakage in the medium, or the failure of the token holder's receiver. The standard requires the station to send any outstanding frames it may have and try to pass the token repeating the above procedure. If it is still not successful, the station abandons its attempts at maintaining the ring and listens for transmissions from other stations.

8.4.3.2 Station Insertion. As was mentioned earlier, the token bus protocol allows for listen-only stations to be connected to the bus. These stations never receive the token and cannot initiate transmissions. They may, however, respond to transmissions from a token holder. The station insertion procedures are defined to allow such stations to join the logical ring if they wish. The basic manner in which this is done is for the token holder to solicit insertion into the ring after transmitting its frames and before passing the token. Because inserting a new station requires some time, a token holder is allowed to solicit insertions only if the token rotation time (i.e., the time it takes the token to make a full circle) is within predetermined bounds. To reduce the contention that would arise when several stations want to be inserted, the process is controlled by allowing only stations whose addresses fall within a certain range to accept an invitation to join the ring.

The insertion procedure depends on the location of the token holder in the logical ring:

Case 1. The token holder's TS is greater than the NS (for all stations except one). The token holder sends a *solicit_successor_1* MAC control frame with a null data field in which SA is the sender's address and DA is the sender's NS. This frame solicits the insertion of any station with an address between TS and NS. One response window, which is one slot-time in length, is opened after the solicit_successor_1 frame. If nothing is heard during the response window, the token holder passes the token, assuming no station within the specified range wants to be inserted. If a transmission is heard during the response window, the token holder will wait until the transmission is complete and the bus is idle.

A station within the address range specified by the solicit_successor_1 frame that wants to be inserted responds with a set_successor MAC control frame (the same frame used to respond to a who_follows frame). Thus, if one set_successor frame is heard by the token holder, it will reset its NS value to the source address of the set_successor and pass the token to that station. However several stations may respond with a set_successor to the insertion solicitation. In such a case, the token holder hears a collision start during the response window and attempts to resolve the collision through a procedure described in Section 8.4.3.3. The outcome of this procedure is to pick, from among the stations sending the colliding set_successor frames, the one with the highest address. This station's address is made into the token holder's NS. Notice that the stations losing the contention resolution process will get an opportunity to get inserted into the logical ring just before the next token pass.

Case 2. The token holder's TS is less than the NS. Exactly one station should have this property. The insertion procedure is very similar to case 1 described above. The difference here is that a *solicit_successor_2* MAC control frame is broadcast with two response windows opened afterward. Stations with addresses less than TS (contained in SA of the solicit_successor_2) may respond with a set_successor during the first response window. Stations with addresses greater than NS (contained in the DA of the solicit_successor_2) may respond with a set_successor in the second response window, but only if they do not hear any transmissions start during the first response window. (Recall that all transmissions are heard by all operational stations.) If a collision occurs between set_successor frames, the contention resolution procedure described in Section 8.4.3.3 will result in identifying and inserting the station with the highest address. The token is then passed to that station.

8.4.3.3 Contention Resolution. This procedure is invoked whenever the token holder hears a collision of set_successor frames starting in a response window. The token holder waits until the transmissions are completed and then transmits a *resolve_contention* MAC control frame. Four response windows are opened after the frame. This frame has a null data field, SA is the sending station address and DA is the broadcast address. Only the stations involved in the set_successor collision are involved in the collision resolution process, which has the objective of allowing the highest address station to transmit successfully.

After the resolve_contention frame is issued the contending stations may start transmission of their set_successor frames in the response windows as follows:

A station may respond in the first window if the two most significant bits (MSB) of its address are 11.

A station may respond in the second, third, or fourth window if the two MSBs of its address are 10, 01, or 00, respectively *and* the station did not hear a transmission start before its window comes up. Otherwise the station refrains from transmission and is eliminated from contention.

If no successful set_successor frame is heard after this procedure, this implies that at least two stations with the same two MSBs desire insertion into the ring. The process is now repeated starting with a transmission of a new resolve_contention frame by the token holder. Only stations that transmitted (i.e., were not eliminated) in the previous round are allowed to transmit in this round. These stations now use the next two bits of their addresses to choose one of the four windows. If still more collisions occur, the procedure is repeated again and again, each time using the next two address bits, until a successful transmission takes place. Note that if, as should be, no two addresses are the same, we are guaranteed that eventually a successful transmission will take place after a maximum of eight rounds (for 16-bit addresses) or 24 rounds (for 48 bit addresses).

8.4.3.4 Station Deletion. A station may remove itself from the logical ring in one of two ways:

1. The station may refuse to respond to a token passed to it. This will result in the station being bypassed as described in Section 8.4.3.

2. When the station has the token, it first sends a set_successor frame to its predecessor (PS) and then passes the token to its successor (NS). This method is more efficient.

8.4.3.5 Ring Initialization. This procedure is triggered by the detection of a prolonged period with no activity on the bus, which can occur if a token holder fails before passing the token. Each station has an *inactivity timer* that counts how long it has been since any transmission was detected on the bus. If this timer expires, a station will try to reinitialize the ring. Notice that since silence is detected by all stations at about the same time (within a maximum propagation delay), all functioning stations on the bus will try to reinitialize the ring. The reinitialization process is carried out in two steps:

1. *Token claiming*. Each station with an expired inactivity timer attempts to claim the token for itself. The procedure (described below) results in the highest address station being declared as the token holder.
2. *Ring reestablishment*. The station winning the token claiming procedure declares itself as the ring, that is, it sets NS = PS = TS, and attempts to insert other stations on the ring. This is the same procedure used when the who_follows procedure fails (see Section 8.4.3.1).

To claim the token, a station transmits a *claim_token* MAC control frame. This frame contains a data field that contains 0, 2, 4, or 6 slot-time octets (recall that a slot-time is defined in octets). The actual contents are arbitrary. The length is based on two bits (starting with the most significant two bits) of the address of the station sending the claim_token frame. Stations with address bits 00, 01, 10, and 11 send a claim_token frame with the length of data field 0, 2, 4, and 6 slot-times, respectively. After completing the transmission, a station waits 1 slot-time and then senses the channel. If transmission is detected, this means that at least one other station (with a higher address) is transmitting a longer claim_token frame. The station then defers to the higher address station and is eliminated from the token-claiming procedure. If silence is detected 1 slot-time after the completion of the transmission of a claim_token frame, the station transmits another claim_token frame with a length determined by the next two MSBs of its address. This process is repeated until all address bits have been used up.

If after transmitting the claim_token frame with a length determined by the last two address bits, silence is detected (after waiting one slot-time) the station has won and may now reestablish the ring. The above procedure will have exactly one winner, as long as no two addresses are the same.

The winner of the claim token procedure now forms a logical ring by itself. That is, it sets TS = NS = PS. Ring reestablishment is accomplished by sending a solicit_successor_2 MAC control frame with DA = SA = TS. This frame will solicit insertion from all operational stations. What follows is exactly the same as in the station insertion procedure described in Section 8.4.3.2; resulting in the identification of the next highest address operational station on the ring. The token is passed to this newly inserted station, which in turn repeats the insertion process. This is done until no more stations are inserted and the ring returns to normal operation.

8.5 DISTRIBUTED QUEUE DUAL BUS

The IEEE 802.6 working group was formed to consider a standard for a metropolitan-area networks (MANs). Initially the working group considered a network of interconnected token rings to provide MAN-type connectivity. However, this approach was discarded after it lost industrial backing, and the search was then on for another candidate. The protocol then called *Queued Packet and Switch Exchange* (QPSX) was identified as a good starting point for the 802.6 working group's activities. The protocol has since been renamed the Distributed Queue Dual Bus (DQDB) protocol.

8.5.1 Distributed Queue Dual Bus Network Architecture

A DQDB network consists of two opposite flowing unidirectional busses, known as the *dual bus pair*. One bus is denoted bus A and the other bus B (see Fig. 8.17). Each node can read both busses and can write by ORing onto the existing information being transferred. Thus a node can observe and interpret activity on the busses and may change 0 bits into 1 bits but never the opposite.

Data being transmitted on both busses is formatted into slots that are generated at the head of both busses at exactly the same rate. This is done by slaving the slot generator at the head of bus B to the slot generator at the head of bus A. Each slot can carry a payload of 52 octets. Slots are initially empty (all 0's) and a payload is written at a node by ORing into an empty slot. To prevent other nodes from writing (and destroying the contents) of a full slot, each slot contains a *busy bit*, which is initially set to 0 by the slot generator and to 1 by a writing node. Nodes are prevented from writing into a slot with a busy bit set to 1. A slot also contains reservation bits used in the operation of the distributed queue described in Section 8.5.5.

8.5.2 Routing in the Distributed Queue Dual Bus

Unlike other 802 protocols, the DQDB MAC layer has to make routing decisions, as it sends information on one bus or the other depending on the destination. There are essentially two different ways for routing:

1. *Bothway transmission* in which all data is transmitted on both busses. This is feasible if efficient use of the bandwidth is not required.

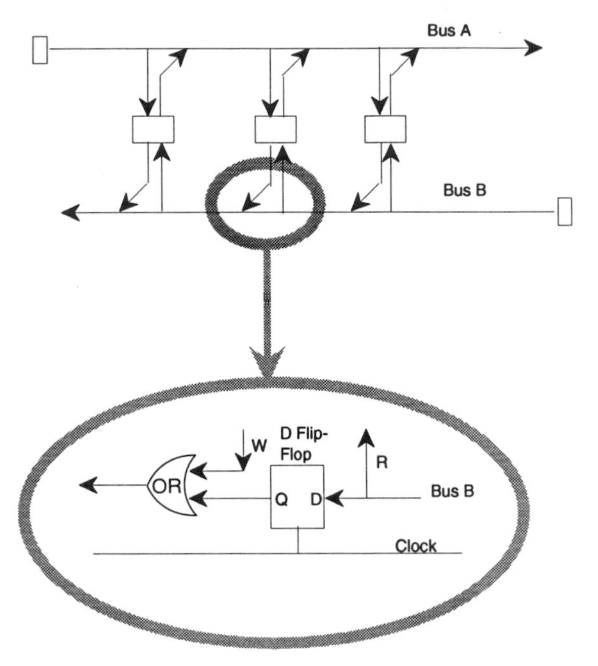

Figure 8.17. DQDB network architecture.

2. *Bus selection tables* in which a table is used to associate bus A or B with each node address in the network.

If tables are used, they need to be maintained since the relative positions of nodes may change over time due to the resiting of nodes. Also when the network is large, the tables may become unacceptably long. It is, therefore, beneficial to consider using *self-learned tables*. In this approach, the tables are maintained by observing traffic on the busses. Whenever a node receives data from source X on a bus, it knows that the opposite bus should be used for traffic destined to X and this information is entered in the bus selection table. Traffic destined to nodes for which no bus has been determined are sent using bothway transmission. If the tables become full, then newly learned addresses may be allowed to overwrite old entries that have not been used recently. Entries in the table may also be purged periodically to avoid determining routes based on old information.

8.5.3 Overview of Medium-Access-Control Layer Services

The DQDB MAC and physical layer standards are designed to operate within the framework of the IEEE 802 LAN standards. In particular, the DQDB protocol will interface directly with a higher LLC layer and deliver its

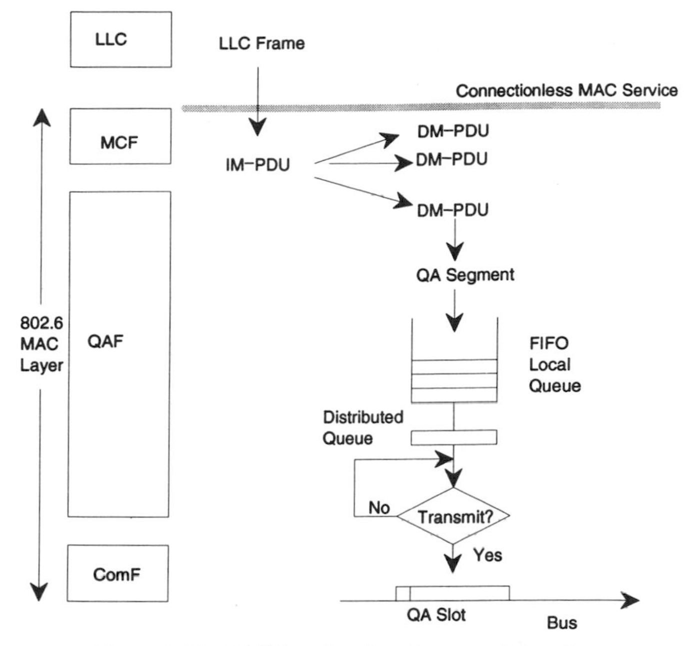

Figure 8.18. DQDB protocol sublayers and data flow.

frames to a remote LLC layer. This is known as the *connectionless MAC service* (CMS). Although other services are defined by the DQDB standard, only how the CMS is provided is discussed here. The CMS is suitable for most data communication applications using the DQDB protocol.

Figure 8.18 shows the subdivision of the MAC layer into three sublayers: the MAC convergence function (MCF), the queued-arbitrated function (QAF), and the common function (ComF). The MCF accepts frames from the LLC layer and adds protocol control information (i.e., a header) to form an initial MAC protocol data unit (IM_PDU). This is divided into units of 44 octets. Other control information is added to each unit to form a derived MAC protocol Data Unit (DM_PDU). The DM_PDUs are then passed to the QAF which adds still more control information to form a queued-arbitrated (QA) segment. A QA segment represents the payload that can be carried by one slot. The QAF maintains a First-in-first-out (FIFO) queue in which it holds the segments awaiting transmission. The position at the head of that queue is part of a network-wide distributed queue.

The ComF serves as a relay for the transfer of slots to the local physical layer and implements several operations and management functions that are required for the correct functioning of the protocols. These functions are defined in a separate layer, because they are also required to implement other DQDB services not discussed here. The use of the ComF layer may be shared between the QAF and any other sublayers requiring such functions.

The QAF accepts QA segments and transmits them (using the ComF as a relay to the physical layer) according to the distributed-queue protocol. This protocol, which is described in detail in Section 8.5.5, is an important part of the DQDB standard. Each node maintains a set of counters in the QAF, which are used to operate a queue of QA segments distributed across the network. Updating of the counters is triggered by segment arrival events as well as the contents of busy and reservation bits in the slots being read.

On the receiving side, QAF uses the services provided by the ComF layer to read QA segments being transmitted. Based on the control information in the segment headers, the appropriate ones are delivered to the MCF. The receiving MCF reads and removes the control information added by the sending MCF and reassembles the DM_PDU's into an IM_PDU which is then delivered to the appropriate LLC layer entity.

8.5.4 Frame Formats

8.5.4.1 Initial MAC Protocol Data Unit. The MCF encapsulates an LLC frame into an IM_PDU as follows: A brief description of the fields contained in the IM_PDU headers and trailer is presented below.

Common PDU header/trailer fields:

- Reserved, 1 octet.
- Beginning/end tag (BEtag), 1 octet, is a value inserted in the header and trailer of the same IM_PDU to insure proper reassembly.
- Buffer allocation size (BASize), 2 octets, is the length in octets of the MCP header, the header extension, and the information field of the IM_PDU; it is used for buffer allocation purposes at a destination node and to insure proper reassembly.

MAC convergence protocol (MCP) header fields:

- Destination/source address, 8 octets each.
- Protocol identification, 1 octet, identifies the user of the MAC layer (PI = 1 indicates LLC as the user).
- Quality of service, 4 bits, indicates the requested delay or loss.
- Header extension length, 4 bits, the length of the header extension field is in units of 4 octets.
- Bridging, 2 octets, is reserved for future use, probably as a counter for the number of bridges traversed or the maximum number of bridges in an extended 802.6 LAN.

Header extension field, 0 to 20 octets, is reserved for future use.

IM_PDU information, 0 to 9188 octets, contains the LLC frame to be transmitted.

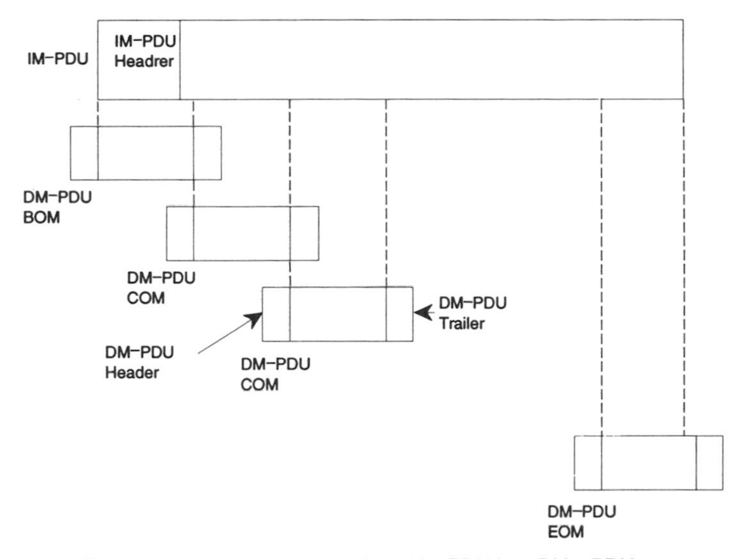

Figure 8.19. Segmentation of an IM_PDU into DM_PDUs.

Pad, 0 to 3 octets, contains sufficient octets (all zeroes) so that the length of the information field plus Pad is an integer multiple of 4 octets.

8.5.4.2 Derived MAC Protocol Data Unit. As mentioned earlier, the MCF segments an IM_PDU into 44 octet units encapsulating each into a DM_PDU. A typical segmentation process is illustrated in Figure 8.19.

The DM_PDU header is two octets long and contains two fields: the segment type and the IM_PDU identifier (MID).

Segment Type. This field is 2 bits long and indicates the position of the DM_PDU in the original IM_PDU as follows:

00 Continuation of message (COM)
01 End of message (EOM)
10 Beginning of message (BOM)
11 Single segment message (SSM) is used when the DM_PDU contains an entire IM_PDU (i.e., the IM_PDU's length is less than or equal to 44 octets).

IM_PDU Identifier (MID). This 14-bit field is used to unambiguously associate a DM_PDU with a single IM_PDU from a single MCF entity. This field is used in the reassembly process. All DM_PDUs derived from the same IM_PDU have the same MID value. Assignment of these identifiers is done through the use of a distributed algorithm. The DM_PDU Trailer contains the following fields:

1. *Payload length* (6 bits). The length in octets of the payload. This is important in an EOM or SSM DM_PDU as the payload may be less than 44 octets and padding characters may be added.
2. Payload CRC (10 bits).

8.5.4.3 Queued-Arbitrated Segment. A QA segment is formed by adding a header to a DM_PDU. This is done by the QAF sublayer (see Section 8.5.3). The QA segment header is 4 octets long and is formatted as follows:

Virtual circuit identifier (VCI), 20 bits, is set to all 1s for the connectionless MAC service.

Payload type, (PT) 2 bits, The standard currently defines only the value 00 to indicate user data.

Segment priority, (SP) 2 bits, use is under study.

Header check sum, (HCS) 1 octet, protects against header errors.

8.5.4.4 QA Slot. As described earlier, QA segments are transmitted by the ORing inside slots generated in the dual bus pair. A slot contains other control information in an 8 bit *access control field*, which contains, among other fields, the following:

Busy. One bit indicates whether the slot contains information (busy = 1) or is empty (busy = 0),

SL_Type. One bit indicates whether the slot can carry a QA segment.*

REQ_3, REQ_2, REQ_1, REQ_0. Each field is one bit long and is set to zero when the slot is generated. These bits can be set by nodes in slots appearing on a bus to request access for a QA segment on the opposite bus. Requests can be made at one of four priority levels with REQ_3 being the highest. The use of these bits is explained in more detail later.

8.5.5 Distributed Queue Operation

As QA segments are generated in the QAF sublayer, they enter the FIFO queue to await transmission (see Fig. 8.18). Queue-arbitrated segments at the head of the FIFO queue are said to be in the distributed queue. Only one segment per node and per bus is allowed in the distributed queue. The QAF now attempts to transmit the segment over the appropriate bus. Figure

*Another type of segment, the prearbitrated (PA) segment, is defined in the DQDB standard. A slot is normally designated at the time it is generated as either a QA or PA slot. We do not discuss PA segments here.

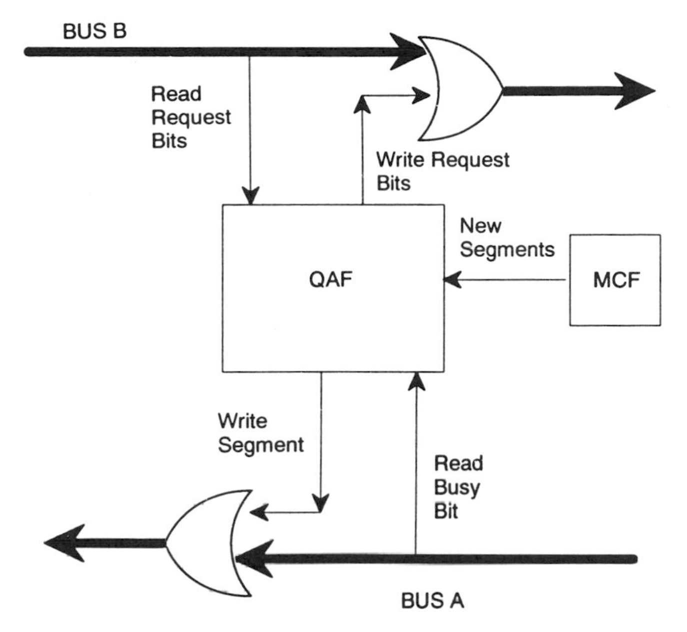

Figure 8.20. Information read and written by QAF.

8.20 shows the information read by the QAF to determine when to transmit a segment on bus A. In the following, we consider only transmissions on bus A. An identical and independent procedure is used for transmissions on bus B.

A segment is transmitted by the ORing onto the bus onto the bus and thus the busy bit has to be read to make sure that the slot is empty. If the protocol allows a node to write into the first available slot that it encounters after entering a new segment in the distributed queue, a busy node close to the head of a bus (i.e., the point where the slots are generated) may capture most of the bandwidth. Because of this, the protocol allows a node to write into an empty slot only if it is the node's turn. Otherwise, the empty slot is allowed to go past the node. For a node holding a segment in the distributed queue to allow an empty slot to go by, it first must be assured that the empty slot will get used by some other node downstream. Also the downstream node's segment must have entered the distributed queue before the segment that the node is holding. Determination of these conditions is made by the node reading the request bits in the slots passing on bus B. (Bus B request bits are used for transmission decisions on bus A.)

Whenever a node has a segment to transmit on bus A, it sets the next available request bit (of the appropriate priority) on bus B. This informs the nodes upstream of the presence of a segment.

8.5.5.1 Determination of a Node's Turn. To determine its turn in the distributed queue, each node maintains a request counter (RQ) and a countdown counter (CD) for each bus. The bus A RQ counter (recall we only consider bus A transmissions) is incremented for every set (= 1) request bit that the node reads on bus B. These bits indicate arrivals of segments in downstream (on bus A) nodes. If the node does not hold a segment for transmission on bus A, the RQ counter is decremented for every empty slot that goes by. This implies that when a node is not holding a segment, the RQ counter counts the number of segments awaiting transmission (i.e., in the distributed queue) in downstream nodes.

If a segment arrives at a node, the RQ counter is copied into the CD counter and the RQ counter is zeroed. Now the RQ counter is still incremented for every request bit read on bus B. However, only the CD counter is decremented for every empty slot that goes by. A node transmits in an empty slot if its CD counter is zero before the arrival of the slot. When a node is holding a segment, the CD counter indicates the number of segments queued downstream (on bus A) that arrived before the node's own segment. The RQ counter indicates the number of segments queued downstream that arrived after the node's own segment. Thus, CD + RQ counts the total number of segments queued downstream.

A node receiving a segment will attempt to inform other upstream nodes of this new arrival. This is done by setting a request bit on bus B. Observe that, in order to set a request bit, a node has to wait for a slot on bus B where the request bit is zero. Thus, it could actually happen that a segment is transmitted before the corresponding request bit is set. Each node, therefore needs to maintain a queue of request bits.

8.5.5.2 Example. The following is an example that illustrates the operation of the distributed queue. To show the important features of the protocol, we assume that the propagation delay on both busses is zero. This implies that slots propagate in zero time and that requests can be heard at the moment a request bit is set. Later we discuss the effect of nonzero propagation delays on the operation of the protocol.

The example is shown in Table 8.1 for a network of seven nodes (labeled 0 through 6) (see Fig. 8.21) and for transmissions by these nodes on bus A only. (Request bits are of course sent on bus B.) The leftmost column describes those events that can either be a slot generation event or a segment arrival event. The other columns labeled RQx and CDx contain the values of the bus A RQ and CD counters for node x where $x = 1$ to 6. Note that node 0 never transmits on bus A and thus does not need to maintain counters for bus A. A CD counter value of "–" means that the node does not contain a segment for transmission. If a slot is used by a node to transmit, the word transmit shows in the node's column in the row corresponding to the slot generation event.

In the example, initially all nodes are empty and thus all the RQ counters are zero and the CD counters have a null value. The first event is an arrival of a segment to node 1. This causes node 1 to send a request bit (on bus B) that increments all the RQ counters in nodes 2–6. Node 1 also copies its RQ

TABLE 8.1 Example of the Use of the Counters to Schedule Transmissions

Event	RQ6	CD6	RQ5	CD5	RQ4	CD4	RQ3	CD3	RQ2	CD2	RQ1	CD1
Initial	0	—	0	—	0	—	0	—	0	—	0	—
Segment to 1	1	—	1	—	1	—	1	—	1	—	0	0
Segment to 3	2	—	2	—	2	—	0	1	1	—	0	0
Segment to 5	3	—	0	2	2	—	0	1	1	—	0	0
Slot 1	2	—	0	1	1	—	0	0	0	—	0	—
											Transmit	
Segment to 6	0	2	0	1	1	—	0	0	0	—	0	—
Slot 2	0	1	0	0	0	—	0	—	0	—	0	—
							Transmit					
Segment to 2	1	1	1	0	1	—	1	—	0	0	0	—
Segment to 4	2	1	2	0	0	1	1	—	0	0	0	—
Slot 3	2	0	2	—	0	1	1	—	0	0	0	—
			Transmit									
Segment to 1	3	0	3	—	1	1	2	—	1	0	0	0
Segment to 5	4	0	0	3	1	1	2	—	1	0	0	0
Slot 4	4	—	0	3	1	1	2	—	1	0	0	0
	Transmit											
Slot 5	3	—	0	2	1	0	1	—	1	—	0	0
									Transmit			
Segment to 6	0	3	0	2	1	0	1	—	1	—	0	0
Slot 6	0	2	0	1	1	—	1	—	1	—	0	0
					Transmit							
Slot 7	0	1	0	0	0	—	0	—	0	—	0	—
											Transmit	
Slot 8	0	0	0	—	0	—	0	—	0	—	0	—
			Transmit									
Slot 9	0	—	0	—	0	—	0	—	0	—	0	—
	Transmit											

counter into its CD counter and sets its RQ counter to zero. The next event is a segment arrival at node 3, which causes a request to propagate to nodes 4, 5, and 6, incrementing their RQ counters. Node 3 also copies its RQ counter into its CD counter and sets its RQ counter to 0. The arrival of a segment at node 5 has a similar effect with the resulting state of the counter shown in the fourth row of Table 8.1.

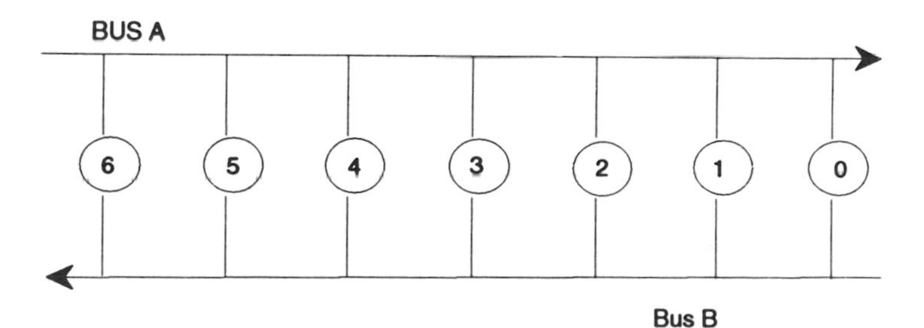

Figure 8.21. Example of a seven node network.

After the three segment arrivals, a slot is generated at the head of bus A and sweeps across the bus. The slot is initially empty. As nodes encounter this empty slot, they decrement their RQ counters (if the node has no segment to transmit) or their CD counters (if the node is holding a segment). The first node that the empty slot encounters with CD = 0 is node 1. Thus node 1, the first node to receive a segment in the example, is the first node that gets to transmit. The example continues with slot events or segment arrival events having similar effects. Observe that in this idealized environment transmission of segments is on a first-come first-served basis.

8.5.6 Unfairness in the Distributed Queue Dual Bus Protocol

8.5.6.1 What's Fair is Fair. Early in the development of the DQDB standard researchers observed that the basic protocol described above could treat nodes unfairly. A fair DQDB network operates in a manner similar to the example presented earlier. In that example, segments arriving into the distributed queue are transmitted on a first-come first-served basis. Under these conditions, if nodes receive multisegment IM_PDUs, the protocol at each node transmits one segment in a round-robin fashion, and bandwidth is shared fairly among all nodes.

A DQDB network using the basic protocol operates fairly as long as the propagation delay along a bus is less than the time required to transmit a 53-octet slot: The propagation delay on a bus is a function of the bus length and the signal propagation speed (a physical constant) and is not affected by data rates. The slot transmission time, on the other hand, is a function of data rates. Table 8.2 shows the ratio of the propagation delay to a 53-octet slot's transmission time for a bus that is 50 km in length at different data rates. Signals are assumed to propagate at 0.6 of the speed of light (180,000 km/second). The right column of Table 8.2 represents the number of slots that can be in transit on the bus at any one time.

The DQDB protocol is intended to operate at speeds exceeding 100 Mbps and to cover a metropolitan area that may extend well over 50 km. Table 8.2 indicates that under these conditions, approximately 100 slots may be in transit on the bus. Therefore, request bits and empty slots do not propagate instantaneously as described in the example of Section 8.55. The effect of this on the behavior of the protocol can best be seen by considering an overloaded network, as describe next.

TABLE 8.2 Number of Slots on a 50 km Bus Segment for Varying Data Rates

Data Rate (Mbps)	Prop. Delay / Slot Time
1	0.65
10	6.55
100	65.51
1000	655.14

8.5.6.2 The Early Bird Gets the Worm. To demonstrate the unfair operation of the basic DQDB protocol, we consider a network with overloaded nodes. An overloaded node may be idle for a while but then all at once receives a large number of segments to transmit. Such a node always has a segment to transmit, for some long period of time, that is, it is always able to bring a new segment into the distributed queue after it transmits a segment. This turns out to be a good model of a node attempting a file transfer over a DQDB network.

Ideally, if several nodes become overloaded with transmissions on the same bus, the protocol ought to give each equal share of the bandwidth. Unfortunately, the percentage of throughput each node is allocated under these conditions is a strong function of the number of slots in transit between the nodes. It also matters which node starts the overload condition first.

Consider an example DQDB network with two active nodes that are 25 slots apart as shown in Figure 8.22. Focusing on data transmission on bus A let us assume that node 1 has been saturated and transmitting for some time (> 25 slots) while node 2 is idle. Node 1 thus fills all the slots while node 2 observes only busy slots passing by its bus A transmitter.

If node 2 now becomes saturated, it queues a segment in the distributed queue and sets a request bit on bus B. This bit requires 25 slots to reach node 1, at which time node 1 lets an empty slot go by. This empty slot takes another 25 slots before reaching node 2, which can now transmit its first segment. Approximately 50 slots pass from the time node 2 becomes saturated until it gets a chance to transmit its first segment. Only at this point can node 2 queue another segment and send another request bit. It again has to wait 50 slots until it receives another empty slot in which to transmit. Whereas we would like nodes 1 and 2 to share bus A bandwidth equally, what actually occurs is that node 2's throughput is approximately 0.02 of node 1's throughput.

A similar situation occurs if node 2 was initially saturated and node 1 becomes saturated later. In an overloaded situation we generally find the following:

1. In the overload condition, all slots are filled by some node.
2. The bandwidth is shared equally only if both nodes experience overload at the same time.
3. Minor differences in the start time of the overload condition (on the order of microseconds) can cause the protocol to allocate bandwidth unfairly.
4. The node that starts to transmit first will capture most of the bandwidth.

8.5.6.3 Don't Try So Hard. The major reason why the DQDB protocol may treat nodes unfairly is that it tries very hard to fill every slot if possible.

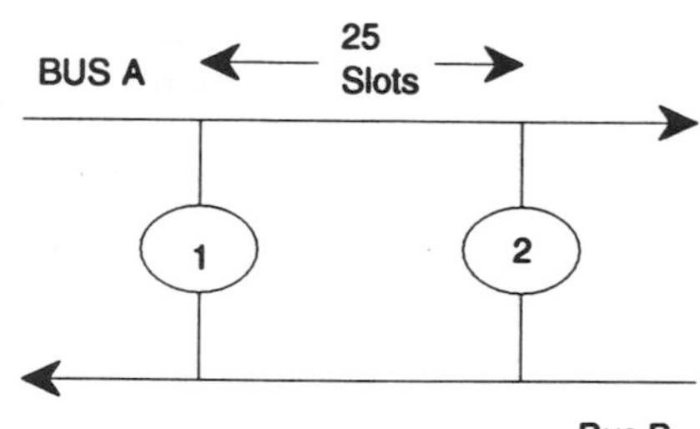

Figure 8.22. Network with two saturated nodes.

The solution to the unfairness problem, thus relies on getting a node to give up some slots that the node thinks it should use. The approach used is known as *bandwidth balancing* and relies on the following concept: Each node uses only a fraction, α, of the available bandwidth, that is, the bandwidth not used by other nodes.

In the DQDB protocol, a node measures available bandwidth by examining empty slots and request bits. In the basic protocol, a node can use all empty slots that have not been requested by downstream nodes. With bandwidth balancing, a node would use only a fraction (say 90 percent) of those slots and let others go by.

To see how this approach helps, consider two overloaded nodes. Node 1, the upstream node, starts first and while it is the only active node it has access to the entire network bandwidth. Because of bandwidth balancing, node 1 uses only α (80 percent in the example) of the available bandwidth. The sequence of events that follows is shown in Figure 8.23.

When node 2 also becomes overloaded, it detects the available bandwidth to be 20 percent (i.e., 20 percent of the slots it receives will be empty). Node 2 thus fills up only 80 percent of those slots, or 16 percent of the total

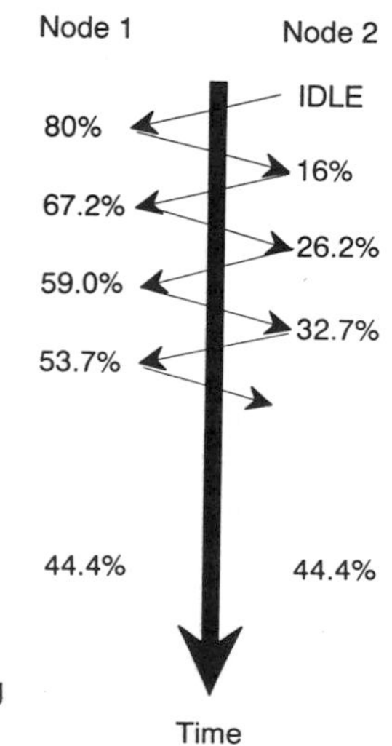

Figure 8.23. Conceptual operation of the bandwidth balancing mechanism.

bandwidth. Node 1 next detects (through the request bits it receives from node 2) that the available bandwidth has just decreased from 100 percent to 84 percent of the slots. It uses only 80 percent of those, that is, it now only fills up 67.2 percent of the slots on the bus. Node 2 detects the drop in node 1's throughput because it receives more empty slots. This sequence of events continues until both nodes reach a steady throughput of 44.4 percent of the slots.*

In the bandwidth balancing example, we make the following observations:

1. Both nodes eventually achieve the same throughput regardless of which node starts first and regardless of the distance between them.
2. Once both nodes reach their steady-state throughput, a total of 88.8 percent of the slots are used with 11.2 percent being wasted. The maximum percentage of wasted slots is 20 percent ($= 1 - \alpha$). This is the case when only node 1 is active.
3. The larger α is, the fewer slots are wasted. However, it requires a longer time to reach steady state for larger α. (The reader should confirm this by repeating the example of Fig. 8.23 using $\alpha = 90$ percent.)

Although we have demonstrated the features of bandwidth balancing using a two-node example, the above observations still hold for a network with any number of overloaded nodes.

8.5.6.4 What Does a DQDB Network Do. Incorporation of bandwidth balancing into the basic DQDB protocol is achieved by defining an additional counter, the bandwidth balancing (BWB) counter. Each node also maintains a parameter known as the bandwidth balancing modulus, BWB-MOD.

The BWB counter is initially set to zero and incremented every time the node transmits a segment provided that the BWB counter < BWB-MOD − 1. If the BWB counter = BWB-MOD − 1, then once a segment is transmitted the counter is reset to 0.

Every time a node's BWB counter is reset to 0, the node increments its RQ counter if it has no segments awaiting transmission. Otherwise, the node increments its CD counter. This means that the node gives up an empty slot it can use every time its BWB counter is reset to 0. The value of BWB-MOD is, thus determined from the desired value of α. (For $\alpha = 90$ percent, BWB-MOD should be set to 9.)

*This steady-state throughput can be found by solving $\gamma = \alpha(1 - \gamma)$, where γ is the steady-state throughput and $\alpha = 0.8$.

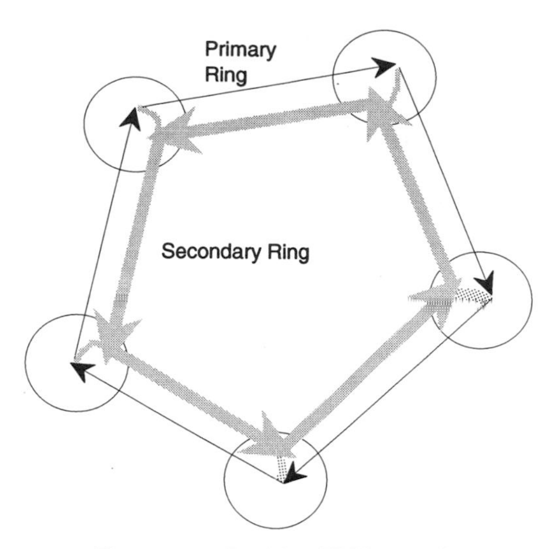

Figure 8.24. Dual ring FDDI network.

8.6 FIBER DISTRIBUTED DATA INTERFACE

The fiber distributed data interface (FDDI) standard, issued by ANSI, defines a high-speed optical fiber token ring network. At the level of our discussion in this chapter, the FDDI standard is essentially similar to the IEEE 802.5 standard. The significant differences are as follows:

The FDDI is designed to operate at 100 Mbps using optical fiber* compared to the IEEE 802.5 token ring's 4 or 16 Mbps over a twisted pair. Because of this, the FDDI standard specifies *fast token release*. In this scheme a new token is issued by a station as soon as it has completed transmission. This is motivated by the fact that at the higher data rates, the ring bit length can be large. Therefore a station waiting for its own transmission to come around the ring before issuing a new token (as is done in the IEEE 802.5 standard) would mean that a significant amount of the ring bandwidth is lost (see Problem 8.10).

The FDDI incorporates a reliability specification into the standard. This enforces a uniform approach to dealing with the potential reliability problems of the ring topology. The FDDI allows for a *dual ring* operation with a primary and a secondary ring as shown in Figure 8.24. Stations with a capability to attach to the two rings are known as *dual attachment stations*. Such a station will perform MAC layer functions on data on only one of the rings (the primary ring). The station will only repeat data received on the other (secondary) ring. Each of the dual attachment stations has the capability to connect the outgoing side of the primary ring to the incoming side of

*The FDDI has also been adapted to operate on twisted pair.

the secondary ring. This capability, coupled with a fault or signal loss-detection capability allows stations to reconfigure around some link and/or station failures. The standard also defines *single attachment stations* that attach to a single ring and have no reconfiguration capabilities. Such stations can be significantly less expensive and are therefore likely to be used in situations where reliability is not the major concern. The FDDI standard also defines single- and dual-attachment concentrators that can then be used to construct FDDI topologies that include both single and dual attachment stations.

SUGGESTED READINGS

The actual standards documents [IEEE89a], [IEEE89b], [IEEE90a], [IEEE90b], and [IEEE90c] provide the only definitive description of the protocols discussed in this chapter. These documents can be hard to understand and often do not provide much motivation or analysis for various design decisions.

A comprehensive textbook discussion of LAN technology can be found in [STAL93]. A book dedicated totally to the token ring technology is [SACK93]. A good discussion of FDDI and concepts for future FDDI-like networking can be found in [ROSS86], [ROSS89], and [ROSS92]. The work in [SHOC82] is a classic discussion of the Ethernet product, which directly led to the development of the IEEE802.3 CSMA/CD standard. Early discussion of the ring architecture and desirable architectural approaches for ring networks can be found in [SALT80], and [SALT83].

PROBLEMS

8.1 A ring network has the following characteristics:

Length of cable connecting stations: 5 km

35 Stations, each incurring 1-bit delay

A data rate of 1 Mbps

Assuming that the signal propagates at 5 μseconds/km, determine the bit length of the ring.

8.2 Explain what might happen in the token bus protocol if Solicit_Successor_1 (used when TS > NS) is used all the time in the insertion procedure (i.e., even when TS < NS).

8.3 Modify the token bus protocol so that if the who_follows procedure fails, the token holder can attempt to find the address of the station that is two hops away.

8.4 In the token bus protocol, a station will be bypassed just because the token passed to it gets lost twice in a row. Modify the protocol so that the station being bypassed is allowed to object if it is alive.

8.5 A token bus logical ring is initialized by mistake as shown in Figure 8.25. While station 12 is holding the token, stations 11, 10, 9, 3, 2, 1, and 0 become operational and want to be part of the ring. Assuming the standard token bus procedures, and assuming that all stations will solicit insertion before passing the token, what will the ring look like after all stations have been inserted.

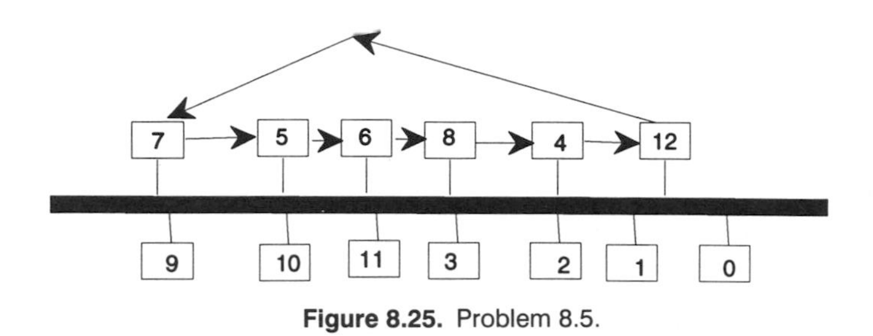

Figure 8.25. Problem 8.5.

8.6 A token bus logical ring is shown in Figure 8.26. Node 10 currently has the token. Assume that the network is cut in the indicated location and assume that after the bus is cut both pieces continue to function electrically. Describe the sequence of events after the cut takes place and show the final configuration of the network.

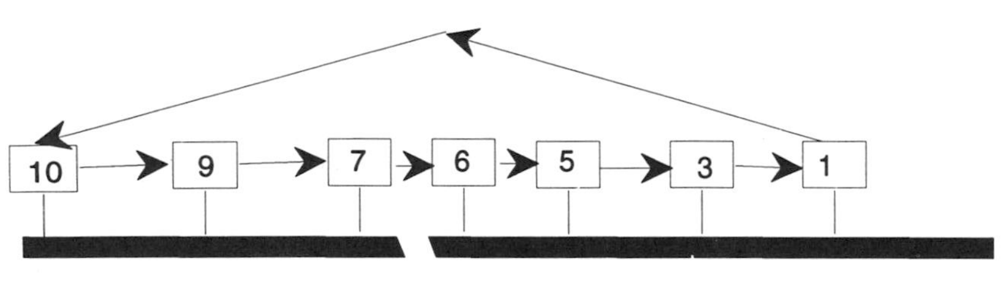

Figure 8.26. Problem 8.6.

8.7 A DQDB network has 10 nodes as shown in Figure 8.27. Node 4 is initially down. Assume that propagation delays are negligible. Also assume that the network is initially idle and bandwidth balancing is not used.

a. What are the values of the bus A RQ and CD counters in all nodes if segments arrive for transmission on bus A to nodes 5, 8, 3, 2, 7, and 6 in that order and before any node gets a chance to transmit.

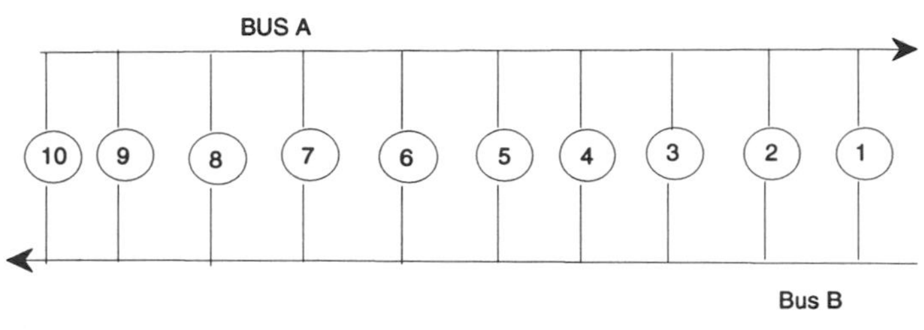

Figure 8.27. Problem 8.7.

b. Node 4 is now repaired (before any transmissions take place) and its RQ counter is initialized to zero. Before any slots appear on the network, node 4 also gets a segment to transmit. Assuming no more segments arrive, determine the order of segment transmissions.

8.8 In a DQDB network there are five active stations, three stations are saturated (i.e., they have a very large number of segments to transmit). The two unsaturated stations have an average arrival rate of 0.1 segments per slot each. The saturated stations use bandwidth balancing with $\alpha = 0.9$, whereas the other two stations do not use bandwidth balancing. Determine the throughput of each saturated station and each nonsaturated station. What is the percentage of wasted slots?

8.9 A DQDB network has six nodes as shown in Figure 8.28. Assuming propagation delays are negligible and each node (except 4) has one segment awaiting transmission. You are told that the following are the values of the bus A RQ and CD counters in each node:

RQ1 = 0, CD1 = 0

RQ2 = 0, CD2 = 1

RQ3 = 1, CD3 = 1

RQ5 = 1, CD5 = 2

RQ6 = 2, CD6 = 2

a. What are the values in RQ4 and CD4.

b. What is the order in which the segments arrived at the stations.

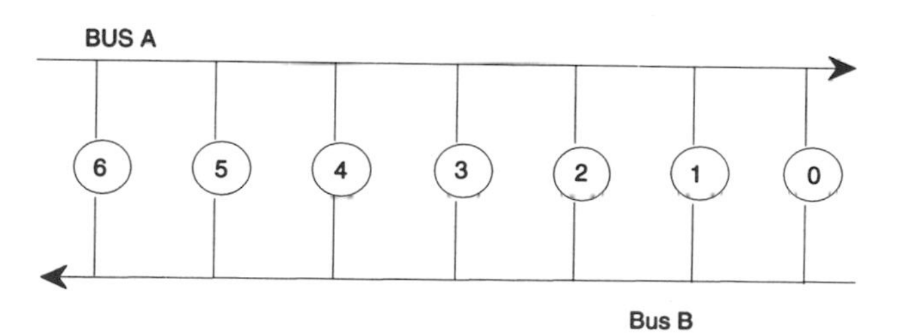

Figure 8.28. Problem 8.9.

8.10 Consider an FDDI ring operating at 100 Mbps with five stations. Assume that each station has one 1000-bit frame to transmit. Calculate the time from when one of the stations receives the token until the time this same station receives the token again assuming the following:

a. Each station releases the token only after the station hears the first bit of its own frame.

b. Each station releases the token as soon as it completes its own frame's transmission.

Assume that the total length of the ring is 20 km and that each station introduces a 1-bit delay. Also assume that signals travel at 0.6 of the speed of light and the token is of negligible length.

8.11 Consider a dual ring FDDI network as shown in Figure 8.24. Show how to reconfigure the network to maintain ring connectivity if

a. One of the links fails

b. One of the stations fails

Can a dual ring handle any combination of link and station failures?

9

NETWORK INTERCONNECTION

9.1 INTRODUCTION

Interconnection of individual networks provides users with an increased level of connectivity, resource sharing, and application-to-application communication potential. This typically requires the installation of additional hardware and software. This chapter presents a discussion of the principles underlying network interconnection technology.

In general, interconnecting networks requires the use of *relays*. Data originating in one network and destined for a receiver on another network must traverse one or more relays. Because interconnected networks may be dissimilar, the relay typically performs any *translation functions* required to make the data originating in one network compatible with the other network.

9.1.1 Relays

We classify relays based on the level of translation that they perform.

Repeaters. These connect two networks at the lowest level, the physical level. A repeater takes bits arriving from one network and repeats them on to another. At most, a repeater might have to translate between two different physical layer formats. This may involve some processing of the transmitted signal such as amplification and noise elimination. Figure 9.1 shows how two local area networks (LANs) may be interconnected with repeaters.

Bridges. These types of relays are generally used to interconnect LANs at the medium access control (MAC) layer. Typical implementations require the interconnected networks to have identical MAC layers. In general, however, bridges may interconnect LANs with different but related MAC layers. Figure 9.2 shows two LANs interconnected with bridges.

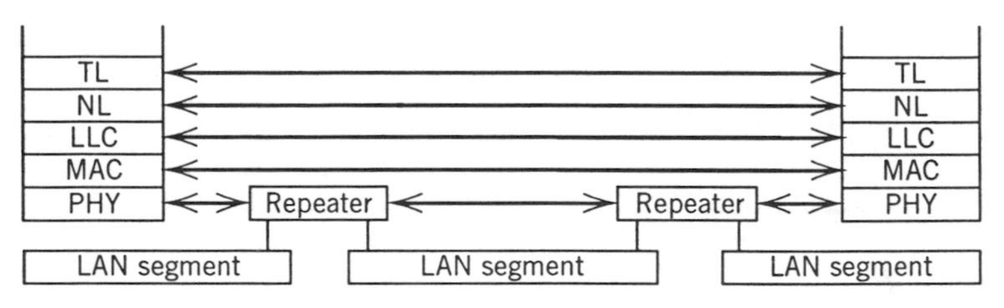

Figure 9.1. Network interconnection with repeaters. TL, NL are the Transport and Network layer protocols, respectively.

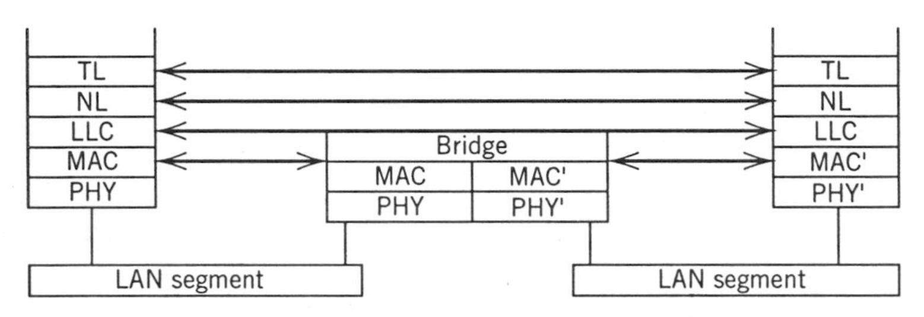

Figure 9.2. Network interconnection with bridges.

Routers. This type of relay connects networks at the network layer and typically performs routing functions much like a switch in a packet switched network. An example of two networks interconnected via routers is shown in Figure 9.3 for the case where the Internet protocol (IP) is used as the network layer and the transmission control protocol (TCP) is used as the transport layer. A network access protocol (NAP) is the set of lower layer protocols required to inject packets into and receive packets from the network. It could be the three layers of X.25 or a MAC layer protocol such as CSMA/CD on top of an appropriate physical layer.

As shown in the figure, a router allows the interconnection of networks that share protocols above the network layer but may have different lower layers (e.g., MAC and/or physical).

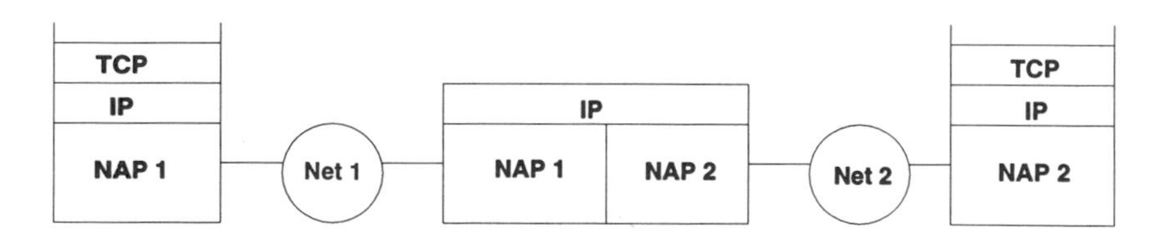

Figure 9.3. Network interconnection with routers.

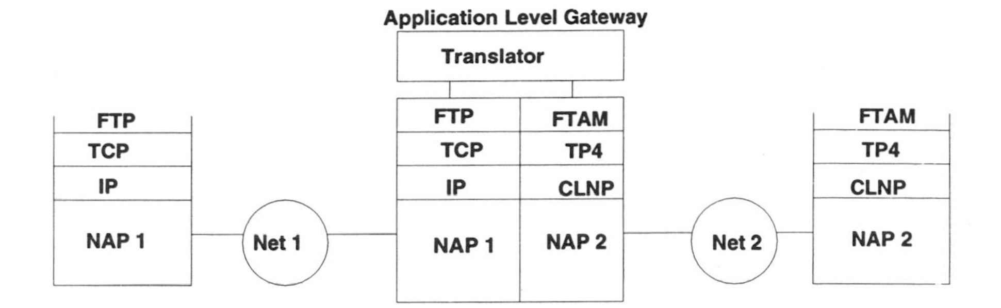

Figure 9.4. FTP / FTAM application layer gateway.

Gateways. These relays interconnect networks at layers higher than the network layer. An example of an application layer gateway (ALG) is shown in Figure 9.4. This particular gateway is used to interconnect the file transfer application used in the Internet suite of protocol; File Transfer Protocol (FTP), to that used in the OSI suite; the File Transfer and Access Method (FTAM). Note that when ALGs are used a different one is required for every application.

In this chapter we discuss bridges and routers in detail, because they are the most common approaches to network interconnection.

9.1.2 Half Relays

One of the problems in network interconnection is assigning administrative responsibility for the relays that interconnect two networks that belong to different organizations. In this situation, it is not clear who should pay for the purchase, maintenance, and communication costs associated with the interconnection. One solution that has been adopted is to split the relaying functions in half and have each organization purchase a *half relay*. Communication between the half relays is carried out over some communication facility such as an X.25 network, integrated services digital network (ISDN) line, or a leased transmission line.

This scenario is shown in Figure 9.5, in which two LANs are interconnected through two half bridges, which are in turn interconnected through a leased transmission facility. Each half bridge (sometimes also known as a *remote bridge*), interfaces on one side with a LAN and on the other side with the transmission line (e.g., a T-1 line). Communication between the two half bridges is carried over a high-level data link control (HDLC) protocol.

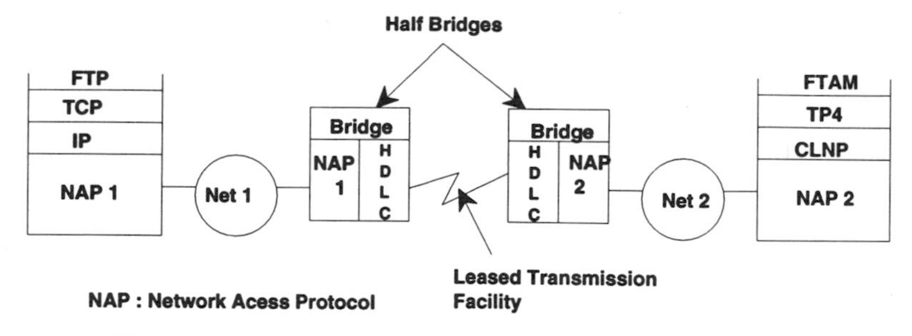

Figure 9.5. Interconnection of two LANs through half bridges.

9.2 INTERCONNECTION THROUGH BRIDGES

There are two basic (and sometimes competing) bridge technologies: transparent bridges and source routing bridges. In this section, we discuss these two approaches.

9.2.1 Transparent Bridges

We say with respect to a source station that a destination is *local* if it is connected to the same LAN and it is *remote* otherwise. The basic principle behind the design of the transparent bridging approach is that a source MAC layer should not be aware of whether a destination is local or remote. In other words, a MAC layer packet should be generated in the same way and should have the same format regardless of the location of the destination. Needless to say, this principle allows for rather simple operation at the source MAC layer. In fact, no change is required to the original (single network) specification of a MAC layer for the use of transparent bridges to be possible. As shown in the remainder of this section this requirement imposes some constraints on how transparent bridges should operate.

9.2.1.1 Getting the Packet to its Destination. The most straightforward approach to achieve the transparency objective is to have the bridges operate in *promiscuous* mode, that is, they examine and process every packet they hear, regardless of address. Each bridge then floods each packet it receives. In other words, each packet received is forwarded on all outgoing ports (except the one the packet was received on). This certainly will get the packet to its desired destination.

However, as with any flooding technique, we need a mechanism to restrict the number of packets. Standard techniques (see Chapter 5) include using a hop count or having bridges remember packets they have seen before. Both of these approaches are not feasible, because they require additional packet

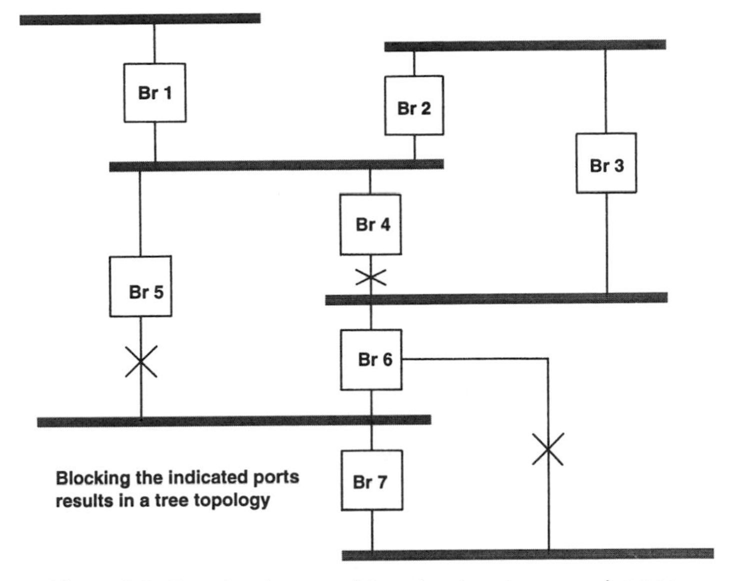

Figure 9.6. Transforming an arbitrary topology to a spanning tree.

header fields (a hop count field or a unique ID field) and thus do not satisfy our transparency objective.

The solution adopted by the IEEE 802.1 committee in its transparent bridging standard restricts the topology of the set of bridge-interconnected LANs to a spanning tree, that is, a tree that spans all LANs. Because the topology has no loops, flooded packets will die by themselves with no need to introduce any additional mechanisms.

Figure 9.6 shows how a spanning tree topology can be obtained by selectively disconnecting bridges from a network. This is done by blocking bridge ports. A blocked bridge port cannot be used for forwarding or receiving data frames.

The tree connectivity solves one problem but introduces another. In a tree network, a single bridge or LAN failure can cause the network to be disconnected. The solution to this dilemma is to allow for arbitrary bridge connectivity but restrict the operation of some bridge ports to that of observers. These bridge ports are not allowed to receive or forward packets. They are, however, in a standby mode and are ready to take an active role in the network if active bridges or ports fail.

9.2.1.2 Tree Configuration Algorithm. The tree configuration algorithm adopted in the IEEE 802 LAN bridging standard is based on the structuring of the tree as shown in Figure 9.7. Each bridge is given a unique identifier (ID). The root bridge is the one with the lowest such ID. Bridges' ports are either a root port, a designated port or a blocked port.

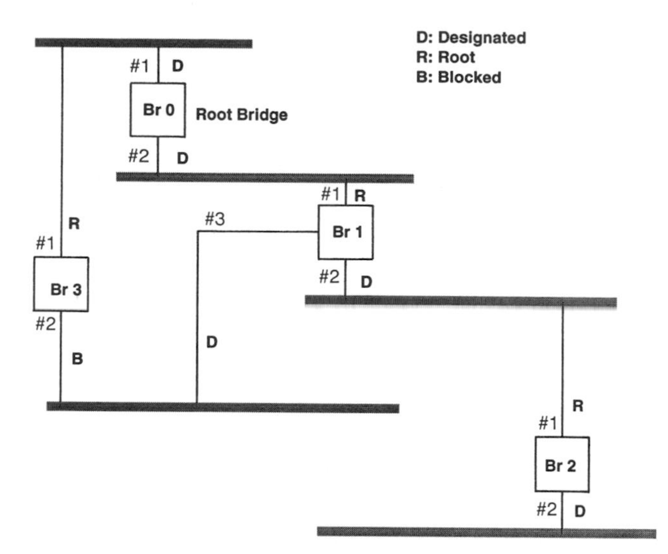

Figure 9.7. Adding structure to a tree topology.

A *root port* is the port in the direction of the shortest path to the root bridge. Each bridge (other than the root bridge) has exactly one such port.

A *designated port* is the port in the direction of the shortest path from a LAN to the root bridge. Each LAN will have exactly one designated port. A bridge may have zero or more designated ports. A particular port cannot be both a root port and a designated port. All ports of the root bridge are designated ports.

A *blocked port* is one that is neither a root port nor a designated port. It is not used to send or receive data frames.

Recall that the main purpose of the tree structure is to provide for the control of the flooded packets. When flooding is indicated, a bridge receiving a normal data frame on a root port or a designated port will forward the packet on all outgoing nonblocked ports, except the port on which the frame was received.

The tree configuration algorithm is a distributed procedure by which all bridges agree on the identity of the root bridge and by which each bridge determines the type of each of its ports. Messages known as a bridge protocol data unit (BPDU) are exchanged among the bridges for that purpose. A bridge will transmit a BPDU according to one of two rules:

If the bridge believes it is the root, it will transmit a BPDU at regular time intervals on all its ports.

If the bridge knows it is not the root, it will forward (after some processing) BPDUs it receives on its root port to all its designated ports. Other BPDUs are discarded.

In both cases above, a BPDU transmitted by a bridge will contain the following information:

The transmitting bridge ID.

The ID of the bridge that the transmitting bridge believes to be the root.

The cost of reaching the root bridge from the transmitting bridge over its root port (e.g., the number of LANs that need to be traversed to get to the root bridge). This cost is 0 if the bridge believes itself to be the root.

Root Bridge / Port Selection. Initially each bridge believes that it is the root. A bridge will drop its claim to being the root the moment it receives a BPDU with a root bridge ID lower than its own. A bridge that is not the root, compares incoming BPDUs on all its ports. Its root port is the one on which **it is receiving BPDUs with the lowest root ID and lowest cost to root, breaking ties in favor of the lowest transmitting bridge ID.**

For example, a bridge (that knows it is not the root) might receive the following three BPDUs:

Port	Root ID	Transmitting Bridge	Cost to Root
1	1	7	20
2	0	6	25
3	7	15	10

The bridge will declare its port 2 to be the root port and bridge 0 to be the root bridge.

As more messages are received, the root port, root bridge ID, and/or cost to root may change. For example if the same bridge above receives a BPDU from bridge 5 on port 1 with cost to root = 5 and root ID = 0, it will now make port 1 as its root port.

A bridge that does not believe it is the root will forward a BPDU that it receives on its root port to all its designated ports. Before forwarding, however, the bridge will change the transmitting bridge ID to its own and increment the cost to reflect the additional cost of the network connected to its designated port.

Designated / Blocked Port Selection. For a bridge that believes it is the root, all ports are designated ports. A bridge that does not believe it is the root will classify a port as designated if it is receiving BPDUs on the port with a root ID that is larger than the ID of its currently believed root or it is receiving BPDUs on the port with a root ID equal to its own believed root ID but with a cost higher than the bridge's root cost plus the cost of the LAN connected to the port. If the cost (as calculated above) and root ID are the same, the port is designated if the BPDU has a higher transmitting bridge ID.

Consider for example the scenario in Figure 9.8. Port 1 on bridge 5 is already identified as the root port to root bridge 0 with a cost of 1 (1 LAN in between bridges 5 and 0). The root ID, cost, and transmitting ID of all

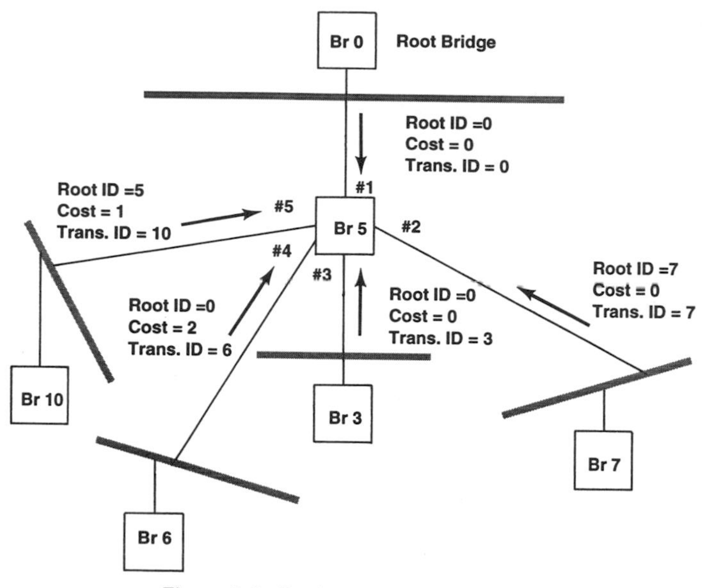

Figure 9.8. Designated port selection.

incoming BPDUs are shown. Ports 2, 4, and 5 are identified as designated, whereas port 3 is blocked.

Recovering from Bridge / Port Failure. Once a tree is initialized, the root bridge continues to periodically issue BPDUs. Other bridges continue to receive BPDUs on their root ports, modify them, and forward them to their designated port. Bridges that do not hear BPDUs on their ports for specified timeout periods will detect failure and will initiate recovery procedures.

For example, if a bridge stops hearing BPDUs on one of its blocked ports, this implies that some designated port must have failed. In this case, the bridge will attempt to claim that its blocked port is now designated. It will succeed if no other port on the same LAN with a lower cost to the root exists. Otherwise, it will return to blocked status.

Example. We illustrate the tree configuration procedure above with a simple example. We consider the network of Figure 9.7 and show how the tree shown in the figure is initialized. To simplify the example we assume synchronous operation of the protocol. That is, all bridges are turned on at precisely the same time, and transmit BPDUs at precisely the same time. These BPDUs, we assume, are received at bridges that are one network away at exactly the same time.

Time 0. When the bridges are turned on, they are all initialized to believe that they are the root, their cost to the root is thus initialized to 0 and all their ports are initialized as designated. Each bridge then transmits a BPDU of the form: transmitting bridge ID, believed root ID, cost to root. This is shown in the following table:

Port:	B0, #1	B0, #2	B1, #1	B1, #2	B1, #3	B2, #1	B2, #2	B3, #1	B3, #2
State	D	D	D	D	D	D	D	D	D
Send	(0, 0, 0)	(0, 0, 0)	(1, 1, 0)	(1, 1, 0)	(1, 1, 0)	(2, 2, 0)	(2, 2, 0)	(3, 3, 0)	(3, 3, 0)

Time 1. As the messages above are received, bridges update their information regarding the root port ID and the cost to root. Bridge 0 continues to believe it is the root, because it is not receiving any message with a lower transmitter ID. Bridges 1 and 3 learn about bridge 0 through the received messages and change their root ID and cost to root accordingly. They also label the ports on which they receive the BPDUs with the lowest transmitting ID as the root port. Bridge 2 behaves similarly except that it hears Bridge 1's messages and therefore thinks bridge 1 is the root. Bridge 0, believing it is the root, continues to issue BPDUs on all its ports. Other bridges forward BPDUs they have received on their respective root ports to designated ports after labeling them with their own transmitting ID and incrementing the cost to root. This is shown in the following table.

Port	B0, #1	B0, #2	B1, #1	B1, #2	B1, #3	B2, #1	B2, #2	B3, #1	B3, #2
Receive	(3, 3, 0)	(1, 1, 0)	(0, 0, 0)	(2, 2, 0)	(3, 3, 0)	(1, 1, 0)	—	(0, 0, 0)	(1, 1, 0)
State	D	D	R	D	D	R	D	R	D
Send	(0, 0, 0)	(0, 0, 0)	—	(1, 0, 1)	(1, 0, 1)	—	(2, 1, 1)	—	(3, 0, 1)

Observe that the bridges do not transmit on their root ports, but they continue to receive and process BPDUs on those ports.

Time 2. As the next round of messages are received the following occurs: Bridge 2 hears a message from bridge 1 with a believed root ID of 0 and cost to root of 1. This changes bridge 2's information regarding who the root is (previously it thought bridge 1 was the root, now it thinks bridge 0 is the root). Bridge 2 also updates its cost to the root to 2 (1 more than the cost of bridge 1). The other significant event is that bridge 3 hears a message from bridge 1 on its designated port. Because the message has the same root bridge ID and the same cost to root as bridge 3's, bridge 3 will block the port as its ID is higher than the transmitter of the message it is hearing (bridge 1). All bridges then transmit another round of messages on their designated ports. All bridges also continue to receive and process messages on their blocked and root ports. The tree is now configured as in Figure 9.7. As long as all bridges and ports are operational, the bridges continue to send and receive BPDUs as shown in this table.

Port:	B0, #1	B0, #2	B1, #1	B1, #2	B1, #3	B2, #1	B2, #2	B3, #1	B3, #2
Receive	—	—	(0, 0, 0)	—	(3, 0, 1)	(1, 0, 1)	—	(0, 0, 0)	(1, 0, 1)
State	D	D	R	D	D	R	D	R	B
Send	(0, 0, 0)	(0, 0, 0)	—	(1, 0, 1)	(1, 0, 1)	—	(2, 0, 2)	—	—

Now let us assume that port #3 of bridge 1 fails and that the network is operational otherwise. Bridge 3 stops hearing messages on its blocked port (port #2) and converts the port to a designated port and starts forwarding BPDUs on it. Not hearing any other BPDUs, the port will remain designated and the tree will be reconfigured around the failure.

9.2.1.3 Learning and Forwarding Databases. The transparent bridging technique as described thus far will work. It is, however, extremely inefficient, because each packet is heard at all LANs regardless of its destination. The obvious solution is to equip each bridge with a routing table (also known as a forwarding database, FDB). Entries in this table are of the form (station address, port number) and indicate which of the bridge's ports should be used to reach a particular destination. An example of this is shown in Fig. 9.9.

Equipped with this information a bridge will forward a packet to the port indicated in the table for the packet's destination address. The packet is discarded if it arrives at the port associated with its destination address in the table. For example, in Figure 9.9 a packet sent from station 3 to station 5 will be forwarded by the bridge. A packet from station 3 to station 1 will be discarded by the bridge.

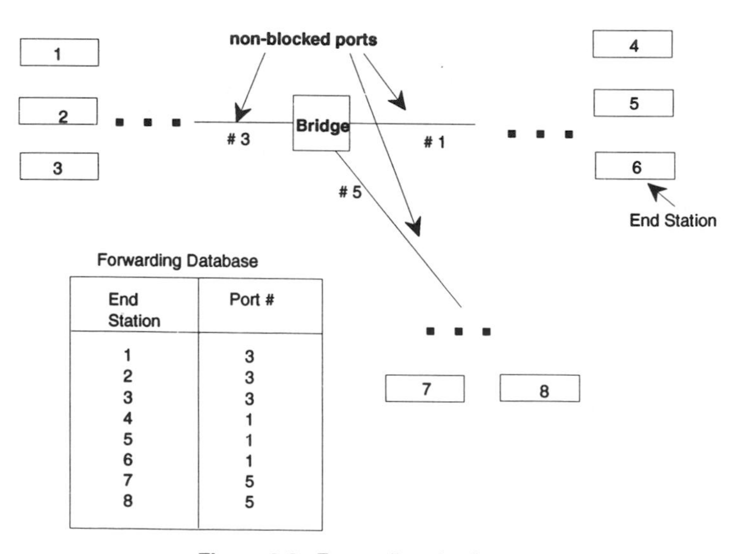

Figure 9.9. Forwarding databases.

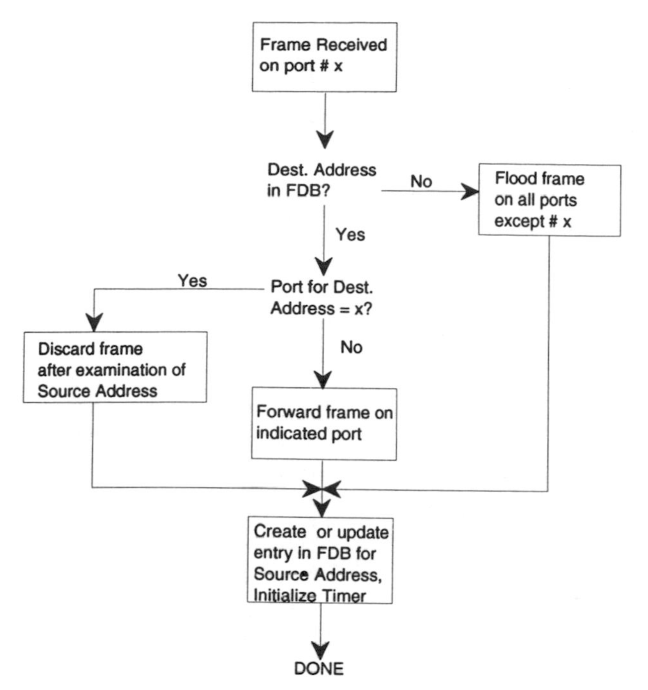

Figure 9.10. Forwarding and learning algorithm.

As with any routing table, the information in the FDB needs to be kept up to date. Changes to the FDB may occur because stations have moved or because the tree connectivity has been changed. Another problem associated with the FDBs is their size. Maintaining information for every station connected to the network is not feasible in large networks. The solution to these problems is to introduce a capability for the bridges to learn (and forget) the location of stations.

This is done as follows: When a bridge receives a packet, it examines the source address. It then inserts or updates an entry for the source address in its FDB indicating the port on which this packet is received. To control the size of the FDB, each entry is associated with a timer that is reset whenever the entry is updated or used. If the timer expires, the associated entry is deleted. When this learning process is used, a bridge may not have routing information for some destinations, and packets arriving for those destinations are flooded. The complete forwarding and learning algorithm employed by the bridges is shown in Figure 9.10.

9.2.2 Source Routing Bridges

One of the major shortcomings of the transparent bridging approach is the requirement to maintain a spanning tree. Whereas the physical topology can be arbitrary, bridges not currently in the tree cannot be used to carry network traffic, which might cause bridges in the tree to experience heavy loads.

The source routing approach was introduced (in part) to address this deficiency. The complexity and other problems associated with the source routing scheme have led to considerable controversy about the relative merits of each. The basic idea behind the source routing scheme is that the MAC layer at the source of the transmission is responsible for specifying (in the MAC frame header) the path that the packet should take to reach its destination. In a network using source routing, LANs and bridges have unique identifiers, and a path is specified in the form of a list:

$$\text{Path} = \{\text{LAN ID1}, \text{Bridge ID1}, \text{LAN ID2}, \text{Bridge ID2}, \ldots, \text{LAN ID}M\}$$

where LAN ID1 and LAN IDM identify the LANs where the source and destination reside, respectively.

Note that including the path in the MAC header has two implications: (1) Headers of existing MAC protocols have to be modified to accommodate source routing and (2) transparency is no longer possible. In fact, the source MAC layer now needs to know the path to the destination.

9.2.2.1 Typical Packet Format. A typical header* for a source routing MAC layer is shown in Figure 9.11. The header contains the following fields (in addition to other standard fields such as source and destination addresses):

L / R Bit†. This is set if the destination is remote, that is, requires bridge forwarding. When the bit is 0, indicating a local destination, the three source routing fields described next need not be present. Since a large portion of traffic is local, this will help reduce the overhead in the MAC packet headers.

Type of Forwarding. This field specifies which type of forwarding is desired for the packet. Three types of packet forwarding are possible:

Single destination forwarding,
All-route broadcast, and
Single-route broadcast.

We will discuss the different forwarding schemes shortly.

*This discussion does not actually conform to any particular standard and is used for clarity of exposition.

†In the IEEE 802 source routing standard, the most significant bit of the source address field is used as the L/R bit. This makes it possible to keep packet formats for local packets unchanged.

Source Address	Destination Address	L/R bit	Type of Forwarding	Path Length	Path Specification	Other Header Info	DATA

Figure 9.11. Source routing MAC header.

Path Length. This field contains the length of the path contained in the header of this packet. Because the length of a path to a destination is variable, this field is necessary for the correct parsing of the header.

Path. This field contains a path description containing a sequence of LAN and bridge IDs as described previously.

9.2.2.2 Packet Forwarding. Assume, for the time being, that a source station knows a path to a remote destination station. The source then needs to indicate on packets transmitted to this destination that it is remote by setting the L/R bit to 1. The type of forwarding on the packet is set to single-destination forwarding. The path and path length should also be included in the appropriate fields.

Bridges (that are still promiscuous) that hear this packet first examine the L/R bit to determine if they may need to forward it. Based on the type of forwarding specified, a bridge then examines the path field searching for its own bridge ID. If a bridge does not find its own ID in the path, the packet is discarded. Otherwise, if it finds its bridge ID in between LANi and LANj, it forwards the packet to LANi if it was received on LANj or vice versa. If the packet arrives from other than LANi or LANj, it is discarded.

9.2.2.3 Path Determination. To determine what path to include in a source routed frame, we could insist that each station maintain a (static) table of paths to all possible destinations. In a moderate size and reconfigurable network, this is not feasible. Another alternative would be to maintain a path server, a central location where all paths are maintained. Whenever a source needs a path it can query the server and obtain a path. The server can maintain and update the global network topology to determine paths. This suffers from the standard centralized routing problems: it is vulnerable to server failures and requires a significant amount of resources (bandwidth and processing power) to make the server access efficient.

A third approach, which is favored by proponents of the source routing approach, is that of path discovery (or path learning). In this approach, mechanisms are defined for a station to obtain in a distributed fashion one or several paths to a desired destination. Once discovered, this path can be maintained in a local cache, discarded when no longer required, and rediscovered when needed again. Many approaches to path discovery are possible. We discuss one such approach below.

The path discovery approach is based on the fact that if a packet from the source is flooded throughout the network, one copy will reach the destination for each different path that exists between the source and destination. In the source routing terminology, this is known as an *all-routes broadcast*.

Path discovery can then proceed as follows. The source sends a packet with all-routes broadcast indicated in the type of forwarding field. Initially the path field of this packet is set to include the source LAN ID. As bridges receive this packet, they forward it after adding their own bridge ID and outgoing LAN ID to the path field. If a bridge finds its own ID or next LAN ID already in the path field, the packet is discarded, because it went through a loop.

Eventually, these packets are received by the destination station (whose address is in the destination field). Each received packet will contain a different specification of a path from the source to the destination. The destination should then choose from among all the paths it is learning about. The chosen path is then sent back to the source using source routing. Alternatively, the destination can send all the paths back to the source for it to make the selection among the different paths.

An alternative path discovery approach uses the "single-route broadcast" type of forwarding. For this, a spanning tree is assumed to be defined in the network (in the same way as it is defined in the transparent bridging scheme).* A single-route broadcast packet traverses the tree and is, therefore, guaranteed to reach all LANs (and hence all destinations). This type of packet contains a zero length path field, and no path is accumulated in it as it traverses the network.

With the use of this type of forwarding, path discovery proceeds as follows: a single-route broadcast packet is sent from the source. This packet reaches the destination and informs it that the source is trying to discover a path to the destination. The destination responds by sending an all-routes broadcast packet, which traverses the network as described above and allows the source to learn various routes to the destination.

An example of the path discovery procedure is shown in Figure 9.12. In the example, a source on LAN 1 sends a single-route broadcast to a destination on LAN 3. The destination responds with an all-routes broadcast. This message accumulates path information as it traverses the network.

9.3 INTERCONNECTION THROUGH ROUTERS

As discussed previously, routers interconnect networks at the network layer. The Internet protocol (IP) is perhaps the best known protocol in this

*In the source routing scheme, this tree is used only to forward single-route broadcasts and is not maintained for use in normal packet delivery. Nevertheless, the fact that the tree needs to be maintained eliminates one possible advantage for the source routing scheme over the transparent bridging scheme.

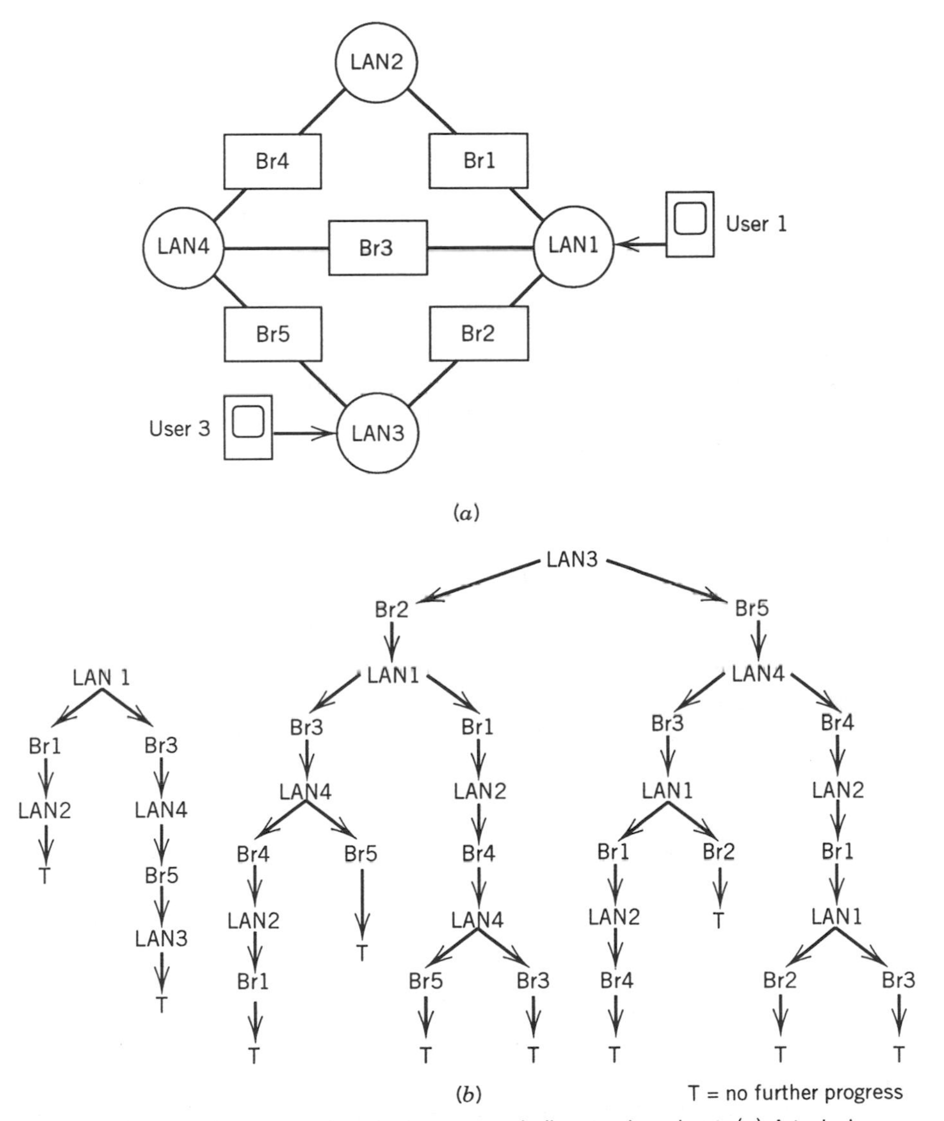

Figure 9.12. Path discovery using single-route and all-routes broadcast. (*a*) A typical example of LAN interconnection. Bridges 2 and 4 are not on the spanning tree. (*b*) Single-route broadcast (from User 1) and all-route broadcast (from User 3) frames propagation corresponding to *a*. T: packet does not make further progress.

category. It is discussed next. We also describe briefly the connectionless network protocol (CLNP), which is an OSI standard protocol.

9.3.1 The Internet Protocol (IP)

9.3.1.1 Service Offered. The Internet protocol is a connectionless network layer protocol used primarily in the context of the Internet which interconnects a very large number of networks worldwide.* It can be argued that the success and widespread use of the Internet is due in part to the design philosophy of IP.

A higher layer protocol that uses IP (or *user* for short), such as the transmission control protocol (TCP) described in Chapter 10, interacts with IP through two operations (or primitives): SEND and DELIVER. The SEND operation is used by the higher layer to pass the information to be transmitted and to instruct IP, through a set of parameters, on the particulars of the desired information delivery. The DELIVER operation is used to pass from IP to a higher layer the information transmitted from a remote host. Again some parameters are passed along with the information to describe delivery particulars.

IP provides a connectionless (or datagram) service to its user. In other words, data given to IP (via the SEND operation) is not guaranteed to be delivered. In IP units of information, called *IP datagrams*, are transmitted. These are an encapsulation of data passed from the higher layer with an *IP header*. IP attempts to provide a *lifetime guarantee*. This promises to the user that if an IP datagram is not delivered within a certain time period, the datagram will never be delivered. This time period known as the datagram *time-to-live* period can be specified in the arguments passed in the SEND operation.

9.3.1.2 A Communication Scenario Using IP. Perhaps the best way to explain how IP works is through an example. Consider the system illustrated in Figure 9.13. A user in station A wants to send some data to another user in station B. As mentioned previously, the user is typically a higher layer protocol and the data it is sending is typically a packet with the higher layer's protocol header encapsulating some information. From the viewpoint of IP, this is irrelevant and the user data is treated in a manner independent of its content.

In the Internet, stations have so-called Internet or IP addresses. A specific address uniquely identifies a particular connection of a station to a network. A station connected to multiple networks will have distinct Internet ad-

*By 1993, the number of hosts connected to the Internet exceeded 1.3 million and is growing at an annual rate in excess of 80%!

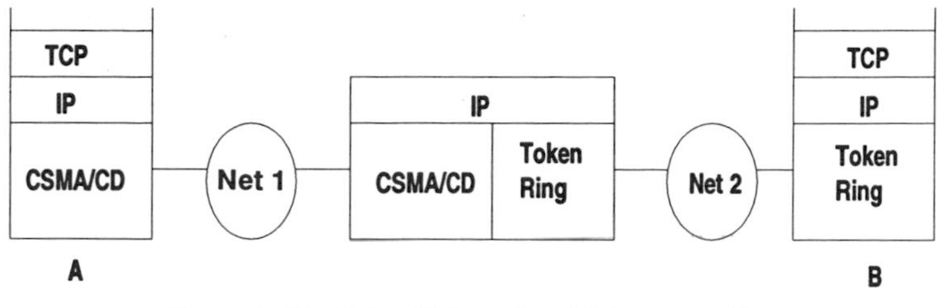

Figure 9.13. Using IP for network interconnection.

dresses for each connection. In our example, we assume station A is connected to a CSMA/CD network (network 1) and station B is connected to a token-ring network (network 2). The connections of station A and B to networks 1 and 2 have Internet addresses I_A and I_B, respectively. A router interconnects networks 1 and 2.

Communication from A to B is initiated when the user at A instructs its local IP to send some data to Internet address I_B. The IP layer in A takes the data given to it and performs the following tasks.

1. Encapsulates the user data in an IP datagram. The datagram will have an IP header that includes among other items, the source and destination Internet addresses. We discuss the details of the IP header shortly.
2. Determines the MAC layer address to be used by the MAC layer in A as its destination address. This requires a mapping between the destination Internet address and a MAC layer address. This mapping could be performed through looking up a local table or could be determined when needed. A protocol known as the address resolution protocol (ARP) may be used for this purpose. In our example, and since the IP datagram needs to get to the router first, we need to map the Internet address I_B to the MAC layer address of the router port connected to network 1.
3. Passes the IP datagram to the MAC layer (CSMA/CD in our example) instructing it to deliver the datagram to the MAC layer address determined in Step 2 above.

After the above steps are complete, station A's MAC layer transmits a CSMA/CD frame on network 1, which encapsulates an IP datagram and in turn encapsulates the user data. This frame reaches the router (as it has the router's MAC layer destination address). The CSMA/CD layer in the router then removes the MAC layer header and trailer and delivers the remainder (IP datagram) to the IP layer at the router. This IP layer examines the destination Internet address in the IP header and determines where this datagram should go next. This can be done by examining a local routing table. In this scenario, the router determines that the IP datagram needs to proceed along the router's port connected to network 2.

Before giving the IP datagram to the network 2 MAC layer, the router's IP layer may need to slightly modify its header (the source and destination Internet addresses remain the same, however). It also needs to determine the destination MAC address to which this packet needs to be forwarded. This is again obtained by mapping the destination Internet address (I_B) into a MAC layer address. In this case, the appropriate address is station B's MAC layer address for its network 2 port.

After transmission across network 2 and removal of the MAC layer header and trailer by B's MAC layer, the datagram is delivered to the IP layer in station B. This layer recognizes itself as the destination of the datagram. The IP datagram is then stripped of its header and the data portion is passed on to the appropriate user. Since IP can typically have multiple simultaneous users, the receiving IP layer needs to determine the identity of the user to which an incoming datagram should be delivered. This is specified in a field in the IP header, called the *protocol field*, and reflects the identity of the sending user (higher layer protocol). Various numbers have been agreed upon to represent the most common users such as TCP.

9.3.1.3 Fragmentation and Reassembly. One of the differences among various network standards and technologies is the maximum allowable packet size. To reduce header/trailer overhead, it is desirable to transmit packets with the largest possible data fields. Other factors, such as the desire for fairness in a CSMA/CD network (long packets will occupy the bus for a long time) or the desire to keep error rates below certain thresholds (the probability of packet error increases with the packet size), limit the size of a packet. These packet size limits can be different for different networks. IP was designed to work in this type of environment by allowing a router to fragment an incoming IP datagram that is too large to be transmitted on the outgoing network.

This fragmentation is accomplished by first stripping the header off the original IP datagram. The datagram's data field is then fragmented into the appropriate number of fragments. A new IP header (derived from the original header as described later) is added to each fragment. Each fragment is now treated as an independent IP datagram and delivered to the destination. Note that fragments of an original IP datagram may be further fragmented by other routers as they proceed toward their destination. Note also that, because each fragment is treated independently, they may reach the destination in a different order than their initial sequence of transmission.

Once an IP datagram is fragmented, it needs to be reassembled at some point before it is delivered to the user. To understand why, consider Figure 9.14, which shows an original IP datagram and the result of its fragmentation. It is clear from the figure that delivering the content of an

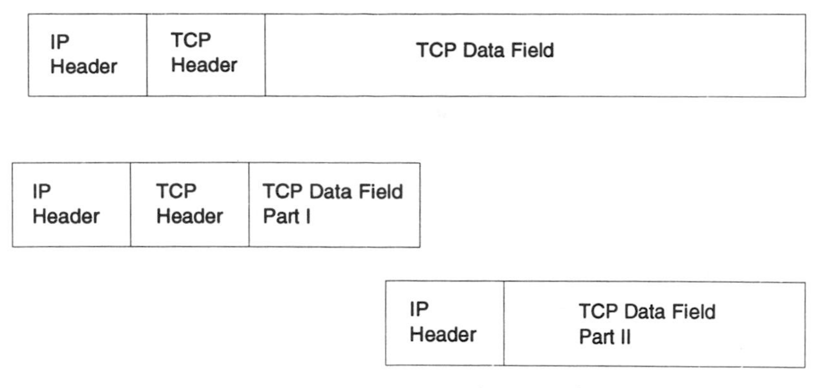

Figure 9.14. IP fragmentation example.

individual fragment to the user (TCP in this case) is meaningless, since each contains only a portion of a TCP packet. The two fragments need to be reassembled at the destination IP layer before they are delivered. Therefore, it is important that the IP headers of the fragments contain enough information to (1) identify the various fragments as belonging to the same original IP datagram, and (2) indicate the order in which these fragments should be reassembled to obtain the original IP datagram.

Each IP datagram contains an identification number (not to be confused with a sequence number). This number, along with the source and destination addresses and the protocol field (identifying the user of IP that issued the datagram), uniquely identifies an original datagram. When fragmented, these fields are copied into the header of each fragment.

Determining the proper assembly sequence is accomplished by maintaining in each header a More Flag and an Offset Field. The More Flag is set to 0 in the last fragment's header and to 1 in all others. The Offset Field indicates the distance in units of 64 bits of the first bit of the datagram (i.e., fragment) from the first bit of the original datagram. All fragmentation, therefore, must be done along 64 bit boundaries.

The table below is an example of the use of these fields in a scenario where an original datagram with a datafield that is 1280 bits long is fragmented to pass through a network with a maximum data field length of 704 bits.

Field	Original Datagram	Fragment 1	Fragment 2
Source	I_A	I_A	I_A
Destination	I_B	I_B	I_B
Protocol	TCP	TCP	TCP
Identification	225	225	225
More Flag	0	1	0
Offset	0	0	11

It should be noted that when an IP datagram is fragmented, there is no guarantee that any or all of the fragments will reach their desired destination.

At the destination, a reassembly process is started whenever a new datagram is received. This is defined as a datagram that does not belong to any reassembly process currently underway. Once started, however, a reassembly process cannot be guaranteed to terminate successfully because fragments may be lost. Therefore, a timer is associated with each reassembly process. If the timer expires before a datagram is completely reassembled, the reassembly is terminated and the fragments collected thus far are discarded.

9.3.1.4 Enforcing Maximum Lifetime. As mentioned earlier, IP attempts to guarantee a maximum time after which undelivered datagrams are never delivered. Each IP datagram contains a *time-to-live* field in its header. This is initialized with some maximum value at the source and decremented as the datagram traverses the network. If the time-to-live field is ever decremented to zero before reaching its destination, it is discarded.

Ideally, the initial value and the process of decrementing the time-to-live field should be done in units of time (e.g., seconds or milliseconds). This, in practice, is very difficult as it requires some form of global timekeeping across the Internet. Instead, the time-to-live field is indicated in number of hops, that is, the number of routers to be traversed. Each router decrements the field by 1. The datagram is discarded by a router that decrements the time-to-live field down to zero.

9.3.1.5 Error / State Reporting. IP embodies an error and state reporting capability that is used to provide feedback about the internal operation and state of the Internet. What is interesting about this capability is that it is provided in a higher layer protocol called the Internet control message protocol (ICMP), which actually uses IP to deliver its messages. ICMP messages are submitted to an IP layer and encapsulated in IP headers before transmission. Strictly speaking, this is a violation of the layering principle. However, it works well in practice.

The ICMP provides the following messages:

Destination unreachable may be caused by a router that does not know how to reach a specified destination or because fragmentation is required, but the *Do Not-Fragment* indicator is set. The message is returned to the source of the datagram, containing the header and the first 64 bits of the datagram that caused the error.

Time exceeded is a time-to-live value that has expired, resulting in a datagram being discarded. The message is returned to the source of the datagram, containing the header and the first 64 bits of the datagram that caused the error.

Parameter problem is a syntactic or semantic error detected in the IP header. The message is returned to the source of the datagram, containing the header and the first 64 bits of the datagram that caused the error.

Source quench provides a form of congestion control by requesting that a source reduce its input rate to the network. (This is a form of a choke packet as discussed in Chapter 6.) The message is returned to the source of the datagram, containing the header and the first 64 bits of the datagram that caused the error.

Redirect advises the source to send future traffic through another router if it is found to provide a shorter route to the destination. The message is returned to the source of the datagram, containing the header and the first 64 bits of the offending datagram that caused the error.

Echo and *echo reply* are a pair of messages exchanged between two nodes to verify that communication is possible.

Timestamp and timestamp reply are a pair of messages exchanged between two nodes to sample the delay characteristics of the Internet. The sender includes a message ID plus the time the message is sent. The receiver appends a receive timestamp and a transmit timestamp to the original message and returns it as the timestamp reply.

9.3.1.6 IP Datagram Format. The format of an IP datagram is shown in Figure 9.15. The length of each field in bits is also indicated in the figure. The following fields have been discussed so far in our discussion of IP in this section: source and destination address, time-to-live, protocol, identifier, and offset. Below is a description of other fields:

Version indicates the version of IP that generated the header.

Internet header length (IHL) indicates the length of the header in units of 32-bit words. Note that the header is at least five words long. The options field can make the header's length variable, and hence the IHL field is needed.

Type of service specifies the quality of service desired by the user generating this datagram. IP, being a datagram protocol, makes no promises that specified quality of service will be met. This field, however, can be used by routers and hosts to establish the eligibility of datagrams for any available special treatment (e.g., transmission priority).

Total length specifies the length of the IP datagram including the header in bytes.

Flags includes the "more" flag, which was discussed previously, and a "do not fragment" flag that prohibits the fragmentation of the datagram.

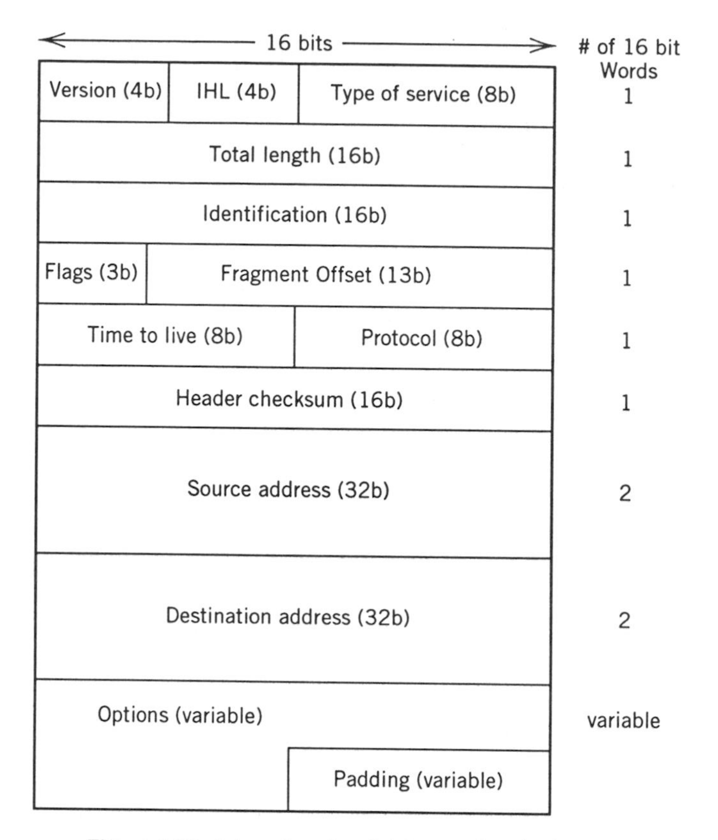

Figure 9.15. Internet protocol datagram header format.

Such a datagram is discarded if fragmentation is required. A third bit in the flags field is unused.

Options and padding specifies options such as security and source routing and padding to insure that the header ends on a 32-bit boundary.

Checksum is an error-detection code that applies only to the header. The concern addressed by this code is the corruption of the header while it is being processed within a router or a host's memory. It is not intended to detect errors in the data that occur during transmission. The assumption is that lower layer protocols (e.g., CSMA/CD) guard against such errors.

It is worth noting at this point that Internet addresses are 32-bits long (as shown in Fig. 9.15). At the time that the IP was developed (in the early 1970s) it was thought that 32 bits would be long enough to accommodate the requirements of the Internet forever. As the Internet has grown to encompass over 1.3 million hosts (as of January 1993) and at its current growth rate (estimated at 80 percent per year), the 32-bit address is perceived to be inadequate. Several proposals are now being considered to develop a next-generation IP with a larger address space (among other improvements). One of these proposals known as TUBA (**TCP** and **UDP*** with **bigger addresses**) involves the use of the OSI's connectionless network protocol described next.

9.3.2 Connectionless Network Protocol

The Connectionless Network Protocol (CLNP) developed by the OSI operates along the same principles discussed for the IP above. In fact, at the level of our discussion they are almost identical with four primary exceptions. Figure 9.16 shows the format of a CLNP datagram.

9.3.2.1 Strict Adherence to Layered Principles. CLNP, being part of the OSI suite of protocols, conforms to very strict guidelines regarding layering and description of functionality and interfaces. The most readily observable feature along these lines is in how error reporting is accomplished with CLNP. Error reports in CLNP are actually carried as CLNP datagrams. A "type" field in the datagram header is used to indicate whether this is a data or error-reporting datagram. An error report (ER) bit is set in the header to indicate whether a datagram should generate an error report.

9.3.2.2 Variable Address Lengths. CLNP allows addresses to be of variable length. An *address length* field is used to indicate how long each address is. It should be noted that, whereas variable length addresses provide complete flexibility, they are not universally accepted as a good solution for two reasons. The first is the overhead required to indicate the length of the address field. The second is that variable length addresses require more processing at routers for parsing of the header, its storage, and manipulation.

9.3.2.3 Elimination of the Fragmentation / Reassembly Overhead when Not Required. The segmentation part containing the identifier, offset, and length of field need not be present if a datagram will not be fragmented, as in cases where the networks being interconnected use identical standards or if the packet will never traverse a router. CLNP (unlike IP) allows for the elimination of the segmentation part. The segmentation permitted (SP) flag is set to 1 if the segmentation part is present and to 0 otherwise.

9.3.2.4 Determination of Higher Layer Protocol. In CLNP, the upper layer protocol that originated or is to receive a datagram is identified in the source or destination address field, respectively. (The protocol identifier field in the CLNP header actually indicates the presence or absence of CLNP,

*UDP stands for the User Datagram Protocol, a connectionless transport protocol used in the Internet.

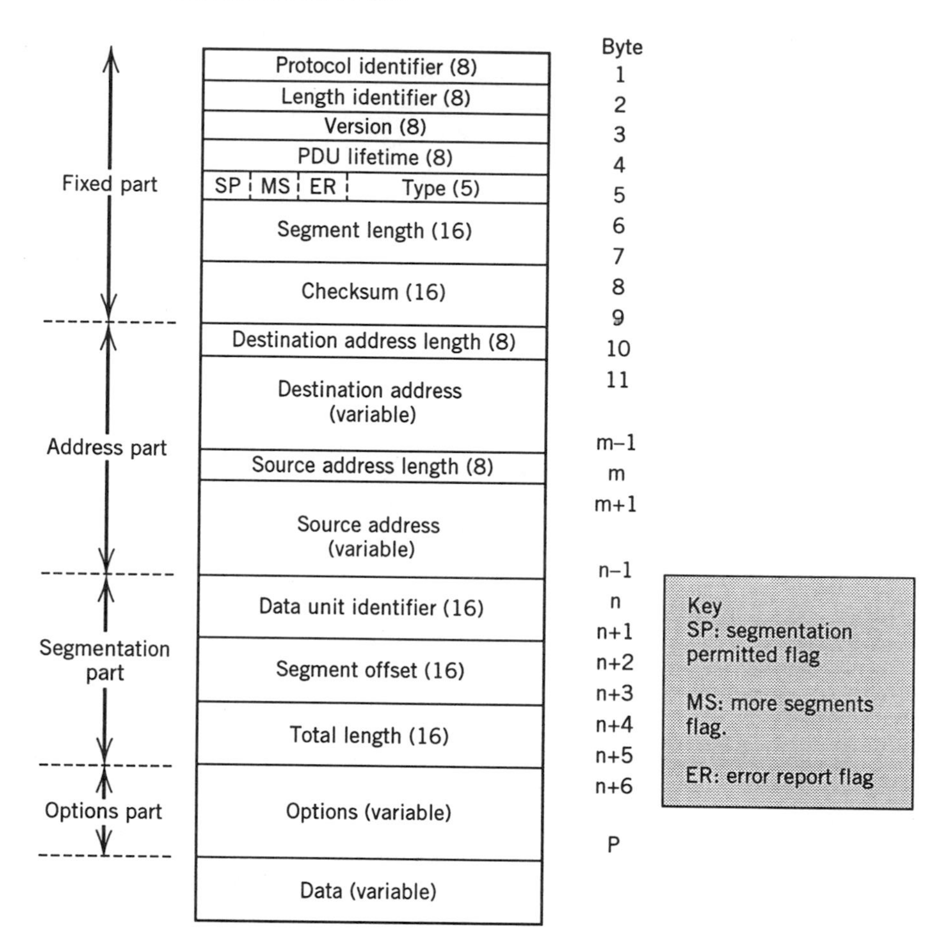

Figure 9.16. The connectionless network protocol datagram format.

because no network layer protocol is required for communication across the same network.)

SUGGESTED READINGS

A general discussion of network interconnection can be found in the excellent text devoted to this topic by Perlman [PERL93]. An extensive annotated bibliography can be found in [BIER90]. The application layer gateway and other interconnection approaches are discussed in [ROSE92], and [ROSE90].

The transparent bridging approach is standardized in [IEEE90a]. Some discussion of this technique can be found in [BACK88]. The original tree configuration algorithm was described in [PERL84]. The source routing technique is standardized in [IEEE91]. Discussion of the relative merits of the two bridging schemes is available in [DIXO88], and [SOHA88]. The work in [VARG90] investigates how LANs with incompatible MAC layers may be bridged together.

An early discussion of the various options of interconnecting networks (at the network layer) can be found in [POST80a]. The IP and ICMP standards are specified in [Pos81b], and [Pos81a] and the IP is discussed in [POST81]. The design decisions leading to the IP fragmentation scheme are discussed in [SHOC71]. The CLNP is standardized in [ISO88]. Discussions of possible replacements for the IP can be found in [CALL92], [KATZ93], [DEER93], and [FRAN93].

Internet standards are documented in request-for-comments (RFC) documents. These documents can be obtained by anonymous ftp from ds.internic.net.

PROBLEMS

9.1 Show the layered architecture in a system consisting of two user stations connected by means of half bridges and an X.25 network. Show the layering in the stations and bridges and depict the message frame formats at different points in the interconnection.

9.2 Figure 9.17 shows LANs interconnected by bridges. The bridges have the indicated bridge IDs and port numbers. Find the tree resulting from using the spanning tree algorithm in this network.

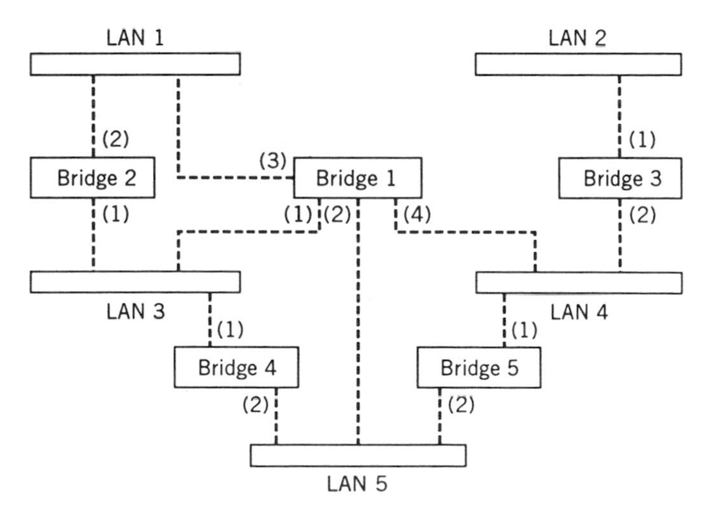

Figure 9.17. Problem 9.2.

9.3 In the process of formulating a spanning tree, a bridge receives the following five BPDUs on its five ports. Find the identity of the designated ports and the resulting BPDU message to be transmitted on these ports if this bridge ID is 14. Repeat if the bridge ID is 4.

Port Number	Believed Root	Cost	Transmitting Bridge ID
Port 1	22	45	28
Port 2	10	39	44
Port 3	9	54	51
Port 4	9	46	36
Port 5	48	0	48

9.4 Figure 9.18 shows LANs interconnected by source routing bridges. Assume bridges 1, 5, and 7 are not part of the spanning tree. Show the single route and the all-route broadcast messages exchanged in order for S1 to determine a path to S6.

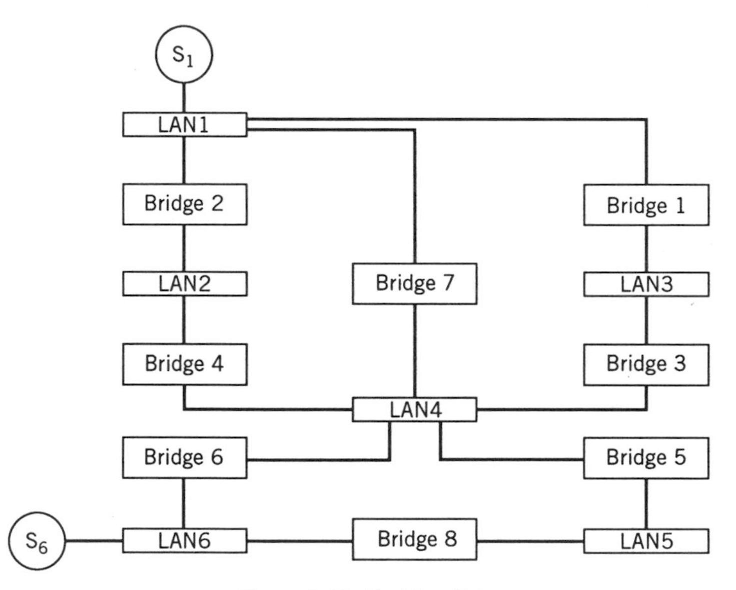

Figure 9.18. Problem 9.4.

9.5 Repeat Problem 9.4 using the assumption that no single-route broadcasts are possible (no tree is maintained). Also assume that the destination in this case returns all paths it learns about. Compare the message traffic with that of Problem 9.4.

9.6 Discuss various possibilities of when a source might undertake a path discovery procedure in the source routing scheme.

9.7 Compare source routing to transparent bridging from the viewpoint of bandwidth efficiency, routing overhead, route finding efficiency, and reliability.

9.8 At a certain router, an IP datagram consisting of 926 data bytes is to be fragmented for transmission on a network with a maximum data field of 256 bytes. The header length of the original datagram is 36 bytes. Assuming that all fragments are of the maximum length (except the last one) find the values of the IHL, total length, fragment offset, and more flags of each resulting fragment.

9.9 Describe a way that a user of IP can determine the current route to a destination using the facilities provided by ICMP.

9.10 IP currently provides for the reassembly of a fragmented datagram at the destination. One alternative to this is to have a datagram reassembled by a router just before exiting the network for which this fragmentation was necessary. Compare these two different approaches.

9.11 One possible way to avoid fragmentation altogether is to insist that all datagrams be of a certain maximum size that will comply with the requirements of any network. Comment on this approach.

10

TRANSPORT PROTOCOLS

10.1 INTRODUCTION

The only remaining piece necessary to complete the data communication picture is the discussion of the end-to-end functions performed by the transport layer. The techniques discussed so far provide us with capabilities to deal with effective medium sharing, reliable single-hop transmission, and routing through a switched network or router-interconnected networks. Even with all these capabilities we are unable to perform two primary tasks:

- ensure that data is delivered reliably to a destination. Despite our best efforts, packets can be lost. For example, they can be corrupted while being stored in a switch or router's memory, or they may circulate in a network for a long time and be dropped because they have exceeded their time to live.
- discriminate between multiple communicating entities on the same machine. Network addresses can typically identify a single machine. But several processes on the same machine may be running simultaneously, with each carrying on its own independent conversation.

Among other things, transport protocols help to perform the above tasks and are thus an important part of any communication architecture. The transport layer provides end-to-end functionality to higher protocol layers, such as a file transfer application. It does not deal with the details of how to get a packet to a destination. Rather, it deals with how to address a particular process in a destination machine and sometimes how to make sure that a packet is actually received and accepted by the intended receiver. Transport layer functionality need only be present at end nodes. Strictly speaking it is not required in routers and switches. Normally, however, both routers and switches will have transport layer functions implemented since they also serve as end nodes of management operations.

10.1.1 Type of Transport Service

A transport protocol provides the rules of exchange between the transport layers at the two ends of a conversation. The network layer services are used to deliver transport protocol exchanges. In other words, a transport layer is said to operate using the services of the network layer. A transport layer may provide a connection-oriented or a connectionless type of service. The latter form provides no assurance that the transport layer will deliver packets to a destination without loss. A connectionless transport layer typically operates on top of a connectionless network layer (such as IP or CLNP).

A connection-oriented transport layer can be built to run on either a connectionless or a connection-oriented network layer (such as layer 3 of X.25). When a connectionless network layer is used, the connection-oriented transport service requires that many elaborate functions be performed by the transport layer. Conversely, if a connection-oriented network layer is used, the connection-oriented transport layer can be rather simple, although it will typically need to enhance some of the services provided by the network layer.

10.1.2 Quality of Service

Higher layers using the transport layer may specify quality of service parameters either at connection establishment time or at data transmission time. It is up to the transport layer to examine its user's requirements and to respond with approval, denial or negotiation for alternate values for the required service parameters. The user's requirements can be expressed in terms of quality of service (QOS) parameters. Important parameters include a specification of end-to-end connection throughput or packet delay. It may also be possible to request a certain level of service priority (among other transport layer connections) and a level of connection security.

One of the factors determining whether a certain specified QOS parameter can be offered is the level of service offered by the network layer. For example, if the network is using 56 kbps transmission lines and the transport layer user requests 1 Mbps throughput, the transport layer cannot meet this requirement.

10.1.3 Multiplexing

Multiplexing can be defined as the sharing of a service provided by an entity at some protocol layer by multiple entities at higher layers. Multiplexing can occur in many protocol layers. The transport layer multiplexing functions allow several users of the transport layer to use a single set of network layer facilities used to convey messages across a network. For example, a transport protocol can be used to allow multiple processes on one machine to establish

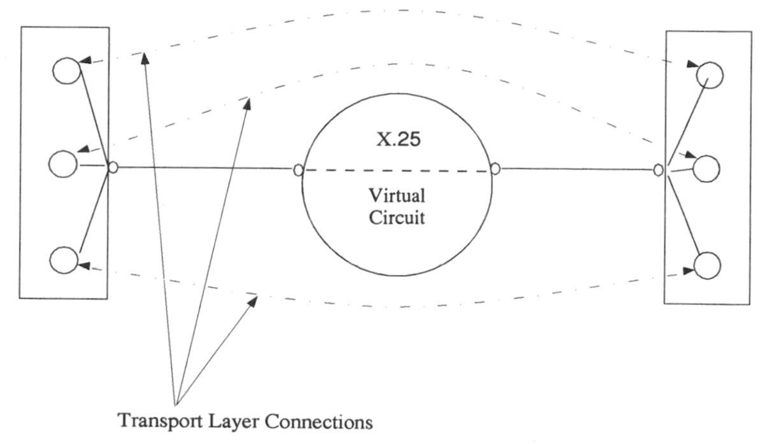

Figure 10.1. Multiple transport connection using a single X.25 Virtual Circuit.

multiple transport layer connections to processes on another machine with all these connections being carried over the same X.25 virtual circuit. This is shown in Figure 10.1. Connectionless transport protocols sometimes have multiplexing as their only function.

10.1.4 Have we seen this before?

The reader may now be thinking that some of this discussion sounds familiar. After all, we argued in Chapter 4 that transmission of frames on data links may be unreliable and that frames may be lost or corrupted. We then proceeded to discuss how data link layer protocols may be used to help overcome this problem. It may seem that the problem of providing end-to-end transport layer functionality, especially where the deficiencies of the network layer need to be overcome, can be solved using the same techniques used in a data link layer protocol such as HDLC, but this is not quite true. Whereas the network layer, much like a transmission link, provides an unreliable conveyance mechanism, there are subtle and important differences between the two. The first is that an unreliable network layer can resequence packets, which can never happen on a transmission link. In addition, a network layer can arbitrarily delay a packet with the packet showing up at its destination long after it was sent, whereas a transmission link will always deliver a packet within a given time governed by the propagation delay and packet transmission time. These two differences are sufficient to require that some mechanisms be used in a transport protocol that are different from those used in a data link layer protocol. As it turns out, though, much of what we have learned about data link layer protocols, such as the use of acknowledgments, the timeout mechanism, windows and retransmissions is very much a part of connection-oriented transport protocols.

10.1.5 The Players

Many transport protocols have been developed and are in use today. In this chapter we focus on the transmission control protocol (TCP) a connection-oriented protocol which is used in the Internet. Many of the principles of the design of transport protocols will be apparent in our discussion of TCP. We also discuss the user datagram protocol, a connectionless protocol used in the Internet. We briefly discuss the OSI transport protocols, particularly how they differ from TCP.

10.2 THE TRANSMISSION CONTROL PROTOCOL

The transmission control protocol (TCP) is a connection-oriented transport protocol designed to work in conjunction with the Internet protocol (IP), which was discussed in Chapter 9. TCP provides its user with the ability to transmit reliably a byte stream to a destination and allows for multiplexing multiple TCP connections within a transmitting or receiving host machine.

Being connection-oriented, communication using TCP requires a connection establishment phase which can be followed by a data transmission phase. A connection is terminated when it is no longer in use. The discussion below will include a description of these three phases.

10.2.1 Service Provided by the Transmission Control Protocol

A higher layer that uses TCP interacts with TCP through operations, or primitives. Such primitives are very similar to subroutine calls and have parameters associated with them. The interface with TCP can be defined in terms of such primitives. The original TCP standard document defines what it calls user commands as well as TCP-to-user messages. It is, however, a loose definition and it is acknowledged in the standard that different implementations of TCP may have different interface specifications.

Examples of the user commands described are:

1. the OPEN command, which can be either active or passive. An active OPEN command requests that a connection be established to a specified destination (destination is specified by a combination of the IP address and the port number described below). A passive OPEN is a "listen" request and allows for the quick acceptance of an incoming connection establishment request.

2. the SEND command which is used to request the transmission of data on an already established connection.
3. the RECEIVE command which is used to allocate a certain size buffer for data reception. In the standard this command returns only after data has been received in the allocated buffer. This would be a "blocking" implementation of the receive function. Other types of implementations include non-blocking ones where a RECEIVE returns immediately and is retried periodically until the allocated buffer is full. Alternatively, data reception could cause an interrupt resulting in the execution of an interrupt handler.
4. the CLOSE command which is used to request the termination of an existing connection.

Issuance of these commands by the user can sometimes trigger the exchange of messages between the local TCP entity and a remote TCP entity. For example, an active OPEN will typically cause a connection request to be issued to the other side. Also these commands may result in some TCP-to-user message being returned. An example of this is a *local-connection-name* that is returned as a result of the OPEN command. This local-connection-name is then used as an argument to SEND, RECEIVE, and CLOSE commands to identify the particular connection.

10.2.2 Transmission Control Protocol Packet Format

We show in Figure 10.2 the format and fields of a TCP packet. The remainder of this section defines the meaning of these fields and illustrates how they are used. It should be noted that *all* TCP messages, regardless of purpose, have the same format. Contents of various fields will change depending upon the purpose of the packet.

Since TCP uses IP, a TCP packet is encapsulated into an IP datagram prior to transmission. The IP packet will indicate the source and destination machines (actually IP addresses indicate network interfaces of a particular machine). Differentiating particular processes within a particular machine is done using the TCP *port* fields. The source port number indicates the sending process and the destination port number indicates the receiving process on the destination machine. In the Internet, some port numbers have been reserved to identify "well known services" such as the file transfer protocol (FTP). Processes implementing these services will always communicate using the same port number on all machines. In TCP packets, the sending and receiving port numbers are indicated using the 16 bit fields shown in Figure 10.2.

TCP is a byte stream protocol in the sense that it looks at the data it is transmitting as a sequence of bytes rather than a sequence of packets. Sequence numbers refer to a byte number. Each data packet carries the

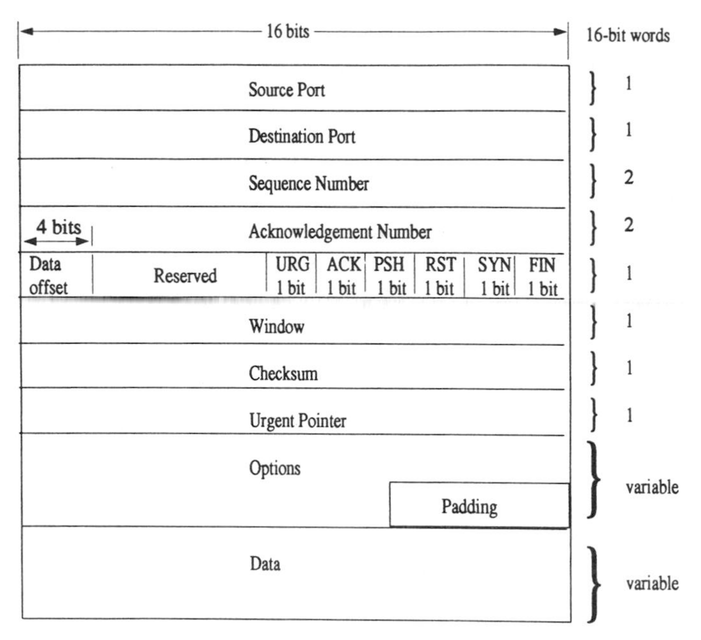

Figure 10.2. TCP Packet Format

sequence number of the first byte in the *data* field. Data is also acknowledged using the sequence number of the byte expected next. Acknowledgments are piggybacked on data packets. (This is much like the HDLC numbering approach except that now bytes are being numbered and multiple bytes can be transmitted in a packet.) In data packets, the sequence number and acknowledgment number are indicated in the corresponding 32-bit fields as shown in Figure 10.2.

The other fields in Figure 10.2 are discussed in the context of the various TCP procedures described next.

10.2.3 Initial Sequence Numbering

Let us assume for the time being that a TCP connection between two entities has been established successfully. The actual connection establishment procedure is described in Section 10.2.4. For now all we need to know is that, as part of the connection establishment procedure, the two ends agree on two initial sequence numbers, one for each direction. This agreement is necessary because it is desirable to have all connections start with different sequence numbers. To understand why, let us consider a situation where all connections start sequence numbering at 0. A packet with sequence number 0 is transmitted but delayed for a long time inside the network. The packet is retransmitted and the connection continues and is ultimately closed. Immediately following this, a new connection between the same two ends is opened (again with initial sequence number 0). Before any data is transmitted on this new connection the original packet with sequence number 0 reappears and is accepted. This clearly will cause problems. However, if the new connection starts using a different sequence number, this old packet would not cause problems, as long as the sequence numbers have not wrapped around to 0. With 32 bit numbers available a packet has to live for a long time before its sequence number is ever reused (see Problem 10.2).

In TCP each end can use a local 32-bit clock to determine the initial sequence number. Note that because clocks on different machines are not synchronized, the initial sequence number in each direction can be different. If a machine crashes it may forget the value of its 32-bit clock. This can cause it to pick sequence numbers that will be confused with old packets that may still be in the network. In such cases, the TCP standard recommends that the machine stay quiet, that is, not attempt to start a connection, for a period of time equivalent to the maximum lifetime of a packet. This is typically on the order of minutes. Recall that IP attempts to enforce a maximum packet lifetime value using the time-to-live field (see Chapter 9).

10.2.4 Connection Establishment

The TCP connection is established by the *three-way handshake*, in which three messages need to be exchanged before a connection is established.

1. A connection request (CR) message is sent by the connection initiator.
2. A connection confirmation (CC) message is sent by the connection destination.
3. An acknowledgment (ACK) message is sent by the connection initiator.

The CR message carries an identification which is echoed in the CC message. This allows the source to match a CC message with a CR message that it has sent. The CC message, in turn, carries its own identification which is echoed by the ACK message. This allows the connection destination to match the ACK with the CC it has sent. An example of this process is shown in Figure 10.3 In the three-way handshake if the ID of a CC message cannot be matched with the ID of a previously transmitted and still outstanding CR message, the CC message is discarded and a REJ message is sent back to the source of the CC message echoing the ID of the CC message.

The three-way handshake can guard against many potential problems related to the possibility of arbitrarily long delays that can be encountered by a packet as it proceeds to its destination (see Problem 10.4 for an example).

In TCP, the CR message has the SYN flag set to 1 (other flags are 0), and the initial sequence number indicated in the *sequence number* field as the ID. The CC message has the SYN and ACK flags set. The ID of the CC message is indicated in the sequence number field and the *acknowledgment* field contains the sequence number in the CR message incremented by 1. Finally, the ACK message, which in TCP can contain data, will have the SYN flag set

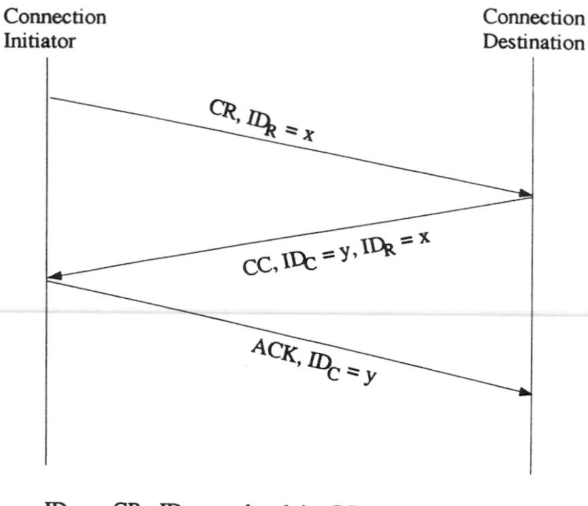

Figure 10.3. Three-Way Handshake.

to 0, the ACK flag set to 1 and the acknowledgement field containing the sequence number in the CC message.

In the REJ message, the reset (RST) flag is set to 1. This message echoes the ID of the rejected message in the *acknowledgement* field and has the ACK bit set to 1.

10.2.5 The Push and Urgent Functions

As was mentioned before, TCP is a byte stream protocol. The user of a TCP protocol delivers a byte stream to TCP. It is up to TCP to make packets out of the byte stream. Typically, the TCP layer will accumulate the bytes in a buffer and send a packet once the buffer is full. In this scenario the user of TCP has no control over when the bytes that it is giving to TCP will actually be sent. A similar operation occurs at the receiver of TCP packets. As data bytes are received they are accumulated into a receive buffer and delivered as the receiving layer sees fit, perhaps once the buffer is full. Again the user has no control of when bytes are actually delivered.

TCP defines the push function to give the user more control. The user is allowed to indicate at the end of a byte stream a push (PSH) flag. This flag forces the sending TCP to send everything it has in its buffers regardless of whether or not they are full. A packet transmitted this way will have the PSH flag in the TCP header set to 1 (see Figure 10.2). This will, in turn, cause the receiving TCP layer to deliver all data received on this connection up to the end of the packet with the PSH flag set. In most actual implementations of the TCP programming interface the user (or higher layer) is not allowed to specify the push function. Rather, it is left up to the TCP protocol itself to determine when the PSH flag is to be set.

Another function provided by TCP is the ability to indicate that some data is urgent. This is done by setting the URG flag to 1. The urgent data is all the bytes appearing before the first *n* bytes where the value of *n* is indicated by the 16-bit *urgent pointer* field. The existence and location of urgent data is also communicated to the receiving user of TCP. The TCP standard does not specify how urgent data is to be treated by the user. The assumption, however, is that the user will have special procedures invoked once notified of the existence and location of urgent data. Urgent data can be used, for example, to carry an interrupt request from a user that requires immediate attention by the receiver.

10.2.6 Error Detection and Recovery

The TCP header contains a 16-bit *checksum* field. This is used to detect errors in TCP packets. Recall that TCP packets are typically transmitted using IP which cannot guarantee error-free transmission of packets. It is, therefore, up to the TCP protocol to detect and recover from errors. The checksum field actually does more than protect against errors in the TCP packet; it also protects against the misrouting of this packet. To do this the checksum is actually calculated over a packet that contains, in addition to the TCP header, a pseudo-header as shown in Figure 10.4. This pseudo-header explicitly contains the source and destination IP addresses, the protocol identifier used in the IP *protocol* field to identify TCP, and the length of the TCP datagram. In order for the checksum to be regenerated correctly at the receiver, it is not sufficient that the TCP packet be error-free. It is also necessary that the IP source and destination addresses and the IP protocol field be the same as when generated at the source of the TCP packet. For example, if for some reason a router were to change an IP datagram's destination, the TCP packet enclosed in the IP datagram will be rejected when it reaches this wrong destination.

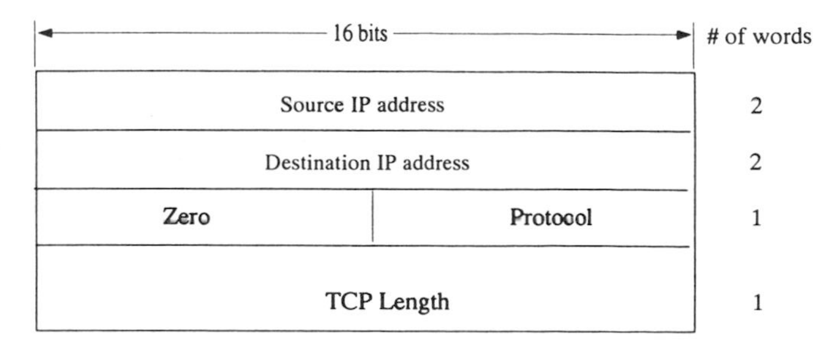

Figure 10.4. Pseudo-Header Used for TCP Checksum Calculation.

This unusual error detection technique is a reflection of the desire to maintain correct transport layer operation no matter how unreliable IP and its routing protocols are.

Packets that are received correctly are acknowledged by the receiver. Acknowledgment numbers are carried in the 32-bit *acknowledgement* field. The ACK bit is set in such packets to indicate that the acknowledgement field should be read. An acknowledgement field with value x indicates that the last byte that was received was $x - 1$ and that the next byte expected is byte number x. For example, a TCP source sends a packet with sequence number 128. If the packet contains 256 bytes and the packet is correctly received, the acknowledgement field value returned should be 384.

For each packet transmitted, TCP maintains a timer. If the timer expires, the packet is retransmitted. One interesting aspect of TCP in particular, and of transport protocols in general, is the need to adjust continuously the threshold at which the retransmit timer expires. The reason for this is that the TCP packets are being carried over a path with an unpredictable delay. Not only is the path changing during the lifetime of the connection but the state of the routers (or switches) is also changing as more or less traffic is being carried by the network.

TCP, therefore, incorporates a round-trip estimation mechanism that is used to determine what should be considered a normal acknowledgment delay for transmitted packets. This round-trip delay estimate is used as the basis to set the retransmit timer threshold.

10.2.7 Flow Control

One of the important functions of TCP is its ability to control the rate of flow of bytes sent out of the transmitting TCP layer in a manner that accommodates the capabilities of the receiver. TCP employs an ACK/CREDIT scheme as discussed in Chapter 6. Acknowledging a packet (or a sequence of bytes) does not imply any permission to send additional bytes. Such permission is given explicitly in the form of credit allocation. In TCP the credit value is indicated in the 16 bit *window* field.

When a TCP layer receives a packet with the acknowledgment field containing a value, a, and the window field containing a value, b, it is interpreted as:

> The sender of this packet has received all data bytes up to and including byte number $a - 1$. Transmission of bytes in the range a to $a + b - 1$, inclusive, is allowed.

10.2.8 Options

The TCP header allows the sender to specify various options in the *options* field. Only one option is currently defined to indicate the maximum packet size that can be received by a TCP layer. This is only communicated in an initial connection request. The *padding* field is used to ensure that the TCP header ends on a 32-bit boundary. The option field has the effect of making the header length variable. The *data offset* field is used to specify the actual length of the entire TCP header in 32-bit words. It is used to correctly parse the TCP header.

10.2.9 Connection Termination

TCP allows either end to terminate a connection. Since neither end is dominant, it makes no sense for one of them to just decide to terminate a connection. Therefore, when the user of TCP decides it no longer wants to send data, it informs its TCP layer, which will issue a TCP packet with the FIN flag set to 1. This packet contains no data but has a sequence number. The receiver of the packet with the FIN flag set to 1 will acknowledge the packet. This has the effect of closing the connection in one direction only. The sender of the FIN flag will not send any more data, but will continue to receive. A connection is completely terminated only after both sides have sent packets with the FIN flag set to 1 and when both these packets have been acknowledged.

10.3 THE USER DATAGRAM PROTOCOL

The user datagram protocol (UDP) is a connectionless transport protocol designed to operate using the service of IP. Its primary functions are error detection and multiplexing. UDP does not guarantee the delivery of packets but will guarantee that if a packet is ever delivered in error such an error will be detected. In addition it allows for the discrimination among multiple processes residing on the same host machine.

The format of a UDP packet is shown in Figure 10.5. The source and destination port fields have the same function as in TCP. In addition the header contains a checksum field used for error detection and a length field specifying the length of the UDP datagram, including the header in octets.

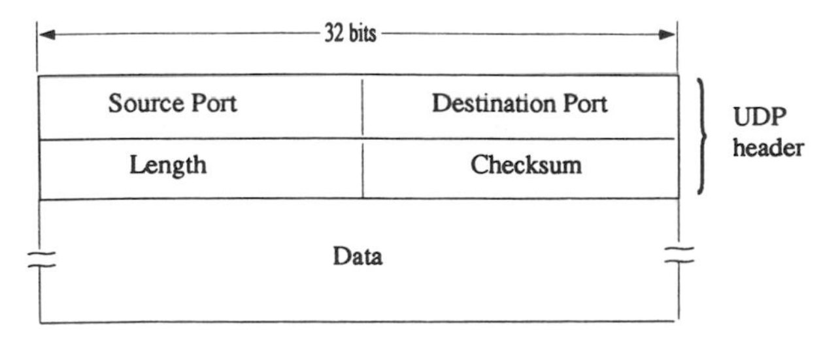

Figure 10.5. UDP Packet Format

The checksum field is calculated in the same way as the TCP calculation described above.

The reader should compare the UDP packet format with that of TCP (shown in Figure 10.2). The difference in complexity in the formats is a good measure of the relative complexity of providing a connection-oriented versus a connectionless transport service.

10.4 OSI TRANSPORT PROTOCOLS

Although the OSI transport protocols are designed to provide a service similar to that of TCP, they approach this problem in a different way. In the following sections, we focus on aspects of the OSI protocols that are significantly different from the TCP approach.

10.4.1 A Set of Protocols

OSI defines five transport protocols known as TP0 through TP4. Each protocol is designed to operate on a different type of network layer service. Also the protocols differ somewhat in the functionality they provide to the user. TP0 and TP2 are designed to operate on a highly reliable network layer. Although they can detect network layer problems they are not equipped to recover from them. The main enhancement of TP2 over TP0 is the addition of the multiplexing and flow control functions. TP1 and TP3 are designed to operate on a network layer that is mostly reliable but occasionally suffers from problems that require recovery by the transport layer. X.25 is an example of such a network layer. Again TP3 provides some enhancements over TP1 including flow control and multiplexing. TP4 is the most powerful of the OSI transport protocols and is designed to provide "top-of-the-line" transport layer services including multiplexing, flow control, and QOS specification. TP4 is designed to operate over even a connectionless network layer such as CLNP (see Chapter 9) and is very similar in functionality to TCP.

10.4.2 Protocol Services

OSI protocol services are defined through a set of primitives that are used by the user of the transport layer (the session layer in the OSI model) to invoke transport layer services. There are three main types of primitives*: CON-

*OSI notation uses a prefix to indicate the service provider for a particular primitive, that is the layer with which this service primitive is defining an interaction. In the case of the transport layer the primitives are prefixed with T_ to become T_CONNECT, T_DISCONNECT and T_DATA and T_EXPEDITED_DATA. We omit this prefix in our discussion.

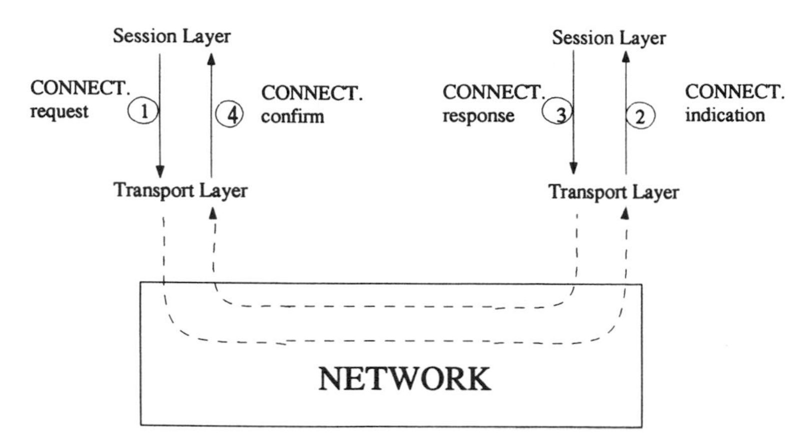

Figure 10.6. CONNECT primitives used in the successful establishment of a connection.

NECT, DISCONNECT, DATA, and EXPEDITED_DATA. These primitives serve the same purpose as the user commands and the TCP-to-user messages defined for TCP and discussed in section 10.2.1*.

1. The CONNECT primitives are used to set up a connection and can be in any of four forms:
   - *CONNECT.request* is used to indicate that the session layer wishes to connect to a remote session layer.
   - *CONNECT.indication* is used to inform a session layer of the desire of a remote session layer to establish a connection.
   - *CONNECT.response* is used to accept a connection request from a remote session layer.
   - *CONNECT.confirm* is used to inform a session layer that a previously issued *CONNECT.request* has been accepted.

Figure 10.6 shows the uses of these primitives.

2. The DISCONNECT primitives are used to close a connection and can be in one of two forms:
   - *DISCONNECT.request* is used to reject a CONNECT.indication or to initiate the closing of a connection that is currently established.
   - *DISCONNECT.indication* is used to locally reject a CONNECT.request or to inform a session layer that an existing transport connec-

*Whereas the TCP interface definition is labelled as "at best, fictional" in the TCP standard, such interface definition is an important part of the OSI transport protocols.

tion is being closed (by the other end of the connection or by the network layer).

3. The DATA primitives are used to request the transmission of data. There are two forms of the DATA primitives:
   - *DATA.request* is used to request that data be transferred on an already established connection.
   - *DATA.indication* is used to inform a session layer of the arrival of data on an existing transport connection.
4. The EXPEDITED_DATA primitives are used to request the transmission of high priority data. (This is analogous to urgent data in TCP.)
   - *EXPEDITED_DATA.request* is used to request that urgent data be transferred on an already established connection.
   - *EXPEDITED_DATA.indication* is used to inform a session layer of the arrival of urgent data on an existing transport connection.

We note that typically the invocation of these service primitives will either cause the transmission of a transport protocol packet (also called a transport protocol data unit, TPDU) or will be the result of the reception of a TPDU.

10.4.3 Packet Formats

Another interesting feature of the OSI transport protocols is the relative complexity of the packet formats defined. Unlike TCP, which defines a single packet format and uses flags to indicate whether control (e.g., connection request) or data packets are being sent, the OSI protocols define various packet structures.

Figure 10.7 shows the format of a connection request and connection confirmation packets. The packet header has a fixed part and a variable part. The fields in the fixed part have the following meaning:

- *Header length* is the length of header (including the variable part) in bytes.
- *Type* indicates which kind of TPDU this is. Codes defining a connection request or a connection confirmation TPDU are used.
- *Credit* is used to give an initial packet credit to the receiver.
- *Source reference* is the number that will be used by this transport layer to identify the connection locally. In other words packets received on this connection should be identified by this number.
- *Destination reference* is the number used by the other end of the connection to identify it. This field is all zeros in the connection request packet since the identifier assigned by the remote end is not known yet.

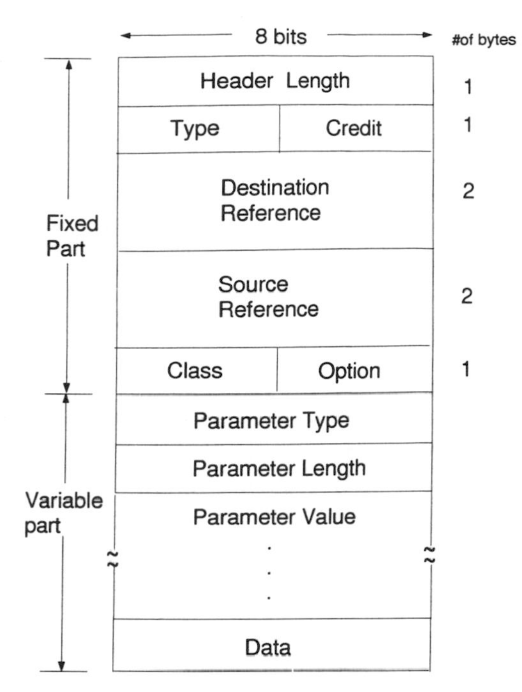

Figure 10.7. Connection request and confirmation packet format.

- *Class* identifies which transport layer class is being requested.
- *Option* specifies the use of short or long sequence number and credit fields (short = 7-bit sequence number and 4-bit credit field, long = 31-bit sequence number and 16-bit credit field)*.

The variable part of the header of the connection request and confirmation packets contains a number of fields. Among them are:

- calling transport address and called transport address
- protocol version number
- checksum (only required for class 4 protocols since they do not assume an error-free network layer)
- quality of service parameters.

Each parameter in the variable part of the header is preceded by an octet indicating its type and another octet indicating its length.

*Recall that longer sequence numbers allow the use of larger window sizes and thus allow efficient operation in long propagation delay scenarios (see Chapter 4).

Both the connection request and confirmation packets allow the inclusion of a short data field within the packet.

Figure 10.8 shows the format of data and acknowledgment packets. In a data packet the credit field is not used and only a destination reference is included. The data packet contains a sequence number which identifies the packet. The acknowledgment packet contains a credit field and an acknowledgment number. The values of the type field identify a packet as data or acknowledgment packets. We make two observations. First, OSI transport protocols do not allow for piggybacked acknowledgments and, second, the sequence numbering in the OSI protocols is done relative to packets and not bytes as is the case with TCP.

The destination reference field is used to identify the particular transport connection to the receiving transport entity. This field is not present in data packets for TP0 and TP1 since they do not allow multiplexing and there will be exactly one transport connection at the destination.

In the OSI protocols, a data TPDU is typically sent as a result of the session layer invoking the DATA.request primitive on an existing connection.

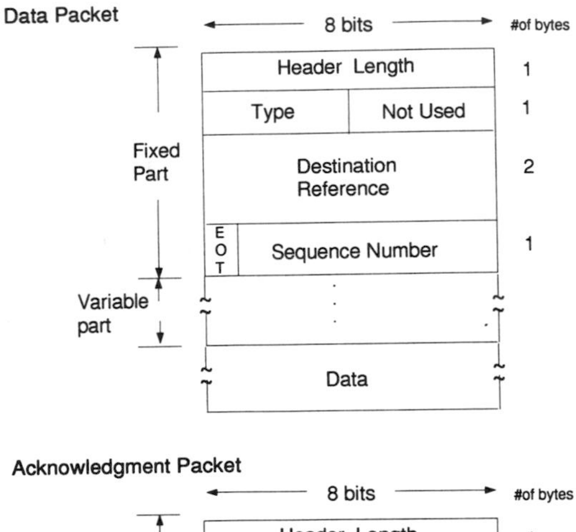

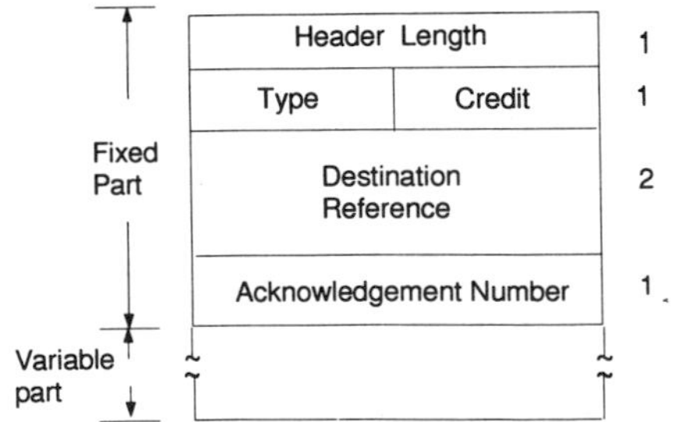

Figure 10.8. Data and Acknowledgment packet formats.

The data that needs to be sent is an argument of this invocation and is known in the OSI terminology as a transport service data unit (TSDU). In order to maintain layer independence, the size of a TSDU is typically decided upon independently from the maximum data packet size that can be transmitted on a transport connection (that is, within a data TPDU). In cases where the TSDU is larger than this maximum size, it will have to be fragmented and transmitted in several data TPDU. The EOT bit in the data packet is set to 0 if a data TPDU is the first or middle fragment of a TSDU and to 1 if it is the last fragment. This information can be used by the receiver to wait until a full TSDU is received before informing the local session layer of data arrival via the DATA.indication primitive.

10.4.4 Other Protocol Mechanisms

OSI transport protocols allow for two- or three-way handshakes for connection establishment. Two-say handshakes are used in TP0 through TP3 since they assume a reliable network layer. TP4, on the other hand, assumes an unreliable network layer and uses a three-way handshake.

Connection establishment involves the transmission of a connection request TPDU and a response with a connection conformation TPDU. In cases where a two-way handshake is used, this is sufficient to establish the connection. For TP4 where a three-way handshake is required, the connection confirmation packet is acknowledged with a data TPDU or an acknowledgment TPDU.

Two other TPDUs, a disconnect request and a disconnect confirm are used to refuse a connection request or to tear down existing connections.

Expedited data is sent using an expedited data TPDU and an expedited data acknowledgment TPDU. Expedited data TPDUs are numbered and only one such TPDU may be outstanding at a time. That is, the transmission of expedited data is performed using a stop-and-wait protocol (see Chapter 4). The requirement is that once an EXPEDITED_DATA.request has been issued by the user, this data should be sent in an expedited data TPDU that is received ahead of any future data TPDUs. In all protocols except TP4 this is not a problem since they are designed to operate with a network layer that maintains packet sequencing. TP4, however, is designed to work with a network layer that may resequence packets. Therefore, in order to maintain the sequencing requirement, the transport layer has to refrain from any normal TPDU transmission until the acknowledgment for an outstanding expedited data TPDU is received.

Error control is needed only in TP4 and is accomplished through timeouts and packet retransmissions. Flow control uses an acknowledgment/credit scheme much like the one used by TCP (with the exception of numbering packets not bytes).

10.5 OSI TRANSPORT SERVICE ON TOP OF THE TRANSMISSION CONTROL PROTOCOL

The OSI transport protocols and TCP have been developed largely by separate communities. Each set of protocols has its proponents and its detractors. It has become clear in recent years that the TCP/IP networking approach is extremely mature with a wealth of experience with usage and implementations. The OSI protocol community, however, has achieved a certain measure of success proposing and implementing higher layer protocols such as the X.400 mail protocol and the FTAM file transfer protocol. These applications are defined with the use of OSI transport protocols in mind.

As was mentioned earlier TCP and TP4 perform basically the same functions. It therefore should be possible for an application that can use TP4 to be modified to use TCP. While it is feasible to modify the application, it requires that each possible application be modified. It is much simpler to just build a layer on top of TCP that provides a TP4 interface (that is an OSI transport layer interface) to an application and then provides the TP4 service using TCP. Since TCP and TP4 are very similar, this layer would not have to do much. Indeed a specification of such a layer has been developed and an implementation is available. With such a layer one can run many OSI applications in a TCP/IP environment, getting the benefit of the enhanced applications without having to abandon a mature TCP/IP network.

The basic structure of a protocol architecture that provides OSI transport service on top of TCP is shown in Figure 10.9. The packetization layer accepts packets from a higher layer and delivers a byte stream to TCP. The transport convergence layer accepts and delivers transport layer service primitives (such as CONNECT.request or DATA.indication). An interesting observation to make is that the interface to the packetization layer looks very much like the interface to a reliable network service. This is because TCP already provides reliable end-to-end delivery of sequenced data. A result of this is that the functions performed by the transport convergence layer are actually exactly those performed by a TP0 protocol.

SUGGESTED READINGS

The TCP protocol is defined in [POST81d]. Numerous changes to the original description and many studies have been conducted regarding TCP. A good starting point to this literature is the paper by Jacobson [JACO88] which describes modifications to the TCP flow control algorithm and the paper by Clark et al [CLAR89] which analyzes TCP processing requirements and the time spent performing each required step. More recently there has been some work on assessing the effect of the operating system environment on the performance of TCP [PAPA93]. UDP is described in [POST80b].

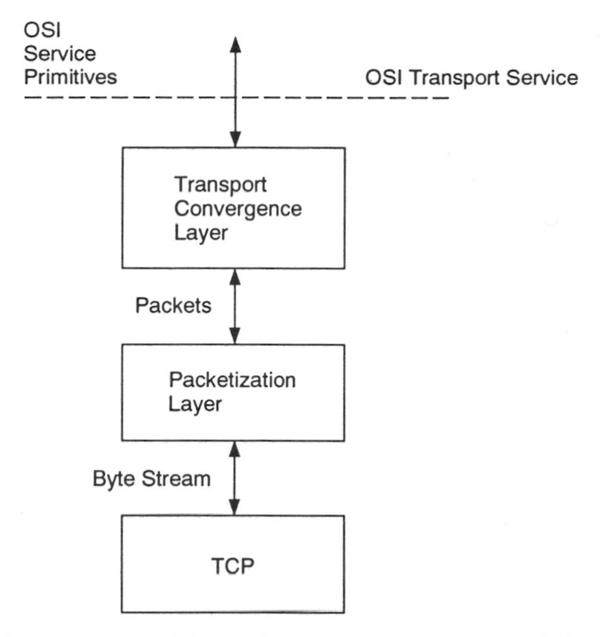

Figure 10.9. Providing an OSI transport service on top of TCP.

The details of the OSI transport protocols can be found in [OSI84a, b, c, d] and [OSI86a]. The provision of OSI transport protocols on top of TCP is described in [ROSE87].

In some instances transport protocol processing can become a bottleneck, particularly in high speed networks. The work in [LAPO91] is a good starting point to this literature. Work in this area includes [LAPO93], [ZITT93], [LIND93], [JAIN90a], and [HAAS91].

PROBLEMS

10.1 Explain the advantages of allowing multiplexing at the transport layer. What other protocols/layers that we have seen allow multiplexing? What would happen if multiplexing was disallowed in all protocol layers?

10.2 Consider a TCP layer using a 32-bit clock to derive its initial sequence numbers. How long will it be before a sequence number can potentially be used again assuming that: (1) the data rate is 1 Mbps? (2) the data rate is 1 Gbps?
What are the implications of this on using a protocol such as TCP when the data rate can be high?

10.3 Explain why the push function is not used on every packet transmitted or every packet received. In other words what are the advantages of having the transmitter or receiver determine when it is time to send or deliver a packet?

10.4 A source, *A*, sends a CR message to a destination, *B*. The CR message is delayed inside the network for a long time. *A* then retransmits a CR (after a timeout) and the connection establishment procedure is successful. After some data is transmitted the connection is closed. Some time later the original CR that had been delayed in the network reappears at *B*. *B* mistakes this for a legitimate new connection request from *A* and responds with a CC echoing the ID in the CR. At about the same time *A* decides to issue a new connection request to *B*. The new CR has a different ID than the previous CR. Explain how the three-way handshake can recover from this situation.

10.5 Figure 10.10 is a portion of a TCP conversation. Each packet is identified by sequence number, acknowledgment number, and window. All frames have a 64-byte data field.

a. Fill in the blank numbers in the conversation.

b. How many more frames can *B* transmit before receiving more credit from *A*?

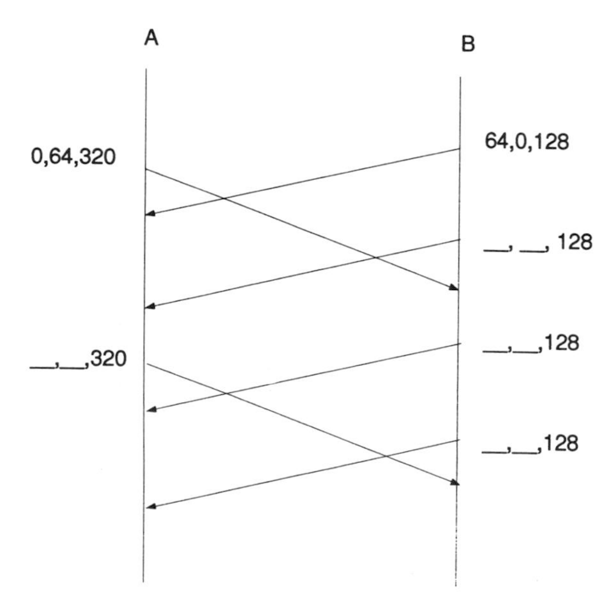

Figure 10.10. Problem 10.5.

10.6 Figure 10.11 shows part of a TCP conversation using the same convention as Problem 10.5. The initial credit given to the two sides is 256. The initial sequence numbers for *A* and *B* are 1001 and 501, respectively. Assuming each packet contains 128 bytes of data, find and correct all mistakes in the figure.

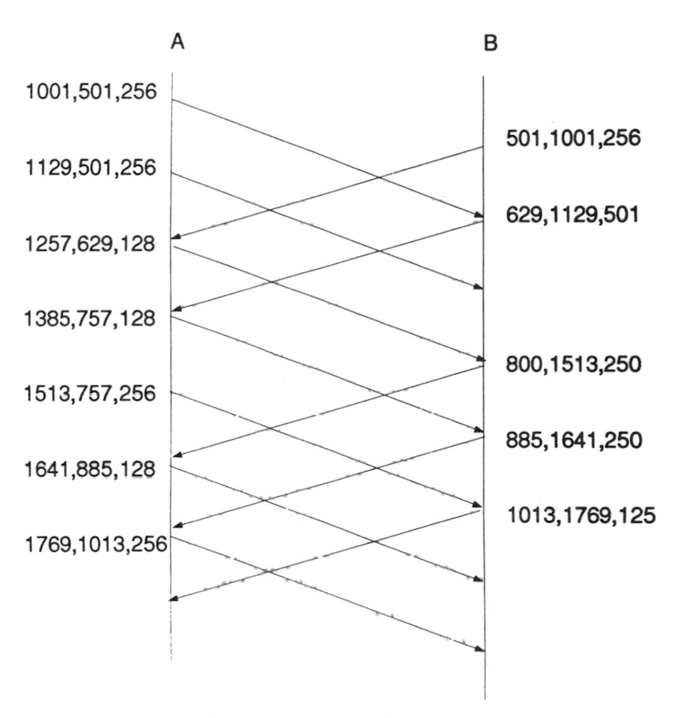

Figure 10.11. Problem 10.6.

10.7 Discuss, in some detail, the reasons that the round trip delay experienced by TCP might change.

10.8 Describe some approaches by which a transmitting transport protocol can determine the round trip delay it is experiencing.

10.9 In TCP, a receiver has the option of transmitting an acknowledgment immediately upon receiving a data packet or waiting for some time to see if this acknowledgment can be piggybacked on a data packet in the reverse direction.

a. Explain the advantages of each scheme.

b. Discuss the implications of waiting to see if the acknowledgment can be piggybacked on the functions that the transmitter and receiver have to perform.

10.10 The fact that OSI transport protocols do not allow for acknowledgment piggybacking can be a source of inefficiency. Explain why this may or may not be the case.

10.11 A multicast transport protocol is one that allows a source to send data reliably to multiple destinations rather than just to one. Discuss the various issues involved in the design of such a protocol. You may assume the existence of a connectionless multicast network layer. Address the following in your discussion:

a. What are the issues involved in connection management?

b. How can such a protocol work if the set of destinations are not known in advance?

c. How is reliable transmission defined in such a protocol and how can it be achieved?

d. What about flow control?

11

INTEGRATED SERVICES DIGITAL NETWORKS

11.1 INTRODUCTION

Since the introduction of the telegraph in 1850s and telephone networks in 1890s, a large variety of dedicated networks (dedicated to a single telecommunications service) have been developed and deployed over the world. Examples of dedicated networks are the telephone network, the telex network, the CATV network, and the packet-switched networks (PSNs). Each of these networks has been designed to meet the technical requirements of a specific service. For example, the public switched telephone networks has been designed using three main technical requirements of voice traffic: approximately 3 kHz of bandwidth, 3 minutes holding time per call, and 3 ccs traffic load (centi-call seconds; 1 Erlang* = 36 ccs busy hour traffic per time). Another example is the CATV network, which has been built around the units of 6-MHz channels that are needed for the video signal. To limit the proliferation of dedicated networks, there have been engineering solutions to integrate emerging new services in the existing dedicated networks. A typical example of an early service integration is the transmission of computer data over analog telephone networks by data modems. However, data transmission over the analog telephone network is limited by the approximately 3-kHz available channel bandwidth, resulting in a low transmission bit rate, as well as the lack of special features (such as low bit-error rate) needed by data customers.

With the development and availability of digital transmission and digital switching, a new and promising approach for an integrated public digital network became a reality. A significant development was the introduction of the 64-kbps digital rate, the basic channel rate of the standard PCM system in the evolving digital telephone network. The 64 kbps digital rate is large

*The traffic load in Erlangs carried by a transmission link is defined as the average number of calls per unit time multiplied by the average call holding time and is a dimensionless quantity. It is also equal to the average number of channels in simultaneous use in the link.

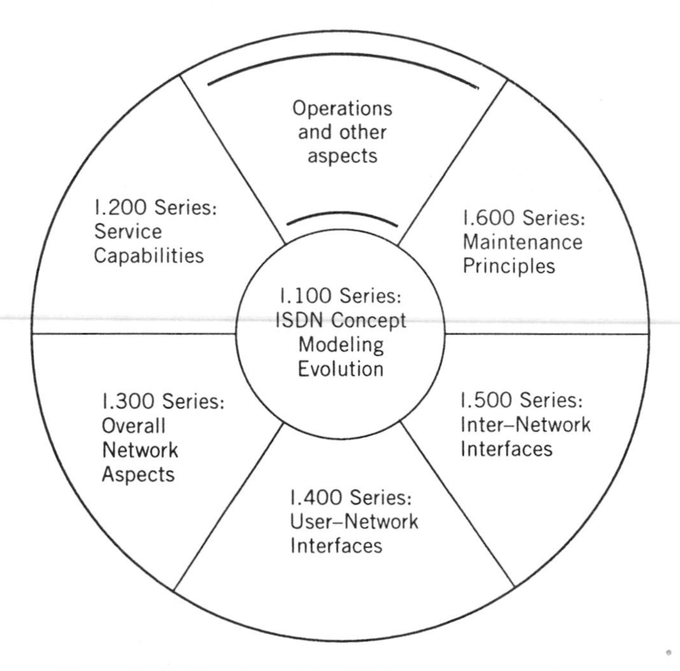

Figure 11.1. Structure of the I-Series recommendations.

enough to allow the integration of nearly all nonvoice services, except for high-bit-rate services, such as real-time video. This led to the envisioning of the integrated services digital network (ISDN) to provide the user with end-to-end digital connectivity to support a wide range of services including voice and nonvoice services over a single connection to the network. The single connection to the network is basically a limited set of standard multipurpose user-network interfaces.

In 1972, the vision of an ISDN emerged in the CCITT recommendation G.705 (CCITT uses the word "recommendation" rather than the word "standard"), which defined the conceptual principles of ISDN. In 1984, the 6th plenary assembly of the CCITT adopted the first *I-series* recommendations, which was a milestone in the development of the ISDN strategy. The I-series provides the principles and the guidelines for the ISDN as well as the detailed specifications of the user–network and internetwork interfaces. The I-series also contains references to other important CCITT standards, such as signaling in the Q-Series and data protocols in the X-series. Figure 11.1 shows the general structure of the I-Series recommendations, which include the following:

I.100 Series — General Concepts. The I.100 series recommendations provide a general introduction to the concept of ISDN. Recommendation I.130 proposes the attribute methods for describing the services available through ISDN. Each ISDN service (such as a voice service, data, etc.) is characterized by specific values assigned to each descriptive attribute. The definition and meaning of each attribute is given.

I.200 Series — Services. The I-200 series recommendations provide a classification and description of telecommunication services supported by an ISDN. Also, they provide the basis for defining the network capabilities required by an ISDN. ISDN services have been classified into: bearer services, teleservices, and supplementary services.

A *bearer service* is a type of telecommunications service that provides the capability for the transmission of signals between user–network interfaces. In other words, the bearer services are various forms of transport with attributes defining the capacity, the type of connection, and data communication protocols to be used. Bearer services are grouped into two groups: circuit-mode bearer services such as a voice service, and packet-mode bearer services such as a data service.

A *teleservice* is a type of telecommunication service that provides the complete capability, including terminal equipment functions, for communication between users according to protocols established between telecommunication administrations. Examples of teleservices are teletex (such as the interconnection of communicating word processors), telefax, videotex, telex, and multimedia communications (such as simultaneous voice and graphics communications).

In ISDN *supplementary services* additional capabilities are provided in conjunction with regular ISDN services (i.e., the bearer services and teleservices). Examples of ISDN supplementary services include call forwarding, call waiting, calling line ID presentation, call transfer, conference calling, and call holding.

I.300 Series — Network Aspects. Here, the functions and overall network aspects are defined. The main recommendations in this series are the I.320-Protocol Reference Model and the I.330-Numbering and Addressing Principles. The I.320 protocol reference model allocates functions in a modular fashion for the telecommunication protocol. The I.330 recommendations define principles for numbering and addressing in ISDNs.

I.400 Series — User – Network Interfaces. This series defines the interface between user equipment and ISDNs. The I.412 recommendation defines the various types of channels along the user–network interface, whereas recommendations I.430 and I.431 define the physical characteristics of the interface. Recommendations I.440–441, also known as Q.920–921, define the data-link-level procedure used on the D-channel (to be defined later). Recommendations I.450–451/Q.930–931 establish a flexible signaling procedure for establishing, modifying, and disconnecting calls.

I.500 Series — Inter-Network Interfaces. This series defines ISDN internetworking with various networks. The I.515 defines parameter exchange for ISDN networking, the I.520 provides general arrangements for network internetworking between ISDNs, and I.530 defines ISDN and the public switched telephone network internetworking.

I.600 Series — Maintenance Principles. This series defines the general maintenance principles of ISDN with applications to ISDN subscriber installation, basic access, primary rate accesses, and statistical multiplexed ISDN basic access.

11.2 ISDN CHANNEL DEFINITIONS

The information carrying capacity of ISDN user–network interfaces is defined in terms of channels that have a fixed bit rate across each interface. The user–network interfaces are located within the customer premises. Only three ISDN user–network interfaces are described:

Basic rate interface has an information-carrying capacity of 144 kbps simultaneously in both directions.

Primary rate interface has an information-carrying capacity of either 1.536 Mbps or 1.984 Mbps simultaneously in both directions.

Each of the three physical interfaces carries overhead bits leading to aggregate transmission rates of 192 kbps, 1.544 Mbps, and 2.048 Mbps, respectively. Each of these interfaces is composed of a number of channels. The various types of channels are the B, D, E, and H channels.

B (bearer) channels are 64-kbps channels that may be used to carry voice, data, facsimile, or image at 64 kbps or rate-adapted to 64 kbps (see below). The B channel may be used to provide access to a variety of different communication modes, such as circuit or packet switched.

D (demand) channels are mainly intended for carrying signaling information to control ISDN services. The D channel may be either 16 or 64 kbps depending on the specific interface. In addition to signaling, they may also carry other information such as packet-switched data.

E channels provide signaling for circuit-switched traffic and are only used at the user–network interface where multiple-access configurations are employed.

H channels are provided for user information at the primary rate interfaces. H0 channels are 384 kbps. There are two H1 channels: H11 at

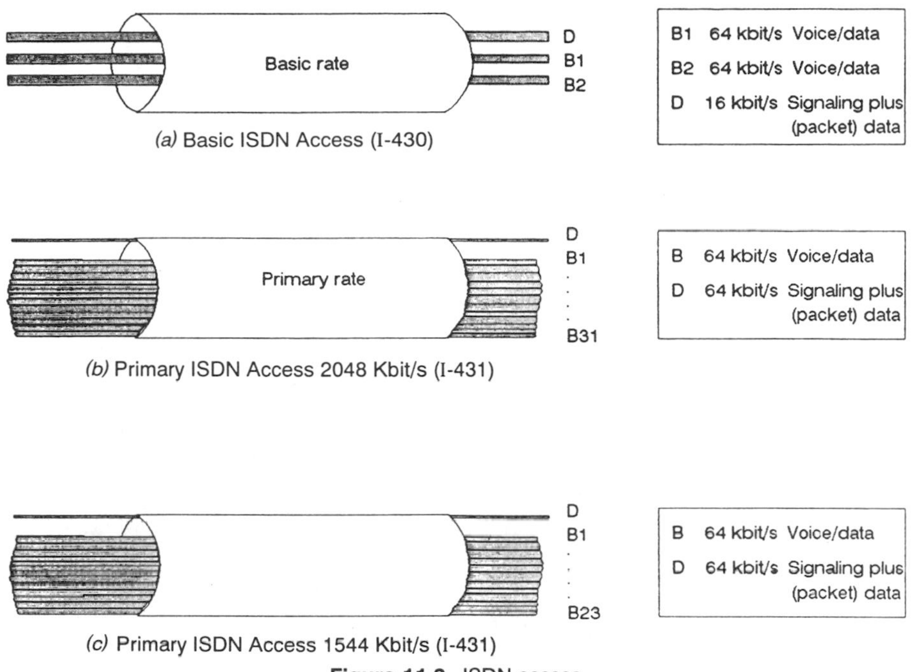

Figure 11.2. ISDN access.

1.536 Mbps, which will provide worldwide compatability, and H12 at 1.92 Mbps, which may be provided in countries using 2.048-Mbps primary rate transmission systems. H channels may be used for fast facsimile, video, high-speed data, and high-quality audio.

The B, D, E, and H channels may be packaged together within the basic and the primary rate interfaces (Fig. 11.2). The channel structure for the basic access interface consists of two B channels (called B1 and B2) and a 16-kbps D channel. Note that the network or the user terminal may not support all of these channels. Thus, the basic access capabilities are D, B and D, or 2B and D.

The channel structures for the 1.544- and 2.048-Mbps primary rates may be structured using B, H0, H1, or combined B and H0 channel interface, with the 64-kbps D channel for signaling as shown in Table 11.1.

11.2.1 Rate Adaptation and Multiplexing

For information streams with rates less than 64 kbps (such as a non-ISDN 9600-bps terminals or non-ISDN 16-kbps compressed voice telephones), they must first be rate adapted to be carried on a B channel. The I-460 Series recommendations describe procedures to adapt information streams of 8, 16,

TABLE 11.1 Primary Rate Interface Structures

Channel Interface	Primary Rate Interface 1.544 Mbps	Primary Rate Interface 2.048 Mbps
B (1 B = 64 kbps)	23B + D	30B + D
H0 (1 H0 = 384 kbps)	4H0[a] or 3H0 + D	5H0 + D
H1	H11[a]	H12 + D
Combined B and H0	*n*B + *m*H0 + D	*n*B + *m*H0 + D

[a]Signaling is provided by the D channel of another interface.
Note: all D channels are 64 kbps.

and 32 kbps to 64 kbps. For example, the first 1, 2, or 4 bits, respectively from a byte (an octet) in a B channel are used with the remaining bits set to binary 1. Also, streams at other rates are first rate adapted to 8, 16, or 32 kbps by the creation of a frame (similar to HDLC), with extra capacity filled in by flags.

Streams of 8, 16, and 32 kbps may be multiplexed into a single B channel, up to the total capacity of 64 kbps. This is done by permitting bits from different streams to be interleaved.

11.3 USER–NETWORK INTERFACE CONFIGURATIONS

A main objective of ISDN is to allow different types of applications and terminals to use the same interface simultaneously or alternately. To assist in specifying the ISDN user–network interface, the functions have been classified into generic groups, as shown in Figure 11.3. The five functions groups are as follows:

Network termination 1 (*NT*1) corresponds to a transmission line termination and includes functions associated with the physical layer such as line maintenance, and interface termination.

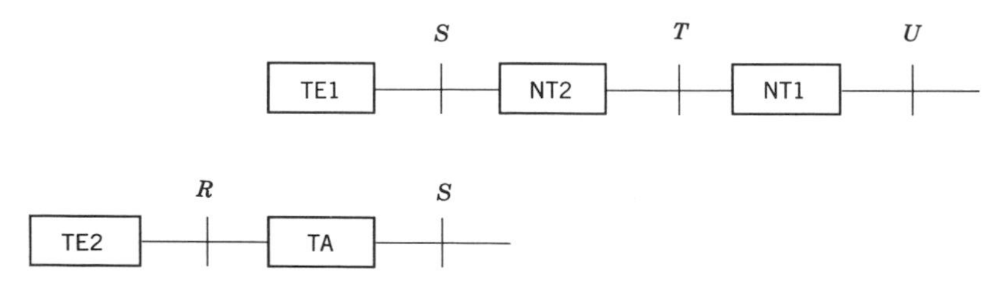

Figure 11.3. Functional groupings and reference points for the ISDN user–network interface.

Network termination 2 (*NT*2) corresponds to intelligent devices that may include physical, link, and layer functions. Network termination 2 devices may perform functions, such as protocol handling, switching, and concentration. Examples of NT2 devices are PBXs, LANs, and controllers to connect terminals within the customer premises to each other and to the network.

Terminal equipment type 1 (*TE*1) corresponds to devices equipped with ISDN interface (i.e., ISDN terminals), such as integrated voice and data terminals, digital telephone, picture-phone, and so on.

Terminal equipment type 2 (*TE*2) refers to terminals not conforming to the ISDN interface standard. Examples are those terminals based on existing recommendations such as X.21 and X.25

Terminal adaptor (*TA*) is needed to allow a TE2 (i.e., a non-ISDN terminal) to be connected to an ISDN user–network interface. This is analogous to the packet assembler and disassembler (PAD) facility allowing asynchronous terminals to interface with an X.25 packet switch.

Besides the functions grouping discussed above, the reference points help to define the ISDN user–network access configuration. The reference points, *points R*, *S*, *T*, and *U* in Figure 11.3, define the separations between the function groups. Reference point *R* (for rate) is the separation point between TE2 and TA. Reference point *S* (for system) corresponds to the interface of individual ISDN terminals. Reference point *T* (for terminal) corresponds to a minimal ISDN network termination at the customer's premises. Reference point *U* (for user) describes the subscriber line interface. The *U* reference point is not recognized in the I-Series recommendations and is valid mostly in the United States, where the regulations are such that some carriers cannot provide the NT1 functions on a customer's premises. In general, the physical location of the user–network interface may be at either reference point, *S* or *T*, according to the type of customer installation.

11.4 ADDRESSING

Terminals connected to an ISDN can be voice, circuit-switched data, packet-switched data, integrated voice/data, or picture-phone terminals. The terminals may have an ISDN interface or it may be connected to the ISDN by the terminal adapter (TA). An ISDN addressing plan provides the network with the ability to recognize any terminal on an ISDN port.

The main features of the ISDN addressing and numbering are as follows;

It is based on the telephone numbering plan. For example, the current telephony country code is the same in the ISDN numbering plan.

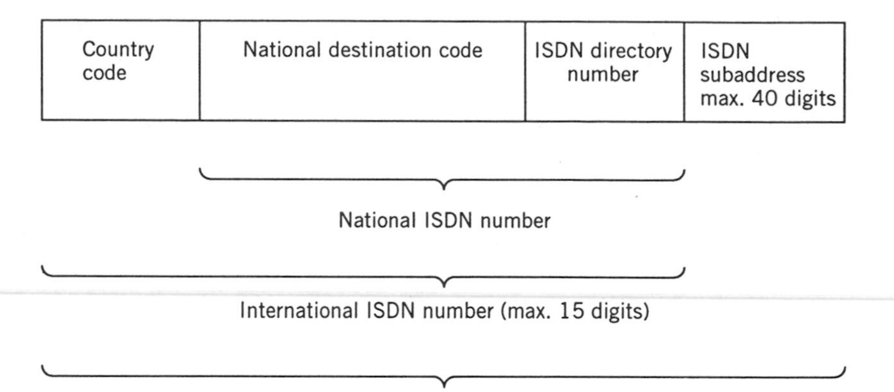

Figure 11.4. Structure of the ISDN address.

It is independent of routing. As an example, it is independent of the transit carrier selection.

It is independent of the service (voice, data, etc.) or the performance of the connection.

It is not alphanumeric.

It should allow for internetworking with only the use of the ISDN number.

Figure 11.4 shows the format of the ISDN address. The format includes an ISDN subaddress, ISDN directory, national destination, and country code. The maximum length of the ISDN address is 55 digits.

11.5 ISDN PROTOCOL LAYERS

The ISDN protocol structure is based on the layered protocol developed by ISO and CCITT, as shown in Figure 11.5. Recommendations I.430 and I.431 define the layer 1 characteristics of the user–network interface to be applied at the *S* or *T* reference points for the basic interface and the primary interface, respectively. Recommendations I.440 (Q.920) and I.441 (Q.921) deal with general aspects and specifications of the data link layer, respectively. Recommendations I.450 (Q.930) and I.451 (Q.931) deal with general aspects and specifications of layer 3, respectively. Below, we describe the three ISDN layers.

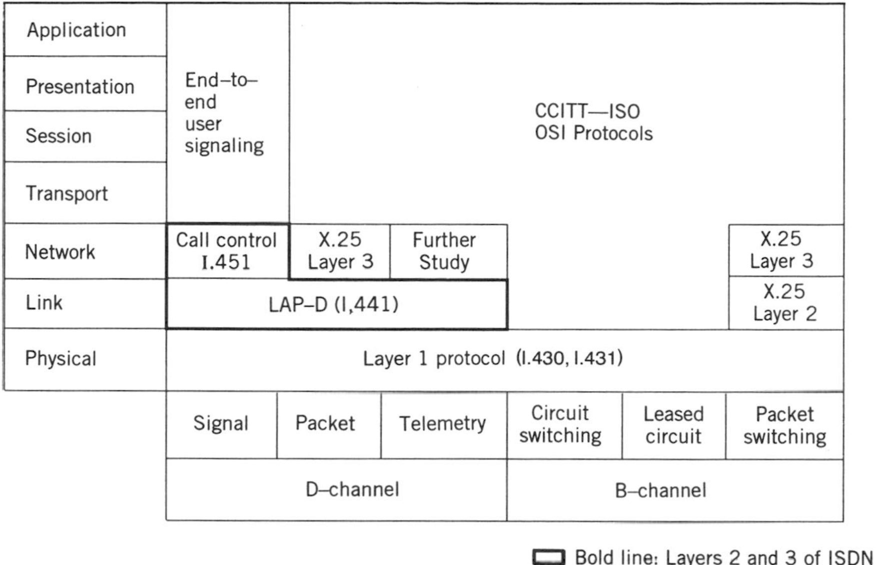

Figure 11.5. Layered protocol structure at the ISDN user – network interface.

11.5.1 Layer 1 — The Physical Layer

11.5.1.1 The Basic Rate Interface. The main function provided at the basic interface is the bidirectional transmission of the two independent B channels (user information at 64 kbps each) and the D channel (signaling or X.25 information at 16 kbps), which are time-division multiplexed over a four-wire interface (*S* or *T* reference points). Frames are transmitted full-duplex using a balanced interface, in which one pair of wires is used for transmitting, and another pair of wires is used for receiving. The I.430 recommendations specifies the ISDN basic rate connector, the so-called RJ-45, as shown in Figure 11.6. It has eight pins, only four of which are usually used. Pins 3/6 and 4/5 are used as the transmit–receive pairs.

Two possibilities are available for transfer of power across the interface. One possibility is to supply power over these transmit–receive pairs, that is, pins 3/6 and 4/5 (power source 1). The other possibility is to supply power over pins 1 and 2 (power source 3) or pins 7 and 8 (power source 2). Power sources 1 and 2 are provided by the NT, whereas power source 3 is provided by the TE.

Both point-to-point and point-to-multipoint modes of operation are supported in the basic rate interface (see Fig. 11.7). In a point-to-point wiring link, in which only one source and one sink are interconnected, the maximum distance between the TE and NT is 1 km because of the interface cable attenuation. In the point-to-multipoint wiring configuration (passive bus) (Fig. 11.7*b*), more than one source can be connected to the same sink or

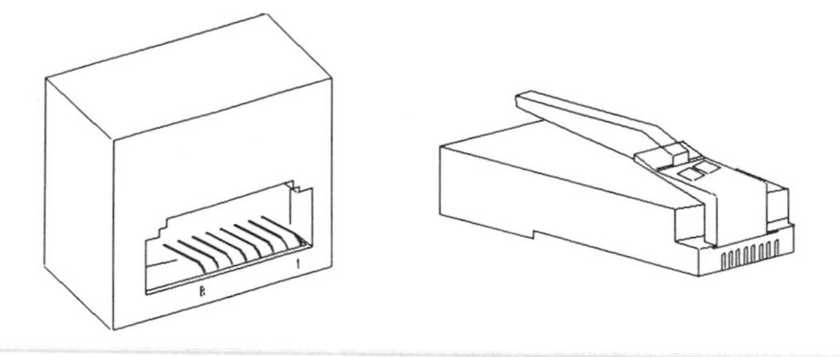

Eight-pole Jack Eight-pole Plug

(*a*) ISDN Basic Rate Connector

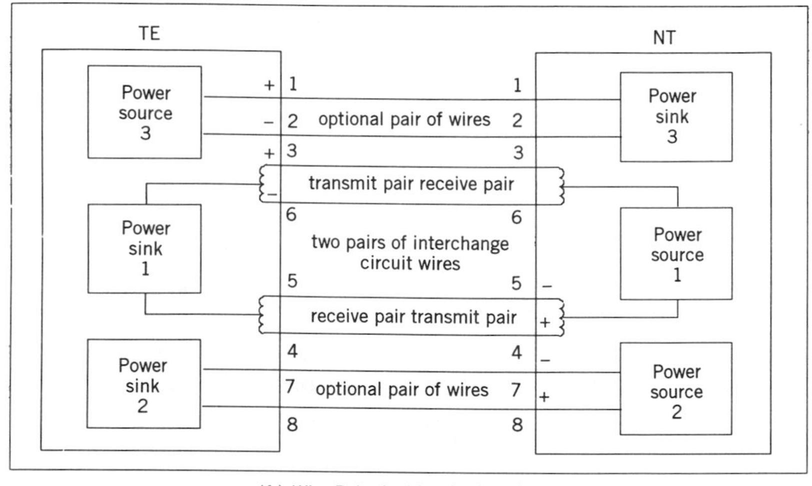

(*b*) Wire Pairs Insider the Interface Cable at S/T Reference Point

Figure 11.6. ISDN basic rate interface.

more than one sink to be connected to the same source. Up to a maximum of 8 TEs may be connected at random points along the interface cable. The maximum distance from the NT is on the order of 100–200 m. The distance can be extended to 500 m, if the terminals are grouped at the far end of the interface cable (see Fig. 11.7*c*), the extended passive bus.

Frame Structure. The digital signal is structured into layer 1 frames of 48 bits (Fig. 11.8). Frames are simultaneously flowing from both the TE and the NT. There are four octets for B1 and B2 channels, four bits for the D channel, and twelve overhead bits for control. The overhead bits are used differently depending on whether frame was transmitted from the TE or the NT. Thus, the frame length is 48 bits and of 250 μsecond duration, resulting in 192 kbps.

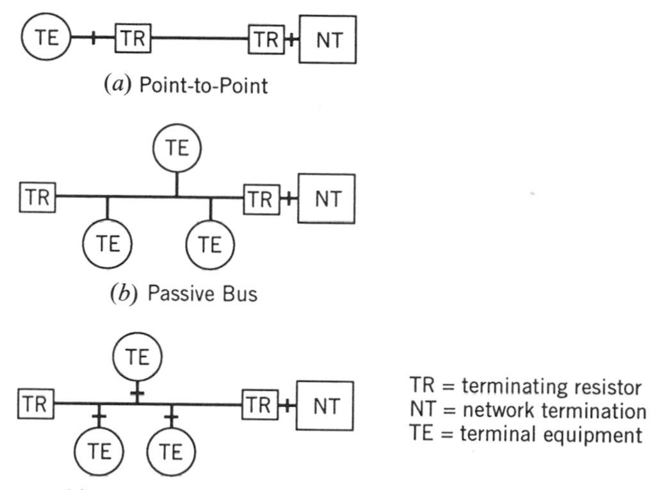

Figure 11.7. Wiring configurations at the basic interface.

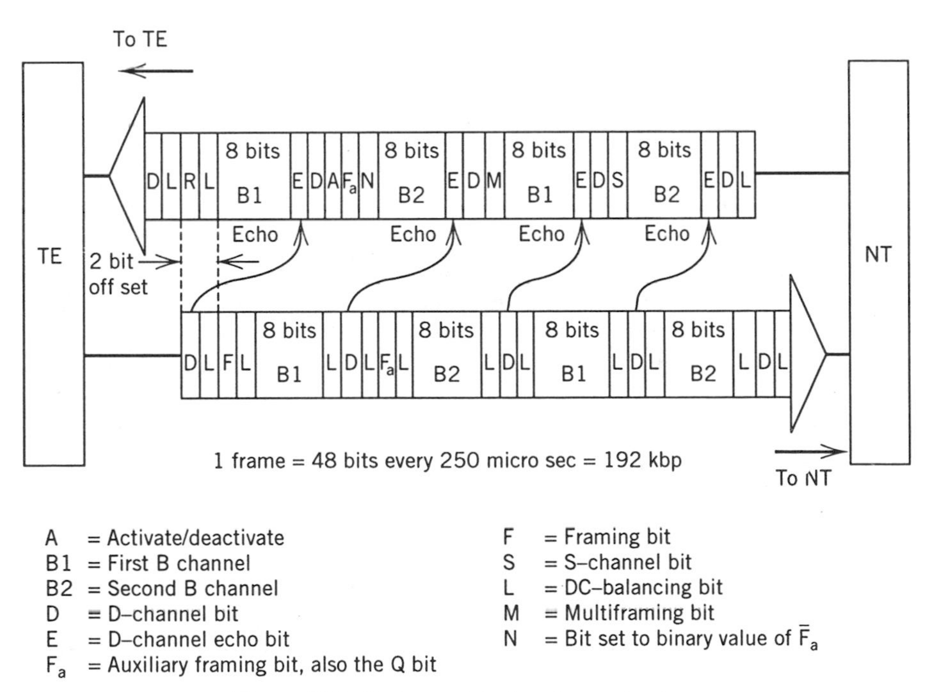

Figure 11.8. Basic rate interface frame structure.

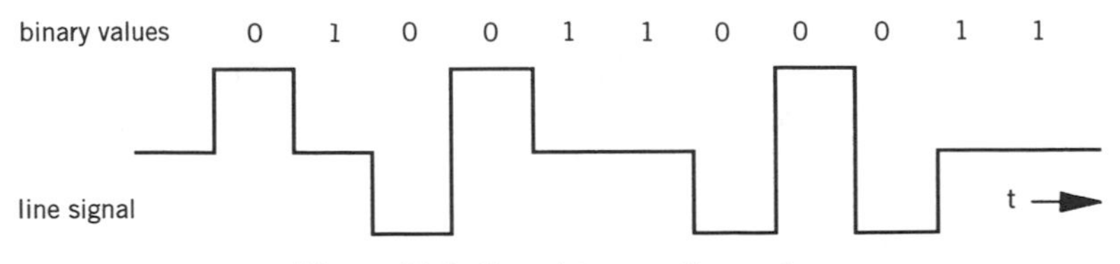

Figure 11.9. Pseudoternary line code.

Because a basic rate interface can support point-to-point or point-to-multipoint configuration, some of the control bits in the frames transmitted to the TE are used to arbitrate the use of the D channel. Frames transmitted by the NT contain four E bits, one A bit, two S bits, one L bit, one F bit, one Fa bit, one N bit, and one M bit. These are explained as follows;

E bits (*Echo*): These bits echo the values of the D-channel bits previously transmitted to the NT. This is used to arbitrate D-channel contention.

A bits (*activate/deactivate*): A-bits allow a terminal to come on line (activate) or place a terminal offline (deactivate).

L bit: This bit is used to DC-balance the whole frame.

M bits: These bits are multiframe bits.

S bits: These bits are used for functions such as rate adaptation.

F bit: This is the framing bit. The *Fa bit* is the auxiliary framing bit and the *N bit* is set to binary value equal to the complement of the Fa bit.

Frames transmitted by the TE to the NT have 8 bits for the bits that are used to provide individual DC balance for each B octet and D channel bits. This is needed to prevent undesired line code violations and to limit the DC content of the signal received by the NT in the passive bus configuration.

Notice that the TE synchronizes its transmission so that there is a 2-bit delay between its transmitted framing bit F and a received framing bit. The TE derives its timing (bit, octet, or frame) from the NT, whereas the NT derives its timing from the network clock.

Once the D channel has been accessed by a terminal, a possible collision with other terminals who have the same priority and accessing at the same time may occur. This situation can be detected and resolved as follows; when the TE transmits a D-channel bit to the NT, the NT echoes the D-channel bit in its next E-bit position. If the transmitted bit is the same as the received echo, the terminal shall continue its transmission. If, however, the received echo is different from the transmitted bit, the terminal shall cease transmission immediately and return to the D-channel monitoring state.

Regarding the line code, the pseudoternary* line code with 100 percent pulse width (Fig. 11.9) is used at the *S* and *T* reference points for the basic

*Considering the passive bus configuration, AMI (alternate mark inversion) line code, see Section 2.4.1, has been selected.

rate interface. The pseudoternary code uses three voltage levels to transmit binary ones and zeros. A binary one is represented by no line signal (zero voltage), whereas a binary zero alternates between positive and negative voltages.

11.5.1.2 Primary Rate Interface. As discussed earlier, there are two defined primary rate interfaces: the 1.544 Mbps North American, South Korean, and Japanese standard (23 B channels at 64 kbps each, one D channel at 64 kbps, plus overhead), and the 2.048 Mbps European standard (30 B channels at 64 kbps, one D channel at 64 kbps, plus overhead). Unlike the basic rate interface, both types of primary rates support only point-to-point configuration at the *S* and *T* reference points. The primary rate interface is also permanently activated, and thus it has no need for the activation/deactivation control bits or for arbitration of the D channel.

The I-431 recommendations describe the layer 1 characteristics of the primary rate interface, which has been largely derived from the existing G.703, G.732, and G.733 recommendations, discussed earlier in Chapter 2. The frame structure for the 1.554-Mbps and 2.048-Mbps interfaces are the same as discussed earlier. For the 1.544-Mbps interface, time slot 24 is assigned to the D channel or E channel when either of these channels is present. The B8ZS (bipolar with eight-zero substitution) line code is recommended. The frame structure for the 2.048-Mbps interface consists of 32 time slots, numbered from 0 to 31. Time slot 0 is assigned to frame alignment; time slot 15 is devoted to the D channel or E channel, when either of these channels is present.

11.5.2 Layer 2—Data Link Layer

Recommendations for ISDN layers 2 and 3 are applied to a variety of interface structures, such as the basic interface structure and the primary interface structure. As mentioned earlier, the protocol structure is based on the layered protocol model developed by the ISO and the CCITT (see Fig. 11.5). Layer 2, commonly known as *LAPD* (*link access procedure for the D channel*) is defined in the recommendations I.440 (General Aspects) and I.441 (Detailed Specification), which, are also known as Q.920 and Q.921, respectively. The LAPD is based on the balanced mode high-level data link control (HDLC), such as the X.25 LAPB, discussed earlier in chapter 4. The B channel protocols for the data link layer are not defined by ISDN standards and thus they can vary depending on the type of service.

LAPD provides layer 2 addressing, flow control, and error detection for the D channel. The flow and error-control procedures are very similar to X.25 LAPB, however LAPD is different in its addressing capabilities. Because the D channel controls all of the B channels on the interface as well as its own data transmission, LAPD has the capability of supporting multiple data link connections simultaneously (note that X.25 LAPB supports only one data link connection over one physical layer connection) at the data link

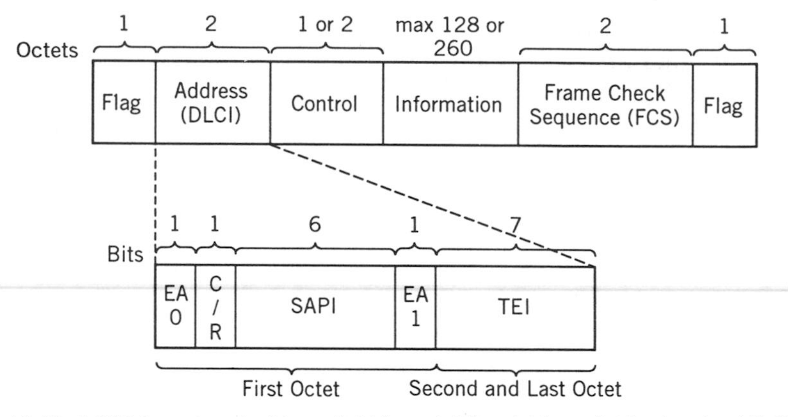

Figure 11.10. LAPD format and address field format. EA = Address field extension bit; C / R, command and response field bit; SAPI, service access point identifier; TEI, terminal endpoint identifier; DLCI, data link control identifier.

layer. Figure 11.10 shows the LAPD frame format and the address field format.

To identify a specific data link connection, the layer 2 address field, called the data link control identifier (DLCI), consists of two subfields, the service access point identifier (SAPI) subfield and the terminal endpoint identifier (TEI) subfield. It is the SAPI and TEI address combination that uniquely addresses or identifies the multiple logical entities on the D channel.

The SAPI identifies a point at which data-link-layer services are provided to layer 3. The SAPI field is 6 bit long, thus allowing for 64 possible service access point identifiers. The SAPI values assigned by Q.921 are shown in Table 11.2.

The TEI identifies a specific connection endpoint (whether it is a logical device or a group of logical devices) within a service access point. The TEI values range from 0 to 127 and are shown in Table 11.3.

TABLE 11.2 SAPI Values

SAPI Values	Layer 2 Service
0	Frame is carrying signaling information and call control procedures
1	Packet mode communications using Q.931 call control procedures
16	Frame is carrying an X.25 packet-mode user information
63	Frame is carrying layer 2 management information
32 through 47	Reserved for national use
all others	Reserved for future standardization

TABLE 11.3 TEI Values

TEI values	User Type
0 – 63	Non automatic TEI assignment user equipment (The TEI determines its own TEI)
64 – 126	Automatic TEI assignment user equipment
127	Group TEI (The NT determines the TEI values) broadcast data link, all 1s.

The TEI may be assigned automatically (by the network) by using a TEI assignment procedure, or it may be assigned at the time of subscription and can be entered into the user equipment by the user or the manufacturer. The network must verify the TEI value entered by the manufacturer to guarantee that the TEI is not already used by another user equipment.

Figure 11.10 shows the two-octet DLCI format. The 6-bit-long SAPI and the 7-bit-long TEI identify a unique data link. The first bit (EA, extended address) in each octet contains the address extension bit; a binary 1 indicates the last octet of the address, otherwise we have a binary 0. The second bit of the address field is the command/response bit.

11.5.3 Layer 3: Network Layer

Recommendations I.450 (General Aspects) and I.451 (Detailed Specification), also known as Q.930 and Q.931, respectively, define the layer 3 protocol. The main purpose of the network layer of the ISDN user–network interface signaling protocol is to establish, maintain, and clear network connections such as

1. Circuit-switched connections using the B channel
2. Packet-switched connections using either the D or B channel
3. User-to-user signaling connections using the D channel.

Below we first discuss the control signaling messages format exchanged over the D channel, and then we discuss the call control procedures for a circuit-switched connection as an example.

11.5.3.1 Layer 3 Message Format. Figure 11.11 shows the layer 3 Q.931 signaling message format. It contains a protocol discriminator, call references, message type, and information field. The protocol discriminator distinguishes messages for signaling (Q.931) from other types of messages. A protocol discriminator of 8 indicates Q.931, whereas a value of 10 indicates X.25 messages. The call reference value identifies the call at the local user–network interface. It does not have end-to-end significance. It is similar in function to the logical channel number (LCN) in the X.25 protocol and

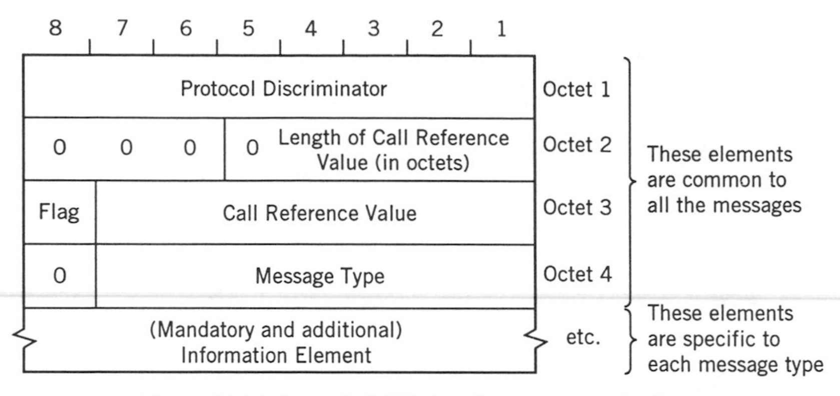

Figure 11.11. Layer 3 Q.931 signaling message structure.

may be one or more octets long. The call reference value is assigned at the beginning of a call and remains fixed for the duration of the call. The call reference flag identifies which end of the data link originated the call. The originator side always sets the call reference flag to 0, and the remote side sets it to 1. The message type field identifies the function of the message being sent, such as SET UP, CALL PROCEEDING, ALERTING, CONNECT, RELEASE. Bit 8 is reserved for possible future use. The information element field has two categories; single octet information element and variable-length information elements.

Circuit-Switched Calls Using the B Channel. Signaling information are transmitted on the D channel to establish a voice call on the B channel. A user initiates a call establishment by transferring a SET UP message across the user–network interface (see Fig. 11.12). The SET UP message is the I-field of the LAPD I-frame. It contains, among other information elements, the terminal number. Upon receiving the SET UP message, the local exchange (ISDN exchange) sends a message (using signaling system number 7, SS7) through the network to establish a route and allocate resources for the requested call. Also, it sends back a CALL PROC (call proceeding) message to the user to indicate that the call is being processed. The CALL PROC message contains the B channel allocated to the call. When the destination's local exchange receives a call, it transfers a SET UP message across the user–network interface. If a multipoint terminal configuration exists, this message will be sent using a broadcast capability at the data link layer. The called terminal responds with a CALL PROC message contained in a LAPD I-frame. The CALL PROC message has local significance and it is optional. ALERTING has end-to-end significance, and it alerts its user to the arrival of a call, for example by ringing a bell similar to the conventional telephone. When the called terminal answers the call (i.e., lifting the handset), the CONNECT message is sent to

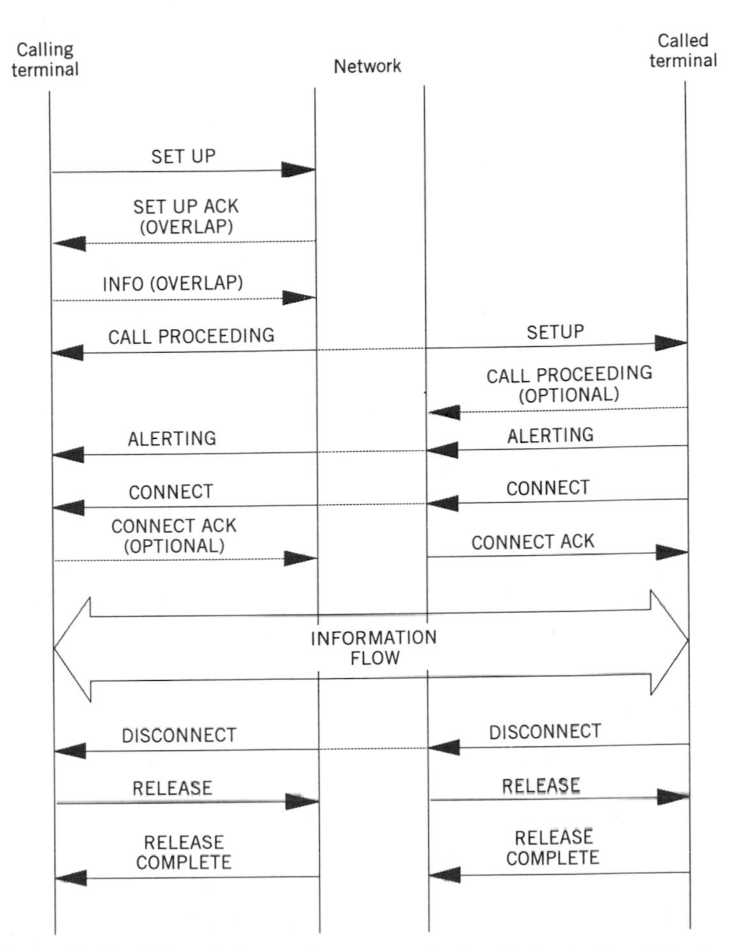

Figure 11.12. Call control procedure; example of a simple circuit-switched call.

the network signifying call acceptance. This completes call setup on the D channel and now the user information is exchanged over the B channel.

Either the user or the network can initiate call termination by sending a DISCONNECT message. The RELEASE and RELEASE COMPLETE messages are exchanged on a link-by-link basis. This releases the call reference number and the B channel for use by the next call.

11.6 SIGNALING

Signaling is basically the process of exchange of information related to the establishment and control of connections across the telecommunication network. Signaling is classified into user–network signaling, interoffice

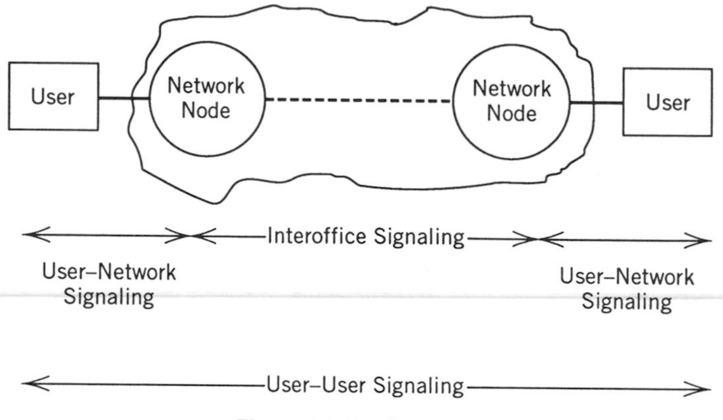

Figure 11.13. Signaling.

signaling, and user–user signaling (Fig. 11.13). User–network signaling is an access signaling and refers to the interaction between the user equipment and the local switch for call establishment. Interoffice signaling is a backbone signaling and refers to the interaction between the network nodes regarding the establishment and control of a call, and the allocation and management of network resources. User–user signaling deals with defining the communications parameters between the end users.

An example of interoffice signaling is the *common channel signaling system number* 7 (SS7). Common channel signaling uses dedicated channels for the exchange of signaling between the switching nodes. Thus, the common channel signaling (CCS) network can be viewed as a packet-switched network (PSN) overlaid on the circuit-switched voice network (Fig. 11.14). The primary components of CCS (or the SS7) are signaling end points (SEPs), or simply signaling points SPs, and signaling transfer points (STPs). The STPs are packet switches and responsible for routing signaling messages from one SEP (i.e., network node) to another. For reliability reasons, it is recommended that SEPs be connected by at least two STPs, known as a mated pair of STPs.

11.7 EVOLUTION TOWARD BROADBAND ISDN

As already stated, I-Series recommendations for ISDN provide two interfaces based on the 64-kbps digital rate: the basic rate interface at 144 kbps and the primary rate interface at 1.5 Mbps or 2.09 Mbps. Such rates can support a wide range of services; however high-bit-rate services such as image and video services cannot be provided on the 64 kbps ISDN, which is sometimes referred to as narrowband ISDN (N-ISDN). The advances in optical fiber, computing switching, and other technologies such as digital signal processing and video encoding techniques have stimulated a significant and rapid

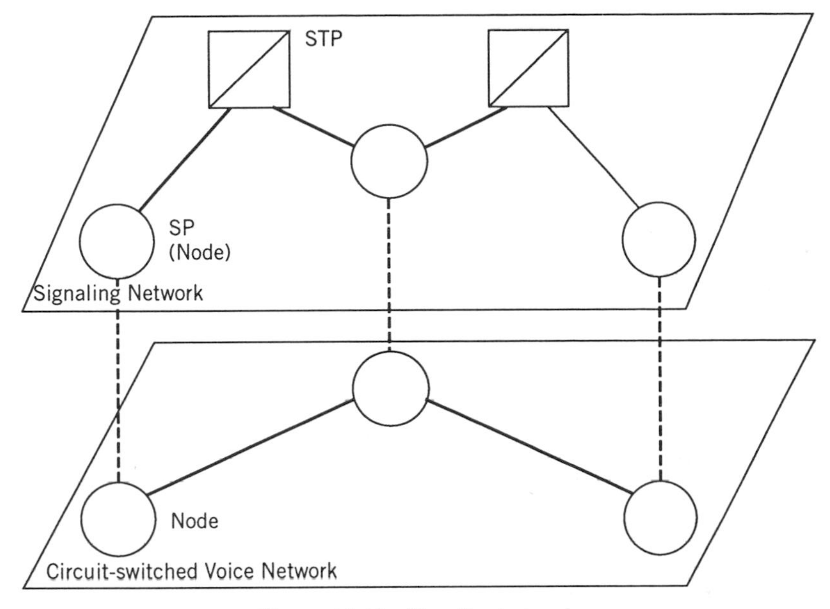

Figure 11.14. Signaling network.

increase in the demand for new communication services and made broadband integrated services digital networks (B-ISDNs) possible and potentially economic in the future. It is the increasing demand for these high-bit-rate services that primarily triggered the evolution of B-ISDN.

There is a wide diversity of services expected to be provided on BISDN. Examples of these services are full-motion video and high-definition television, image, videotelephony, videoconferencing, videotex, video surveillance, data, electronic mail, data transactions, voice, video and voice mail, LAN interconnection, and high-speed data communications. Figure 11.15 shows the ranges of services expected on broadband networks. All of these services demand high-speed transmission and switching within the network, and many require new signaling capabilities well beyond that of the current ISDN Q.931.

The need for services employing bit rates greater than 2 Mbps was clearly seen when the I-Series recommendations were written. Thus, CCITT formed the Task Group on Broadband Aspects of ISDN in 1985, to study other H channels and interface types above 2 Mbps. Today, the framework of broadband ISDN is clearer and much progress has been achieved in terms of meeting the goal of broadband ISDN.

The new synchronous digital hierarchy (SDH), with its counterpart in North America, the Synchronous Optical Network (SONET), specifies a worldwide universal *Network Node Interface* (*NNI*). In 1988, CCITT approved recommendations G.707, G.708, and G.709, which specified the bit rate,

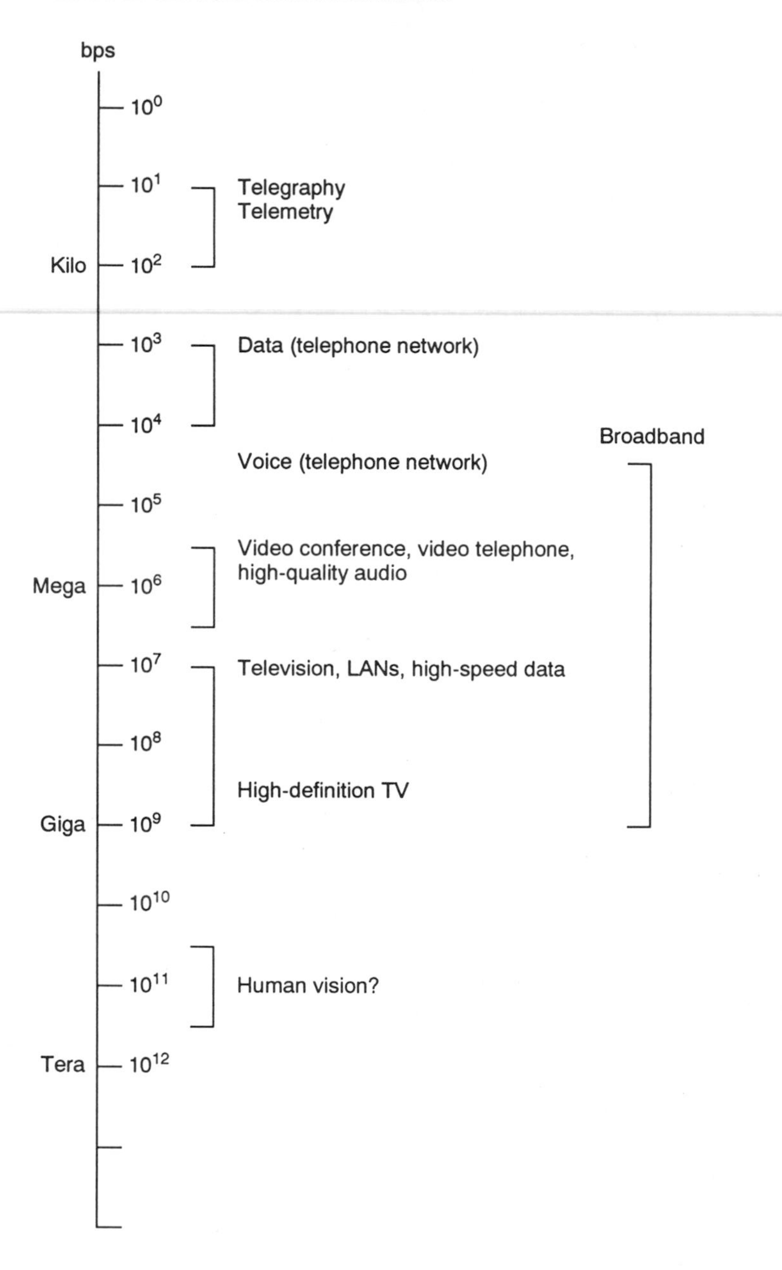

Figure 11.15. Ranges of services expected on broadband networks.

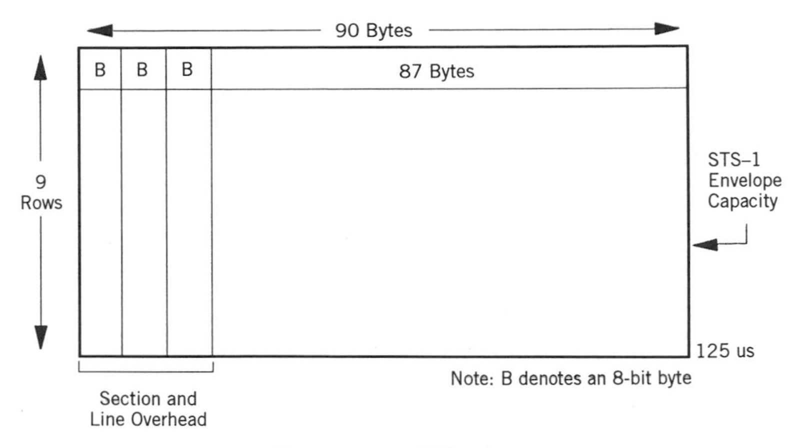

Figure 11.16. STS-1 frame.

format, multiplexing structure, and payload mapping for the NNI of the SDH.

Also CCITT B-ISDN standards have chosen the asynchronous transfer mode (ATM) as the transport technique for B-ISDN. CCITT standards for B-ISDN call the switching and multiplexing aspects the transfer mode. The ATM is a high-bandwidth, low-delay, packet-like switching and multiplexing technique. Below we discuss both the SONET and the ATM.

11.7.1 SONET

Single-mode fiber is becoming the medium of choice for high digital transport; however many proprietary fiber optic transmission systems cannot interconnect with each other. SONET specifies standard optical signals, a synchronous frame structure for the multiplexed digital traffic, and operations procedures. The synchronous transport signal-level 1 (STS-1) is the first level and the basic building block of the SONET signal hierarchy. The STS-1 has a bit rate of 51.84 Mbps. Figure 11.16 shows the STS-1 frame structure, which is $9 \times 90 = 810$ bytes. (Note that the frame is drawn as 90 columns and nine rows of 8-bit bytes. The order of transmission of the bytes is row by row, from left to right.) One entire frame is transmitted every 125 μseconds, which is equivalent to 8000 frames per second, leading to the transmission rate of 51.84 Mbps. Out of the 810 bytes, 27 bytes (the first three columns in the STS-1 frame) are allocated for various network transport functions, and the other 783 bytes are allocated to the synchronous payload envelope (SPE), which carries information. When the STS-1 is converted into an optical signal, this basic signal is then called an optical carrier-level 1 (OC-1) signal.

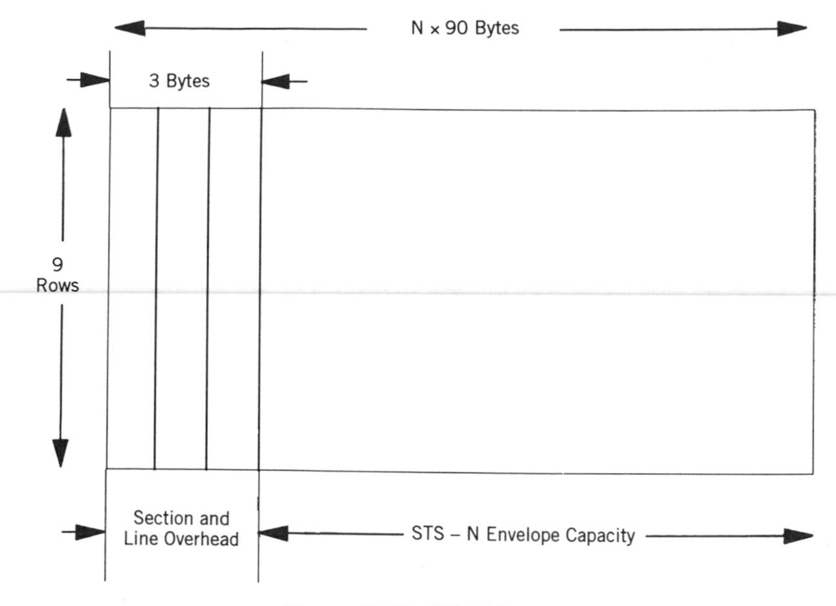

Figure 11.17. STS-*N* frame.

The OC-1 is the lowest level optical signal to be used by SONET equipment and network interfaces.

Higher levels of the SONET hierarchy are achieved by synchronously multiplexing N frame STS-1s to form an STS-N frame (see Fig. 11.17). Similarly the STS-N signal is converted to an optical carrier-level N (OC-N) signal. The OC-N has a transmission rate exactly N times that of an OC-1. Certain values of N are preferred; Table 11.4 shows the OC-N levels allowed by the American national standard. For SDH (CCITT standards), the lowest rate is 155.52 Mbps designated STM-1, which corresponds to STS-3.

TABLE 11.4 Allowable Levels of the SONET / SDH Hierarchy

OC Level	SONET Designation	CCITT Designation	Payload rate (Mbps)	Line Rate (Mbps)
OC-1	STS-1		50.112	51.84
OC-3	STS-3	STM-1	150.336	155.52
OC-9	STS-9	STM-3	451.008	466.56
OC-12	STS-12	STM-4	601.344	622.08
OC-18	STS-18	STM-6	902.016	933.12
OC-24	STS-24	STM-8	1202.688	1244.16
OC-36	STS-36	STM-12	1804.032	1866.24
OC-48	STS-48	STM-16	2405.376	2488.32

11.7.2 Asynchronous Transfer Mode

The selection of the transfer mode was crucial in the BISDN recommendations. The transfer mode must have the capabilities to handle the followings:

Both narrow-band and broadband rates;

Both stream traffic (such as voice and video) and bursty traffic (such as interactive data)

Unforeseen demands; and

Delay and/or loss sensitive quality requirements.

Neither circuit mode nor packet mode can meet all of these requirements. Two similar approaches have emerged; asynchronous transfer mode (ATM) and fast-packet technology. Both terms are widely used as synonyms, because they are very similar. Fast-packet technology applies to both transmission and switching systems (Chapter 12 is dedicated to fast-packet switching). The ATM has the advantages of two different types of digital multiplexing; one is the packet multiplexing and the other is asynchronous time division multiplexing (ATDM). ATM is similar to synchronous TDM in the sense that it uses a slotted time format; however, it does not allocate time slots on a fixed per call basis, but rather on a dynamic basis. In synchronous TDM, the position in a frame implicitly determines the association within the information stream. However, in ATM the association with an information stream is made explicit through the header or label on the data unit. The multiplexed information flow is organized in fixed size data units called cells. Fixed size cells, as opposed to variable size cells, reduce protocol processing, and enable hardware switching. Although SONET provides physical transport at data rates that are multiples of 51.84 Mbps, ATM allows the SONET payload to be flexibly allocated to a wide range of applications with varying bandwidth needs.

Table 11.5 shows the 1992 CCITT recommendations on B-ISDN, and Figure 11.18 shows the BISDN ATM protocol structure. The protocol structure (described in the I.321 recommendations) consists of three separate planes: the user plane, the control plane, and the *management plane*. The *user plane* provides for user information flow along with its associated control (such as flow control and error recovery). The *control plane* includes the call and connection control functions. It handles the signaling functions necessary for call and connection setup, supervision, and release. The *management plane* provides the plane management and layer management functions. The layer management performs layer-specific management functions while the plane management performs management and coordination functions related to the complete system.

Both the user plane and the control plane have a layered architecture as in OSI. The layers are the physical layer, the ATM layer, the ATM adaptation layer (AAL), and the higher layers.

TABLE 11.5 1992 CCITT Recommendations on BISDN

Recommendation	Title
I.113	Vocabulary of terms for broadband aspects of ISDN
I.121	Broadband aspects of ISDN
I.150	BISDN ATM functional characteristics
I.211	B!SDN service aspects
I.311	BISDN general network aspects
I.320	BISDN protocol reference model (1988)
I.321	BISDN protocol reference model and its application (1991)
I.327	BISDN functional architecture
I.35b	BISDN performance
I.361	BISDN ATM layer specification
I.362	BISDN ATM adaptation layer (AAL) functional description
I.363	BISDN ATM adaptation layer (AAL) specification
I.413	BISDN user – network interface
I.432	BISDN user – network interface physical layer specification
I.610	Operation, administration, and maintenance principles of BISDN access

The *physical layer* allows either the use of the SDH format or a cell-based format at the user–network interface (UNI). The UNI is defined in recommendations I.413 and I.432. The reference configuration at the UNI for narrowband ISDN (NISDN) is also applicable to B-ISDN. Two interface bit rates are defined; 155.52 Mbps and 622.20 Mbps. The 155.54 Mbps data rate can handle one or more video channels depending on the video coding technique and the video quality. The 622.08 Mbps data rate can handle multiple video channel distribution, such as in videoconferences.

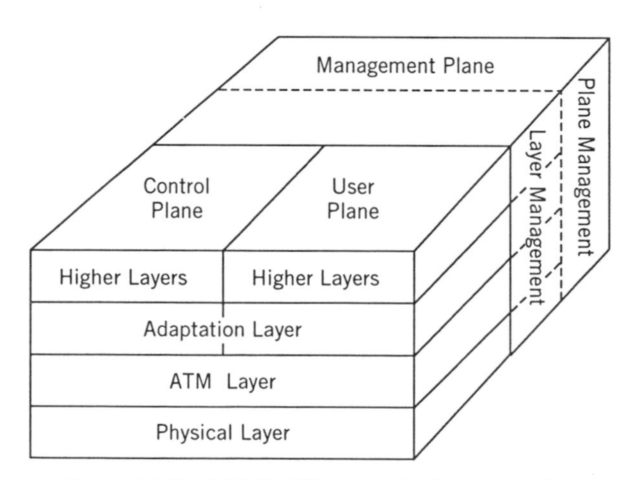

Figure 11.18. BISDN ATM protocol reference model.

There are two approaches to the definition of the ATM underlying transmission media. The first approach adopts the SDH (e.g., SONET) whereas the second approach favors the asynchronous time division (ATD). The ATD approach stems from the same principle as the ATM. It does not require a specific frame structure across the UNI. However, it does require a framing pattern within the ATM cell stream. Transmission overheads are carried in the cells and therefore limits the total usable capacity by the offered load.

The SONET approach can support both ATM and synchronous transfer mode (STM) payloads, which is useful only in the earlier stages of BISDN implementation. SONET's cross-connection transport service is not as flexible as the virtual path identifier approach (see the discussion on ATM Layer below) employed in pure ATM networks, and statistical bandwidth allocation per link capacity is not supported.

Because SONET requires more circuit switching, it can support some classes of traffic that may be more suitable for circuit switching such as broadcast television. However, the ATD approach, combined with efficient variable-bit-rate coding methods and network flow control techniques, can deliver the same service without circuit switching capabilities. The SONET provides the capability to transmit a combined stream of several ATM streams, for example, four ATM streams at 155 Mbp/s can be combined into a single 622 Mbp/s (STS-12) SONET stream.

The ATM layer provides cell transfer capabilities and is common to all services. Figure 11.19 shows the ATM cell header structure at both the UNI and the network–network interface (NNI). The cell size is 53 bytes long; the header size is 5 octets, and the information field is 48 octets. The primary role of the header is to identify cells belonging to the same information stream (same call or connection).

Information transfer is connection oriented. The ATM uses two connection concepts; the virtual channel and the virtual path. A *virtual channel* provides a logical connection between end users. A *virtual path* defines a collection of virtual circuits traversing the same path in the network. The cell header consists of the following

Generic flow control (GFC) field is a 4-bit field defined at the UNI to assist the user in controlling their traffic flow according to a certain quality of service. There is no GFC at the NNI.

Virtual channel identifier (VCI) is 12 to 16 bits at the UNI and 16 bits at the NNI. It identifies a particular end-to-end switched connection and deals with the switching functions of cells belonging to a certain logical connection. The value of the VCI may change as the cell traverses the network.

Virtual path identifier (VPI) is 8 to 12 bits at the UNI and 12 bits at the NNI. It consists of a bundle of virtual channels (see Fig. 11.20) that are

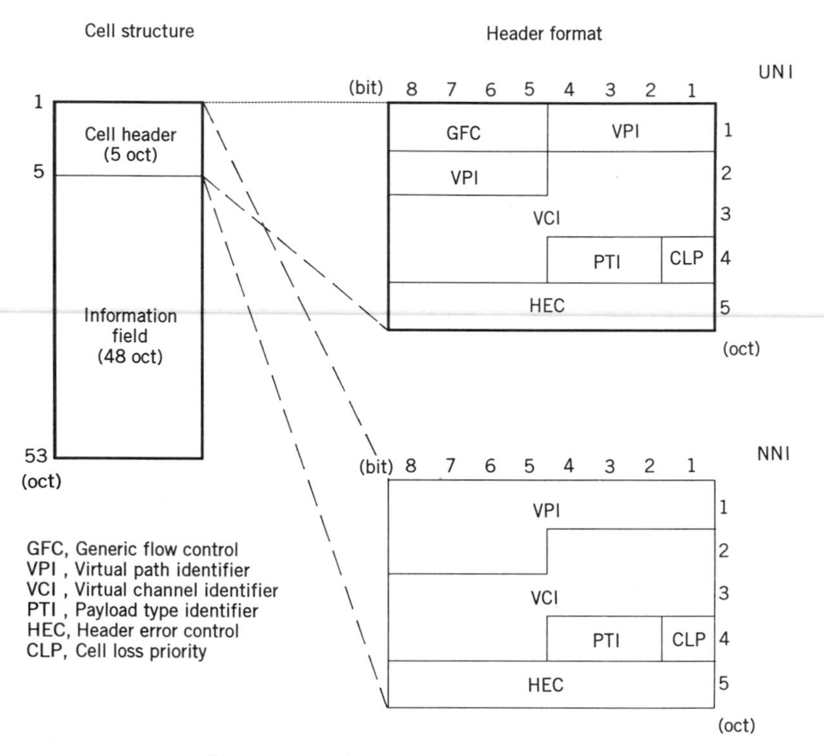

Figure 11.19. ATM cell header structure.

transported on the same physical media, from one end to the other. It deals with the cross-connection functions of the cells. The VPI emulates the functions of the trunk concept in circuit switching, and it greatly enhances the concept of dynamic routing and resource management according to the traffic-required quality of service; this subject is discussed further below.

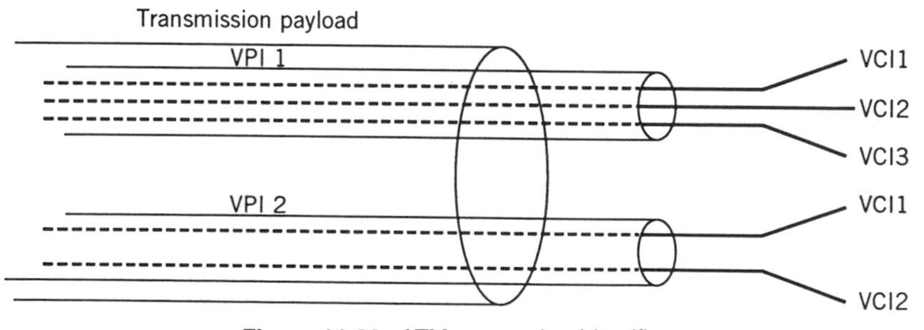

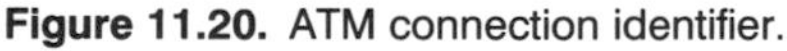
Figure 11.20. ATM connection identifier.

Payload type (*PT*) is a 3-bit field and the payload type identifier (PTI) is used to distinguish network information from user information.

Cell loss priority (*CLP*) is a 1-bit field and is used to indicate the loss priority by the end point and to indicate selective cell discarding in network switches. Users can set the CLP bit to indicate a lower priority cell (CLP value = 1), which may be subject to discarding depending on the network traffic conditions. If the CLP is not set (CLP value = 0), the cell has normal priority for traffic congestion control.

Header error control (*HEC*) is a one-byte field used for error detection and correction on the header. Thus, the transmitter calculates the HEC value for the first four octets of the cell header, and inserts the results into the HEC field, the last octet of the header. The generator polynomial is $x^8 + x^2 + x + 1$.

The concept of virtual path provides the tool to implement an efficient and flexible traffic control functions at the network level. As mentioned earlier, a virtual path is a logical connection between the virtual path terminators, which is composed of a bundle of virtual channels, also known as virtual circuits (see Fig. 11.20). Thus virtual paths define the cross-connection functions across the network, whereas virtual channels (or circuits) are concerned with switching and connection-establishment functions. The virtual path concept has the following characteristics:

1. A predefined route is associated with each virtual path in the physical network.
2. A virtual path is identified using a label attached to the cell (the VPI). Each VPI has a local significance over the link, it does not provide a global significance over the network. The reasons behind this assignment method are to keep the VPI length small because there is an upper limit to the number of multiplexed paths per link, to avoid management of virtual paths in a centralized manner, and to provide flexibility in assigning virtual paths per physical link.
3. Each virtual path is assigned a certain bandwidth that determines the number of virtual channels it can support. The bandwidth allocated can be either deterministic or statistical. In the deterministic case the virtual path is called a labeled deterministic path (LDP), whereas in the statistical case it is called a labeled statistical path (LSP). The LSP has several advantages over the LDP, such as the ability to exercise path bandwidth control, improvement of the transmission efficiency by exploiting the statistical multiplexing gain, and optimum allocation of bandwidth according to the traffic characteristics of each call.
4. Virtual paths are statistically multiplexed on the physical link on a cell multiplexing basis.

The virtual path concept has also provided several advantages that have contributed to the realization of flow control functions at the network level:

1. Elimination of the transit node processing per call setup, thus simplifying the node functions and providing fast switching per cell. No processing is required at the transit nodes when the path capacity is allocated or changed. Also, all switching fabrics need only simplified functions.
2. Separation of the logical transport network from the physical transmission network, thus providing flexibility in performing traffic management. For example, we can change the virtual path capacity without affecting the physical interface structure. This feature will greatly simplify the network architecture.
3. Direct multiplexing of virtual paths with different capacities while using simple hardware and software. There is no need to employ hierarchical multistage multiplexers as with the STM case. Dynamic path routing and dynamic path bandwidth allocation are now simple to implement.
4. Statistical bandwidth allocation for the calls per virtual path provides efficient utilization of the total link capacity.

ATM Adaptation Layer (AAL). This layer supports higher layer functions of the user and control planes and is service dependent. In other words, the AAL supports information transfer protocols that are not based on ATM, and it thus represents the link between many B-ISDN services and ATM. For example a 64 kbps voice connection is an application to be used over ATM. In this case, the digitized voice samples (the output bits of the pulse code modulation) must be assembled into cells for transmission and reproduced smoothly and with a constant flow of bits at the receiver. Since a variety of services are expected to use AAL capabilities, four classes of service are defined (see Table 11.6): A, B, C, and D. The criteria used for their classifications are bit rate, connection mode, and timing relationship between the source and destination. Class A service is a constant bit rate and

TABLE 11.6 BISDN Service Classes

Criteria	Class A	Class B	Class C	Class D
Bit rate	constant	variable	variable	variable
Connection mode	connection-oriented	connection-oriented	connection-oriented	connection-less
Timing relationship between source and destination	required	required	not required	not required
Examples	circuit switched voice	VBR video	X.25	LAN interconnection

connection-oriented service, and a timing relationship is required. An example of class A service is circuit-switched traffic, such as voice and constant-bit-rate video services. Class B service is a variable-bit-rate and connection-oriented service, in which timing is required. A good example of a class B service is variable-bit-rate (VBR) video. Such VBR video may be used in video teleconferencing in which the bit rate varies depending on the scene activity. Both C and D class services are variable bit rate, but timing is not required. However, class C is connection-oriented, whereas class D is connectionless. Data transfer applications are examples of class C and D services. Hence, data services that are connection-oriented, such as X.25 data services and signaling services, are class C services, and those that are connectionless, such as LAN interconnection, are class D services. To support the four classes of service, four AAL types are available; type 1, 2, 3/4,* and 5. AAL types 1 and 2 correspond to service classes A and B respectively. Both AAL type 3/4 and type 5 can be used for class C and D.

The AAL for each type is subdivided into two sublayers: the segmentation-and-reassembly (SAR) sublayer and the convergence sublayer (CS). The prime functions of the SAR sublayer are the segmentation of user data units into ATM cells and reassembling of cells into information blocks. The convergence sublayer is service dependent and provides for error detection, flow control, and other functions.

SUGGESTED READINGS

A number of text books are now available on ISDN, such as [RONA88], [STAL89], and [VERM90]. [DECI86] is an excellent reference for ISDN. [CCIT84] provides CCITT standards for ISDN. [HEWL90] is a good reference for ISDN testing equipment and protocol analyzers. [JABB91], [MITR91b], [MODA90], [KEAR90], [LAZA94] and [BOLO94] are detailed references for common channel signaling SS7.

Regarding broadband ISDN, [CCITT88a, b, c] are the CCITT Blue Book for the new synchronous digital hierarchy (SDH). [ACAM94], [WU91], and [MCQU90] are good references on BISDN, SDH, and ATM. [DEPR90], [SOHR91], [DAY91], [ATMF93], and [PARU93] are good references on ATM. [JOHN93] provides a survey of ATM technology. [HABI92], [DEPR94], [DUB094], and [ZARR94] discuss multimedia traffic. [MIKI90] offers an international perspective on the evolution toward SDH. [BALL89] and [IEEE93a] provide a detailed look at SONET. [TANT94a & b] discusses different topics in high speed networks.

*The distinction between type 3 and type 4 is so minor that they were merged into one single 3/4 type.

PROBLEMS

11.1 **a.** Can the D channel carry voice traffic? Why?

b. In the ISDN basic rate interface, what is the maximum number of possible voice channels? and what is the maximum number of possible data channels?

11.2 What are the minimum bit rate and the corresponding structure of the ISDN user network interface needed for each of the following terminals;

a. A personal computer with an output rate of 16 kbps

b. A video teleconferencing terminal with an output bit rate of 192 kbps (compressed video).

11.3 The D channel access control allows for a number of terminals (TEs) connected to the NT in the passive bus configuration to gain access to the D channel in an orderly fashion. Assume two TEs (TE1 and TE2), that have the same priority, access the D channel at the same time. TE1 bit stream is 10110011 and TE2 bit stream is 10111011. Show how the collision is detected and resolved.

11.4 **a.** In the basic rate interface, ISDN standards specify the procedure for activation and deactivation of terminals from the network side and only activation of the NT from the terminal side. Explain why deactivation of the NT by the terminals has not been specified.

b. Explain why in the primary rate interface there is no need for the activation or deactivation control bit or the E bits.

11.5 Summarize the differences between X.25 LAPB and ISDN LAPD protocol functions and format.

11.6 Two TEs (TE1 and TE2) are connected on the same physical line to the network. Two logical links are established on the D channel for each of the two TEs, and TE1 had an additional logical data channel on the B channel. There is also a broadcast logical channel. Choose appropriate SAPI and TEI values for the logical connections on the D channel.

11.7 To emphasize the need for B-ISDN, obtain the transmission time for a 50 Mbyte file at the following speeds:

a. 9.6 kbps modem

b. 64 kbps ISDN B channel

c. 1.54 Mbps full ISDN primary rate or the T1 rate

d. 155 Mbps B-ISDN access line

11.8 Compare the protocol overhead percentage for the following protocols;

a. ATM

b. SONET

c. ATM over SONET

d. Conventional LAN such as FDDI

11.9 A voice connection is an example of a service that is to use the ATM adaptation layer for compatability with ATM. Provide another example for a service that needs to be adapted to ATM format for information transfer and discuss.

11.10 Flow control schemes that rely on the end-to-end exchange of control messages in order to regulate traffic flow would not work in high speed networks. Provide an example to support this argument. [Hint: consider two ATM nodes which are 100 km apart and are connected by a 100 Mb/s Link].

12

SWITCHING TECHNIQUES AND FAST PACKET SWITCHING

12.1 INTRODUCTION

There are several forms of switching technique, the most widely known being circuit and packet switching. Other techniques are multirate circuit, fast circuit, burst, and fast packet switching.

By the end of the last century, the idea of communication switching became important with the proliferation of the telephone. Today, one may call over extensive networks of compatible switching nodes and transmission facilities to reach any of over 500 million telephone sets. As discussed earlier in Section 1.5, the most popular technique for switching voice is circuit switching. Circuit switching provides for dedicated paths between two users for the duration of a call and it is suitable for delay-sensitive services such as voice. Telephone switching has changed over the years from manually operated switchboards through numerous types of electromechanical systems to the current variety of stored-program electronic-switching systems.

Figure 12.1 shows the evolution of switching systems. The first experimental digital system was proposed at Bell Laboratories in 1959. The No. 4 Electronic Switching System (4ESS) is a toll and tandem system that was first placed in service by the Bell system in Chicago in 1976. As mentioned in Chapter 1, the first experimental packet-switching system was setup under the sponsorship of the Advanced Research Projects Agency (ARPA) in 1966. Research on asynchronous transfer mode (ATM) switches and fast packet switches is underway, and commercial products are already available. Photonic switching is still in its infancy and is expected to mature in the next century.

12.2 DIGITAL CIRCUIT SWITCHING

Circuit switching can be accomplished by two common techniques; space-division switching and time-division switching. In space-division switching,

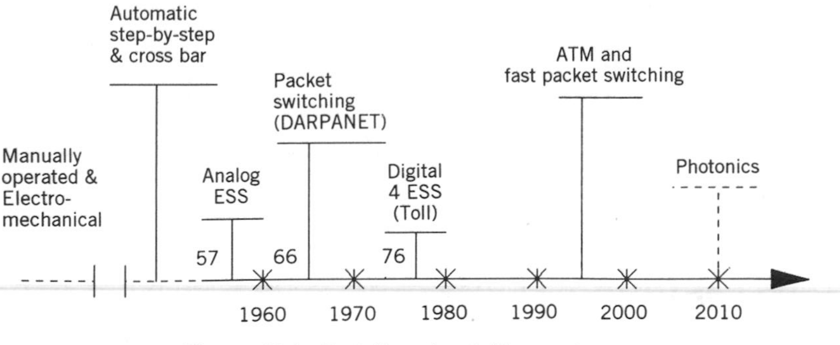

Figure 12.1. Evolution of switching systems.

each connection occupies a physically distinct and identifiable path. All earlier switching systems were space-division switches. In general, the trend in recent years has been toward time-division because of the widespread use of digital systems. However, both time- and space-division stages can be incorporated into the switch architecture.

12.2.1 Space-Division Switching

Figure 12.2 shows examples of a single-stage space-division switch. A simple rectangular $n \times m$ crossbar switch is shown. The switch can perform three basic functions: a concentration function if $n > m$, an expansion function if $n < m$, or a distribution function if $n = m$.

The crossbar switch connects input line to an output line by closing the appropriate crosspoint. The number of crosspoints grows with n^2 (for a square matrix) making it prohibitive for large n, thus it is more economical to divide the switch into a number of stages. Figure 12.3 shows an example of a three-stage space-division switch. The N input lines are divided into N/n

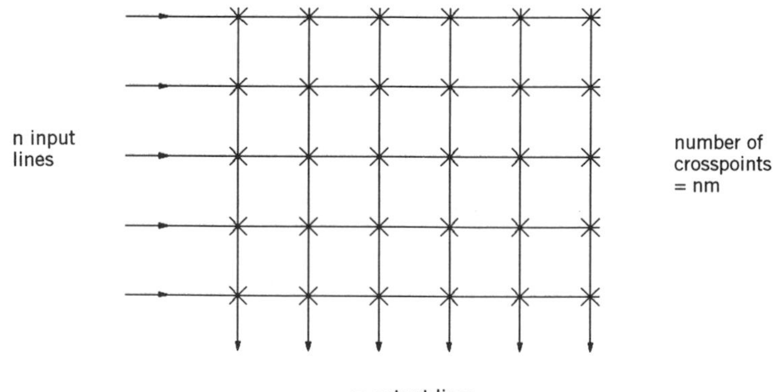

Figure 12.2. Single-stage space-division switch.

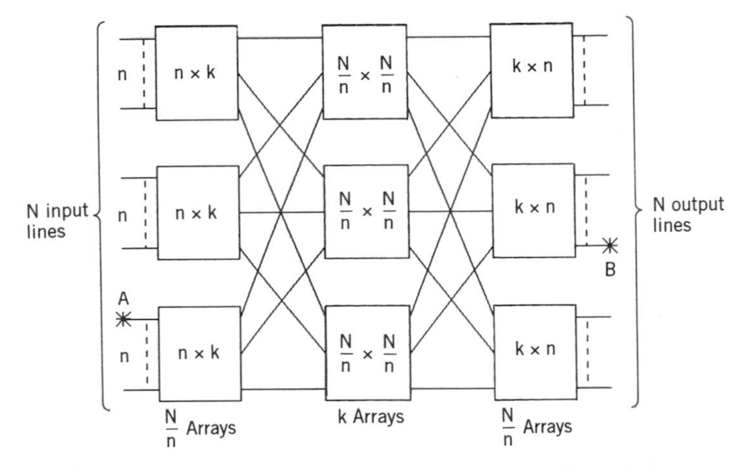

Figure 12.3. A symmetrical three-stage space-division switch.

groups of n lines each and fed into the first stage matrix. Thus there are N/n crossbar matrices, each having n input lines and k output lines. Usually n is greater than k.

The second stage consists of k switch modules, the dimension of each is $(N/n) \times (N/n)$. The third stage is N/n matrices, each being $k \times n$ rectangular matrix structure. It should be noted that there is *more than one path between every input and output line*, thus the multistage network adds the advantage of increasing the reliability of the switch.

Another advantage of this multistage network is the reduction in the number of crosspoints. However, a multistage space-division switch may be internally blocked. It should be noted that a single-stage crossbar matrix (with $m \geq n$) is nonblocking. A crossbar matrix is internally nonblocking if there is always a path available between every input and output. Figure 12.4 shows an example of blocking in a two-stage switch. Input line 1 is blocked from reaching output line 4 (input line 2 has already established a path with output line 3). However, blocking can be eliminated by increasing the value of k to 4. It should be clear that if the value of k is increased instead to 3, the probability of blocking is reduced. C. Clos of AT&T Bell Labs has obtained the minimum value of k, and hence the minimum number of crosspoints for a nonblocking three-stage switch. Let N_x be the total number of crosspoints in a three-stage switch, as shown in Figure 12.3. In the first and last stages, there are nk crosspoints in each of N/n arrays. Thus, the number of crosspoints in the first and last stages is $2(nk)(N/n) = 2kN$ crosspoints. Also, there are $(N/n)^2$ crosspoints in each of k arrays in the second stage; that is $k(N/n)^2$. Thus,

$$N_x = 2kN + k(N/n)^2 \tag{12.1}$$

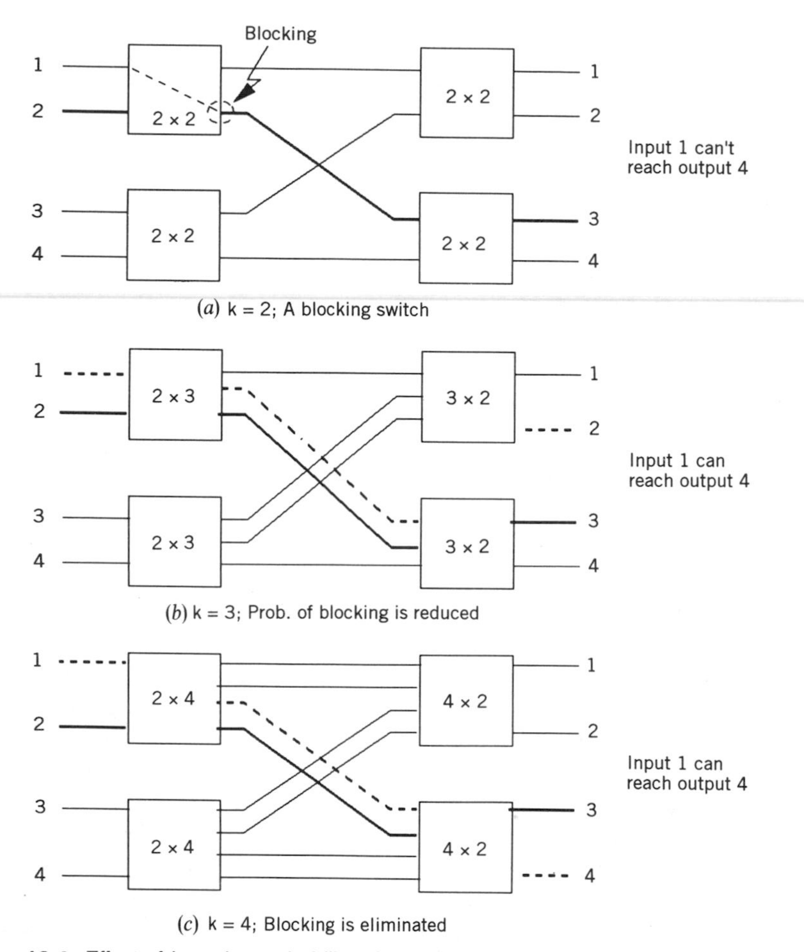

Figure 12.4. Effect of k on the probability of blocking in a two-stage space-division switch.

Now, we determine the minimum value of k for a nonblocking three-stage switch. Consider Figure 12.3; for input line A to reach output line B, there are k possible paths. The worst case for blocking occurs when $(n - 1)$ of the inputs to the matrix used by A in the first stage are occupied by other sources and when $(n - 1)$ other sources occupy $(n - 1)$ lines of the output matrix used by B. Thus, a minimum of $(n - 1) + (n - 1) + 1$ paths must be available between A and B.

Thus a three-stage space-division switch will be nonblocking if

$$k \geq 2n - 1 \tag{12.2}$$

substituting Eq. 12.2 into Eq. 12.1, we get

$$N_x = 2N(2n - 1) + (2n - 1)(N/n)^2 \tag{12.3}$$

By differentiating N_x with respect to n and equating to zero, we obtain the optimum value of n, that is, n_{opt}. For N large, we get

$$n_{\text{opt}} \sim (N/2)^{1/2} \tag{12.4}$$

Substituting into Eq. 12.3, we obtain the optimum number of crosspoints in a nonblocking three-stage switch.

$$N_{\text{xopt}} = 4N\left[(2N)^{1/2} - 1\right] \tag{12.5}$$

For example, for a three-stage switch, $N_{\text{xop}} = 7680$ for $N = 128$; whereas a single-stage switch would require 16,384 crosspoints.

The design of the switch can be further refined if we allow rearrangement of the existing connections and thus release a portion of an existing path to allow the setup of a new connection. Such rearrangeable switches will be nonblocking with n secondary stage switches, rather than $2n - 1$. This type of switch is called rearrangeably nonblocking.

12.2.2 Time-Division Switching

By the late 1950s, time-division transmissions became a reality. Circuit switching of digital signals may be provided by space-division switching. Unlike space division switching for analog signals, where connections are maintained continuously for the duration of each call, space-division switches for time-division networks are used to interconnect (or permute) time-division transmission lines for each time slot period. However, to switch digital signals from individual time slots on one input line, the digital signals may be required to be placed in the same or a different time slot on other output lines. This shifting or interchanging of time slots, referred to as time slot interchange (TSI), is essential to time-division switching. Typically, TSI involves the use of memory to hold the digital samples (i.e., signal on one time slot) for no more than one frame, so that they may be sent out in a different time slot (or channel). Figure 12.5 shows the operation and implementation of TSI. Basically, digital samples in incoming time slots are written into sequential locations of the memory. However, digital samples for outgoing time slots are read from the addresses obtained from the address store. As shown in Figure 12.5, data store address i is read during outgoing time slot j and vice versa. Thus, the data store memory is accessed twice during each time slot (a write and a read is required for each channel entering and

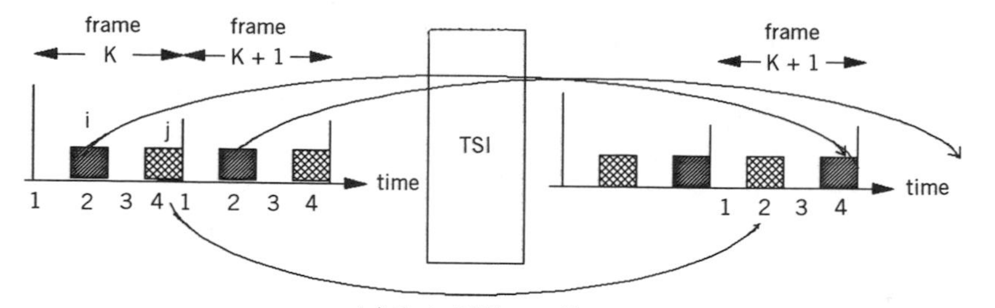

(a) Typical TSI operation

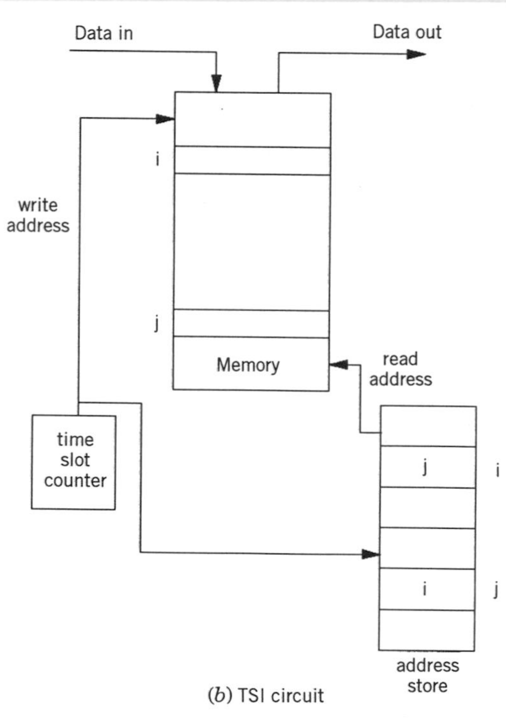

(b) TSI circuit

Figure 12.5. Time slot interchange (TSI).

leaving the TSI memory). If we define t_c as the memory access time in microseconds, then the maximum number of channels $C_{\max}$ that can be supported by the memory switch is given by

$$C_{\max} = \frac{125}{2t_c}$$

where 125 is the frame time in microseconds for the 8 kHz sampled voice.

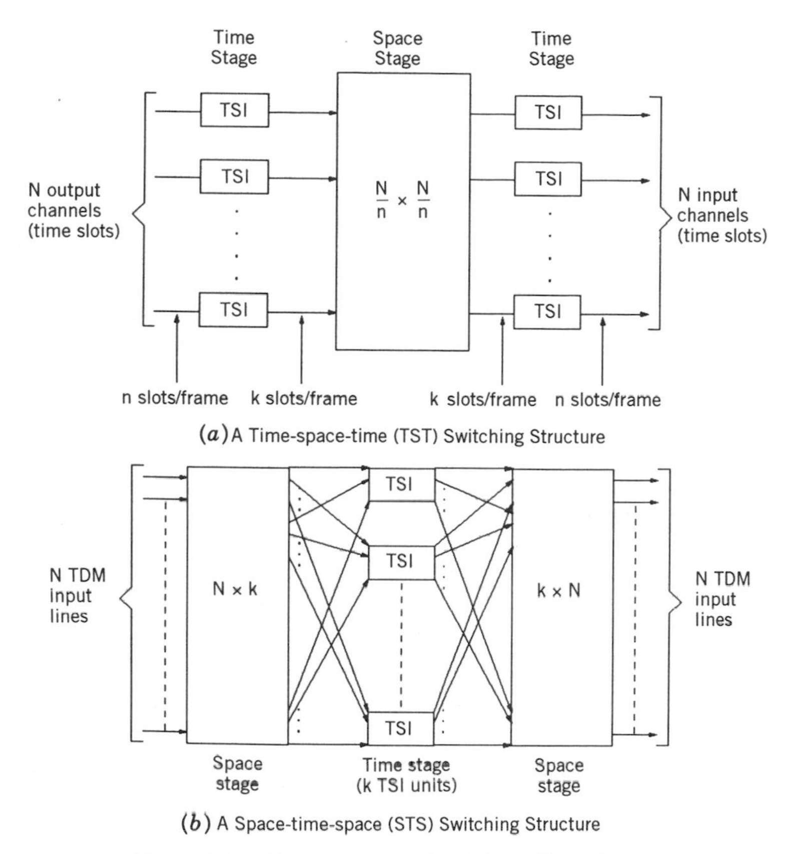

(*a*) A Time-space-time (TST) Switching Structure

(*b*) A Space-time-space (STS) Switching Structure

Figure 12.6. Basic structure of switch configurations.

Thus, the limitation of memory speed limits the size of a time switch and hence some amount of space division is necessary in large switches.

There are a variety of network configurations that can be used to accomplish switching operations in both space and time dimensions. Figure 12.6 shows two basic structures. The first structure, consists of a space stage between two time stages and is referred to as a time-space-time (TST) switch. The second structure is referred to as a space-time-space (STS) switch. As an example, the AT&T No. 4 ESS switch is an STS switch with four stages in the space switch (i.e., a TSSSST). Figure 12.6a shows, as an example, the structure of the TST switch that is similar to the three-stage space-division switch in Figure 12.3. There are N input channels (time slots) using N/n links, with every link time-multiplexing n channels. Thus, each TSI contains n time slots at its input and has k output channels per frame ($k > n$). A proof, similar to the case of the three-stage space-division nonblocking switch, can show that $k = 2n - 1$ is required for the TST to be a nonblocking switch.

12.3 ADVANCES IN SWITCHING TECHNIQUES

Current packet switches are used exclusively for data applications, and they employ both virtual circuits and datagram techniques. The switching capacity of current conventional packet switches ranges from 1000 to 4000 packets per second with average delay at the switch (processing delay) of 20 to 50 mseconds. They are designed to handle narrowband data packets. The switching functions are typically performed by means of software processing on a general-purpose computer or a set of special-purpose processors.

Advances in the field of very large-scale integrating (VLSI) and optics technologies have led to completely new concepts in the design and architecture of high-performance switching fabrics to accommodate a wide range of bandwidths. The essential goal today is to build packet switches that can handle traffic rates of the order of 100,000 to 1,000,000 packets per second per input line. These broadband switches will be flexible to provide dynamic bandwidth, as opposed to the fixed allocation of today's switches. They will also support all ranges of interactive voice, data, and video services, with very low switching delays. With low probability of error in optical fiber, end-to-end error control is only envisioned in the broadband network, as opposed to hop-by-hop error control in today's environment. They also have the potential of providing multicast and broadcast connections, as opposed to the point-to-point connections of today's networks. These features are summarized in Table 12.1 below.

The first generation of these switches uses complementary metal oxide silicon (CMOS) VLSI technology and the laboratory prototypes may typically

TABLE 12.1 Comparison Between Today's Packet Switches and Future Broadband Packet Switches

Characteristics	Today's Switches	Future Switches
Interface line speed	Low, 64 kpbs	High, 150 Mbps
Switch throughput (capacity)	Thousands of packets / sec	Millions of packets / sec
Bandwidth allocation	Fixed	Dynamic
Switching time	50 msec – 100 msec	$<$ 10 msec
Routing function implementation	Software	Hardware and software
Basis of architecture	General or special processor	VLSI technology with special architectures
Error control	Performed by the switch on a hop-by-hop basis	Performed on end-to-end basis
Services	Narrowband datapacket	Voice, data, video
Connections	Point-to-point	Point-to-point, multicast, broadcast

handle 32 × 32 lines running at 150 Mbp/s per line and some are even reaching 1 Gbps per line. Most of the existing approaches of high-performance switching fabrics employ a high degree of parallelism, distributed control, and the routing function is performed at the hardware level.

A fast packet switch is basically a box (Fig. 12.7), with N inputs and N outputs, which routes the packets arriving at its inputs to their destination outputs. In Figure 12.7*a*, we write the source address i and the destination address j between parentheses for every packet. Throughout this chapter, we assume that each packet is destined to a single output port. There is no coordination among arriving packets as far as their destinations are concerned, and thus more than one packet arriving in the same slot may be destined to the same output port. We call such an event an *output conflict* or *external blocking*. Note that the assumption here is that the switch under consideration is internally nonblocking, otherwise the conditions in Figure 12.7 might cause internal blocking. When output conflict occurs, the switch should deliver only one of the requests destined for the output port. Figure

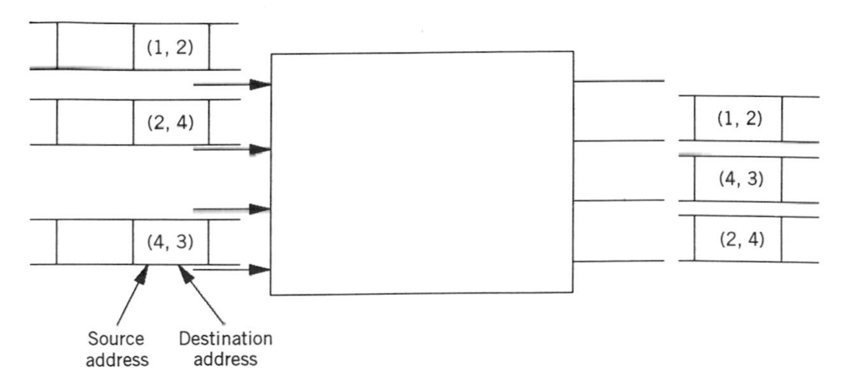

(*a*) A Packet Switch

(*b*) Output conflict in a Packet Switch

Figure 12.7. Packet switch operation.

12.7*b* shows an example of output conflict. As seen in the figure, two packets (1, 3) and (2, 3) are destined to the same output port 3 at the same time slot. The packet (1, 3) loses the contention and packet (2, 3) is delivered to output port 3. The blocked packet (1, 3) would try again during the next time slot. Thus, a mechanism is needed for resolving output conflict, and buffering of packets within the switch must be provided.

Thus, a packet switch is a box that provides two basic functions: *switching* (or equivalently, routing) and *buffering*. There are two other functions that may at times be required. One function is the *multicast* function. In some applications, it may be necessary for a packet originating at a source node in the network to be destined to more than one destination. Thus, a packet switch may have the capacity to replicate a packet at several of their output ports as requested. The other function is the *priority function*. In this case, the switch should have the capability to differentiate among packets according to priority information provided in them, and to give preference to higher priority packets.

Several architectural designs for fast packet switches have emerged in recent years. These can be classified into the following three categories

Shared-memory type
Shared-medium type
Space-division type

In the following sections, we give detailed descriptions of the most common switch architectures within each category. To simplify the discussion, we make the following assumptions: all lines (N input lines and N output lines) have the same transmission capacity, and all packets are of the same size. Also, we assume that the time axis to be slotted with the slot size is equal to the transmission time of a packet and that the operation of the switch is synchronous.

Traffic patterns affect the switch performance. The traffic pattern has two components: The first is the packet arrival process at the input lines, and the second is the destination request distribution for arriving packets. We assume that the packet arrival process is a Bernoulli process with parameter ρ and is independent from all other input lines. That is, in any given time slot, the probability that a packet will arrive on a particular input is ρ. Thus, the average utilization of the input line is ρ and equals the average utilization of the output line for a stable system. Each packet has equal probability $1/N$ of being addressed to any given output (i.e., a uniform traffic pattern), and successive packets are independent.

12.3.1 Shared-Memory Fast Packet Switches

Shared-memory fast packet switches are very similar in design concept to today's conventional packet switches. The switch's memory is shared by all

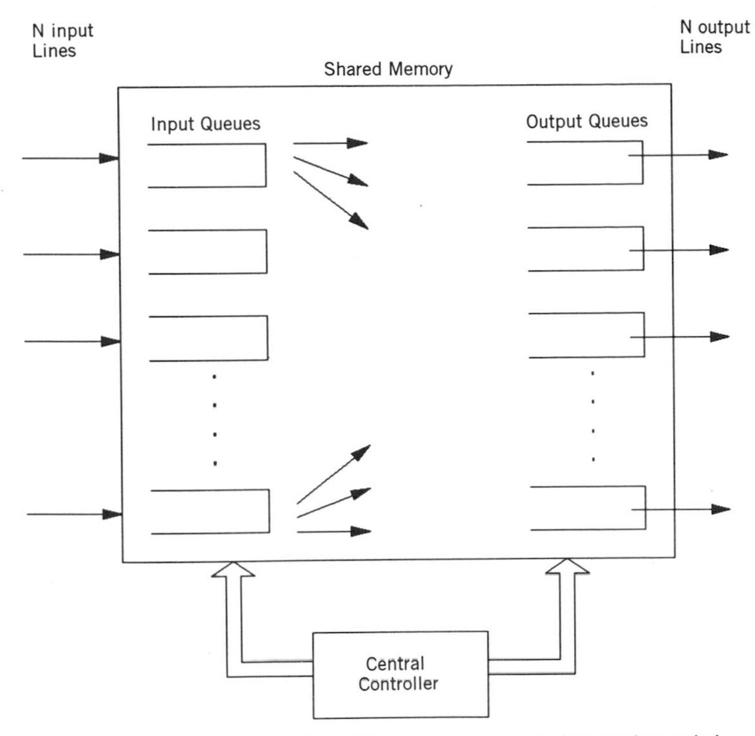

Figure 12.8. Conceptual design of the shared-memory fast packet switch.

input and output lines. Arriving packets on all N input lines are multiplexed and stored in the common memory and then organized into N separate output queues, one for each output line as in Figure 12.8. Thus, the central controller must have the capability to sequentially process N incoming packets and select N outgoing packets in each time slot. The memory bandwidth, access time, and size are important factors in the switch design. The memory bandwidth should be large enough to handle all input and output packets simultaneously. If we define V as the port speed, say in bits per second, then the memory bandwidth should be at least $2NV$. For example, for a switch with 64 input lines, and a line speed of 100 Mbps, the memory bandwidth is 1.28 Gbps. The memory access time should be small enough to allow N times access for the input queues and N times access for the output queues during one time slot. A good example of the shared-memory fast packet switch is the Prelude switch developed in France.

There is no coordination between the packet arrivals at the input queues and their requested destination and the finite size of the memory. Therefore, some packets may not be accepted by the switch and are lost. Thus, *the maximum acceptable packet loss probability* becomes a design constraint on the memory size. The way the memory is shared among the output queues affects the memory size to a great extent. The memory size is also a function of the switch size N, the offered load ρ, and the traffic pattern. Below, we discuss the impact of output queueing on the memory size requirements.

12.3.1.1 Output Queueing. Here, we consider two basic approaches for sharing the output queues. The first approach is based on equally partitioning the output memory between the output lines, which is referred to as *complete partitioning*. The second approach pools all output memory into one completely shared buffer rather than have a separate buffer for each output. We will refer to this approach as *full sharing*. In complete partitioning, a packet destined to a given output is lost if the buffer allocated to that output is full, whereas in full sharing a packet is lost only if the entire memory is full.

Complete Partitioning of the Output Buffer. Here, the output memory is divided into N separate queues, each one of size b packets allocated to a particular output line. Define the random variable A as the number of packet arrivals at a given output queue during a given time slot.* Because the probability of one packet arrival at a given output during a given time slot is ρ/N, it follows that A has the binomial probabilities

$$a_k = \text{Prob}[A = k] = \binom{N}{k}\left(\frac{\rho}{N}\right)^k\left(1 - \frac{\rho}{N}\right)^{N-k}, \quad K = 0, 1, \ldots, N \quad (12.6)$$

For a large size switch, that is, when $N \to \infty$, the arrival process becomes Poisson and we have

$$a_k = \text{Prob}[A = k] = \rho^k \frac{e^{-\rho}}{k!}, \quad k = 0, 1, \ldots, \infty \quad (12.7)$$

Let n be the number of packets in the given output queue at the end of the mth slot. We model n by a finite-state, discrete-time Markov chain with state probabilities as shown in Figure 12.9*a*. If there are $(n + 1)$ packets in the output buffer in the previous time slot, one packet will be transmitted in the current slot and n packets remain in the queue given zero packet arrival. This event is represented by the state transition probability a_0 in Figure 12.9*b*. On the other hand, if there are $(n - i)$ packets in the previous time slot, $0 < i < n$, then $(i + 1)$ packet arrivals will be needed to reach state n, because one packet will be transmitted during the current slot. This event is

*Note that more than one packet may arrive at a given time slot on different input lines and destined to the same output.

represented by the transition probability a_{i+1}. We should also examine the boundary conditions. State $n = 0$ is reached from state $n = 1$ with zero packet arrivals (with probability a_0) and from state $n = 0$ with zero or one packet arrival (with probability $a_0 + a_1$). Regarding state $n = b$ (the top part of Fig. 12.9*b*), if there are $(b - i)$ packets in the output queue in the previous slot, the arrival of $(i + 1)$ packets or more will fill the queue up to b packets and any extra packets will be discarded. This event is given by the transition probability

$$\sum_{k=i+1}^{N} a_k, \quad 0 \leqslant i \leqslant b$$

We define P_n as the steady-state probability of having n packets at a given output, $0 \leqslant n \leqslant b$. Then we can write the Markov chain balance equations, i.e.,

$$P_n = a_0 P_{n+1} + \sum_{i=0}^{n} a_{i+1} P_{n-i}, \quad 0 < n < b \tag{12.8a}$$

$$P_0 = a_0 P_1 + (a_0 + a_1) P_0 \tag{12.8b}$$

$$P_b = \sum_{i=0}^{b} \left[\sum_{k=i+1}^{N} a_k \right] P_{b-i} \tag{12.8c}$$

and

$$\sum_{n=0}^{b} P_n = 1 \tag{12.9}$$

Equations 12.8 and 12.9 can be solved numerically to obtain P_n. The probability that the output queue is idle is the probability of having zero packets in the output queue and of zero packets arriving. Thus, the normalized switch throughput ρ_0, which is also the output line utilization is given by

$$\rho_0 = 1 - P_0 a_0 \tag{12.10}$$

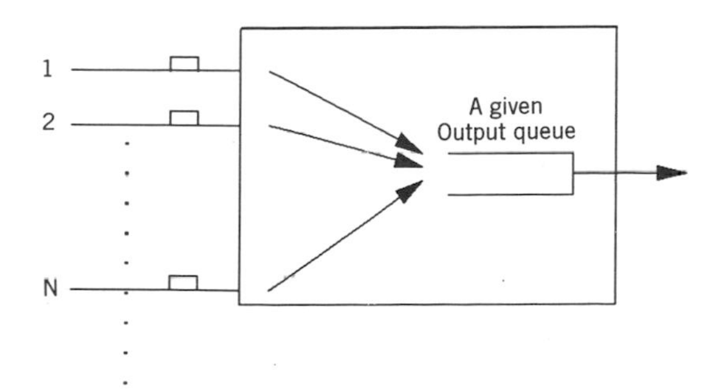

Figure 12.9*a*. Queueing model for a given output queue.

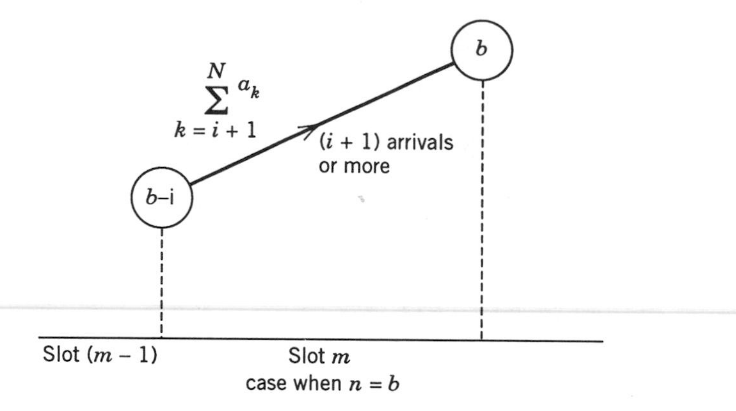

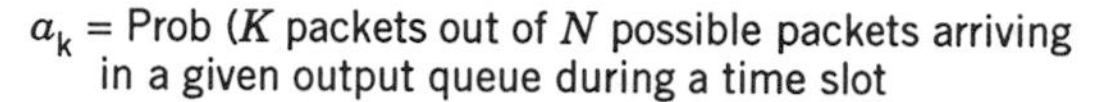

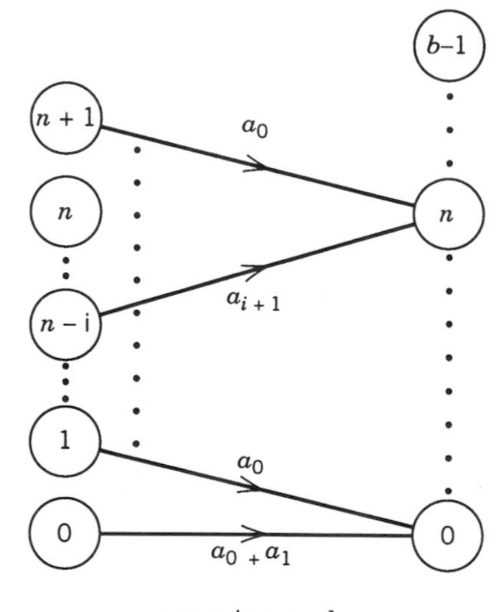

Figure 12.9*b*. State transition diagram for the number of packets in a given output queue.

Note also that ρ_0 equals the arrival rate ρ multiplied by the probability that a packet succeeds in passing through the switch. Therefore, the probability of a packet loss is equal to 1 minus the probability of a Packet Success or

$$P_L = 1 - \frac{\rho_0}{\rho} \tag{12.11}$$

Figure 12.10 shows the packet loss probability for the completely partitioned memory case as a function of the output buffer size b (in packets) for a various number of users N and offered load $\rho = 0.9$. These curves are useful

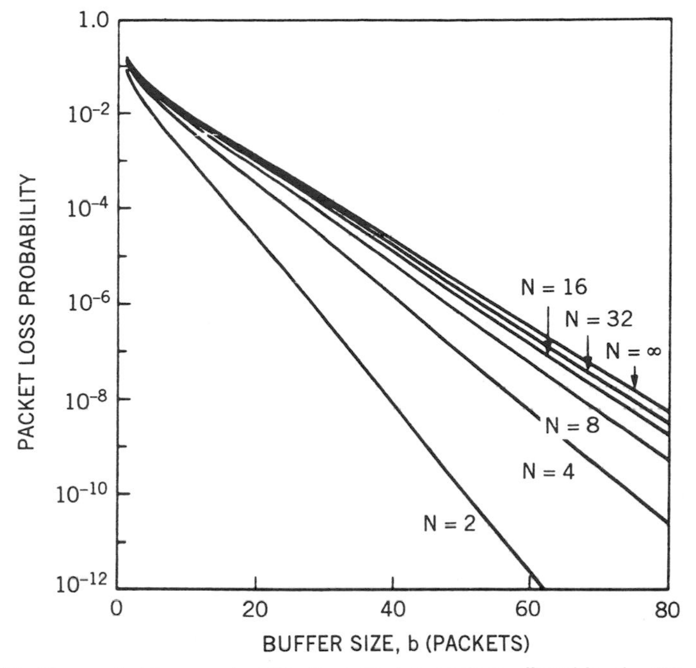

Figure 12.10. The packet loss probability for output queueing; offered load = 0.9. *Source* From [HLUC88] with permission of the IEEE.

in switch design. For example, to design a switch with $N = 16$ lines, given that the probability of a packet loss should be less than 10^{-7} at an offered traffic $\rho = 0.9$, the output buffer size per line should be approximately 60 packets. Note that, for a given b, the loss rate increases with N. For example, at 90 percent load and 40 packet buffers, for $N = 4$ the loss probability is approximately 10^{-6}, and for $N = 16$, the loss probability is approximately 10^{-5}. Note also that the $N = \infty$ curve is a good approximation for finite $N > 32$. In Figure 12.11, for $N = \infty$, we plot the packet loss probability as a function of the output queue size b for various values of ρ. At 80 percent offered load, with $b = 28$, the packet loss probability is under 10^{-6} for arbitrarily large N.

12.3.1.2 Full Sharing. In full sharing, rather than have a separate buffer for each output, all memory is pooled into one completely shared buffer. All queues share the entire memory, and a packet is lost only if the entire memory is full. With complete sharing, the loss rate performance improves dramatically, thus requiring overall a much smaller size memory than in the complete partitioning case.

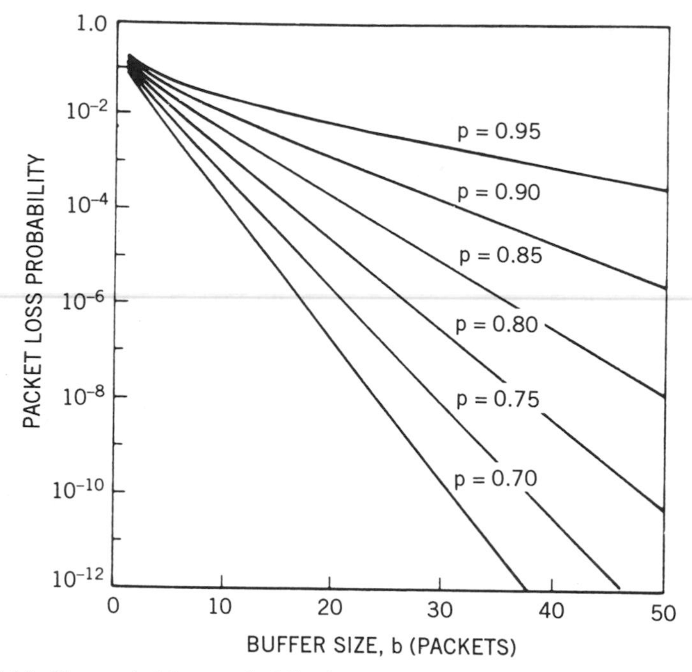

Figure 12.11. The packet loss probability for output queueing for various values of offered load.
Source From [HLUC88] with permission of the IEEE.

12.3.2 Shared-Medium Fast Packet Switches

Shared-medium fast packet switches employ a bus or a ring topology as the switching medium. This approach has been used in many of today's switches. One current example of a switch that uses the bus as the shared medium is the Datakit packet switch. Another example is the broadcast time-division bus that arises in the design of some circuit switches such as digital private branch exchange (PBX). Hence, the technology of the shared medium is well understood. In addition they provide flexibility in terms of the access protocol and distribution of traffic. An essential factor in realizing the shared-bus architecture is how to implement the high-speed bus and the allocation of memory. One way to increase the speed is to use multiple rings or multiple buses in a single or multiple structure. In this type of switch, all packets arriving on the input lines are synchronously multiplexed onto the common high-speed medium, which has a bandwidth equal to N times the rate of a single input line (Fig. 12.12). Each output line is connected to the bus by an interface consisting of an address filter and an output buffer. In general, the shared-bus architecture results in separate output queues with no sharing.

Examples of the shared-medium type of fast packet switches are the packetized automated routing integrated system (PARIS) switch by IBM, the

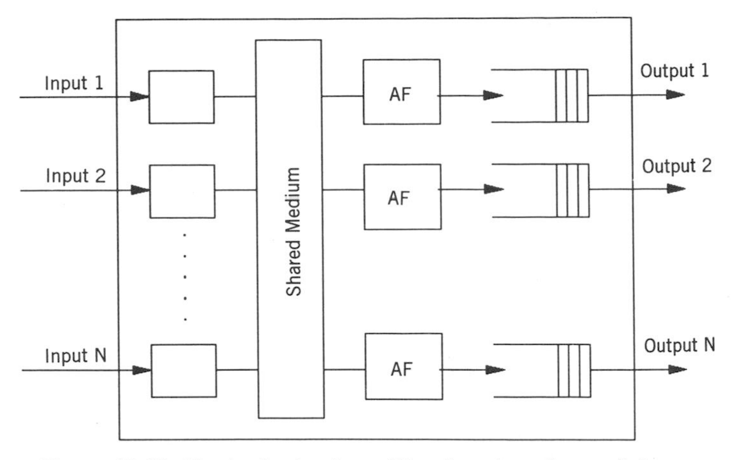

Figure 12.12. The basic structure of the shared-medium switches.

asynchronous transfer mode output buffer modular (ATOM) switch by NEC, and the synchronous composite packet switching (SCPS) architecture.

12.3.3 Space-Division Fast Packet Switches

A space-division packet switch is a box with N inputs and N outputs that routes the packets arriving on its inputs to the appropriate output. This is achieved by the establishment of multiple concurrent paths from the inputs to the outputs.

Thus, a space-division packet switch is different from the shared-memory and the shared-bus architecture, where traffic from all input lines is multiplexed into a single stream with a bandwidth N times the bandwidth of a single line. The header of each arriving packet to the switch input lines contains the routing information needed to establish input–output paths.

There are three main categories of space-division switches: (1) crossbar fabrics, (2) Banyan-based fabrics, and (3) fabrics with N^2 disjoint paths.

Some space-division switches suffer from internal blocking (referred to as a blocking switch), which makes it impossible for all required paths to be set simultaneously. In such a situation, packets may have to be buffered within the switch until appropriate connections are available. Contrary to the shared-memory and shared-medium switches, the location of the buffer is not possible at the outputs. The previous discussion on the output queues for the shared-memory switches is also valid for the nonblocking space-division packet switches. Below we discuss the impact of the input queues on the performance of space-division packet switches.

12.3.3.1 Input Queues. Input queueing consists of placing a separate buffer at each input to the switch (see Fig. 12.13). A packet arriving at an

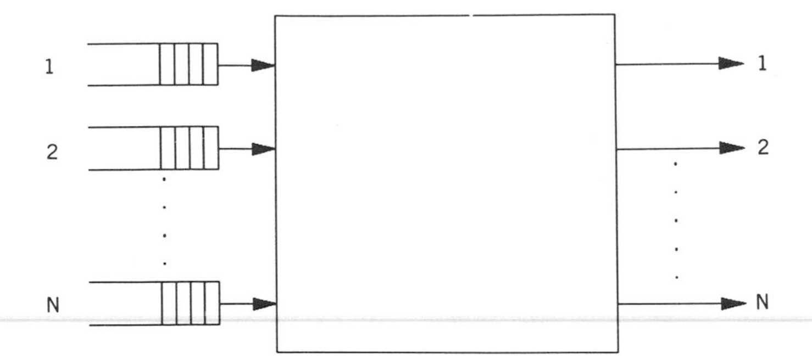

Figure 12.13. Input buffering in a space-division packet switch.

input line first enters the first-in first-out (FIFO) buffer where it awaits access to the switch fabric. At the beginning of every time slot, only the head of the line (HOL) packets contend for access to the switch outputs. If k packets at the head of the FIFO input are addressed to a particular output, one is chosen at random with probability $1/k$. The other $k - 1$ packets must wait until the next time slot. Note that there may be packets, queued behind the unsuccessful $(k - 1)$ packets, and, consequently, blocked from reaching possibly idle outputs on the switch. This results in reducing the switch throughput. As we see below the maximum throughput, for large N, is 0.586. We use the same assumptions and notation as defined previously in the analysis of the output queues in Section 12.3.1.

12.3.3.2 Heavy-Load Analysis. To determine the maximum throughput, we study the heavy load case (i.e., $\rho = 1$). Whenever a packet is transmitted through the switch, a new packet immediately replaces it at the head of the input queue. We define $B(i)$ as the number of blocked packets, destined to output i, at the heads of input queues at the end of the mth time slot. Also, define $A_m(i)$ as the number of packets moving to the heads of the input queues during the mth slot and destined for output i. Hence, referring to Figure 12.14,

$$B_m(i) = \max[0, B_{m-1}(i) + A_m(i) - 1] \tag{12.12}$$

Let F_m represent the total number of packets transmitted through the switch during time slot m. Because it is a heavy-load situation, F_m equals the total number of input lines N minus the total number of those lines whose HOLs are blocked. Hence,

$$F_m = N - \sum_{i=1}^{N} B_m(i) \tag{12.13}$$

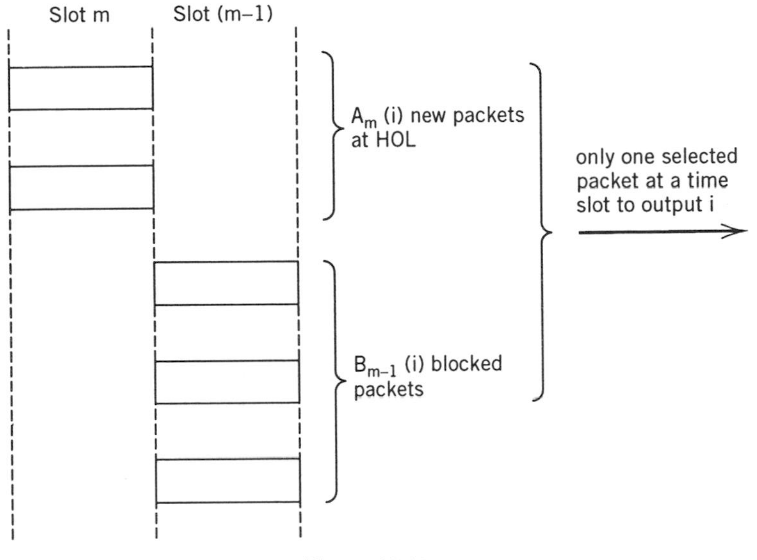

Figure 12.14.

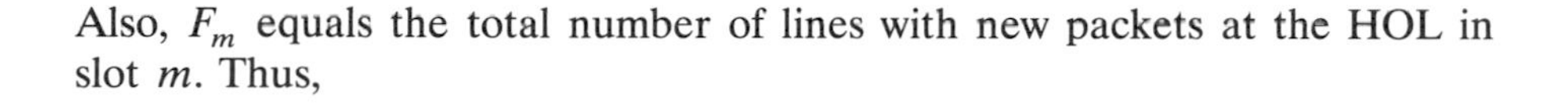

Also, F_m equals the total number of lines with new packets at the HOL in slot m. Thus,

$$F_m = \sum_{i=1}^{N} A_{m+1}(i) \tag{12.14}$$

The utilization of the output line ρ_0 (which is the same as the switch throughput) is given by

$$\rho_0 = \overline{F}/N \tag{12.15}$$

where, $\overline{F}$ is the average steady-state number of input queues with free HOL. Using Eq. 12.13 and taking the steady-state average of both sides, we get

$$\overline{F} = N - \overline{B}(i) \tag{12.16}$$

Since a new packet arrival at the HOL has an equal probability $1/N$ of being addressed to any given output and total f_{m-1} packets transmitted through the switch during slot $(m - 1)$, $A_m(i)$ has a binomial probability distribution. That is, the probability of having k packets moving to the HOL during slot m and destined for output i, Prob$[A_m(i)]$, is given by

$$\text{Prob}[A_m(i) = K] = \binom{F_{m-1}}{K}\left(\frac{1}{N}\right)^K \left(1 - \frac{1}{K}\right)^{F_{m-1}-K}, \quad K = 0, 1, \ldots, F_{m-1} \tag{12.17}$$

Notice that as $N \to \infty$, we can heurestically argue that $\overline{F}$ becomes large and the steady-state number of packets $A(i)$ moving to the head of free-input queues at each time slot and destined for output i follows a Poisson distribution with a rate $\overline{F}/N = \rho_0$. Also, by examining Eq. 12.12, although $B_m(i)$ does not represent the occupancy of any physical queue, we find that it has the same form as the basic queueing relationship for a single-server queueing system with a deterministic service rate. Note that the service discipline of this artificial queue is not FIFO but random selection. However, since the random-selection service discipline does not depend on the packet service time (which is fixed in this case), the queue occupancy distribution remains the same as the FIFO case. Hence with the argument of Poisson arrival and the similarity with the single-server queueing system, we can use the results for the steady-state queue size for $M/D/1$, for $N = \infty$. Thus,

$$\overline{B}(i) = \frac{\rho_0^2}{2(1 - \rho_0)} \tag{12.18}$$

Using Eqs. 12.15 and 12.16, we get, for large N,

$$\overline{B}(i) = 1 - \rho_0 \tag{12.19}$$

From Eqs. 12.18 and 12.19, we get

$$\rho_0 = 2 - \sqrt{2} = 0.586 \tag{12.20}$$

That is, the maximum achievable switch throughput is 0.586 when the switch is saturated and $N = \infty$.

It should be noted that a more rigorous analysis for finite N shows that the above result is valid for N as small as 8. This is shown in Figure 12.15. However, the switch throughput can be increased by dropping packets (see Problem 12.5).

The above analysis assumes that each input queue is served on a FIFO basis. By relaxing this assumption, the switch throughput can be increased beyond 0.586. This can be achieved by allowing packets behind the HOL to be considered. For example, at the beginning of each time slot, we allow the first w packets in each input queue to contend sequentially for access to the switch outputs. Packets at the heads of the queues contend first. Because of

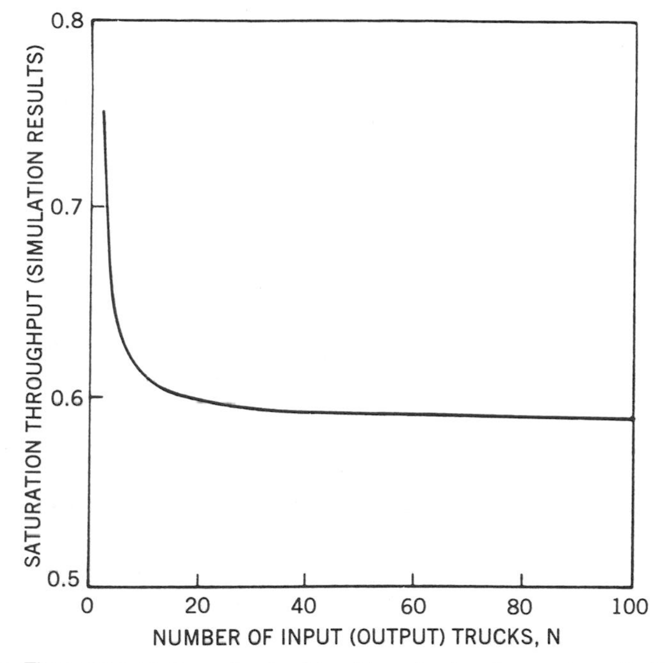

Figure 12.15. The maximum throughput achievable using input queueing with FIFO buffers. *Source* From [HLUC88] © 1988 IEEE.

the output conflicts, those inputs that did not succeed in transmitting the first packets in their input queues contend with their second packets for access to any remaining idle outputs, and so on up to w times. Clearly, $w = 1$ (i.e., a window size of 1) represents the case of input queueing with FIFO buffers.

Table 12.2 shows the effect of the window size w on the maximum throughput achievable for various switch sizes N. Note that the throughput increases with increasing values of w, but with little improvements beyond $w = 4$.

TABLE 12.2 The Maximum Throughput Achievable with Input Queueing for Various Switch Sizes *N* and Window Sizes *w*

	Window Size *w*							
N	1	2	3	4	5	6	7	8
2	0.75	0.81	0.89	0.92	0.93	0.94	0.95	0.96
1	0.66	0.76	0.81	0.85	0.87	0.89	0.91	0.92
8	0.62	0.72	0.78	0.82	0.85	0.87	0.88	0.89
16	0.60	0.71	0.77	0.81	0.81	0.86	0.87	0.88
32	0.59	0.70	0.76	0.80	0.83	0.85	0.87	0.88
64	0.59	0.70	0.76	0.80	0.83	0.85	0.86	0.88
128	0.59	0.70	0.76	0.80	0.83	0.85	0.86	0.88

Source: From [HLUC88] with permission of the IEEE.

Now we discuss the three categories of space-division packet switches below.

12.3.3.3 Crossbar Space-Division Switches. This type of switch was originally designed for circuit switching, as discussed earlier in this chapter. However, they could also support fast packet switching.

12.3.3.4 Banyan-based Space-Division Switches. To better understand this type of switch, we start with a discussion of interconnection networks.

Interconnection Networks. A typical switch or interconnection network consists of a number of switching elements and interconnecting links. Interconnection functions are realized by controlling the switching elements. A single-stage network is a switching network (box) with N inputs and N outputs, as shown in Figure 12.16*a*.

A multistage switch (or network) has many stages of interconnected switches and is characterized by three features: the switch element, the network topology, and the control structure. Many switch elements are used in a multistage network. Each switch element is essentially an interchange device with two inputs and two outputs. Figure 12.16*a* shows four functions of a switch element: bar (straight), cross (exchange), upper broadcast, and lower broadcast. In many cases, only the top two functions (the bar and the cross) of Figure 12.16 are used.

We can add another capability to the basic 2×2 switching element: the capability to arbitrate between conflicting requests. If both inputs require the same output line, then only one of them will be connected and the other will be blocked or rejected (in some cases it can even be buffered).

The 2×2 basic switching box (with the bar and cross states only) can be used to construct 1×2^n demultiplexer. This can be built in the form of a binary tree with n stages comprising a total of $N - 1$ switching elements, where $N = 2^n$. Figure 12.17 shows an example for a 1×8 demultiplexer tree. In such a tree, there is a unique path from the root to each of its leaves. It is possible to design the 2×2 switching element to be in the bar state or the cross state according to the value of a single bit of the destination address. If this bit value is a binary 0, the packet is routed to the upper output, and if the bit value is a binary 1, the packet is routed to the lower output, regardless of which input to the switching element the packet arrived at. This capability, of routing the packet through the switch fabric based on the address field contained in the packet header, is called *self-routing* or *digit-controlled routing*.

For example, referring to Figure 12.17, if a packet happens to arrive at A and is destined to a destination with binary address (d_2, d_1, d_0), then the root node is controlled by bit d_2, the switching elements in the second stage are controlled by bit d_1, and the switching elements in the last stage are controlled by bit d_0. Clearly, A or B can be connected to any one of the

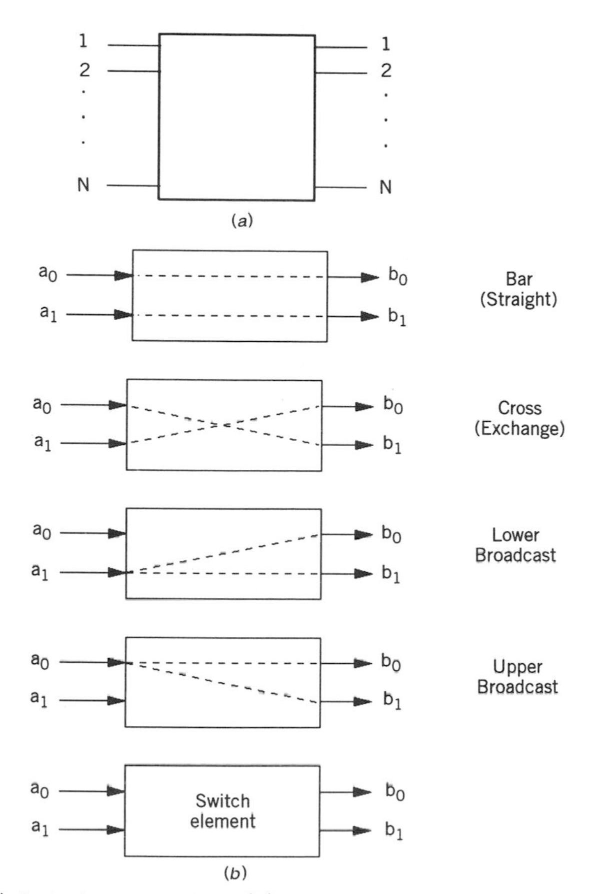

Figure 12.16. (*a*) A single stage switch. (*b*) A 2 × 2 switch element with its four possible functions.

eight output lines. The dotted path shows the route for a packet destined from A to 100. In general, if the destination address is k digits corresponding to the k stages of the network, each digit controls the switching elements in the corresponding stages.

The procedure to build the 1×2^n demultiplexer tree can be extended to build a $2^n \times 2^n$ multistage network (see Fig. 12.18). By adding an additional switching element at the first stage of the 2 × 8 multiplexer in Figure 12.18*a*, we obtain the 4 × 8 multiplexer as shown in Figure 12.18*b*. Adding another switching element at the first stage and two more switching elements at the second stage, we obtain the 6 × 8 multiplexer in Figure 12.18*c*. Finally, adding an additional switching element at the first stage, we obtain the 8 × 8 multistage network (Fig. 12.18*d*). It should be noted that there are many possible forms other than the form shown in Figure 12.18*d* for 8 × 8 multistage networks.

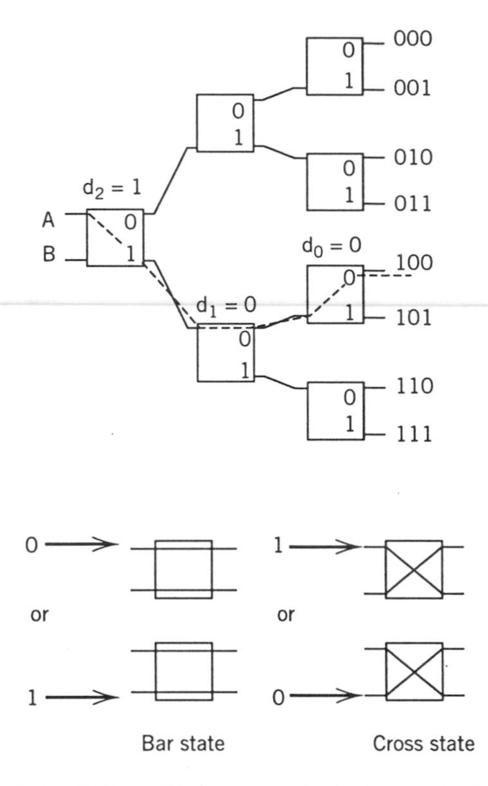

Figure 12.17. 1×8 Demultiplexer using 2×2 switching elements.

In general, a multistage network consists of n stages to connect N input lines to N output lines, where $N = 2^n$. Each stage may use $N/2$ switch elements. The interconnection patterns from stage to stage determine the network topology. Each stage is connected to the next stage by at least N paths. The cost of paths of a size N multistage network is proportional to $N \log_2 N$. The network delay is proportional to the number of stages n. Two basic control structures are used to determine how the switch element functions will be set. The first control structure is the individual stage control. It requires n sets of control signals to set up the functions of all n stages of the switch elements, because the same control signal is used to set all switch elements in the same stage. The alternative control structure is the individual switch element control, which uses a separate control signal to set the states of each switch element. Thus providing for flexibility in setting up the connection paths, but requires $nN/2$ control signals. A multistage network can be classified into three classes: blocking, rearrangeably nonblocking, and nonblocking.

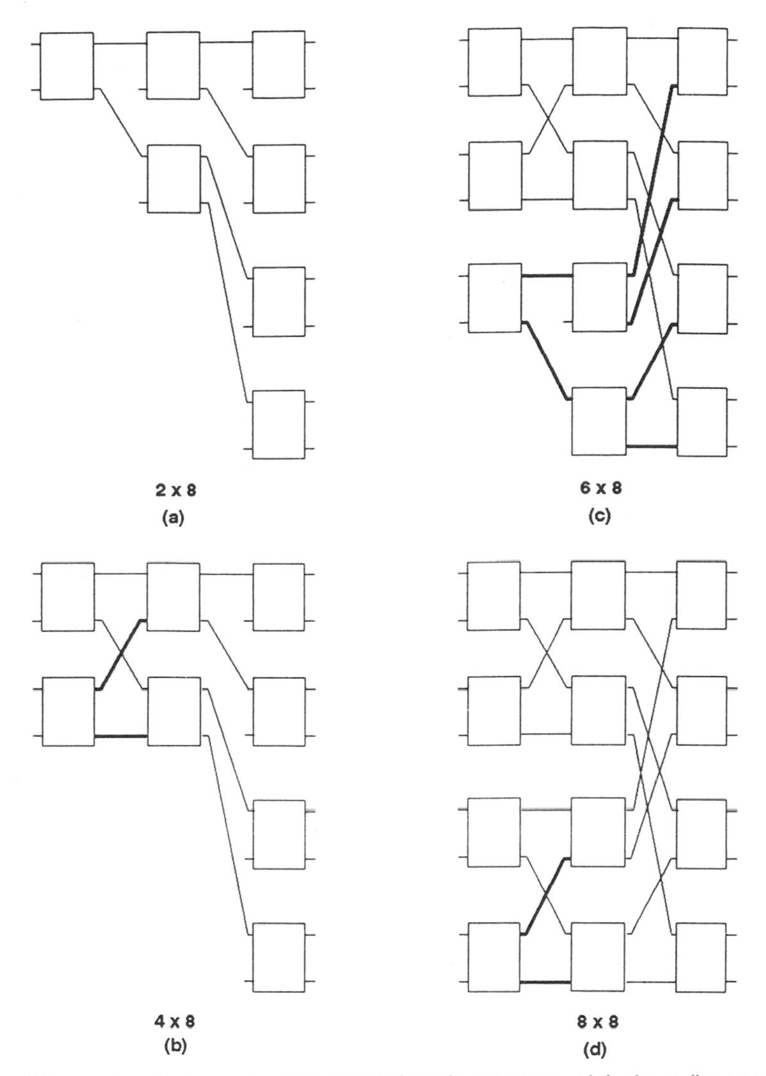

Figure 12.18. Constructing a multistage network to interconnect eight input lines to eight output lines using binary switches.

In blocking networks, conflicts between simultaneous connections of different stages may result in the use of the interconnection links. Example of a blocking network is the Omega network. A rearrangeably nonblocking network can perform all possible connections between inputs and outputs by rearranging its existing connections so that a connection path for a new input–output pair can always be established. For example, the Benes network is a rearrangeably nonblocking network. A nonblocking network can handle all possible connections without blocking. Examples of nonblocking networks are the Clos switch and the crossbar switch.

The Banyan and the Delta Networks. A Banyan network is a class of the multistage interconnection network. The main characteristic of a Banyan network is that there is exactly one path between any input and any output of the switching network. The number of switching elements required to connect N inputs to N outputs equals to $(N/2)\log_2 N$ for a binary Banyan. A crossbar switch, by contrast, requires N^2 elements.

Delta networks are a subclass of the Banyan network that self-route (i.e., they are digit controlled). We can construct an 8×8 Delta network using a 1×2^n demultiplexer. For each additional input, superimpose a demultiplexer tree on the partially constructed tree. The already existing links can be used as part of the new tree or add extra links and switching elements if needed. An example is shown in Figure 12.19, with the addition of the next tree in heavy lines.

The only rule to follow during the construction of the network is that if a switching element has its inputs coming from other switching elements, then both inputs must come from the upper lines of preceding-stage switching elements or both must come from the lower lines of other preceding-stage switching elements. It should be noted that for the above 8×8 network only 12 modules are necessary to build this network. Examples of Delta networks are the Omega networks, the indirect binary n-cube, the flip, the reverse baseline, and the shuffle-exchange.

Before we give a formal definition of the Delta network, we need to describe the switching element in a general way. Let an $a \times b$ switching element be a box that has the capability to connect any of its a inputs to any one of the b outputs.

Label the inputs $0, 1, \ldots, a - 1$ and the outputs $0, 1, \ldots, b - 1$. An input line is connected to the output labeled d if the control digit (address digit) supplied by the input is d, where d is a base-b digit. Also, an $a \times b$ switching element arbitrates between conflicting requests by accepting some and rejecting others.

Now, a Delta network is defined as an $a^n \times b^n$ switching network with n stages, consisting of $a \times b$ switching elements. The interconnection or link pattern between stages is such that there exists a unique path of constant length from any source to any destination. Furthermore, a packet's movement through the network can be controlled by the destination address included in the packet. Also, in a Delta network, no input or output line of any switching element is left unconnected. Thus, in Figure 12.19, the network of (d) is a Delta network, but not the networks of a, b, or c.

The construction of an $a^n \times b^n$ Delta network follows the same procedures described above for the $2^n \times 2^n$ Delta network. Clearly, many link patterns are available for an $a^n \times b^n$ Delta network. A regular link pattern that is appropriate for an $a^n \times b^n$ Delta network is the a-shuffle link pattern. First let us explain the shuffle function. The *q-shuffle* is better understood by using the example of a deck of qr playing cards. Divide the deck of qr cards into q piles of r cards each; top r cards in the first pile, next r cards in the

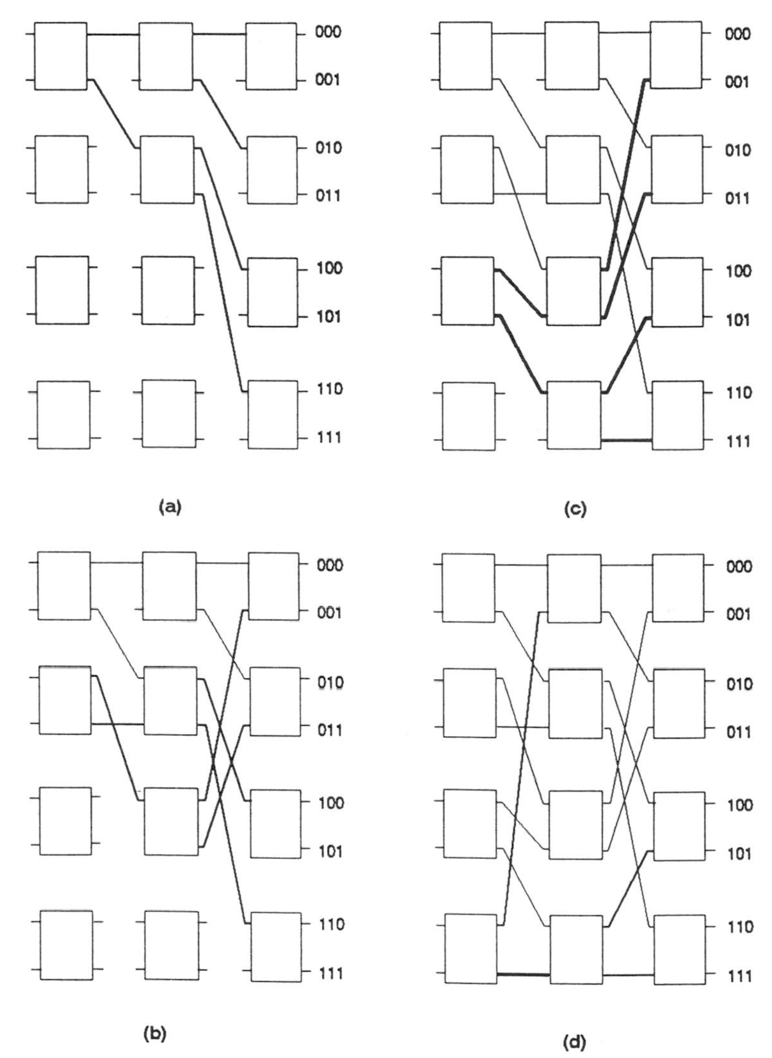

Figure 12.19. Construction of an 8 × 8 Delta network.

second pile, and so on. Now pick the cards, one at a time from the top of each pile; the first card from the top of pile one, the second card from the top of pile two, and so on in a circular fashion until all cards are picked up.

This new order of cards represents the q-shuffle of the previous order. Note that the q-shuffle is an inverse permutation of the r-shuffle of qr objects. Also, the 2-shuffle is called the *perfect shuffle*. Figure 12.20 shows an example of the 4-shuffle of 12 lines (or objects). Figure 12.21 shows the construction of a $2^3 \times 2^3$ Delta network using the interstage pattern as the 2-shuffle. In many cases, the switching element is of size $b \times b$, with n stages,

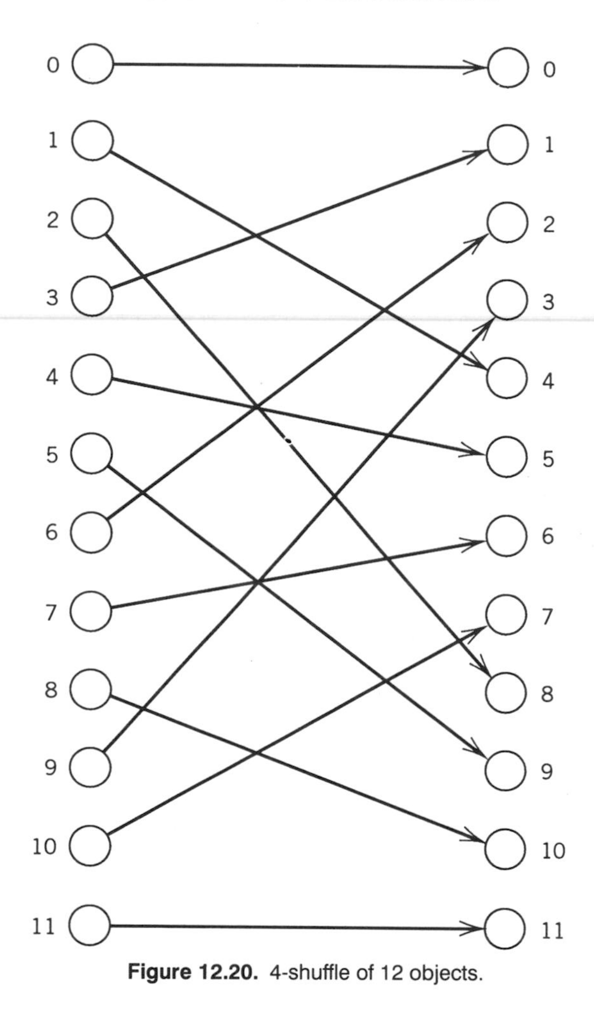

Figure 12.20. 4-shuffle of 12 objects.

N input ports, and N output ports, where $N = b^n$. The destination address is k digits long on base b. The Delta network is sometimes referred to as Delta-b network.

Since the Banyan and Delta network provide a unique path between any input and output pair, these networks are fault prone. It is desirable to provide multiple paths between each input and output pair to keep the network operational in the presence of a few faults. Problem 12.9 discusses possible solutions to increase the reliability of the Delta network.

Delta networks are blocking networks (i.e., packets can collide with each other and get lost). Blocking in a Delta network may occur by internal blocking, in which packets destined for different outputs may compete for a particular link inside the network. It may also occur by output blocking, in which packets compete for the same output port (see Problem 12.10).

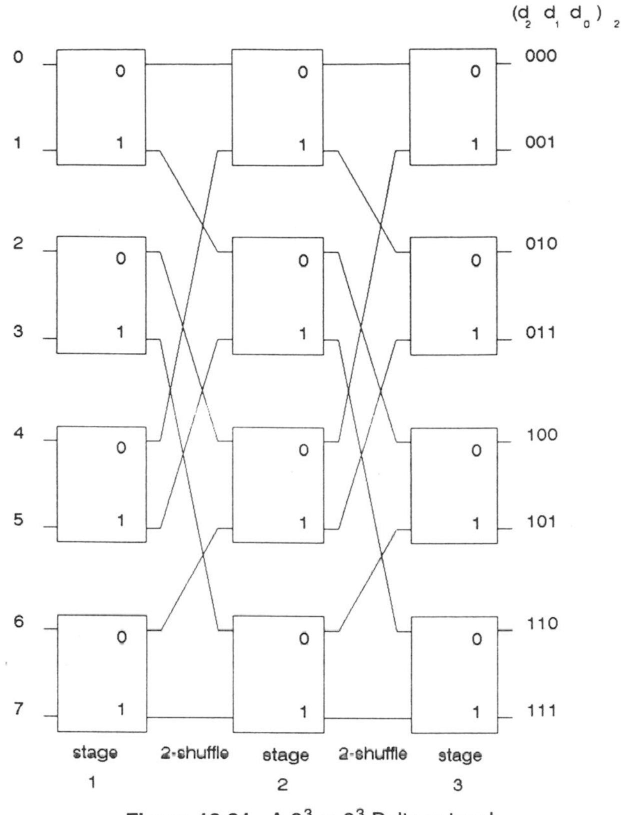

Figure 12.21. A $2^3 \times 2^3$ Delta network.

Because of blocking, Delta networks have a limited throughput that is much lower than that obtained for the internally nonblocking switches, such as crossbar switches. Simulation results have shown that the throughput for Delta networks degrades with increasing the switch size N. For example, for $N = 32$ the maximum throughput is 0.4, and for $N = 1024$ it is 0.26, assuming a uniform traffic pattern.

The basic approach in the design of Banyan networks attempts to overcome blocking and thus improve the switch performance in terms of throughput and packet loss probability. One approach is called the buffered Banyan in which buffers are placed at the points of conflicts. Another approach attempts to avoid completely the internal blocking problem. This can be achieved by first sorting the packets based on their destination addresses and then routing them through the Banyan network. This type of network is called the sort-Banyan network or the Batcher-Banyan network, since a popular sorter network (the bitonic sorter) has been introduced by Batcher. Below, we discuss these two approaches.

BUFFERED-BANYAN NETWORKS. In a buffered-Banyan network, each switching element is a 2 × 2 crossbar switch with packet buffer on each of its two input lines. If we assume that the buffer size is one packet and both buffers at a switching element have a packet and are going to the same output of the switching element, one of the packets will be chosen randomly and the others will stay in the buffer. For a packet to be able to move forward, either the buffer at the next stage is empty or there is a packet in the buffer, and that packet is able to move forward.

The buffered-Banyan performs reasonably well in terms of blocking probability and delay for balanced load. However, for a bursty traffic pattern with long bursts, it may become heavily congested. A solution to this problem is to randomize traffic at the input so as to distribute it across the entire network. Thus, a distribution network is placed in front of the interconnect network.

SORT-BANYAN NETWORKS. The basic idea of preventing internal blocking in a sort-Banyan network is to implement a sorting network in front of a Banyan network to generate a strictly increasing order of destination addresses for the Banyan network (i.e., a compact and monotone sequence). Figure 12.22*a* shows the internal blocking in a Banyan network. Two packets, one destined to 001 and the other destined to 000, arrive simultaneously at input lines 0 and 5, respectively. Conflict occurs in the second stage as shown in the figure. A solution to this problem is shown in Figure 12.22*b*, in which a sorting network and a shuffle-exchange interconnect pattern are inserted in front of the Banyan network. Thus, the packet destined to output line 000 enters the banyan network at input line 0, whereas the packet destined to output line 001 enters the Banyan switch at input line 2. The paths for both packets are now disjoint (nonoverlapping) paths and hence there are no conflicts.

However, if the two packets are destined for the same output, they will collide within the Banyan network (i.e., output conflict). There, the sort-Banyan, as presented so far, resolves the internal blocking, but it does not solve the output contention problem. Different approaches were proposed to implement the sort-Banyan network with solutions to overcome the output contention.

The starlite switch developed by AT&T and the sunshine switch proposed by Bellcore are examples of the sort-Banyan switches. Below we discuss the starlite switch.

STARLITE SWITCH. To overcome the output conflict in a sort-Banyan network, the starlite switch uses a trap network between the sort and the Banyan network (Fig. 12.23). The trap network detects packets with the same destination addresses at the output of the sort network and separates them from the packets with distinct addresses. The packets with repeated addresses are fed back into the sort network in the next time slot. Care must be

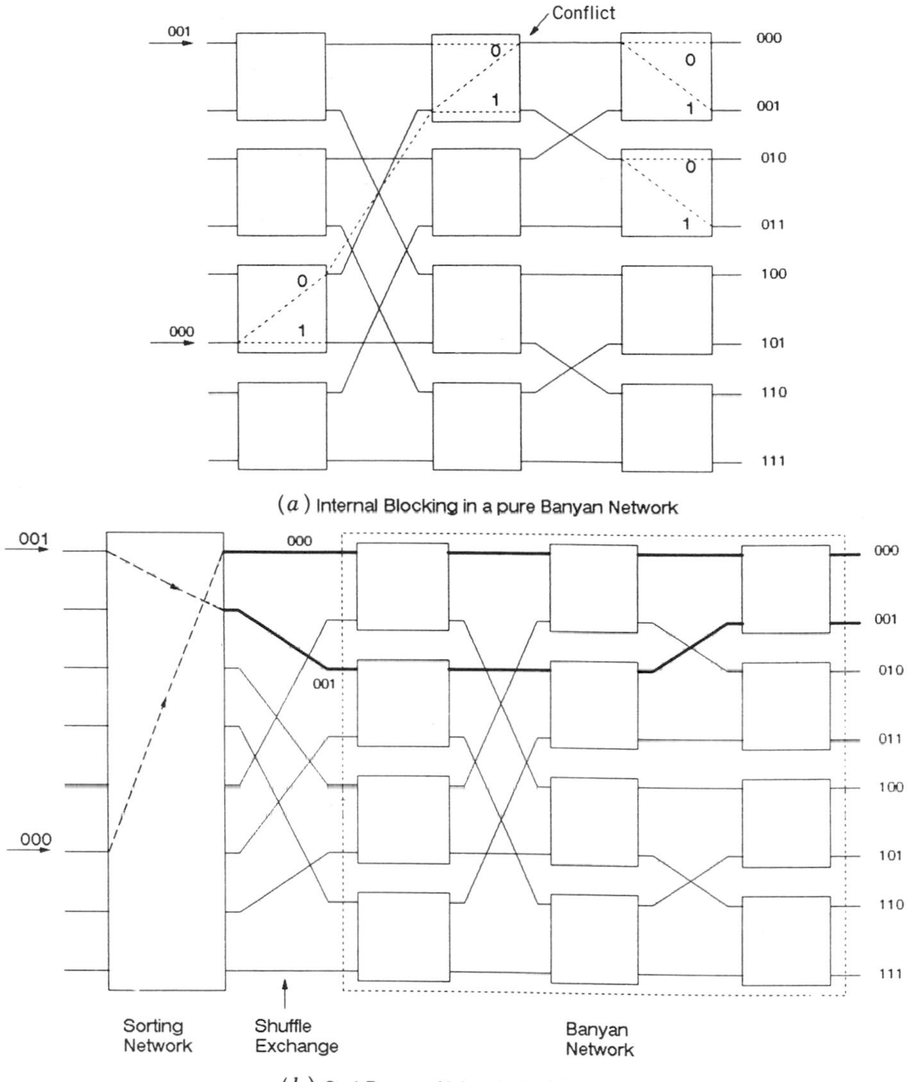

(b) Sort-Banyan Network (no internal blocking)

Figure 12.22. Basic architecture of a sort-Banyan network.

exercised, because the recycled packets through the switch may be delivered out of sequence. This problem can be solved by including a priority for the old packets over the new ones in the sort network. The purpose of the concentrator network shown in Figure 12.23 is to reduce the number of the input lines to the sort network. This is possible since it is likely that a significant number of users will be idle or a significant percentage of the active users will be idle.

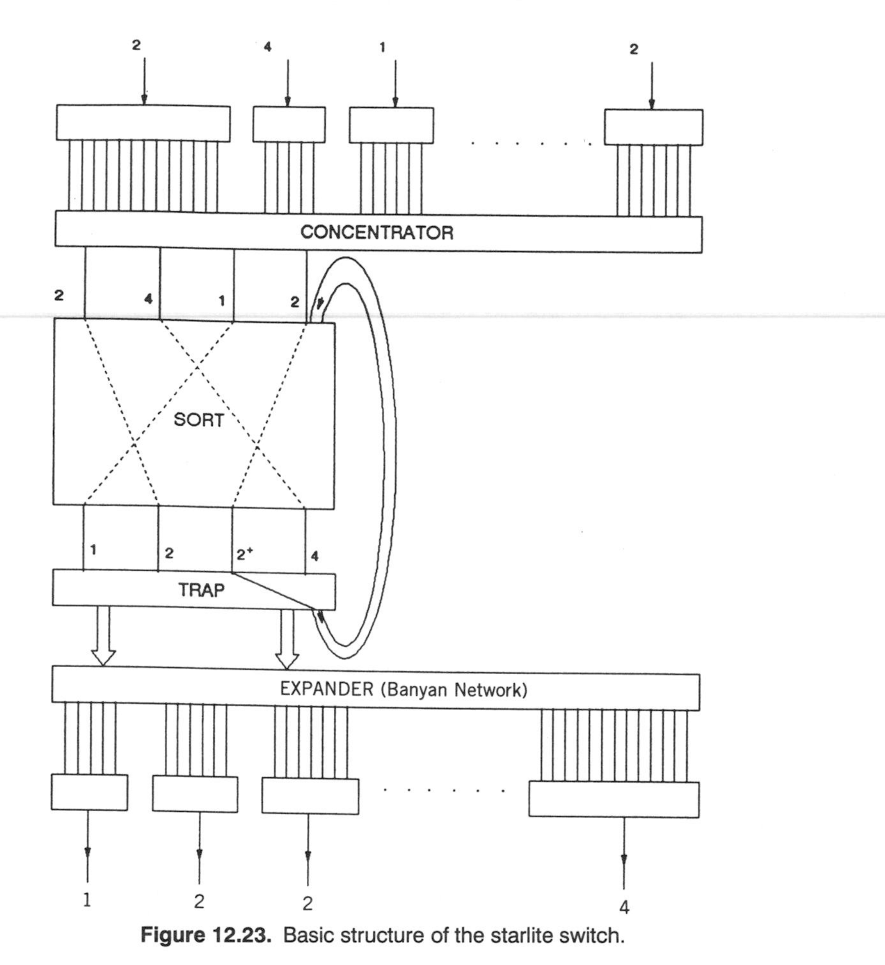

Figure 12.23. Basic structure of the starlite switch.

12.3.3.5 Switching Fabrics with Disjoint-Path Topology. The buffered-Banyan and the Batcher-Banyan are based on multistage interconnection networks comprised of small switching elements. The third type of space-division switches is based on a fully interconnected topology. Every input has a nonoverlapping direct path to every output so that no blocking or contention may occur internally. In addition, they employ output queueing to resolve the output port contention. Two good examples of switching fabrics with disjoint-path topology are the knockout switch and the integrated switch fabric. Below, we give a description of the knockout switch.

Knockout Switch. Proposed by AT&T Bell Laboratories, the knockout switch is geared toward a pure packet-switched environment. The knockout switch is designed for fixed-length packet, and the knockout II is designed for

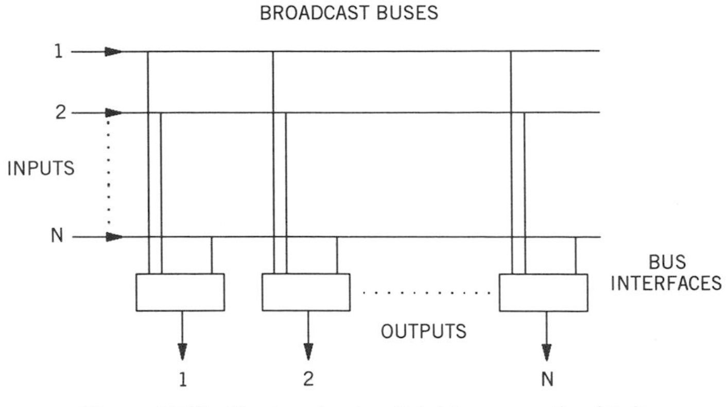

Figure 12.24. The knockout switch interconnection fabric.

a variable-length packet. The knockout switch uses one broadcast input bus from every input port to all output ports (see Fig. 12.24).

Each output port has access to the packets arriving on all inputs. This is achieved through the bus interface, which can receive packets from each input line. Thus, this simple structure provides several important characteristics. Since, each input has a direct path to every output, no internal blocking occurs. The only congestion in the switch occurs at the interface to each output, and thus the interface design should attempt to minimize the impact of this output conflict.

In addition to the above, note that each bus is undirectional and driven only by one input. Thus, higher transmission rates can be achieved on the buses, and the design is more tolerant of faults when compared to a shared parallel bus accessed by all inputs.

The knockout switch lends itself to broadcast and multicast communications. Finally, the switch is modular: The N broadcast bus can be on an equipment backplane with the circuitry for each of the N input–output pairs placed on a single plug-in circuit card. Hence, the switch can grow modularly from 2×2 up to $N \times N$ by adding additional circuit cards.

In Figure 12.25, one of the output bus interfaces is shown in more detail. The bus interface has three major components. The first component is the set of N packet filters, each interfacing a bus line. The packet filter recognizes the address of each packet on the broadcast bus and passes those with that output address to the next component, which is the concentrator. The $N \times L$, with $L \ll N$, concentrator selects up to L packets out of those accepted by the filters. If more than L packets are addressed to the same output line in a given slot, only L are received into the shared buffer and the remaining ones are lost. The shared buffer is composed of a shifter and L separate FIFO buffers. The L FIFO buffers are completely shared and, thus, are equivalent to a single FIFO queue with L inputs and one output.

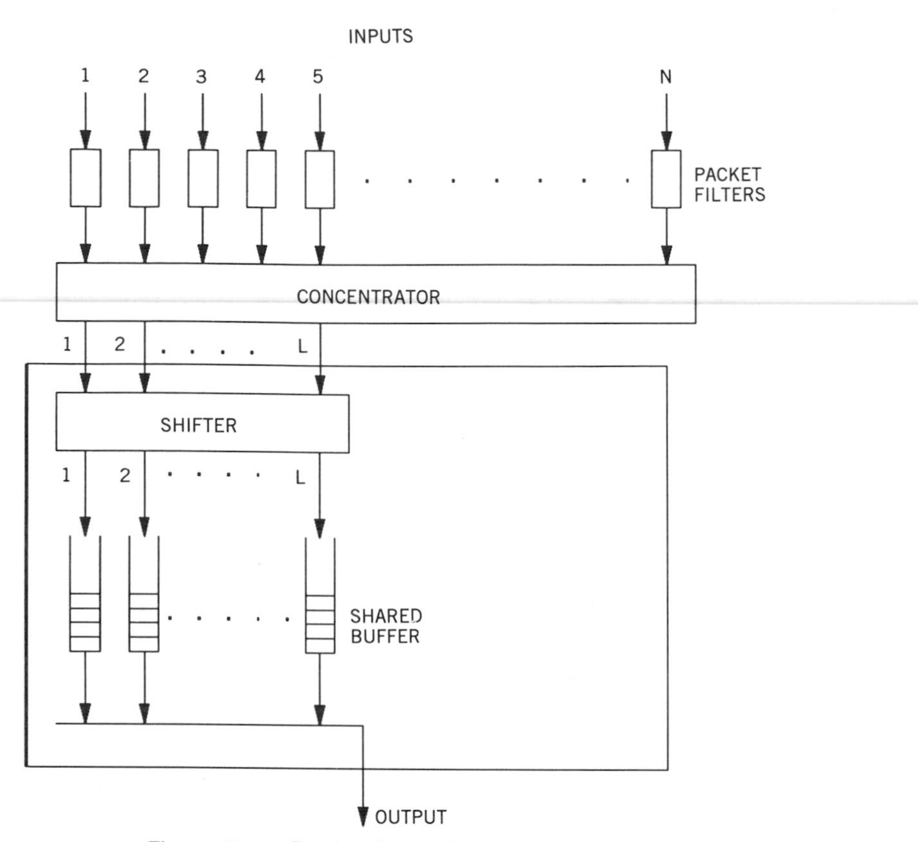

Figure 12.25. Bus interface — the knockout switch.

Figure 12.26 shows the packet format. The destination address is at the beginning of each packet, followed by a single activity bit. The destination address length is $\log_2 N$ bits, and the activity bit indicates the presence (logic 1) or absence (logic 0) of a packet in the time slot.

Determination of Packet Loss Probability in the Knockout Switch. The main idea behind the knockout switch design is that within any packet-switching network, packet loss is inevitable from transmission line errors, network failure, buffer overflows, etc. Clearly, recovery is made possible by various retransmission protocols. Hence, the fact that packets can be dropped within the N to L concentrator should not be alarming as long as the probability of packet loss from output congestion is kept below the loss expected from the other network sources mentioned above.

Let ρ be the probability that a packet independently arrives in a time slot at each input, and each packet is equally destined for each output with probability $1/N$. Then the probability of k packets arriving in a time slot all

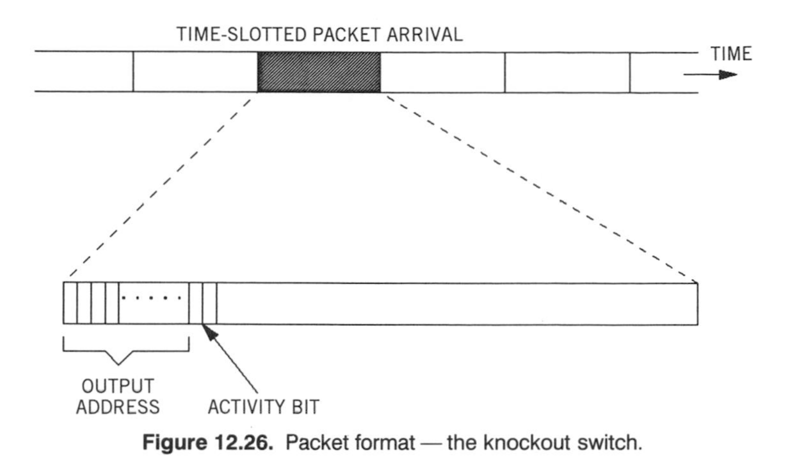

Figure 12.26. Packet format — the knockout switch.

destined for a given output

$$P_k = \binom{N}{k}\left(\frac{\rho}{N}\right)^k\left(1 - \frac{\rho}{N}\right)^{N-k}, \quad k = 0, 1, \ldots, N \tag{12.21}$$

A packet is dropped in the $N \to L$ concentrator if the number of packets arriving in a time slot destined for the same output exceeds the number of concentrator outputs L.

Since the average number of packets lost equals the probability of packet arrival times the packet loss probability P_L, we obtain the probability of a packet loss

$$\begin{aligned} P_L &= \left(\frac{1}{\rho}\right) \times \text{Ave. number of packets lost} \\ &= \left(\frac{1}{\rho}\right) \times \sum_{k=L+1}^{N} (k - L) P_k \\ &= \left(\frac{1}{\rho}\right) \times \sum_{k=L+1}^{N} (k - L) \binom{N}{k}\left(\frac{\rho}{N}\right)^k\left(1 - \frac{\rho}{N}\right)^{N-k} \end{aligned} \tag{12.22}$$

Taking the limit as $N \to \infty$, we get

$$\text{Prob[packet loss]} = P_L = \left(1 - \frac{L}{\rho}\right)\left(1 - \sum_{k=0}^{L} \frac{\rho^k}{k!} e^{-\rho}\right) + \rho^L \frac{e^{-\rho}}{L!} \tag{12.23}$$

Using Eqs. 12.22 and 12.23 we show in Figure 12.27, the probability of packet loss versus the number of outputs on the concentrator L for $N = 16, 32, 64,$

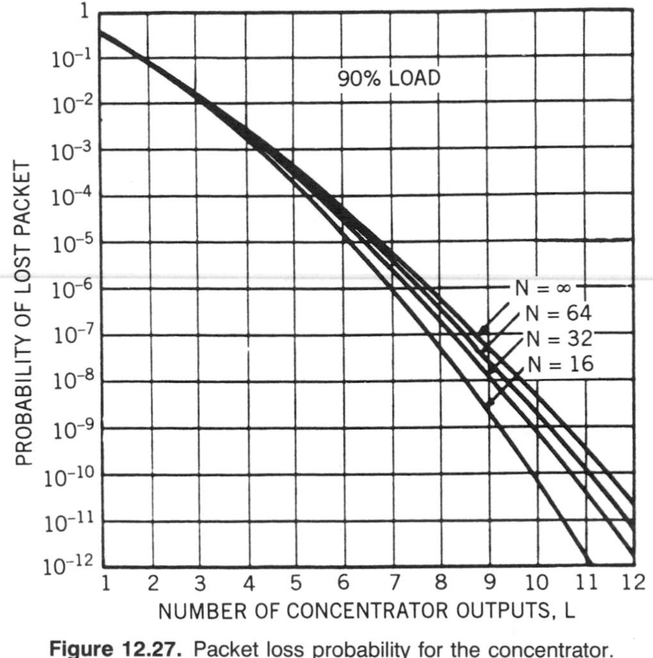

Figure 12.27. Packet loss probability for the concentrator.
Source From [YEH87] © 1987 IEEE.

and ∞ at an offered load of $\rho = 0.9$. For a probability of packet loss to be less than 10^{-6} for large N, only eight concentrator outputs are required. Thus, the number of separate buffers needed to receive simultaneously arriving packets with a relatively low probability of packet loss is reduced from large N to $L = 8$. It was assumed above that any packet is equally destined to any of the output lines. However, if the traffic pattern is nonuniform, L has to be high (up to 20) to sustain the same packet loss rate.

The $N \times L$ Concentrator Design. Figure 12.28 shows a 2×2 contention switch used as the basic building block of the concentrator. The two outputs are labeled "winner" and "loser." The activity bit (equals 1 for an arriving packet) controls the 2×2 contention switch. If only one input (left or right input) has an arriving packet (activity bit = 1), it is routed to the winner (left) output. To achieve this, the switch examines the activity bit for only the left input. If it is 1, the left input is routed to the winner output and the right input is routed to the loser output.

Figure 12.29 shows a block diagram for an 8 to 4 concentrator made of these 2×2 switch elements. The box marked D is a 1-bit delay element for synchronization purposes. The concentrator follows a certain scheme to select a maximum of L packets from the N incoming lines. The scheme is similar to a tournament. In the first round of tournament, N players

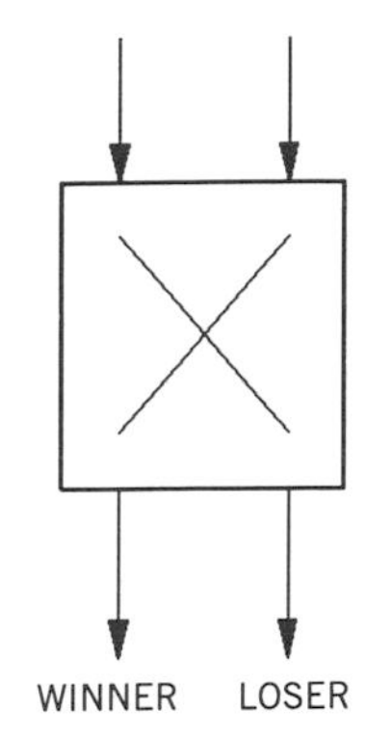

(a) The 2 × 2 contention switch

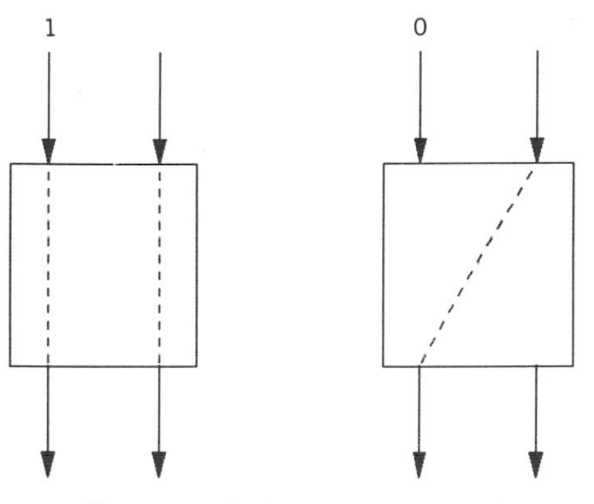

(b) States of the 2 × 2 contention switch

Figure 12.28. The 2 × 2 switching element used in the knockout switch.

compete, $N/2$ winners are determined, and they advance to the second round, which results in $N/4$ winners and so on. Similarly the losers compete with each other. Note that whenever there is an odd number of players in a round, one player must wait and compete in a later round in the section. This is achieved by using the delay element. As an example, Figure 12.29 shows four packets arriving at input lines 1, 2, 4, and 5 (the activity bits on these lines are equal to one). Packets on lines 1 and 2 (p_1 and p_2) compete, and p_1 is a winner. Packets on line 4 (p_4) proceed to the second round and lose to p_1. Packets on line 5 (p_5) proceed to round 2 and then 3 to lose to p_1. The dotted lines show the path for every packet. A packet losing L times is knocked out of the competition and is lost. This event occurs only if more than L packets arrive in any given slot. Fortunately, the probability, for $N \geq 8$, of packet loss is very small.

The shared buffer is the next block after the concentrator (Fig. 12.25). The use of L separate FIFO buffers provides the capability of storing up to L packets within a single time slot.

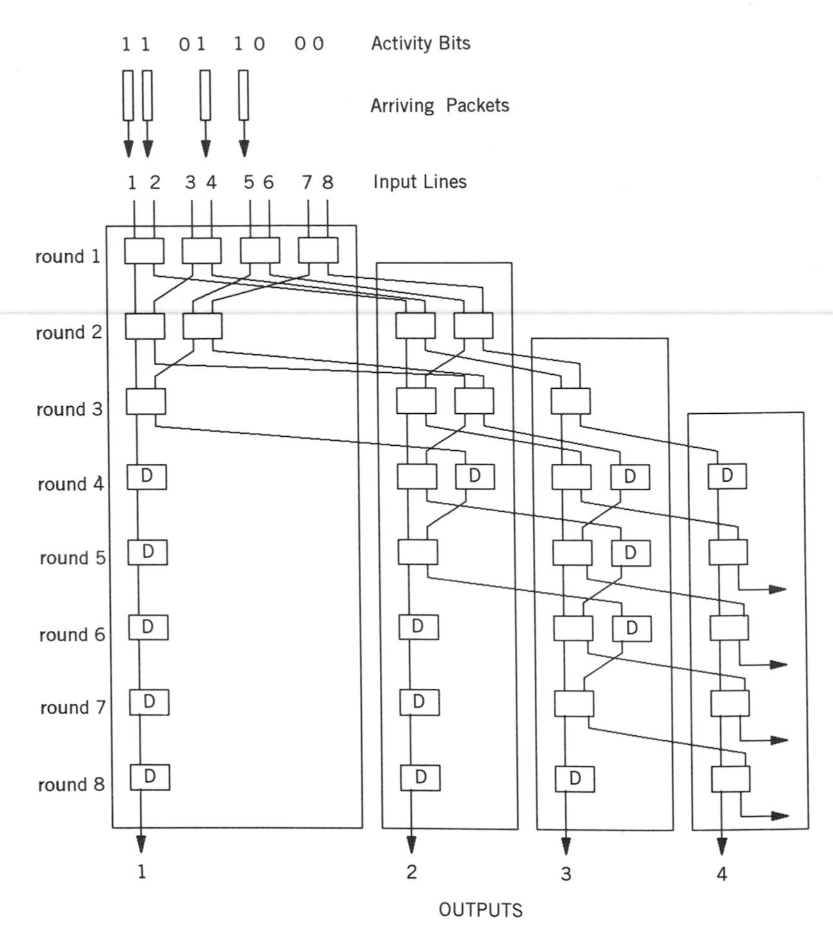

Figure 12.29. The 8-input / 4-output concentrator.

12.4 PHOTONIC SWITCHING

With the wide-scale introduction of optical fibers, it is expected that early in the next century essentially all telecommunications signals, both long distance and local loop, will be carried by optical fibers. The switching system that interconnects a large collection of these fiber-optic cables can be divided into three main categories. In the first category, the switching system includes an optical-to-electrical conversion (o/e) followed by an electronic switch, such as a fast packet switch. The electronic switch is then followed by an electrical-to-optical conversion (e/o). The second category is all optical (or photonic) switches that avoid the need (and associated cost, power, etc.) for o/e or e/o conversion. The optical signal is switched directly and no electrical conversion is required. In this category, the control of the network is electronically implemented. These photonic switches can switch, under elec-

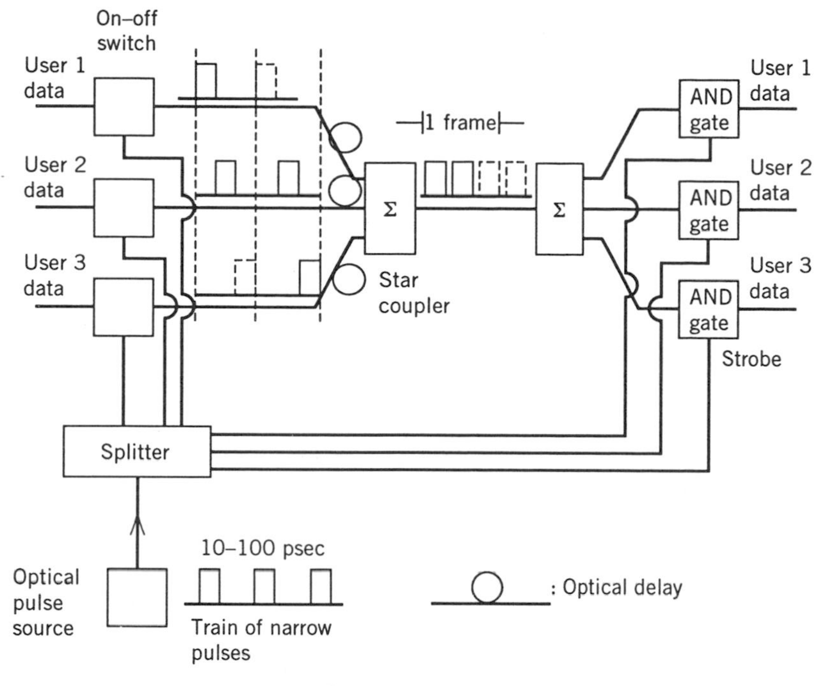

Figure 12.30. Optical time-division multiplexing.

trical control, extremely wideband signals (e.g., gigabits per second) at potential reconfiguration times on the order of less than 1 nsecond. The final category of switching systems is similar to the previous class in that no o/e or e/o conversion takes place. The difference is that the control of this class of photonic switch is optical. Current research on optical switches is focused on applications in space-, time- and wavelength- (frequency) division switching systems. Two important concepts in the design of optical switches are optical time-division multiplexing and wavelength-division multiplexing. Below, we give a brief description of each.

12.4.1 Optical Time-Division Multiplexing

Figure 12.30 shows an optical time-division multiplexing approach. The optical source generates very narrow optical pulses (say of width 1 to 10 psecond, corresponding to bandwidths of 100–1000 GHz). The narrow pulses are split into N paths. For each path, the narrow pulses are modulated by the user's data. Thus, a narrow pulse will either be passed or blocked, corresponding to whether the user's bit is a logical 1 or a logical 0. Delay is inserted into each path such that successive paths are offset in time by one narrow pulse, thus time multiplexing the narrow bits associated with each user.

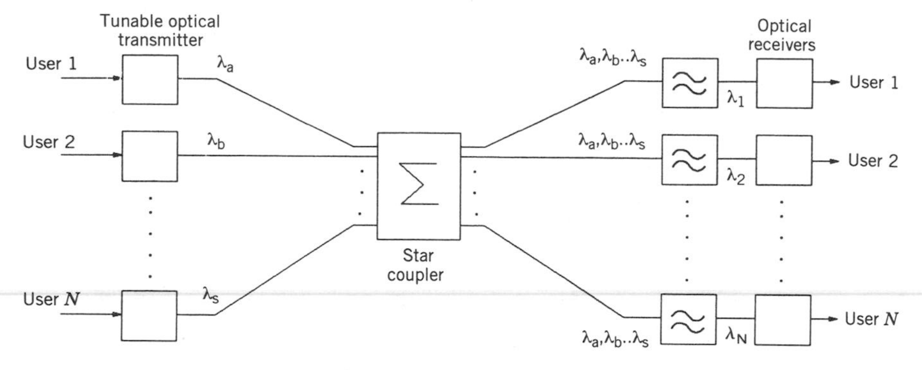

Figure 12.31. Optical wavelength-division multiplexing.

Now, the composite signal is split prior to reaching the receiver and is fed into optical AND gates. Optical AND gates have two inputs: the composite signal and a suitably delayed replica of the original narrow pulse stream. The output of the AND gates is processed by an optical receiver to electronically regenerate the desired packet.

12.4.2 Wavelength-Division Multiplexing

Figure 12.31 shows the basic concepts of optical wavelength-division multiplexing (WDM). Each receiver is assigned a unique wavelength and a transmitter wishing to access that receiver tunes its transmitter to that receiver's wavelength and sends its packet. A passive star coupler linearly combines all simultaneously transmitted packets. The assignment of different wavelengths to each individual packet preserves their individual identity. Packets addressed to the same receiver do collide, because they are assigned the same wavelength of the intended receiver. However, a proper multiple-access algorithm (see Chapter 7) can be devised.

The main drawback of this approach is the need for a bank of transmitting lasers, one for each receiver. The lasers should be rapidly tunable (i.e., able to tune in tens of nanoseconds) over the broad optical band. Another approach is to assign a unique wavelength to each transmitter. However, the receiver will select, from the multitude of WDM signals, on a packet-by-packet basis, the corresponding wavelength at each point in time. A side channel for signaling is needed to inform the receiver of the appropriate channel to which it must tune. Instead, a rapidly tunable optical filter may be used.

SUGGESTED READINGS

[CLOS53] presents the analysis of nonblocking switching networks. [BENE64] discusses rearrangeable networks. [MCDO83], [NUSS79], [REY86], [ABBO84], and [JOEL79], are good references for digital switching techniques.

[TURN86] presents motivations and directions for high-speed switching. [HLUC88] and [KARO87] are excellent references for the impact of buffer allocation on the design of fast packet switches. [DEVA88] discusses the architecture of the Prelude switch. The Paris switch is discussed in [CIDO88]. Further readings on the Delta network can be found in [PATE81] and [DIAS84]. [JENG83] provides a detailed analysis of buffered-Banyan networks and [RATH88] for ATM switching network. The Starlite switch is described in [HUAN84], and the Batcher-Banyan is described in [BATC68]. The knockout switch is described in [YEH87]. [TOBA90] and [AHMA89] are excellent surveys of fast packet switching. [HUI90] presents a mathematical treatment of switching and traffic theories. [ELHA92] discusses improving switch performance in satellite environment.

[MIDW88], [ACAM89], [BRAC89], [GREE92a & b], [CHEU90], [HINT93], and [BALD87] are good references on optical switching and systems.

PROBLEMS

12.1 **a.** Obtain the minimum value of N (number of lines) at which a three-stage nonblocking spacc-division switch has a number of crosspoints that is less than that of a corresponding single-stage switch.

b. Draw the blocking diagram for a three-stage Clos switch with $N = 100$, showing the number of arrays (i.e., matrices) in each stage.

12.2 Let Q_m be the number of packets in a given output queue of a packet switch at the end of the mth time slot and A_m be the number of packet arrivals during the mth time slot. b is the buffer size for every output line. Prove that

$$Q_m = \min\{\max(0, Q_{m-1} + A_m - 1), b\}$$

12.3 A fast packet switch of the shared-memory type employs the complete partitioning approach. It has three input and three output lines. The total memory size is 2. Assume the probability of a packet arrival in a slot $\rho = 0.6$.

a. Determine a_k the probability of k packet arrivals to a given output during a time slot.

b. Obtain P_n, the probability of having n packets at a given output, $0 \preccurlyeq n \preccurlyeq 2$.

c. What is the packet loss probability?

d. Repeat for $\rho = 0.9$.

12.4 Depending on the particular internal fabric of the space-division packet switch, some of these switches are blocking switches. Explain why it is not possible to buffer packets at the output of a blocking space-division switch.

12.5 Consider a space-division packet switch in which we eliminate the input queues. Whenever k packets are addressed to a particular output in a time slot, only one can be transmitted over the output trunk. The remaining $(k - 1)$ packets are dropped from the switch.

a. Prove that the switch throughput ρ_0 is given by

$$\rho_0 = 1 - \left(1 - \frac{\rho}{N}\right)^N$$

where N is the total number of input lines and ρ is the input line utilization.

b. What is the probability that an arbitrary packet will be dropped from the switch p_{dropping}?

c. For $N = \infty$, what is the threshold value for ρ at which the switch throughput ρ_0 is larger when we drop packets than when we queue them on input trunks? Explain why.

12.6 Is the interconnection network shown in Figure 12.32 a Delta network? Explain why or why not.

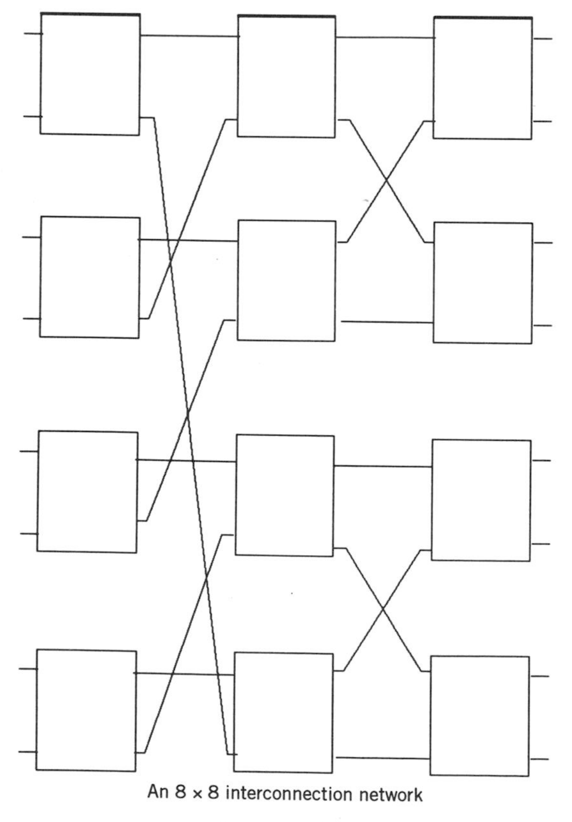

An 8 × 8 interconnection network

Figure 12.32.

12.7 Consider an $a^n \times b^n$ Delta network composed of N $(a \times b)$ switching elements, prove that the total number of $a \times b$ switching elements N is given by

$$N = \begin{cases} \dfrac{(a^n - b^n)}{(a - b)}, & a \neq b \\ nb^{n-1}, & a = b \end{cases}$$

Obtain N for a 16×16 Delta network composed of N 2×2 switching elements.

12.8 For a $4^2 \times 3^2$ Delta network using the 4-shuffle as the interstage link pattern, obtain the number of switching elements N and draw that network showing the output line addresses.

12.9 To provide multiple paths between each input and output pair to keep the Delta network operational in the presence of few faults, three possible strategies can be used: (1) add an extra stage of switches to create two disjoint paths from any input link to each output link (these networks are called augmented networks), (2) have multiple Delta networks in parallel, (3) use multiple links for each switch connection.

a. Draw a 8×8 augmented network and show an example of two disjoint paths.

b. Draw block diagrams for strategies (2) and (3).

12.10 Draw a 8×8 Delta-2 network using the 2-shuffle as the interstage link pattern.

a. Assume two packets simultaneously arrive at input lines 0 and 3. Each packet is destined to output lines 3 (binary address 011). Draw the path for each packet and show, if any, the type of conflict that occurs.

b. A packet arrives in a given time slot at input line 0 and is destined to output line 3. At the same time slot, another packet arrives at input line 4 and is destined to output line 2. Draw the path for each packet and show, if any, the type of conflict that occurs.

12.11 Two packets arrive simultaneously at a $2^3 \times 2^3$ Delta network. The first packet arrives at input line 2 and is destined to output line 011. The second packet arrives at input line 3 and is destined to output line 010. Assume the interstage pattern is the 2-shuffle.

a. Draw the Delta network and show the paths for both packets indicating any conflicts.

b. Draw the sort-Banyan network for the above network showing the paths for both packets.

12.12 Draw a 4×2 knockout concentrator. Assume at a given time slot that there is a packet arriving at each of the four input lines to the concentrator. Draw dotted lines to show the state of the 2×2 switching element and the winning packets.

12.13 Repeat the above problem assuming a packet arrives at each of input lines 1 and 4 only.

12.14 The packet arriving at the left input of the knockout concentrator always has the highest priority (i.e., it is always a winner) and the lowest priority packet is the one arriving at the Nth input line (the worst-case input).

a. Prove that the packet loss probability for the worst-case input ρ_ω, as $N \rightarrow \infty$, is given by

$$\rho_\omega = 1 - \sum_{K=0}^{L-1} \frac{\rho^K}{K!} e^{-\rho}$$

where ρ is the offered load.

b. Compare (a) to the packet loss probability averaged over all inputs as given by Eq. 12.23.

c. Suggest how to improve the probability of packet loss for the worst-case input.

APPENDIX 1A

TELECOMMUNICATIONS STANDARDS ORGANIZATIONS

This appendix lists some of the major organizations involved in telecommunications standards. We group them into three categories; international, regional, and national.

1A.1 INTERNATIONAL ORGANIZATIONS

1A.1.1 International Telecommunications Union (ITU)

The ITU is the specialized agency of the United Nations responsible for telecommunications. The ITU traces its origins back to 1865. Its members are national governments; there are presently 181 member countries (as of 1993). Recognized private operating agencies such as AT&T, scientific and industrial organizations, such as IBM, and international organizations such as the ISO, can participate in the technical committees. The ITU headquarters are in Geneva, and it comprises four permanent organs: the General Secretarial, the International Telegraph and Telephone Consultative Committee (CCITT)* and the International Radio Consultative Committee (CCIR).

The United States Department of State oversees U.S. participation in the CCIR and CCITT as the official U.S. member in the ITU. It appoints the head of the U.S. delegation and accredits the delegates from various U.S. companies. The CCIR activities are concentrated on radio communications (spectrum use, satellite, microwave, mobile services) and television (including high definition television). The CCITT's jurisdiction is over all other aspects of telecommunications. It carries out studies and issues recommendations on technical, operational, and traffic questions relating to all public-service telecommunications-telegraphy, telephony, data transmission, telex, and related telecommunications services other than those directly connected with

*As of 1993, CCITT became ITU Telecommunication Standardization Sector (ITU-TSS)

radio communications. The results of the studies are published as recommendations. The process of adopting a recommendation usually proceeds in a 4-year cycle (or a 2-year cycle for an accelerated procedure).

1A.1.2 International Standards Organization (ISO)

Founded in 1946, the ISO is a nongovernmental organization that operates on a voluntary basis. Experts from manufacturers, consumers, governmental, and general-interest groups contribute to national positions that are presented to the ISO by a member body. Although the ISO is a nongovernmental organization, more than 70 percent of the ISO member bodies are governmental standards institutions or organizations incorporated by public law. The U.S. member is the American National Standards Institute (ANSI). As of 1988, the ISO had 73 full members from the national standards organizations. Full members are entitled to participate and exercise full voting rights on any technical committee and can participate in the administration of the ISO. The ISO had 14 correspondent members that do not actively participate in the technical work and have no voting powers.

1A.1.3 International Electrotechnical Commission (IEC)

Established in 1906, the IEC is a nongovernmental organization that has over 43 member countries. Activities cover the areas of electronic components, test methods, electrical apparatus, wire, cable and connectors, measuring instruments, and so on.

1A.2 REGIONAL ORGANIZATIONS

A number of regional international organizations are in existence today. Examples are the Conference of European Post and Telecommunications Administration (CEPT), the European Telecommunications Standards Institute (ETSI), Pacific Telecommunications Council (PTC), Group Special Mobile (GSM), the Inter American Telecommunication Conference (CITEL) for Latin America, and the Arab Telecommunication Union. Other important European organizations include European Telecommunication and Professional Electronics Industry (previously known as European Committee of Telecommunications and Electronic Industries, ECTEL), the European Computer Manufacturers Association (ECMA), the European Committee for Standardization (CEN), the European Economic Commission (EEC), and the European Committee of Electrotechnical Standardization.

1A.2.1 The European Telecommunication Standards Institute (ETSI)

ETSI is a pan-European standards organization established in 1988. It includes members from within and outside the European Community.

1A.2.2 The Japan Standards Association

This nonprofit institution was established in 1945 to promote industrial standardization and quality control in industry. The *Telecommunication Technology Council* (*TTC*) of Japan acts as a clearinghouse for data on international standards and technical specifications.

1A.3 NATIONAL ORGANIZATIONS

1A.3.1 American National Standards Institute (ANSI)

Founded in 1918, ANSI is a nonprofit organization with the main goal of coordinating voluntary standards activities in the United States. Its members include industrial companies, professional societies, trade associations, consumer groups, and governmental and regulatory bodies.

ANSI also represents the United States in the ISO and the IEC. Its role with respect to the ISO is analogous to that of the State Department for the CCITT and the CCIR.

1A.3.2 Institute of Electrical and Electronics Engineer (IEEE)

Over the years, the IEEE, one of the largest professional societies in the world, has produced a number of telecommunications-related standards, primarily through the efforts of the IEEE Communications Society Technical Committees. The IEEE is an ANSI-accredited organization.

1A.3.3 Electronic Industries Association (EIA)

Established in 1924, the EIA is an ANSI-accredited organization of electronic equipment manufactures.

1A.3.4 Standards Committee T1

The Standards Committee T1 is sponsered by the *Exchange Carriers Standards Association* (*ECSA*) and, as of 1984, its membership includes 90 exchange carriers, manufacturers, vendors, government agencies, user groups, and so on. It is ANSI-accredited.

1A.3.5 Government Organizations

The Federal Communications Commission (FCC) is an "independent agency" formed by the U.S. Congress in 1934 to regulate wire and radio communications. Other important government organizations are the National Institute of Standards and Technology (NIST, formerly the National Bureau of Standards), the Department of Defense (DOD), and the National Telecommunication and Information Administration (NTIA) which is part of the Department of Commerce.

APPENDIX 2A

CHARACTER CODES

Examples of character codes are the Morse code, the Baudot (5 bits per character code), the Binary-coded Decimal (BCD), which is a 6-bits-per-character code; the Extended Binary-Coded Decimal Interchange Code (EBCDIC), which is an 8-bits-per-character code; and the American Standard Code for Information Interchange (ASCII), which is a 7-bits-per-character code. Another code is the International Alphabet Number 5 (IA5) which is the standard code defined by the CCITT and recommended by the ISO. In practice, IA5 is almost exactly the same as the ASCII code. The following is a description of these transmission codes.

MORSE CODE

Morse Code, invented by Samuel Morse in 1837, is perhaps the earliest code. It consists of a series of dots and dashes of unequal length. The dash is three times the length of the dot. Besides the dash and the dot, Morse code defines also the space. Hence, the Morse code is a ternary code and not a binary code. The space between dot and dashes in a single letter is one unit of time, whereas between letters it is three time units, and between words it is six time units. The Morse code takes into account the relative frequency in the English language to optimize the time to transmit a message (Table 2.A1). For example the letter T (encoded as –) occurs more frequently than the letter Y (encoded as –·––). In this aspect, Morse code is an intuitive precursor of the information theory and coding theory. Morse code is a variable length code. However the difficulty of recognizing the words, makes it impractical for data and computer communications. One reason is that the coded character is not encapsulated between a start and a stop bit.

TABLE 2.A1 Morse Code

Symbol	International Morse
A	·–
B	–···
C	–·–·
D	–··
E	·
F	··–·
G	––·
H	····
I	··
J	·–––
K	–·–
L	·–··
M	––
N	–·
O	–––
P	·––·
Q	––·–
R	·–·
S	···
T	–
U	··–
V	···–
W	·––
X	–··–
Y	–·––
Z	––··

BAUDOT CODE

The Baudot Code, developed by Emil Baudot, encodes character into an equal number of bits (fixed length code), in this case five bits per character. Hence, the Baudot Code allows up to 32 unique character codings. By the use of two shift characters; letter shift and figure shift, it was possible to encode more numeric and control characters. Table 2.A2 shows the Baudot Code. For example, to transmit the letter T, a letter shift of bit pattern 11111 is first transmitted followed by bit pattern 00010; while to transmit number 5, a figure shift of 11011 is followed by bit pattern 00010. The Baudot Code is used in the Telex network for sending messages and it may still find limited application with radio amateurs and some communication systems for the deaf, as well as being used in most printing telegraphy.

BCD / EBCDIC CODES

The binary coded decimal (BCD) code is one of the earliest codes that converts the alphanumeric characters into binary representation. BCD is a

TABLE 2.A2 Five-Level Baudot Code

		Bit Selection				
Letters	Figures	1	2	3	4	5
A	–	1	1			
B	?	1			1	1
C	:		1	1	1	
D	$	1			1	
E	3	1				
F	!	1		1	1	
G	&		1		1	1
H				1		1
I	8		1	1		
J		1	1		1	
K	(	1	1	1	1	
L	)		1			1
M	.			1	1	1
N	,			1	1	
O	9				1	1
P	0		1	1		1
Q	1	1	1	1		1
R	4		1		1	
S		1		1		
T	5				1	
U	7	1	1	1		
V	;		1	1	1	1
W	2	1	1			1
X	/	1		1	1	1
Y	6	1		1		1
Z	"	1				1
FUNCTIONS						
Carriage Return					1	
Line Feed			1			
Space				1		
Letters Shift		1	1	1	1	1
Figures Shift		1	1		1	1

6-bits-per-character code. Hence it can represent up to 64 different character. Table 2.A3 shows the code format for BCD.

The extended binary coded decimal interchange code (EBCDIC) is an extension of the BCD and it is an 8-bits-per-character code. EBCDIC is a IBM proprietary code, and it is used with most IBM terminals and computers. EBCDIC allows, in addition to encoding letters, numbers, and punctuation marks, the encoding of a large number of control characters that are needed to establish the handshaking between the communicating devices. Also, because of the large number of possible different characters coding ($2^8 = 256$ combinations) a number of combinations are not used in EBCDIC. Table 2.A4 shows a typical implementation of EBCDIC for the IBM 3270 terminal.

TABLE 2.A3 Binary Coded Decimal System

Bit Position						
b_6	b_5	b_4	b_3	b_2	b_1	Character
0	0	0	0	0	0	A
0	0	0	0	1	1	B
0	0	0	0	1	0	C
0	0	0	1	0	1	D
0	0	0	1	0	0	E
0	0	0	1	1	1	F
0	0	0	1	1	0	G
0	0	1	0	0	1	H
0	0	1	0	0	0	I
0	1	0	0	0	1	J
0	1	0	0	1	0	K
0	1	0	0	1	1	L
0	1	0	1	0	0	M
0	1	0	1	0	1	N
0	1	0	1	1	0	0
0	1	0	1	1	1	P
0	1	1	0	0	0	Q
0	1	1	0	0	1	R
1	0	0	0	1	0	S
1	0	0	0	1	1	T
1	0	0	1	0	0	U
1	0	0	1	0	1	V
1	0	0	1	1	0	W
1	0	0	1	1	1	X
1	0	1	0	0	0	Y
1	0	1	0	0	1	Z
1	1	0	0	0	0	0
1	1	0	0	0	1	1
1	1	0	0	1	0	2
1	1	0	0	1	1	3
1	1	0	1	0	0	4
1	1	0	1	0	1	5
1	1	0	1	1	0	6
1	1	0	1	1	1	7
1	1	1	0	0	0	8
1	1	1	0	0	1	9

ASCII CODE (AND IA5)

The most common code in use is the American Standard Code for Information Interchange (ASCII), which is defined in ANSI X3.4-1968 and is shown in Table 2.A5. The code is basically identical to the CCITT Alphabets and the ISO standard 646.

ASCII is a 7-bits-per-character code; thus 128 different characters can be represented. With this large number of different characters, ASCII allows for a range of additional control, and graphical characters. Table 2.A5 shows the ASCII code. Although the ASCII code is a 7-bits-per-character code, the

TABLE 2.A4 EBCDIC Code Implemented for the IBM 3270 Information Display System

	Bits 0 1	0 0				0 1				1 0				1 1			
	Bits 2 3	00	01	10	11	00	01	10	11	00	01	10	11	00	01	10	11
Bits 4567	HEX 0 / HEX 1	0	1	2	3	4	5	6	7	8	9	A	B	C	D	E	F
0000	0	NUL	DLE			SP	&	-									0
0001	1	SOH	SBA					/		a	j			A	J		1
0010	2	STX	UEA		SYN					b	k	s		B	K	S	2
0011	3	ETX	IC							c	l	t		C	L	T	3
0100	4									d	m	u		D	M	U	4
0101	5	PT	NL							e	n	v		E	N	V	5
0110	6			ETB						f	o	w		F	O	W	6
0111	7			EBC	EOT					g	p	x		G	P	X	7
1000	8									h	q	y		H	Q	Y	8
1001	9		EM							i	r	z		I	R	Z	9
1010	A					C	!		:								
1011	B					.	$	.	#								
1100	C		DUP		RA	.	*	%	@								
1101	D		SF		NAK	(	)	-	'								
1110	E		FM			.	:		.								
1111	F		ITB		SUB			.									

TABLE 2.A5 U.S. ASCII Code

				$b_7 \rightarrow$	0	0	0	0	1	1	1	1
				$b_6 \rightarrow$	0	0	1	1	0	0	1	1
				$b_5 \rightarrow$	0	1	0	1	0	1	0	1
Bits				Column	0	1	2	3	4	5	6	7
b_4	b_3	b_2	b_1	Row								
0	0	0	0	0	NUL	DLE	SP	0	@	P	‘	p
0	0	0	1	1	SOH	DC1	!	1	A	Q	a	q
0	0	1	0	2	STX	DC2	"	2	B	R	b	r
0	0	1	1	3	ETX	DC3	#	3	C	S	c	s
0	1	0	0	4	EOT	DC4	$	4	D	T	d	t
0	1	0	1	5	ENQ	NAK	%	5	E	U	e	u
0	1	1	0	6	ACK	SYN	&	6	F	V	f	v
0	1	1	1	7	BEL	ETB	'	7	G	W	g	w
1	0	0	0	8	BS	CAN	(	8	H	X	h	x
1	0	0	1	9	HT	EM	)	9	I	Y	i	y
1	0	1	0	10	LF	SUB	*	:	J	Z	j	z
1	0	1	1	11	VT	ESC	+	;	K	[	k	{
1	1	0	0	12	FF	FS	,	<	L	\	l	¦
1	1	0	1	13	CR	GS	–	=	M	]	m	}
1	1	1	0	14	SO	RS	.	>	N	^	n	~
1	1	1	1	15	SI	US	/	?	0	_	o	DEL

encoded characters are, in most cases, stored and transmitted using 8 bits per character, i.e., one byte. There are a number of ways to choose the value of the eighth bit. It could be left arbitrary and no use made of it. In some cases, it can always be set to 1 so that it can be used as a timing source. In most cases, the eighth bit is a parity bit used to detect errors. The parity bit is set, to binary 1 or 0, such that the total number of 1s in each byte is an odd number (in this case it is referred to as odd parity), or an even number (in this case it is referred to as even parity). Error detection is discussed in detail in Section 2.9.

APPENDIX 3A

THE MOMENT-GENERATING AND CHARACTERISTIC FUNCTIONS

In probability theory higher moments of a random variable provides some useful vehicles for evaluating the mean, variance, and distribution functions of continuous as well as discrete random variables. The moment-generating function (MGF) or probability generating function of a discrete random variable Y is given by

$$G_Y(z) \triangleq E(z^Y) = \sum_{i=-\infty}^{\infty} z^i P_Y(i)$$

where $P_Y(i)$ is the probability density function (pdf) of Y, and z is a complex variable; $G_Y(z)$ can be looked upon as the z transform of Y. From $G_Y(z)$, it easily follows that,

$$G_Y(1) = \sum_{i=-\infty}^{\infty} P_Y(i) = 1$$

The first and second moments of Y (mean and mean square value) are easily obtained by successive differentiation as follows:

$$\left.\frac{dG_y(z)}{dz}\right|_{z=1} = \sum_{i=-\infty}^{\infty} iz^{i-1}P_Y(i)\big|_{z=1} = \sum_{i=-\infty}^{\infty} iP_Y(i) = \overline{Y}$$

$$\left.\frac{d^2G_y(z)}{dz^2}\right|_{z=1} = \sum_{i=-\infty}^{\infty} i(i-1)z^{i-2}P_Y(i)\big|_{z=1} = \sum_{i=-\infty}^{\infty} i^2P_Y(i) - \sum_{i=-\infty}^{\infty} iP_Y(i)$$

$$= \overline{Y}^2 - \overline{Y}$$

and

$$\sigma_Y^2 = \overline{Y^2} - \overline{Y}^2 = \left.\frac{d^2G_Y(z)}{dz^2}\right|_{z=1} + \left.\frac{dG_Y(z)}{dz}\right|_{z=1} - \left\{\frac{dG_Y(z)}{dz}\right\}^2_{z=1}$$

As examples, we take the (zero-one) distribution $P_Y(0) = q$ and $P_Y(1) = (1 - q) = p$, it easily follows that

$$G_Y(z) = q + pz, \quad E(Y) = p, \quad E(Y^2) = p \quad \sigma_Y^2 = pq$$

The power of the (MGF) becomes evident when we deal with sums of random variables, such as the binomially distributed random variable, which is the sum of n (zero-one) distributed random variables (x_i), that is,

$$Y = x_1 + x_2 + \cdots + x_n$$

It is easily seen that if $x_{1,}, \cdots, \mathrm{x}_n$ are independent random variables then,

$$G_Y(z) = E\{(z)^Y\} = E\{Z^{x_1}\} \cdot E\{Z^{x_2}\} \ldots E(Z^{x_n})$$

$$= G_{x_1}(z) \cdot G_{x_2}(z) \cdot \ldots \cdot G_{x_n}(z)$$

and in the case of n identically distributed random variables

$$G_Y(z) = \prod_{i=1}^{n} G_{x_i}(z) = [G_x(z)]^n$$

For the binomially distributed Y variable, we finally obtain

$$G_Y(z) = \prod_{i=1}^{n} G_{x_i}(z) = [G_x(z)]^n$$

At this point, it becomes clear that the pdf of Y can also be obtained from its MGF, for this distribution and can be expanded,

$$G_Y(z) = q^n + nq^{n-1} \cdot pz + n(n-1)q^{n-2}P^2z^2 + \cdots + \binom{n}{i} p^i(q)^{n-i} z^i + \cdots$$

So the coefficient of z^i is nothing but the aforementioned $P_Y(i)$.

It is a straighforwardly seen that,

$$\overline{Y} = np, \quad \overline{Y}^2 = n^2p^2 + npq, \quad \sigma_Y^2 = npq$$

For a Poisson distributed random variable Y.

$$P_Y(j) = \lambda^j e^{-\lambda}/(j!), \quad j = 0, 1, \dots$$

It is an easy exercise to prove that

$$G_Y(z) = e^{-\lambda(1-z)}$$

$$\bar{Y} = E_Y = \lambda, \quad \sigma_Y^2 = \lambda$$

While for a geometrically distributed random variable Y,

$$P_Y(j) = pq^{j-1}, \quad j = 1, 2, \dots$$

$$G_Y(z) = \frac{pz}{1 - qz}, \quad \bar{Y} = \frac{1}{p}, \quad \sigma_Y^2 = \frac{q}{p^2}$$

For a continuous random variable Y, the corespondence is the characteristic $\phi_Y(w)$ function defined as the Fourier transform or, alternately, the Laplace transform $\phi_Y(s)$ of the pdf of Y,

$$\phi_Y(\infty) = E(e^{+j\omega y}) = \int_{-\infty}^{\infty} e^{j\omega y} P_Y(y)\,dy, \quad j = \sqrt{-1}$$

or

$$F_Y(s) = E(e^{-sy}) = \int_{-\infty}^{\infty} e^{-sy} P_Y(y)\,dy$$

Following the same steps as in the discrete case, one can easily see that,

$$\phi_Y^{(0)} = \int_{-\infty}^{\infty} P_Y(y)\,dy = 1 \text{ (as it should be)}$$

$$\left.\frac{d\phi_Y(\omega)}{jd\omega}\right|_{\omega=0} = \int_{-\infty}^{\infty} \frac{(+jy)}{j} | e^{j\omega y} |_{\omega=0} P_Y(y)\,dy = \int_{-\infty}^{\infty} y P_Y(y)\,dy = \bar{Y}$$

In general the mth moment is found to be,

$$\left.\frac{d^m \phi_Y(\omega)}{d\omega^m}\right|_{\omega=0} = j^m \overline{Y^m}$$

Alternately, we can define the Laplace transform (e.g., for an exponentially distributed random variable), i.e., if,

$$P_Y(y) = \mu e^{-\mu y}$$

then,

$$F_Y(s) = \frac{\mu}{s + \mu}$$

$$\bar{Y} = \left. \frac{-dF_Y(s)}{ds} \right|_{s=0} = \frac{1}{\mu}$$

$$\overline{Y^2} = (-1)^2 \left. \frac{d^2 F_Y(s)}{ds^2} \right|_{s=0} = \frac{2}{\mu^2}, \quad \sigma_Y^2 = \frac{1}{\mu^2}$$

One can also find $\phi_Y(\omega) = \mu/(\mu - j\omega)$ and arrive at the same values for the moments.

Similar to the discrete case, the pdf $P_Y(y)$, can be found once $\phi_Y(\omega)$ or $F_Y(s)$ are obtained, by means of inversion, that is

$$P_Y(y) = \int_{-\infty}^{\infty} e^{sy} F_Y(s)\, ds$$

or

$$P_Y(y) = \frac{1}{2\pi} \int_{-\infty}^{\infty} e^{-j\omega y} \phi_Y(\omega)\, d\omega$$

For sums of continuous random variables (i.e., $Y = c_1 + c_2 + \cdots + c_n$), it is easy to see that

$$\phi_Y(\omega) = \prod_{i=1}^{n} \phi_{c_i}(\omega)$$

APPENDIX 3B

REVIEW OF MARKOV CHAIN THEORY

A stochastic process $\{Y_n, n = 0, 1, \ldots\}$ taking only integer values on the positive real time is called a discrete process. Moreover it becomes a discrete Markov process whenever state changes take place at equally spaced intervals of time and,

$$P_{ij} = P\{Y_{n+1} = j/Y_n = i, Y_{n-1} = i_{n-1}, \ldots\} = P\{Y_{n+1} = j/Y_n = i\}$$

that is, the probability of the next state $P(Y_{n+1})$ depends only on the most recent state $(Y_n) = i$ and the transition probability P_{ij}. The transition probabilities satisfy

$$P_{ij} \geq 0, \quad \sum_{j=1}^{\infty} P_{ij} = 1, \quad i = 0, 1, \ldots$$

and the transition probability matrix is given by

$$\mathbf{P} = \begin{bmatrix} P_{00} P_{01} P_{02} \cdots \\ P_{10} P_{11} P_{12} \cdots \\ P_{j0} P_{j1} P_{j2} \cdots \end{bmatrix}$$

After the elapse of $(n + m)$ time units starting from some initial time m, the n-step transition probabilities are defined as

$$p_{ij}^{(n)} = P\{X_{n+m} = j/X_m = i\}, \quad n \geq 0, i, j \geq 0$$

Before proceeding further the following definitions are in order:

Homogeneous Markov chains are those having P_{ij} independent of time n.

Irreducible Markov chains are these where every state can be reached from every other state, that is, $P_{ij} > 0 \in (i, j)$.

An absorbing state (i) is defined by $P_{ii} = 1$.

A *periodic Markov chain* is that, where for each state i, there is an integer $d \geq 2$ such that $P_{ii}^{n} \neq 0$ except when n is a multiple of d. If $d = 1$ the chain is aperiodic.

The following Chapman-Kolomogrov equation enables one to evaluate $P_{ij}^{(n)}$. For simplicity we display this in the case of homogenous Markov chains.

$$P_{ij}^{(n)} = \sum_{l=0}^{\infty} P_{il}^{n-1} P_{lj}, \quad i, j, n, m \geq 0$$

Repeated application of the last equation for $n = 1, 2, \ldots$ yields $\mathbf{P}_{ij}^{(n)} = \mathbf{P}^{n}$, that is, the state transition matrix raised to the nth power transition probability. The probability of finding a system in state i, at the nth step is defined as,

$$(q_i)^{(n)} = P(X_n = i)$$

A probability distribution (π_i) is called a stationary (equilibrium) distribution for the Markov chain if when it is selected as the initial state (i.e. $q_i^{(0)} = \pi_i$) we obtain $q_i^{(n)} = \pi_i$ for all n. To obtain π_i we solve the following set of simultaneous equations.

$$\pi_i = \sum_{j=0}^{\infty} \pi_j P_{ji}, \quad i \geq 0$$

The following lemma gives an equivalent definition.

Lemma 1. For an irreducible and aperiodic Markov chain, the following limit exists and is independent of the initial states probabilities

$$\pi_i = \lim_{n \to \infty} \left(\pi_i^{(n)} \right)$$

Now two cases might arise: (a) $\pi_i = 0 \in i \geq 0$, that is, the chain has no stationary (equilibrium) distribution. An example of this case is the $M/M/1$ system in the case where $(\lambda > 1)$.

(b) $\pi_i > 0 \in i \geq 0$; in this case, the quantities are uniquely determined from.

$$\pi_i = \sum_{j=0}^{\infty} \pi_j P_{ji}$$

$$\sum_{i=0}^{\infty} \pi_i = 1$$

These are called the Global Balance equations.

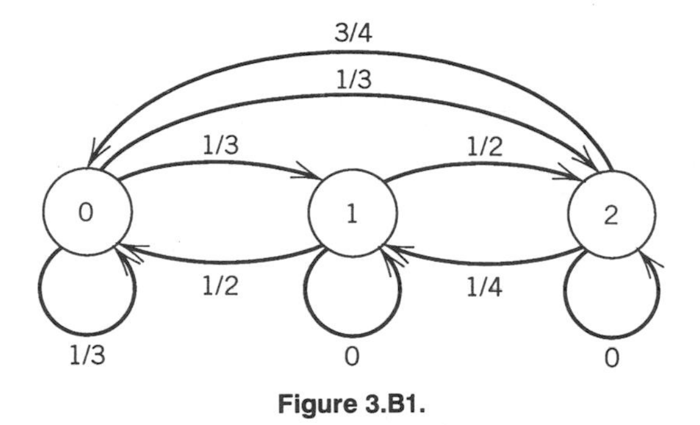

Figure 3.B1.

For example, the irreducible, aperiodic Markov chain in Fig. 3.B.1 is defined by

$$\bar{\pi} = \begin{bmatrix} \pi_0 \\ \pi_1 \\ \pi_2 \end{bmatrix}^T \begin{bmatrix} \frac{1}{3} & \frac{1}{3} & \frac{1}{3} \\ \frac{1}{2} & 0 & \frac{1}{2} \\ \frac{3}{4} & \frac{1}{4} & 0 \end{bmatrix}$$

$$\begin{aligned} \tfrac{1}{3}\pi_0 + \tfrac{1}{2}\pi_1 + \tfrac{3}{4}\pi_2 &= \pi_0 \\ \tfrac{1}{3}\pi_0 + \quad 0 \quad + \tfrac{1}{4}\pi_2 &= \pi_1 \\ \tfrac{1}{3}\pi_0 + \tfrac{1}{2}\pi_1 + \quad 0 \quad &= \pi_2 \end{aligned}$$

Also the condition, $(\pi_0 + \pi_1 + \pi_2) = 1$ is needed to complete this set of linearly dependent equations (make them independent).

Solving the fourth equation together with any two of the first three equations yields the steady-state equilibrium solution, that is,

$$\pi_0 = 0.488, \quad \pi_1 = 0.233, \quad \pi_2 = 0.279$$

One could have arrived at the same conclusion using

$$\bar{\pi} = \bar{\pi}_0 \lim_{n \to \infty} P^n$$

and successive matrix multiplications, where $\bar{\pi}_0$ is some initial state ($[100, \ldots]^T$, for example), finally yields a distribution $\bar{\pi}$ with equal values for different n; usually few multiplications would provide convergence.

Continuous-time discrete state Markov chains are random processes $\{X(t)\}$ that differ from their discrete time counterparts in the following aspects:

A. Set of times at which the states are measured are not fixed now but are exponentially distributed with parameter λ.

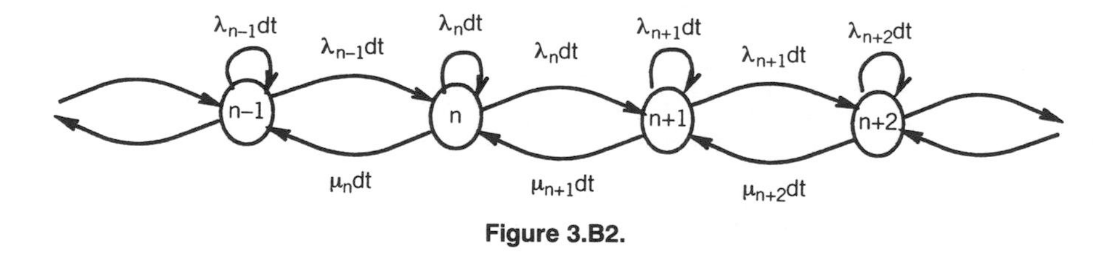

Figure 3.B2.

B. For continuous-time Markov chain, the one-step transition probability matrix elements are replaced by infinitesimal rates, that is,

$$\overline{Q} = \begin{bmatrix} \lambda_{00} & \lambda_{01} & \lambda_{02} \cdots \\ \lambda_{10} & \lambda_{11} & \lambda_{12} \cdots \\ \lambda_{i0} & \lambda_{i1} & \lambda_{i2} \cdots \end{bmatrix}$$

C. The stationary (equilibrium) states probabilities are obtained by solving the Global Balance equations

$$\overline{\pi}^T \overline{Q} = 0, \quad \sum_{i=0}^{\infty} \pi_i = 1$$

rather than $\overline{\pi}^T \overline{P} = \overline{\pi}$, $\Sigma_{i=0}^{\infty} \pi_i$ as in the discrete Markov chain case.

D. Examples of continuous-time discrete state Markov chains appear in Figures 3.1*c* and 3.3. Note that all branches denote rates and that no self-loops exist. If one wishes to translate these rate-dependent state diagrams to probability-dependent diagrams, the rates should be multiplied by the incremental times (dt) and self-loops should be added (Fig. 3B.2)

REFERENCES

[ABBO84] G. F. Abbot, Digital space—a technique for switching high-speed data signals. *IEEE Communications*, 22(4):32–39, 1984.

[ABRA70] N. Abramson, The ALOHA system—another alternative for computer communications. *Proceedings of the Fall Joint Computer Conference*, 1970.

[ABRA77] N. Abramson, the throughput of packet broadcasting channels. *IEEE Transactions on Communicationss*, 25:117–128, 1977.

[ABRA85] N. Abramson, Development of the ALOHANET. *IEEE Transactions on Information Theory*, 31:119–123, 1985.

[ACAM89] A. Acampora and M. Karol, An overview of lightwave packet networks. *IEEE Network*, 3(1):29–40, 1989.

[ACAM94] A. S. Acampora, *An Introduction to Broadband Networks*. New York: Plenum, 1994.

[AHMA89] H. Ahmadi and W. E. Denzel, A survey of modern high-performance switching techniques. *IEEE Journal on Selected Areas in Communications*, 7(7):1091–1103, 1989.

[AIDA94] S. Aidarous and T. Plevyak, eds., Telecommunications Network Management into the 21st Century; Techniques, Standards, Technologies, and Applications. New York, N.Y.: IEEE Press, 1994.

[AMMA92] M. H. Ammar and L. R. Wu, Improving the performance of point-to-multipoint ARQ protocols using destination set splitting. *Proceedings of INFOCOM'92*, Florence, Italy, pp. 2262–2271, May 1992.

[AMMA93] M. H. Ammar, S. Y. Cheung, and C. Scoglio, Routing multipoint connections using virtual paths in an ATM network. *Proceedings of INFOCOM'93*, pp. 98–105, IEEE, 1993.

[AMMA94] M. H. Ammar and G. Rouskas, On the performance of protocols for collecting responses over a multiple access channel. *IEEE Transactions on Communications* [in press].

[ATMF93] ATM Forum, *ATM User-Network Interface Specifications*, Version 2.4 (Draft). The ATM Forum, August 1993.

[ATT80] AT & T, *Telecommunications Transmission Engineering*. Bell System Center for Technical Education, 1980.

[BACK88] F. Backes, Transparent bridges for interconnection of IEEE 802 LANS. *IEEE Network*, 2(1):5–9, 1988.

[BALD87] J. J. Baldini, guest ed., Photonic switching. *IEEE Communications*, 25(5): 1987.

[BALL89] Y. Ballart, SONET: Now it's the Standard Optical Network. *IEEE Communications*, 27(3): 1989.

[BARA64] P. Baran, On distributed communications networks. *IEEE Transactions on Communications Systems*, 12(3): 1–9, 1964.

[BARR91] J. J. Barret and E. F. Wunderlick, LAN interconnect using X.25 network services. *IEEE Network*, 5:12–16, 1991.

BART87] T. C. Bartee, Ed. *Data Communications, Networks, and Systems*. Indianapolis: Howard W. Sams, 1987.

[BATC68] K. E. Batcher, Sorting networks and their applications. *Proceedings Spring Joint Computer Conference*, AFIPS, 1968, pp. 307–314.

[BENE62] V. F. Benes, On pre-arrangeable three-stage connecting networks. *Bell System Technical Journal*, 43:1492, 1962.

[BERG91] A. W. Berger, Performance analysis of a rate-control throttle where tokens and jobs queue. *IEEE Journal of Selected Areas in Communications*, 9(2):165–170, 1991.

[BERN90]· F. Bernabei, C. Calabro, and M. Listanti, A fully distributed routing control scheme in an ATM switch. *Proceedings of International Conference on Communications*, IEEE, 1990, pp. 766–770.

[BERT80] H. V. Bertine, Physical level protocols. *IEEE Transactions on Communications*, 28(4): 1980.

[BERT81] H. V. Bertine, Physical level interfaces and protocols. *Data Communications Interfacing and Protocols*. New York: IEEE, 1981, 2–1 to 2–18.

[BERT92] D. Bertsekas and R. Gallager, *Data Networks*. Englewood Cliffs, N.J.: Prentice-Hall, 1992.

[BIER90] E. W. Biersack, Annotated bibliography on network interconnection. *IEEE Journal on Selected Areas in Communication*, 8(1):22–41, 1990.

[BLAC88] U. Black, *Physical Level Interfaces and Protocols*. Washington, D.C.: Computer Society Press, 1988.

[BLAC89] U. Black, *Data Networks; Concepts, Theory and Practice*. Englewood Cliffs, N.J.: Prentice-Hall, 1989.

[BOCH80] G. V. Bochmann, A general transition model for protocols and communication services. *IEEE Transactions on Communications*, 28(4):643–650, 1980.

[BOCH90] G. V. Bochmann, Deriving protocol converters for communications gateways. *IEEE Transactions on Communications*, 83(9):1298–1300, 1990.

[BOLO94] V. A. Bolotin, P. J. Kuhn, C. D. Pack, and R. A. Skoog, guest eds., Common channel signaling networks: performance, engineering, protocols, and capacity management. *IEEE Journal on Selected Areas in Communications*, 12(3), 1994.

[BRAC89] C. Brackett, guest ed., Lightwave systems and components. *IEEE Communications*, Special Issue, October 1989.

[BRAD89a] R. Braden, ed., Requirements for Internet Hosts-Applications Support. *RFC 1123*, Internet Engineering Task Force, October 1989.

[BRAD89b] R. Braden, ed., Requirements for Internet Hosts-Communications Layers. *RFC 1122*, Internet Engineering Task Force, October 1989.

[BUDR86] Z. L. Budrikis et al., QPSX: a queue packet and synchronous circuit exchange. *ICCC*, Munich, Germany, B3-3, 1986.

[BURG89] F. Burg and N. Iorio, Networking of networks, internetworking according to OSI, *IEEE Journal on Selected Areas in Communications*, 7(7):1131–1142, 1989.

[BUTT91] M. Butto, E. Cavallero, and A. Tonietti, Effectiveness of the "leaky bucket" policing mechanism in ATM networks. *IEEE Journal on Selected Areas in Communications*, 9(3):335–342, 1991.

[BUX83] W. Bux, Local area subnetworks: a performance comparison. *IEEE Transactions on Communications*, 29(10):1465–1473, 1983.

[BUX85] W. Bux and D. Grillo, Flow control in local-area networks of interconnected token rings. *IEEE Transactions on Communications*, 33(10):1058–1066, 1985.

[BUX87] W. Bux et al., Interconnection of local area networks. *IEEE Journal on Selected Areas in Communications*, 5(9):Special Issue, 1987.

[CALL92] R. Callon, TCP and UDP with bigger addresses: a simple proposal for internet addressing and routing. In *RFC 1347*, 1992.

[CALV90] K. L. Calvert and S. S. Lam, Formal methods for protocol conversion. *IEEE Journal on Selected Areas in Communications*, 8(1):127–143, 1990.

[CAPE79] J.I. Capetanakis, Tree algorithms for packet broadcast channels. *IEEE Transactions on Information Theory*, 25(5):505–515, 1979.

[CARL80] D. E. Carlson, Bit-oriented data link control procedures. *IEEE Transactions on Communications*, 28(4):455–467. 1980.

[CCIT84a] CCITT, *Control procedures for Teletex and Group* 4 *Facsimile Services*, Recommendation T.62, 1984.

[CCIT84b] CCITT, vol. VIII, Fascicle VIII.3, Geneva, Switzerland, Recommendation X.75 and X.121, 1984.

[CCIT84c] CCITT, *Final report on the work of study group vii during the study period 1981–1984*, Recommendation X.25, 1984.

[CCIT84d] CCITT, *Network-independent basic transport service for the telematic services*, Recommendation T.70, 1984.

[CCIT88a] CCITT, *Synchronous Digital Hierarchy Bit Rates*, Recommendation G.707, 1988.

[CCIT88b] CCITT, *Network Node Interface for the Synchronous Digital Hierarchy*, Recommendation G. 708, 1988.

[CCIT88c] CCITT, *Synchronous Multiplexing Structure, Recommendation G.* 709, 1988.

[CCIT88d] The International Telegraph and Telephone Consultative Committee, Data Com. Networks (OSI) Models and Notation, Service Definition, vol. VIII, Fascicle VIII.4, Recommendation X.214 (ISO 8072) Transport Service Definition, pp. 278–302, Nov. 1988.

[CCIT88e] The International Telegraph and Telephone Consultative Committee, Data Communication Networks (OSI) Protocol Specifications and Conformance

Testing vol. VIII, Fascicle VIII.5, Recommendation X.224, X.225 (ISO.8073), Transport Protocol Specifications, pp. 36–128, Nov. 1988.

[CEGR75] T. Cegrell, A routing procedure for the tidas message-switching network. *IEEE Transactions on Communications*, 23(6):575–585, 1975.

[CERF90] V. Cerf and R. Kahn, ARPANET maps; 1969–1990, Review ACM, 20(5):81–110, 1990.

[CHEU90] N. M. Cheung, K. Nosu, and G. Winzer, guest ed., Dense wavelength division multiplexing techniques for high capacity and multiple access communication systems. *IEEE Journal on Selected Areas in Communications*, Special Issue, August 1990.

[CHIU89] D -M. Chiu and R. Jain, Analysis of the increase and decrease algorithms for congestion avoidance in computer networks. *Computer Networks and ISDN Systems*, 17:1–14, 1989.

[CHOI86] T. Y. Choi, Formal techniques for the specification, verification, and construction of communication protocols. *IEEE Communications* 23:45–62, 1985.

[CHUA93] J. C-I, Chuang, J. B. Anderson, T. Hattori, and R. W. Nettleton, guest eds., Wireless personal communications: part I and II, *IEEE Journal on Selected Areas in Communications*, Special Issues, August 1993 and September 1993.

[CIDO88] I. Cidon, I. Gopal, G. Grover, and M. Sidi, Real-time packet switching: a performance analysis. *IEEE Journal on Selected Areas in Communications*, 6(9):, 1988.

[CLAP91] G. H. Clapp, LAN interconnection across SMDS. *IEEE Network Magazine*, 5(5):25–32, 1991.

[CLAR89] D. D. Clark, V. Jacobson, J. Romkey, H. Salwen. An analysis of TCP processing overhead. *IEEE Communications*, June 1989:23–29.

[CLAR93] R. J. Clark, M. H. Ammar, and K. C. Calvert, Multi-protocol architectures as a paradigm for achieving inter-operability. *Proceedings of INFOCOM'93*, March 1993, San Francisco, pp. 136–143.

[CLOS53] C. Clos, A study of nonblocking switching networks. *Bell System Technical Journal*, 32(2):406–424, 1953.

[COHE69] J. W. Cohen, *The Single Server Queue*. New York: Wiley, 1969.

[COME88] D. Comer, *Internetworking with TCP/IP: Principles, Protocols, and Architecture*. Englewood Cliffs, N. J.: Prentice-Hall, 1988.

[CONA80] J. W. Conard, Character-oriented data link control protocols. *IEEE Transactions on Communications*, 18(4):445–454, 1980.

[COOP81] R. B. Cooper, *Introduction to Queueing Theory*. New York: Macmillan, 1981.

[COOV92] E. R. Coover, *Systems Network Architecture (SNA) Networks*. New York: IEEE Computer Society Press, 1992.

[DAIG86] J. N. Daigle and J. D. Langford, Models for analysis of packet voice communication systems. *IEEE Journal on Selected Areas in Communications*, 4(6):847–855, 1986.

[DAIG92] J. N. Daigle, *Queueing Theory for Telecommunication*. Reading, MA: Addison Wesley, 1992.

[DALA78] Y. K. Dalal and R. M. Metcalfe, Reverse path forwarding of broadcast packets. *Communications of the ACM*, 21(12):1040–1048, 1978.

[DAVI72] D. Davies, The control of congestion in packet switching networks. *IEEE Transactions on Communications*, 20(3):546–550, 1972.

[DAVI80] D. H. Davis and S. A. Gronemeyer, Performance of slotted aloha random access with delay capture randomized time of arrival. *IEEE Transactions on Communications*, 28(5):703–710, 1980.

[DAY91] A. Day, International standardization of BISDN, *IEEE Magazine of Lightwave Telecommunications Systems*, 13–20, 1991.

[DECI86] M. Decina, W. S. Gifford, R. Potter, and A. A. Robrock, guest eds., Special issue on integrated services digital network: recommendations and field trials-I. *IEEE Journal on Selected Areas in Communications*, 4(3):1986.

[DEER93] S. Deering, SIP: Simple internet protocol. *IEEE Network*, 7(3):16–29, 1993.

[DEPR90] M. DePrycker, *Asynchronous Transfer Mode*. Ellis Horwood Ltd., Englewood Cliffs, N.J.: Prentice-Hall, 1990.

[DEPR94] M. DePrycker, and A. D. Gelman, guest eds., Video on demand. *IEEE Communications*, 32(5):67–109, 1994.

[DEVA88] M. Devaut, J.Cochennec, and M. Servel, The Prelude ATD experiment: assessment and future prospects. *IEEE Journal on Selected Areas in Communications*, 6(9):1528–1537, 1988.

[DIAS84] D. M. Dias and M. Dumar, Packet switching in $N \log N$ multistage network. *IEEE GLOBECOM'84*, Nov. 1984., Atlanta, Georgia, pp. 114–120.

[DIXO88] R. C. Dixon and D.A. Pitt, Addressing, bridging, and source routing. *IEEE Network*, 2(1):25–32, 1988.

[DUBO94] E. Dubois, guest ed., Digital video communications, *IEEE Communications*, 32(5):37–66, 1994.

[DZIO90] Z. Dziong, J. Choquette, K-Q. Liao, and L. Mason, Admission control and routing in ATM networks. *Computer Networks and ISDN Systems*, 20:189–196, 1990.

[EHRL79] N. Ehrlich, The advanced mobile phone service. *IEEE Communications*. March 1979.

[ELHA92] A. K. Elhakeem, S. Bohm, M. Hachicha, and H. Mouftah, Analysis of a new multi-access switching technique for multibeam satellites in a prioritized ISDN environment, *IEEE Journal on Selected Areas in Communications*, 10(2):378–391, 1992.

[ELHA94a] A. K. Elhakeem, D. L. Schilling, W. P. Baier, and M. Nakagawa, guest eds., Code-division multiple access networks (I). *IEEE Journal on Selected Areas in Communications*. Special Issue, May 1994.

[ELHA94b] A. K. Elhakeem, D. L. Schilling, W. P. Baier, and M. Nakagawa, guest eds., Code-division multiple access networks (II). *IEEE Journal on Selected Areas in Communications*. Special Issue, June 1994.

[ENG87] K. Y. Eng, M. G. Hluchyj, and Y. S. Yeh, A knockout switch for variable length packets. *IEEE Journal on Selected Areas in Communications*, 5(9):1426–1435, 1987.

[EPHR87] A. Ephremides, J. E. Wiesethier, and D. J. Baker, A design concept for reliable mobile radio networks with frequency hopping signaling. *Proceedings of the IEEE*, Special Issue, January 1987.

[FLET78] J. G. Fletcher and R. W. Watson, Mechanisms for a reliable timer based protocol. *Computer Networks*, 2(4/5):271–290, 1978.

[FOLT83] H. C. Folts and R. des Jardins, guest eds., Open systems interconnection (OSI), *Proceedings of the IEEE*, Special issue, 17(12):December 1983.

[FOLT90] H. C. Folts, A powerful standard replaces the old interface standby. In Data Communications. New York: McGraw Hill, 1980.

[FRAN93] P. Francis, A near-term architecture for deploying pip. *IEEE Network*, 7(3):30–37, 1993.

[GERL80] M. Gerla and L. Kleinrock, Flow control: a comparative survey, *IEEE Transactions on Communications*, 28(4):553–574, 1980.

[GLEN85] A. B. Glenn and G. E. Keiser, Guest eds., Lightwave technology. *IEEE Communications*, Special Issue, 23(5) 1985.

[GOOD87] D. J. Goodman and A. M. Saleh, The near/far effect in local aloha radio communications. *IEEE Transactions on Vehicular Technology*, 36(1):19–27, 1987.

[GOPA84] I. S. Gopal and J. M. Jaffe, Point-to-multipoint over broadcast channels. *IEEE Transactions on Communications*, 32:1034–1044, 1984.

[GREE80] P. E. Green, Jr., ed., Special issue on computer network architectures and protocols. *IEEE Transactions on Communications*, 28(4):413–424, 1980.

[GREE84] P. Green, Computer communications: milestones and prophecies, in [SCHW84].

[GREE86] P. E. Green, Protocol conversion. *IEEE Trans. on Comm.*, 34:257–268, 1986.

[GREE88] P. E. Green, Jr., ed., *Network Interconnection and Protocol Conversion*. New York: IEEE Press, 1988.

[GREE92a] P. Green, An all-optical computer network: lessons learned, *IEEE Network*, 6(2):56–60, 1992.

[GREE92b] P. E. Green, *Fiber Optic Communication Networks*. Englewood Cliffs, N.J.: Prentice-Hall, 1992.

[GROE84] I. Gorenback, Conversion between the TCP and ISO transport protocols as a method of achieving interoperability between data communications systems. *IEEE Journal on Selected Areas in Communications*, 24(2):288–296, 1986.

[HAAS91] Z. Haas, A protocol structure for high speed communication over broadband ISDN," *IEEE Communications*, Vol 5(1):64–70, January 1991.

[HABE87] D. E. Haber, W. Steinlin, and P. Wild, Silk: an implementation of a buffer insertion ring, in *Advances in Local Area Networks*. New York: IEEE Press, 1987.

[HABI91] I. Habib and T. Saadawi, Controlling flow and avoiding congestion in broadband networks. *IEEE Communications*, 29(10):46–53, 1991.

[HABI92] I. Habib and T. Saadawi, Multimedia traffic characterization. *IEEE Communications*, 30(7):48–54, 1992.

[HAHN91] E. L. Hahne, C. R. Kalamanek, and S. P. Morgan, Fairness and congestion control on a large ATM data network with dynamically adjustable networks. In *Teletraffic and Data Traffic in a Period of Change, ITC-13*. New York:Elsevier Science, 1991, pp. 867–872.

[HALS92] F. Halsall, *Data Communications, Computer Networks, and Open Systems*. Reading, Mass.: Addison-Wesley, 1992.

[HAMM80] R. W. Hamming, *Coding and Information Theory*, Englewood Cliffs, N. J.: Prentice-Hall, 1980.

[HAMM86] J. L. Hammond and P. J. P. O'Reilly, *Performance Analysis of Local Computer Networks*. Reading, MA: Addison-Wesley, 1986.

[HAMN88] M. Hamner and G. Semsen, Source routing bridge implementation. *IEEE Network*, 33–36, 1988.

[HAYE78] J. F. Hayes, An adaptive technique for local distribution. *IEEE Transactions on Communications*, 26(8):1178–1186, 1978.

[HAYES84] J. F. Hayes, *Modelling and Analysis of Computer Communications Networks*. New York: Plenum, 1984.

[HEMR88] C. F. Hemrick, R. W. Klessig, and J. M. McRobert, Switched multi megabit data service and early availability of LAN technology. *IEEE Communications*, 26(4):9–14, 1988.

[HERT78] F. Hertweck, E. Raubold, and F. Vogt, X.25 based process-process communication. *Proceedings of the Symposium on Computer Network Protocols*, Liège, Belgium, pp. 1–22, February 1978. [also in Computer Networks Journal, 2(4/5), 1978.]

[HEWL88] Hewlett Packard, *X.25 Network Performance Analyzer*, 1988.

[HEWL90] Hewlett Packard, *ISDN: selecting the right test tools*. 1990.

[HINT93] H. S. Hinton, *An Introduction to Photonic Switching Fabrics*. New York: Plenum, 1993.

[HLUC88] M. G. Hluchyj and M. J. Karol, Queueing in high performance packet switching. *IEEE Journal on Selected Areas in Communications*, 6(9):1587–1597, 1988.

[HOBE80] V. L. Hoberecht, SNA function management. *IEEE Transactions on Communications*, 28(4):594–603, 1980.

[HONG91] D. Hong and T. Suda, Congestion control and prevention in ATM networks. *IEEE Network*, 5(4):10–16, 1991.

[HOPP87] A. Hopper and R. C. Williamson, Design and use of an integrated Cambridge Ring. In [KUMM87].

[HUAN84] A. Huang and S. Knauer, Starlite: a wideband digital switch. *GLOBECOM'84*, Nov. 1984, Atlanta, Ga., pp. 121–125.

[HUI90] J. Y. Hui, *Switching and Traffic Theory for Integrated Broadband Networks*. Boston: Kluewer Academic, 1990.

[IEEE84] IEEE, 100 years of communications progress. *IEEE Communications*, Special Issue, 22(5), 1984.

[IEEE89A] IEEE, Logical link control. *IEEE 802.2*, 1989.

[IEEE89b] IEEE, Token ring access method. *IEEE 802.5*, 1989.

[IEEE90a] IEEE, CSMA/CD access method and physical layer specification. *IEEE 802.3*, 1990.

[IEEE90b] IEEE, Distributed queue dual bus (DQDB) subnetwork of a metropolitan area network (MAN). *IEEE 802.6*, Dec. 1990.

[IEEE90c] IEEE, Token passing bus access method and physical layer specification. *IEEE 802.4*, 1990.

[IEEE90d] IEEE, Media access control bridges. *IEEE 802.1D*, 1990.

[IEEE91] IEEE, Source routing appendix to token ring access method. *IEEE 802.5*, 1991.

[IEEE93a] IEEE, *IEEE Communications*, Feature topic: SDH/SONET. 31(9), 1993.

[IEEE93b] IEEE, Feature topic: The future of Internet protocol, *IEEE Network*, 7(3), 1993.

[ILYA85] M. Ilyas and H. T́. Mouftah, Performance evaluation of computer communications networks, *IEEE Communications*, 23:18–29, 1985.

[INTE83] Internet Protocol, Military Standard, MIL-STD-1777, U.S. Department of Defense, May 20, 1983.

[ISO88] ISO, Protocol for providing connectionless-mode network service. In *ISO 8743*, 1988.

[IVAN89] F. Ivanak, ed., *Terrestrial Digital Microwave Communications*. Norwood, MA: Artech House, 1989.

[JABB91] B. Jabbari, Common channel signaling system number 7 for ISDN and intelligent networks. In [WU91].

[JACK57] J. R. Jackson, Networks of waiting lines. *Operations Research*, 5:518–521, 1957.

[JACK63] J. R. Jackson, Jobshop-like queueing systems. *Management Science*, 10:131–142, 1963.

[JACO88] V. Jacobson, Congestion avoidance and control. In *Proceedings of ACM SIGCOMM. ACM*, 1988, pp. 314–329.

[JAFF86] A. Jaffari, T. Saadawi, and M. Schwartz, Blocking probabilities in distributed circuit switched CATV system. *IEEE Transactions on Communications*, 34(10):977–984, 1986.

[JAIN90a] N. Jain, M. Schwartz, T. Bashkow, "Transport protocol processing at Gbps rates," ACM Computer Communications Review, Vol 20:188–199, September 1990.

[JAIN86] R. Jain, A timeout-based congestion scheme for window flow-controlled networks. *IEEE Journal on Selected Areas in Communications*, 7(4):1162–1167, 1986.

[JAIN90] R. Jain, Congestion control in computer networks: issues and trends. *IEEE Network*, 4(5):24–30, 1990.

[JAIN91] R. Jain, Myths about congestion management in high speed networks. Technical Report DEC-TR-726, Digital Equipment Corporation, January 1991.

[JAIS68] N. K. Jaiswal, *Priority Queues*. New York: Academic Press, 1968.

[JENG83] Y. Jeng, Performance analysis of a packet switch based on single-buffered banyan network. *IEEE Journal on Selected Areas in Communications*, 1(6): 1014–1021, 1983.

[JOEL79] A. E. Joel, Jr., Circuit switching: unique architecture and applications. *IEEE Computer*, 12(6):10–22, 1979.

[JOEL84] E. A. Joel, The past 100 years in telecommunications switching. In [IEEE84].

[JOHN93] T.J. Johnson, ATM networking gear: welcome to the real world. *Data Communications*, 1993.

[JONE88] W. B. Jones, Jr., *Introduction to Optical Fiber Communication Systems*. Holt, Rinehart and Winston, 1988.

[KAHN72] R. Kahn and W. Crowther, Flow control in a resource sharing computer network. *IEEE Transactions on Communications*, 20(3):539–546, 1972.

[KAMA94] A. E. Kamal, and B. W. Abeysundara, A survey of MAC protocols for high-speed LANs. In [TANT94a].

[KAMO79] F. Kamoun and L. Kleinrock, Stochastic performance evaluation of hierarchical routing for large networks. *Computer Networks*, 3:337–353, 1979.

[KAMO81] F. Kamoun, A drop and throttle flow control policy for computer networks. *IEEE Transactions on Communications*, 29(4):444–452, 1981.

[KARO87] M. Karol, M. Hluchyj, and S. Morgan, Input versus output queueing on a space-division packet switch. *IEEE Transactions on Communications*, 35(12): 1347–1356, December 1987.

[KATZ93] D. Katz and P.S. Ford, Tuba: replacing IP with CLNP. *IEEE Network*, 7(3):38–47, May 1993.

[KAUL92] A. Kaul, Design of a gateway for satellite-based ACARS messaging. Proceedings of the 14th International Communications Satellite Systems Conference and Exhibit, March 1992, Washington, D.C., pp. 78–83.

[KAVE84] M. Kavehard and P. McLane, Performance of direct sequence spread spectrum for indoor wireless digital communications. *Communications of the ACM*, July, 1984.

[KEAR90] T. J. Kearns and M. C. Mellon, The role of ISDN signaling in global network. *IEEE Communications*, 28(7):36–43, 1990.

[KEIS89] B. E. Keiser, *Broadband Coding, Modulation and Transmission Engineering*. Englewood Cliffs, N.J.: Prentice-Hall, 1989.

[KHAN89] A. Khanna and J. Zinky, The revised ARPANET routing metric. *Proceedings of SIGCOMM'89 Symposium*, 1989, ACM, pp. 45–56.

[KLEI75a] L. Kleinrock and H. Opderbeck, Throughput in the ARPANET-protocols and measurement. *Proceedings of the 4th Data Communications Symposium*, Quebec City, Canada, October 1975, pp. 6:1–11.

[KLEI75b] L. Kleinrock and F. Tobagi, Packet switching in radio channels; part I-carrier sense multiple-access modes and their throughput-delay, characteristic. *IEEE Transactions on Communications*, 23(12):1400–1416, 1975.

[KLEI75c] L. Kleinrock, *Queueing Systems*, Vol. I and II. New York: Wiley, 1975.

[KLEI75d] L. Kleinrock and S. S. Lam, Packet-switching in a slotted satellite channel, *IEEE Transactions on Communications*, 23(4):410–423, 1975.

[KLEI77] L. Kleinrock and F. Kamoun, Hierarchical routing for large networks. *Computer Networks*, 1(3):155–174, 1977.

[KLEI80a] L. Kleinrock and F. Kamoun, Optimal clustering structures for hierarchical topological design of large computer networks. *Networks*, 10:221–248, 1980.

[KLEI80b] L. Kleinrock and M. Gerla, Flow control: a comparative survey. *IEEE Transactions on Communications*, 28(4):553–574, 1980.

[KNIG83] K. G. Knightson, The transport layer standardization. *Proceedings of the IEEE*, 71(12):1394–1396, 1983.

[KOBA78] H. Kobayashi, *Modelling and Analysis: An Introduction to System Perfor-*

mance Evaluation Methodology. Reading, MA: Addison-Wesley, 1978.

[KUMM87] K. Kummerle, F. Tobagi, and J. Limb, eds. *Advances in Local Area Networks*. New York: IEEE Press, 1987.

[KURT88] R. L. Krutz, *Interfacing Techniques in Digital Design with Emphasis on Microprocessors*. New York: Wiley, 1988.

[LAM79] S. Lam and M. Reiser, Congestion control in store and forward networks by input buffer limits. *IEEE Transactions on Communications*, 27(1):127–134, 1979.

[LAM80] S. S. Lam, A carrier-sense multiple access protocol for local area networks. *Computer Networks*, 4:21–32, 1980.

[LAM84] S. S. Lam, ed. *Tutorial: Principles of Communication and Networking Protocols*, Silver Spring, Md.: IEEE Press, 1984.

[LAM88] S. S. Lam, Protocol conversion. *IEEE Trans. on Software Engineering*, 14(3):353–362, 1988.

[LAPO91] T. F. LaPorta and M. Schwartz, Architecture, features and implementation of high speed transport protocols. *IEEE Network*, 4(4):14–22, 1991.

[LAPO93] T. F. LaPorta and M. Schwartz. The multistream protocol: a highly flexible high speed transport protocol. *IEEE Journal on Selected Areas in Communications*, 11(4):519–530, 1993.

[LAZA94] A. A. Lazar, K. H. Tseng, K. S. Lim, and W. Choe, A scalable and reusable emulator for evaluating the performance of SS7 networks. *IEEE Journal on Selected Areas in Communications*, 12(3):395–404, 1994.

[LEE93] W. C. Y. Lee, *Mobile Communications Design Fundamentals*. New York: John Wiley, 1993.

[LEIN85] B. M. Leiner, R. Cole, J. Postel, and D. Mills, The DARPA Internet protocol suite. *IEEE Communications*, 23(3):29–34, 1985.

[LEIN87] B. M. Leiner, D. L. Nielson, and F. A. Tobagi, eds., Packet radio networks. *Proceedings of the IEEE*, Special Issue, January 1987.

[LENZ92] L. Lenzini and R. P. Zeletein, From the earth to the sky and back. *IEEE Journal on Selected Areas in Communications*, 8(1):107–118, 1990.

[LIN84] S. Lin, D. J. Costello, Jr., and M. J. Miller, Automatic Repeat-Request-Error-Control Schemes. *IEEE Communications*, 22(12), 1984.

[LIND73] W. Lindsey and M. Simon, *Telecommunications System Engineering*. Englewood Cliffs, N.J.: Prentice-Hall, 1973.

[LIND93] B. Lindgren, R. Krupczak, M. H. Ammar, K. Schwan. Parallelism and configurability in high performance protocol architectures. *Proceedings of the International Conference on Network Protocols, IEEE*, September 1993:6–13.

[LIU82] M. T. Liu, W. Hilal, and B. H. Groomes, Performance evaluation of channel access protocols for local computer networks. *Proceedings of the COMPCON Fall'82 Conference*, 1982, pp. 417–426.

[MAHM89] S. A. Mahmoud, S. S. Rappaport, and S. O. Ohrvik, eds., Portable and mobile communications, *IEEE Journal on Selected Areas in Communications*, Special Issue, 7(1):1989.

[MALA91] C. Malamud, *Analyzing DECnet/OSI Phase V*, New York: Van Nostrand Reinhold, 1991.

[MAXE90] N. F. Maxemchuk and M. El-Zarki, Routing and flow control in high speed, wide area networks, *Proceedings of the IEEE*, 78(1):204–221, 1990.

[MAXE93] N. F. Maxemchuk, Dispersity routing on ATM networks. *Proceedings of IEEE INFOCOM* 93, 1993, pp. 347–357.

[MAYN85] J. Mayne, *Linked Local Area Networks*, First Edition, New York: John Wiley, 1985.

[MCCO88] J. McConnel, *Internetworking Computer Systems, Internetworking Networks and Systems*, Englewood Cliffs, N.J.: Prentice-Hall, 1988.

[MCDO83] McDonald, J. C., Ed., *Fundamentals of Digital Switching*. New York: Plenum Press, 1983.

[MCFA79] R. McFarland, *Protocols in Computer Internetworking Protocol*. Silver Spring, Md.: IEEE Press, 1984.

[MCNA88] J. E. McNamara, *Technical Aspects of Data Communications*. Digital Equipment, 1988.

[MCQU78] J. M. McQuillan, G. Falk, and I. Richer, A review of the development and performance of the ARPANET routing algorithm. *IEEE Transactions on Communications*, 26(12):1802–1811, 1978.

[MCQU80] J. M. McQuillan, I. Richer, and E. C. Rosen, The new routing algorithm for the ARPANET. *IEEE Transactions on Communications*, 28(5):253–261, 1980.

[MCQU90] J. M. McQuillan, Broadband networks: the end of distance? *Data Communications*, 1990.

[MEHM88] M. Mehmet Ali, J. F. Hayes, and A. K. Elhakeem, Traffic analysis of a star local area network. *IEEE Transactions on Communications*, 36:703–713, 1988.

[METC76] R. M. Metcalfe and D. R. Boggs, Ethernet; distributed packet switching for local computer networks. *Communications of the ACM*, 19:7, 395–404, 1976.

[METZ76] J. J. Metzner, On improving utilization in ALOHA networks. *IEEE Transactions on Communications*, 447–448, 1976.

[MIDW88] J. E. Midwinter and P. W. Smith, guest eds., Photonic Switching. *IEEE Journal on Selected Areas in Communications*, Special Issue, 6(7), 1988.

[MIKI90] T. Miki and C. A. Siller, guest eds., Global deployment of SDH-compliant networks. *IEEE Communications*, Special Issue, August 1990.

[MILL87] J. M. Miller and S. V. Ahmed, *Digital Transmission, Systems, and Networks*. Rockville, Maryland: Computer Science Press, 1987.

[MITR91a] D. Mitra and J. Seery, Dynamic adaptive windows for high speed data networks with multiple paths and propagation delays. *Proceedings of INFOCOM '91*. New York: IEEE, 1991, pp. 39–481.

[MITR91b] N. Mitra and S. Usiskin, Relationship of the signaling system no. 7 protocol architecture to the OSI reference model. *IEEE Network*, 26:29–37, 1991.

[MODA90] A. R. Modarresi, R. A. Skoag, and S. M. Boyles, Guest eds., The role of signaling system no. 7 in the global information age network. *IEEE Communications*, 28(7), 1990.

[MORE86] J. P. Morenoy, D. Porter, R. P. Pitkin, and D. R. Oren, the DECnet/SNA gateway product–a case study in gross vendor networking. *Digital Technical Journal*, Digital Equipment Corporation, Hudson, MA, No. 3, Sept. 1986.

[MORG89] L. W. Morgan and G. D. Gordon, *Communications Satellite Handbook*. New York: John Wiley, 1989.

[NAGE87] S. R. Nagel, Optical fiber- the expanding medium. *IEEE Communications*, 25(4), 1987.

[NAKA90] K. Nakao, S. Obara and S. Nishiyama, Feasibility study for worldwide video telex internetworking. *IEEE Journal on Selected Areas in Communications*, 8(1):80–92, 1990.

[NBS83a] National Bureau of Standards, *Specification of a Transport Protocol for Computer Communications, Volume 3: Class 4 Protocol*, Gaithersburg, Md., 1983.

[NBS83b] National Bureau of Standards, *Specification of a Transport Protocol for Computer Communications, Volume 5*: Guidance for the Implementor, Gaithersburg, Md., 1983.

[NUSS79] E. Nussbaum, guest ed., Digital Switching. *IEEE Transactions on Communications*, Special Issue, 1979.

[OSI84a] International Organization for Standardization (ISO), *Information Processing Systems-Open Systems Interconnecting-Transport Protocol Service Definition*, Standard 8072. Geneva, 1984.

[OSI84b] International Organization for Standardization (ISO), *Data Communication: High-Level Data Link Control Procedures-Consolidation of Elements of Procedures*, Standard 4335, 2d ed. Geneva, 1984.

[OSI84c] International Organization for Standardization (ISO), *Data Communication: High-Level Data Link Control Procedures-Consolidation of Classes of Procedures*, Standard 7809. Geneva, 1984.

[OSI84d] International Organization for Standardization (ISO), *Information Processing Systems*: *Open Systems Interconnection-Basic Reference Model*, Standard 7498. Geneva, 1984.

[OSI86] International Organization for Standardization (ISO), *Information Processing Systems: Open Systems Interconnection-Protocol for Providing the Connectionless-Mode Transport Service*, *Standard* 8602. Geneva, 1986.

[PAPA93] C. Papadopoulos, G. M. Parulkar. Experimental evaluation of SUNOS ICP and TCP/IP protocol implementation, *IEEE/ACM Transactions on Networking*, 1(2):199–216, 1993.

[PARU93] G. Parulkar, guest ed., Local ATM Networks. *IEEE Network*, Special Issue, vol. 7(2), 1993.

[PATE81] J. H. Patel, Performance of processor-memory interconnections for multiprocessors. *IEEE Transactions on Computers*, 30(10):771–780, 1981.

[PERL84] R. Perlman, "An Algorithm for Distributed Computation of a Spanning Tree," *Proceedings of the 9th Data Communications Symposium*, 1984.

[PERL88] R. Perlman, A. Harvey, and Varghese, G., *Choosing the appropriate ISO layer for LAN interconnection*. IEEE Network, 2(1):81–86, 1988.

[PERL93] R. Perlman, *Interconnections: Bridges and Routers*, Addison-Wesley, Reading, Mass., 1993.

[PICK82] R. L. Pickholtz, D. L. Schilling, and L. B. Milstein, Theory of spread spectrum communications-a tutorial. *IEEE Transactions on Communications*, 30:855–884. May 1982.

[PICK87] R. L. Pickholtz, Modems, multiplexers, and concentrators. *Data Communications, Networks, and Systems* T. C. Barter, ed. Indianapolis: Howard W. Sams, 63–117, 1987.

[PITT87] D. A. Pitt and J. L. Winkler, Table-free bridging. *IEEE Journal on Selected Areas in Communications*, 5(9):1454–1462, 1987.

[POST80a] J. B. Postel, Internetwork protocol approaches. *IEEE Transactions on Communications*, 28(4):604–611, 1980.

[POST80b] J. Postel, User datagram protocol, *RFC 768*, August 1980.

[POST81a] J. Postel, Internet control message protocol, *RFC* 792. Los Angeles: University of Southern California Information Science Institute, 1981.

[POST81b] J. Postel, ed., Internet Protocol. *RFC 791*. USC-Information Science Institute, 1981.

[POST81c] J. Postel, C. Sunshine, and D. Cohen, The ARPA Internet protocol. *Computer Networks*, 5(4):261–271, 1981.

[POST81d] J. Postel, ed., Transmission control protocol, *RFC 793*, September 1981.

[POST82] J. Postel, Internet control message protocol. *Internet protocol Transition Workbook*, eds. John Postel and Elizabeth Feinler, Menlo Park, CA: SRI International, 1989.

[POST93] J. Postel, ed., *Internet official protocol standards*, *RFC 1500*. Internet Architecture Board, August 1993.

[PROA94] J. G. Proakis and M. S. Salehi, *Communication Systems Engineering*. Englewood Cliffs, N.J.: Prentice Hall, 1994.

[PROC83] IEEE, Open systems interconnection (OSI). *Proceedings of the IEEE*, Special Issue, 71(12), 1983.

[PROS62a] R. T. Prosser, Routing procedures in communication networks-part I, pp. Random Procedures. *IRE Transactions on Communication Systems*, Vol. 10(12):322–329, 1962.

[PROS62b] R. T. Prosser, Routing procedures in communication networks-part II, pp. Random Procedures. *IRE Transactions on Communication Systems*. Vol. 10(12):329–335.

[PRYC88] M. D. Prycker, Definition of network options for the Belgian ATM broadband experiment. *IEEE Journal on Selected Areas in Communications*, 6(9):1538–1544, 1988.

[PURS87] M. B. Pursley, The role of spread spectrum in packet radio networks. In *Proceedings of the IEEE*, Special Issue, January, 1987.

[RAMA88a] V. Ramaswami, A Stable recursion for the steady state vector in Markov chain of the M / G / 1 type, Stochastic Models, Vol. 4:183–188; 1988.

[RAMA88b] K. K. Ramakrishnan and R. Jain, A binary feedback scheme for congestion avoidance in computer networks with a connectionless network layer. *Proceedings ACM SIGCOMM*, ACM, 1988, pp. 303–313.

[RATH88] E. P. Rathgeb, T. H. Theimer, and M. Huber, Buffering concepts for ATM switching networks. *Proceedings of IEEE GLOBECOM'88*, pp. 39.31–39.35.

[RATH91] E. P. Rathgeb, Modelling and performance comparison of policing mechanisms for ATM networks, *IEEE Journal on Selected Areas in Communications*, 9(3):325–334, 1991.

[REY86] R. F. Rey, Technical ed., *Engineering and Operations in the Bell System*. Murray Hill, N. J.: AT & T Bell Laboratories, 1986.

[ROBE78] L. G. Roberts, The evolution of packet switching. *Proceedings of the IEEE*, 66:1307–1313, 1978.

[RONA88] J. Ronayne, *Integrated Services Digital Networks: From Concept to Application*. New York: John Wiley, 1988.

[ROSE80] E. C. Rosen, The updating protocol of ARPANET's new routing algorithm. *Computer Networks*, Vol. 4:11–19, 1980.

[ROSE87] M. T. Rose, E. Dwight Case, ISO transport service on top of the TCP-version 3, *RFC 1006*, May 1987.

[ROSE90] M. T. Rose, Transition and coexistence strategies for TCP / IP and OSI. *Journal on Selected Areas in Communications*, 8(1):57–66, 1990.

[ROSE92] M. T. Rose, *The Open Book*. Reading, MA: Addison-Wesley, 1992.

[ROSS86] F. E. Ross, FDDI-a tutorial. *IEEE Communications Magazine*, 24(5):10–17, 1986.

[ROSS89] F. E. Ross, An overview of FDDI: the fiber distributed data interface. *IEEE Journal on Selected Areas in Communications*, 7(7):1043–1051, 1989.

[ROSS92] F. E. Ross and R. L. Fink, Overview of FFOL: FDDI follow-on LAN, *Computer Communications*, 15(1):5–10, 1992.

[RUBI79] I. Rubin, Message delays in FDMA and TDMA communications channels. *IEEE Transactions on Communications*, 27:769–778, 1979.

[RUBI90] I. Rubin and J. Baker, Media access control for high speed local area and metropolitan area communications networks. *Proceedings of the* IEEE, 78(1): 168–203, 1990.

[RUDIS89] H. Rudin and K. Sabnani, eds., Architecture and protocols for computer networks: the state of the art. *IEEE Journal on Selected Areas in Communications*, 7(7), 1989.

[RYAN85] J. Ryan, guest ed., Telecommunications standards. *IEEE Communications* Special Issue, 1985.

[RYBC80] A. Rybezynski, X.25 interface and end-to-end virtual circuit service characteristics. *IEEE Transactions on Communications*, 28(4), 1980.

[SAAD81] T. N. Saadawi and A. Ephremides, Analysis, stability, and optimization of slotted ALOHA with a finite number of buffered users. *IEEE Transactions on Automatic Control*, 28(3):681–689, 1981.

[SAAD85] T. N. Saadawi and M. Schwartz, Distributed switching algorithm for data transmission over two-way CATV, *IEEE Journal on Selected Areas in Communications*, 3:323–329, 1985.

[SACH88] S. R. Sachs, Alternative local area network access protocols. *IEEE Communications Magazine*, 26(3):25–45, 1988.

[SACK93] G. C. Sackett, *IBM's Token-Ring Networking Handbook*. New York: McGraw-Hill, 1993.

[SALT80] J. H. Saltzer and K. T. Pogran, A star-shaped network with high maintainability. *Computer Networks*, 4:239–244, 1980.

[SALT83] J. H. Saltzer, K. T. Pogran, and D. D. Clark, Why a ring? *Computer Networks*, 7:223–231, 1983.

[SCHW77] M. Schwartz, *Computer Communication Network Design and Analysis*. Englewood Cliffs, N.J.: Prentice-Hall, 1977.

[SCHW80] M. Schwartz and T. E. Stern, Routing techniques used in computer communications networks. *IEEE Transactions on Communications*, 28(4):265–278, 1980.

[SCHW87] M. Schwartz, *Telecommunication Networks: Protocols, Modelling and Analysis*, Reading, MA: Addison-Wesley, 1987.

[SCHW90] M. Schwartz, *Information Transmission, Modulation, and Noise*. New York: McGraw-Hill, 1990.

[SCIE91] Communications, computers and networks. *Scientific American*, Special Issue, 265(3), September 1991.

[SEFE88] W. M. Sefert, Bridges and routers. *IEEE Network*, 2:57–64, 1988.

[SHER92] M. H. Sherif and D. K. Sparrel, Standards and innovation in telecommunications. *IEEE Communications*, 30(9):22–29, 1992.

[SHOC79] J. F. Shoch. Packet fragmentation in inter-network protocols. *Computer Networks*, 3(1):3–8, 1979.

[SHOC82] J. F. Shoch, Y. K. Dalal, and D. D. Redell. Evolution of the Ethernet local computer network. *IEEE Computer*, 13(8):1–27, 1982.

[SKLA88] B. Sklar, *Digital communications*: *fundamentals and applications*. Englewood Cliffs, N.J.: Prentice-Hall, 1988.

[SOHA88] M. Soha and R. Perlman, Comparison of two LAN bridge approaches. *IEEE Network*, 2(1):37–43, 1988.

[SOHR91] K. Sohraby, Guest ed., Congestion control in high speed networks. *IEEE Communications Magazine*, Special Issue, 29(10), 1991.

[SPRA91] J. D. Spragins, J. L. Hammond, and K. Pawlinkowski, *Telecommunications, Protocols and Design*. Reading, MA: Addison-Wesley, 1991.

[SPRO81] D. E. Sproule and F. Mellor, Routing, flow, and congestion control in the datapac network. *IEEE Transactions on Communications*, 28(4):386–391, 1981.

[STAL87] W. Stallings, *Handbook of Computer-Communications Standards, Volume 1:* The Open System Interconnection (OSI) Model and OSI-Related Standards. New York: Macmillan, 1987.

[STAL89] W. Stallings, *ISDN, An Introduction*. New York: Macmillan, 1989.

[STAL94] W. Stallings, *Data and Computer Communications*, Fourth ed., New York: Macmillan, 1994.

[STAL93] W. Stallings, *Local and Metropolitan Area Networks*, 4th ed., New York: Macmillan, 1993.

[STEE92] R. Steele, guest ed., PCS: the second generation, *IEEE Communications*, Special Issue, 30(12), 1992.

[STUC85] B. W. Stuck and E. Arthurs, *A Computer Communications Network Performance Analysis Primer*, Englewood Cliffs, NJ: Prentice-Hall, 1985.

[STUD83] P. Von Studnitz, Transport protocols: their performance and status in international standardization. *Computer Networks*, 7:27–35, 1983.

[SUNS77] C. Sunshine, Source routing in computer networks. *ACM Computer Communications Review*, 7(1):29–33, 1977.

[SUNS78] C. A. Sunshine and Y. K. Dalal, Connection management in transport protocols. *Computer Networks*, 2:454–473, 1978.

[TAJI77] W. D. Tajibnapis, A correctness of a topology information maintenance protocol for a distributed computer network. *Communications of the ACM*, 20(7):477–485, 1977.

[TANE88] A. S. Tanenbaum, *Computer Networks*, 2nd Ed., Englewood Cliffs, N.J.: Prentice-Hall, 1988.

[TANT94a] A. Tantawy, ed., *High Performance Networks: Technology and Protocols*. Boston: Kluwer, 1994.

[TANT94b] A. Tantawy, ed., *High Performance Networks: Frontiers and Experience*. Boston: Kluwer, 1994.

[TARR94] A. Tarraf, I. Habib, and T. Saadawi, A novel neural network traffic enforcement mechanism for ATM networks. *IEEE Journal on Selected Areas in Communications*, 6(12):1088–1096, 1994.

[TAUB86] H. Taub and D. L. Schilling, *Principles of Communications Systems*. New York: McGraw-Hill, 1986.

[TCPM83] Transmission Control Protocol, Military Standard, MIL-STD-1778, U.S. Department of Defense, May 20, 1983.

[TOBA80a] F. A. Tobagi, Multiaccess protocols in packet communication systems. *IEEE Transactions on Communications*, 28(4):468–488, 1980.

[TOBA80b] F. A. Tobagi and V. B. Hunt, Performance analysis of carrier sense multiple access with collision detection. *Computer Networks*, 4:245–259, 1980.

[TOBA84] F. A. Tobagi, R. Binder, and B. Leiner, Packet radio and satellite networks. *IEEE Communications Magazine*, 22(11), 1984.

[TOBA90] F. A. Tobagi, Fast packet switch architectures for broadband integrated services digital networks. *Proceedings of the IEEE*, 78(1):133–167, 1990.

[TODD93] T. D. Todd, T. Liu, and D. P. Taylor, A robust broadband headend switching algorithm. *IEEE Transactions on Communications*, 41(9): 1993.

[TOML75] R. S. Tomlinson, Selecting sequence numbers. *Proceedings of the ACM SIGCOMM/SIGOPS Interprocess Communications Workshop*, Santa Monica, CA, March 1975.

[TOWN93] J. K. Townsend, A. F. Elrefaie, H. Meyr, and M. Pent, guest eds., Computer-aided modeling, analysis, and design of communications links. *IEEE Journal on Selected Areas in Communications*, Special Issue, 11(2), April 1993.

[TURN86] J. S. Turner, New directions in communications (or which way to the information age?). *IEEE Communications*, 24:8–15, 1986.

[VADD92] S. V. Vaddiparty, K. H. Price, and G. P. Heckert, MilSatCom intersatellite link architecture. *Proceedings of the 14th International Communications Satellite Systems Conference and Exhibit*, The American Institute of Aeronautics and Astronautics AIAA, Washington, D.C., pp. 338–348, March 1992.

[VARG90] G. Varghese and R. Perlman, Transport interconnection of incompatible local area networks using bridges. *IEEE Journal on Selected Areas in Communications*, 8(1):42–48, 1990.

[VERM90] P. K. Verma ed., *ISDN Systems; Architecture, Technology, and Applications,* Englewood Cliffs, N.J.: Prentice-Hall, 1990.

[VINT83] C. Vinton and C. Edward, The DoD Internet architecture model. *Computer Networks*, 7, 1983.

[WALR91] J. Walrand, *Communication networks: A first course*. Boston, MA.: Asken Associates Inc., IRWIN, 1991.

[WATS81] R. W. Watson, Timer-based mechanisms in reliable transport protocol connection management. *Computer Networks*, 5:47–56, 1981.

[WATS87] R. W. Watson and S. A. Mamrak, Gaining efficiency in transport services by appropriate design and implementation choices. *ACM Transactions on Computer Systems*, 5:97–120, May 1987.

[WU85] W. W. Wu, *Elements of Digital Satellite Communications*, Vol. I, II, and III. Rockville, Maryland: Computer Science Press, 1985.

[WU91] W. Wu, A. Livne, and J. Griffiths, guest eds., Integrated services digital networks. *Proceedings of the IEEE*, Special Issue, 79(2), 1991.

[YAZI92] S. Yazid and H. T. Mouftah, Congestion control methods for BISDN. *IEEE Communications*, 30(7):42–47, 1992.

[YEE93] J. R. Yee and F. Y. Lin, Real-time distributed routing and admission control algorithm for ATM networks, *Proceedings of INFOCOM 93*, pp. 792–801, IEEE, 1993.

[YEH87] Y. Yeh, M. G. Hluchyj and A. S. Acampora, The knockout switch: a simple, modular architecture for high-performance packet switching. *IEEE Journal on Selected Areas in Communications*, 5(8): 1274–1283, 1987.

[ZARR94] P. N. Zarros, M. J. Lee, and T. N. Saadawi, Statistical synchronization among participants in real-time multimedia conference. *Proceedings of INFOCOM'94*, Toronto, Canada, 1994.

[ZIMM80] H. Zimmermann, OSI reference model-the ISO model of architecture for open systems interconnection. *IEEE Transactions on Communications*, 1980.

[ZITT93] M. Zitterbart, B. Stiller, A. Tantawy, "A Model for flexible high-performance communication subsystem," *IEEE Journal on Selected Areas in Communication*, 11(4): May 1993.

INDEX